Corrosion of Reinforcement in Concrete Construction

Corrosion of Reinforcement in Concrete Construction

Edited by

C. L. Page
Department of Civil Engineering, Aston University, Birmingham, UK

P. B. Bamforth
Taywood Engineering Ltd, Southall, Middlesex, UK

J. W. Figg
Ove Arup and Partners, London, UK

THE ROYAL
SOCIETY OF
CHEMISTRY
Information
Services

Papers presented at the Fourth International Symposium on 'Corrosion of Reinforcement in Concrete Construction', held at Robinson College, Cambridge, UK, 1–4 July 1996.

Special Publication No. 183

ISBN 0-85404-731-X

© SCI 1996

A catalogue record for this book is available from the British Library

Published by The Royal Society of Chemistry,
Thomas Graham House, Science Park, Cambridge
CB4 4WF

Printed by Hartnolls Ltd, Bodmin, Cornwall, UK

Introduction

The papers in this volume were presented at the Fourth International Symposium on Corrosion of Reinforcement in Concrete Construction, organised by the Construction Materials Group of the Society of Chemical Industry in association with the European Federation of Corrosion, the Institute of Concrete Technology, the Institute of Corrosion, the Institute of Materials and NACE International. The Organising Committee was made up of the following individuals:

> J W Figg *(Chairman)*, Ove Arup and Partners
> P B Bamforth, Taywood Engineering Ltd
> J Buekett, Consultant
> R N Cox, Building Research Establishment
> D Harrop, BP Research International
> J W Oldfield, Cortest Laboratories Ltd
> C L Page, Aston University
> D B Storrar, Highways Agency
> K W J Treadaway, Building Research Establishment
> B S Wyatt, Tarmac Global

The Symposium took place from 1-4 July 1996 at Robinson College Cambridge and the trends set by previous meetings in the series, held in 1978, 1983 and 1990, were maintained in that the number of accepted papers exceeded those of all the earlier Symposia as did the extent of international participation. This reflects the worldwide growth of interest in the topic of reinforcement corrosion, which is generally accepted to be the single most significant cause of unsatisfactory performance of concrete structures.

Over the six years that have elapsed since the Third International Symposium was held in 1990, there have been important advances in several areas of the field, which have come about partly as a consequence of increased international collaboration. The move towards harmonisation of European standards has been particularly influential in this respect and recent activities of the European Federation of Corrosion Working Party on Corrosion in Concrete and those of the COST 509 Programme on Corrosion and Protection of Metals in Contact with Concrete are represented in a number of the contributions that appear in these Proceedings. In total the volume contains sixty-four refereed papers with authors from seventeen countries represented. The topics included embrace virtually all aspects of corrosion of reinforcement in concrete and its control. It is hoped that the Proceedings will therefore serve as a useful source of reference for both the research community and practising engineers working in this field.

The editors wish to express their sincere thanks to all members of the panel of referees who gave freely of their time and expert knowledge in reviewing papers. They wish also to acknowledge the substantial contribution made by Anne Borcherds of the Society of Chemical Industry who provided much valuable advice on matters of presentation. Without such support, the preparation of this volume in time for distribution at the Fourth International Symposium would have been impossible.

C L Page
P B Bamforth
J W Figg

Contents

MECHANISMS AND MACROCELL INTERACTIONS

NUMERICAL ANALYSIS OF GALVANIC INTERACTION IN REINFORCEMENT CORROSION

J. Gulikers and E. Schlangen

Delft University of Technology
Department of Civil Engineering
Stevinweg 1
2628 CN Delft, The Netherlands

1 INTRODUCTION

The corrosion of steel in concrete proceeds by means of an electrochemical mechanism which involves both microcell and macrocell corrosion. A microcell is developed when anodic and cathodic regions are formed alternately along the surface of the same reinforcement bar in very close proximity to each other. In such a situation, only the mixed corrosion potentials of the innumerable anodes and cathodes can be measured. In a macrocell, separate anodic and cathodic regions are formed some distance apart, and both the anodic and cathodic potentials can be measured.

Macrocell corrosion is likely to play an important role in reinforcement corrosion. The large dimensions of concrete structures and local differences in exposure conditions will undoubtedly promote differential electrochemical behaviour in the embedded steel. This is accompanied by potential differences along the reinforcement which in turn result in galvanic interaction on a large scale. The overall result is that part of the steel reinforcement will suffer an accelerated corrosion attack, which often leads to premature failure. Galvanic macrocell interaction generally arises from differences in aeration (oxygen), alkalinity (carbonation) or salt concentration (chlorides). In particular so-called active/passive corrosion cells are considered to be of great practical importance since the driving voltage usually amounts to several hundreds of millivolts. Active/passive galvanic cells are very common in chloride contaminated and patch repaired concrete structures. Although galvanic interaction is an integral part of reinforcement corrosion, its effects are often ignored or underestimated, despite the fact that numerous laboratory investigations have demonstrated its practical relevance.[1-5]

Galvanic testing usually leads to considerable difficulties in interpreting the measurements. Under ideal conditions the galvanic current can be converted directly into the corrosion rate. However, practical situations are usually far from ideal and in these cases the measured galvanic current can severely underestimate the true corrosion rate.[6]

In practical situations a numerical analysis can be of great help in understanding the mechanism and relating galvanic currents to the actual corrosion rates. The analysis uses electrochemical parameters as an input instead of employing artificial devices like current or voltage sources. This enables the analysis of galvanic coupling of any number of steel bars exhibiting different electrochemical behaviours. The magnitude of the corrosion rate prior to and following galvanic coupling can be compared to give an acceleration coefficient. This method allows the quantitative determination of the influence of all parameters of importance in reinforcement corrosion, i.e. concrete resistivity, cover depth, steel surface area ratios, and the distance between steel bars. Due to the complexity of galvanic interaction, development of a general correlation between experimental data and the results of numerical analysis was not attempted. Instead, a few cases were examined, and some experimental observations were made based on these studies.

2 ELECTROCHEMICAL THEORY OF GALVANIC CORROSION

Galvanic corrosion can be defined simply as the corrosion which occurs as a result of one piece of metal being in electrical contact with another in an ion conducting corrosive environment. The corrosion process is stimulated by the potential difference that exists between the two metals, the more noble metal acting as a cathode where an oxidising species is reduced, the more active metal, which preferentially corrodes, acting as the anode of the galvanic couple. The presence of these so-called macro-corrosion cells in reinforced concrete structures may have a significant effect on the local corrosion activities.

The electrochemical interaction between regions of steel with different rest potentials can be explained graphically by superimposing the potential-current (E-I, or polarisation) curves for the individual anodic and cathodic components of the galvanic macro-corrosion cell, on the same graph. For the simple case of a two component galvanic cell, this situation is depicted in a so-called Evans-diagram (see Figure 1). From the points of intersection of the cathodic and the anodic polarisation curves with the galvanic current, the potentials of the cathodic and anodic steel areas, denoted E_{gal}^a and E_{gal}^c respectively, and the magnitude of the galvanic current I_{gal} can be determined. In this graph E_{corr}^a and E_{corr}^c refer to the original steady state corrosion potentials of the anodic and cathodic components, respectively, prior to galvanic coupling.

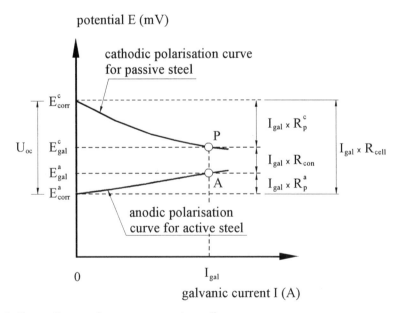

Figure 1 *Evans-diagram for macro-corrosion cells*

In a simplified way the steady-state current, I_{gal}, flowing in this galvanic macro-corrosion cell may be expressed mathematically as:

$$I_{gal} = \frac{E_{corr}^c - E_{corr}^a}{R_p^a + R_{con} + R_p^c} = \frac{U_{oc}}{R_{cell}} \tag{1}$$

The magnitude of the galvanic current, I_{gal}, is dependent not only on the initial potential difference $U_{oc} = (E_{corr}^c - E_{corr}^a)$ prior to galvanic coupling, but also on electrochemical (kinetic hindrance) and geometrical (cathodic to anodic area ratio) parameters. The resistances associated with the cathodic, electrolytic, and anodic retardations are represented by R_p^c, R_{con}, and R_p^a, respectively.

As a result of galvanic coupling the potential of the anodic component is shifted (polarised) towards more positive values, which generally results in an acceleration of the anodic iron dissolution rate while the cathodic reaction rate is simultaneously decreased. The difference in anodic and cathodic reaction rate of the anodic component is then equal to the galvanic current I_{gal} which is supplied by the cathodic component of the galvanic couple exhibiting the opposite behaviour.

Practical situations, however, are more complex as they usually involve a multi-component galvanic macrocell. The resulting galvanic interaction between the reinforcement bars can then be studied only by a numerical analysis which takes into account the specific electrochemical behaviour of the bars and the distribution of concrete resistivity. This permits a quantification of the local reinforcement corrosion rates as affected by galvanic macro-corrosion cells.

3 ANALYTICAL APPROACH

The numerical method used is directed towards solving the governing equations for the potential distribution and current fluxes throughout the galvanic corrosion system, taking into account the appropriate boundary conditions. The galvanic potential distribution in reinforced concrete is characterised by high electrolytic resistivity, ranging from 1×10^3 to 1×10^7 Ωcm. The anodic and cathodic current density at any point on the steel surface embedded in the concrete can be predicted from the electrochemical potential in the electrolyte adjacent to that point. The method models the polarisation behaviour of the cathodic and anodic parts of the galvanic corrosion cell, and the electrochemical properties of the concrete electrolyte. The analytical technique employed is capable of including general and arbitrary geometries and the effects of non-linear polarisation behaviour of the steel electrodes, for multi-component galvanic circuits.

For simplicity the analysis is used to calculate the electrochemical potential distribution in a 2-D situation, i.e., in a concrete cross section containing several steel bars, under steady state conditions.

The equation governing the potential distribution at points in the concrete electrolyte induced by a galvanic system can be expressed by Ohm's law:

$$I = \frac{U}{R_{con}} \tag{2}$$

where I denotes current (A), U potential difference (V), and R_{con} the concrete resistance (Ω). At the steel/concrete interface where polarising currents may enter or leave the system, electrical continuity has to be maintained. The cross sections of the steel electrodes participating in a galvanic circuit are considered to be infinite, coplanar and to exhibit polarisation characteristics under activation control with Tafel behaviour. In the following, it is therefore assumed that the relationship between polarising current, I, and the potential, E, of a steel electrode embedded in concrete, characterised by an initial corrosion current, I_{corr}, and corrosion potential, E_{corr}, can be described as:

$$I = I_{corr}\left[\exp\left(\frac{\ln(10)\,(E - E_{corr})}{b_a}\right) - \exp\left(-\frac{\ln(10)\,(E - E_{corr})}{b_c}\right)\right] \tag{3}$$

where b_a and b_c are the Tafel slopes for the anodic and cathodic reactions occurring at the steel surface. Evaluation of this relationship shows that the resistance of the steel/concrete interface to polarisation, R_p, varies with potential. In general, the resistance decreases as the shift from the free corrosion potential (overvoltage) increases.

In order to comply with the empirically derived value of the Stern-Geary constant B = 26 mV for actively corroding steel,[7] values of 90 mV/decade and 180 mV/decade were chosen for the anodic (b_a) and cathodic (b_c) Tafel slopes, respectively. The actual corrosion current density, i_{corr}, may range from 0.1 μA/cm^2 to values possibly greater than 100 μA/cm^2.[8]

For non-corroding steel ideally passive behaviour is assumed, characterised by $b_a = \infty$. As a consequence it is assumed that the passive steel is capable of supporting cathodic reactions only when the potential is shifted towards less noble values, whereas the anodic current density is restricted under all conditions to the initial corrosion current density for passive steel, i_{pas}. According to Hansson,[8] passive steel exhibits corrosion currents, probably of the order of 10 to 100$\times$10^{-9} A/cm^2. However, from curve fitting procedures performed on cathodic polarisation curves of passive stainless steel embedded in a chloride contaminated concrete environment, it was concluded that the average corrosion current density for passive steel, $i_{pas} \approx 2\times10^{-9}$ A/cm^2, and the average value for the cathodic Tafel slope, $b_c \approx 190$ mV/decade. Throughout the following numerical analyses the latter values will be used, together with a steady state potential value, $E_{corr} = +100$ mV. Prior to galvanic coupling it is assumed that the corrosion currents do not enter the concrete electrolyte and consequently no potential gradients are present.

4 DEMONSTRATION PROBLEM

Numerical analysis of galvanic reinforcement corrosion was performed on a concrete cross section containing two layers of reinforcement steel. Both the top and bottom layers consisted of four steel bars each, placed in parallel with a centre-to-centre distance of 62 mm, and a clear cover depth of 25 mm. The top layer consisted of four mild steel bars $\varnothing$6 mm, whereas the bottom layer contained four stainless steel bars $\varnothing$12 mm. Figure 2 shows the cross section of the specimen, and the position of the steel bars. The dimensions and reinforcement geometry are based on laboratory specimens designed to simulate actual conditions of galvanic corrosion, e.g. the cross section of a bridge deck. Chloride ions and moisture enter from the top surface and promote corrosion of the top layer of the reinforcement. The bottom layer of reinforcement, being surrounded by a drier concrete environment and therefore having more access to oxygen, will act as the cathodic component, supporting anodic

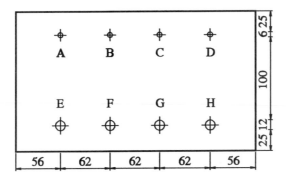

Figure 2 *Cross section of laboratory specimen (measurements in mm)*

dissolution of the top layer anodes. Corrosive environmental conditions were achieved by adding an excess amount of 4% chloride by weight of cement to the mixing water. Consequently the bars in the top layer are in an active state of corrosion whereas the four stainless steel bars remain essentially passive.

The cross section of this type of concrete specimen is modelled as a lattice with discrete resistor elements. This type of modelling was initially developed for investigations into the fracture of concrete[9] and modified for use with resistive elements. The triangular network thus obtained comprises 6500 nodes interconnected by 19171 elements. The electrochemical behaviour of the steel/concrete boundaries is described by interface elements, evenly distributed along the circumference of each steel bar. This results in the implementation of 6 and 16 interface elements for a steel bar $\varnothing$6 mm and $\varnothing$12 mm, respectively. For convenience, unit length, i.e. 1000 mm, is assumed for the depth of the specimen and the length of the eight steel bars contained within.

The solution to problems involving galvanic corrosion is obtained using the conjugate gradient method in an iterative calculation process.[10]

5 RESULTS OF NUMERICAL SIMULATIONS

In the following examples the influence of cathodic to anodic surface area ratio, concrete resistivity and initial corrosion rate on galvanic current, corrosion potential and corrosion rate will be demonstrated. In addition a practical situation is simulated, involving four actively corroding and four passivated steel bars.

5.1 Influence of Surface Area Ratio and Concrete Resistivity

Both the surface area ratio between the anodic and cathodic elements making up a galvanic cell, and the concrete resistivity are considered of special practical relevance in studies of the magnitude of macro-corrosion activity.

Numerical simulations were performed using a stepwise increase of the passive to active steel area ratio P/A, with P/A = 2.0, 4.0, 6.0, and 8.0, respectively. This was accomplished by subsequent coupling of 1, 2, 3, and 4 passive bars to one of the actively corroding steel bars (rebar B). In all cases the initial condition of the corroding steel was characterised by $E_{corr} = -400$ mV and $i_{corr} = 2.0 \times 10^{-6}$ A/cm^2 ($I_{corr} = 377.0 \times 10^{-6}$ A). The steady state condition

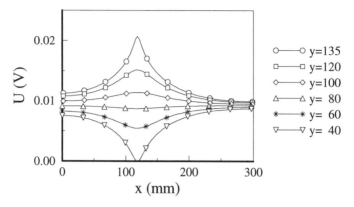

Figure 3 *Potential distribution in the concrete specimen at different heights; P/A = 2.0*

of the passive steel was described by E_{corr} = +100 mV and i_{corr} = 2×10^{-9} A/cm^2 (I_{corr} = 0.754×10^{-6} A). Thus a voltage of 500 mV is available to drive the galvanic corrosion cell. The concrete resistivity throughout the specimen was maintained at 10000 Ωcm.

For the galvanic circuit P/A = 2.0, which was achieved by coupling rebars F and B, the potential distribution in the concrete specimen is shown in Figure 3 for six different heights between the bars. The overall voltage over the concrete necessary to induce ionic current flow between the steel bars is approximately 26 mV, i.e. only 5.2% of U_{oc}. Approximately 59.4% of the total galvanic current flows through a concrete area with a width of 100 mm approximately halfway between the coupled steel bars. However, the concrete resistance plays a minor role in the galvanic corrosion process, and hence it is concluded that the limited size of the specimen has no significant effect on the magnitude of the galvanic current. The potential of the anodic component (rebar B) changes by 12 mV, which results in an increase of nearly 39% in the corrosion rate. By far the largest part of the driving voltage is used to activate the passive steel (rebar F). A cathodic overvoltage of 462 mV or 92.4% of the driving voltage, U_{oc}, is necessary to produce the galvanic current of 204 µA.

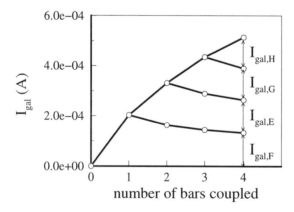

Figure 4 *Influence of the number of coupled passive steel bars on the galvanic current*

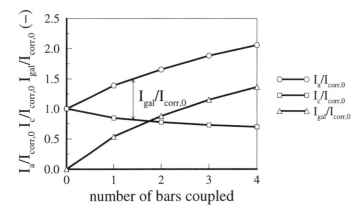

Figure 5 *Relative changes in anodic, cathodic and galvanic current with number of bars coupled*

A stepwise increase in the passive (cathodic) to active (anodic) surface area ratio is accompanied by an increase in the galvanic current, which is subsequently 204, 331, 434 and 514 µA. Due to the exponential relationship between current and potential for the passive steel, there is a less than linear increase in galvanic current, I_{gal}. This is shown in Figure 4, which also shows the individual contributions of the cathodic steel bars. The corresponding acceleration of the corrosion rate of the anodic rebar, B, amounts to 38.9, 65.5, 88.1 and 106.0%, respectively.

In the last situation, with P/A = 8.0, the corrosion rate of rebar B is more than doubled. The relationship between the number of passive steel bars coupled and the associated change of the anodic, cathodic, and galvanic currents, relative to the initial corrosion rate of rebar B, $I_{corr,0}$, is illustrated in Figure 5. Whereas the anodic current, I_a, increases steadily, the cathodic current, I_c, of rebar B decreases, the difference being supplied by the galvanic current, I_{gal}.

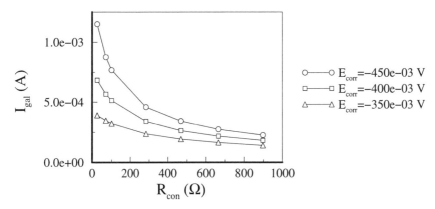

Figure 6 *Influence of concrete resistance on galvanic current for three different corrosion potentials of rebar B*

The influence of concrete resistivity, ρ_{con}, is further investigated for the case P/A = 8.0 (galvanic circuit B-E-F-G-H), ranging from 27 Ωm to 898 Ωm. This range includes resistivity values of practical relevance (portland and blast furnace slag cement concrete under 80% and 98% R.H. conditions). From Figure 6 it would appear that the influence of concrete resistivity on galvanic action is rather pronounced, the higher the concrete resistivity the lower the magnitude of the galvanic current. However, the control exerted by the passive steel remains dominant, gradually decreasing from 89.6% for P/A = 2.0 to 67.8% for P/A = 8.0. In the latter case the control exerted by the concrete resistance amounts to approximately 30%. In Figure 6 the influence of differences in the driving voltage, U_{oc}, is also presented. The driving voltage is altered by changing the corrosion potential, E_{corr}, of rebar B by 50 mV in both positive and negative directions. A significant influence is observed, particularly for low concrete resistivities (portland cement). By keeping the corrosion rate at I_{corr} = 377.0 μA and the concrete resistivity at ρ_{con} = 100 Ωm, the galvanic current increases by 49.0% for U_{oc} = 550 mV, whereas for U_{oc} = 450 mV a decrease of 37.2% is observed. This non-linear dependence of galvanic current on driving voltage can be explained by the fact that the effective passive steel resistance against cathodic polarisation, R_p^c, is strongly dependent on the shift in potential. The higher the driving voltage, the less the resistance will be.

As a result of galvanic coupling, the passive steel is subjected to a strong polarisation in the cathodic direction. Consequently, the potential value of the passive steel is shifted towards lower (more negative) values. For the four cases investigated (P/A = 2.0, 4.0, 6.0 and 8.0), the corrosion potential of the passive steel bars shifted several hundreds of millivolts from their initial values of +100 mV, to values ranging from −326.2 to −365.1 mV as measured by placing a reference electrode on the bottom concrete surface. As a consequence a potential survey will erroneously identify the passive steel as being in an active state of corrosion.

5.2 Simulation of Practical Situation

The large dimensions of actual reinforced concrete structures may result in multi component galvanic interaction. Both passivated and actively corroding sections of the steel reinforcement mesh will participate in the macro-corrosion cell. In the following example all eight steel rebars present in the concrete cross section will be coupled. The four rebars in the top layer exhibit active corrosion, presenting different corrosion rates and corrosion

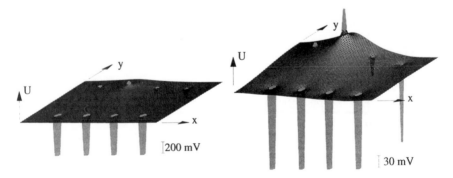

Figures 7a and 7b   *Potential changes of corroding and passive steel bars due to galvanic interaction*

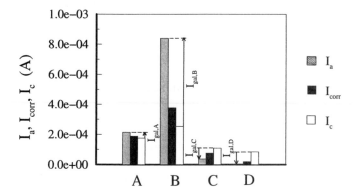

Figure 8 *Changes in anodic and cathodic currents of the corroding steel bars due to galvanic interaction*

potentials. The following combinations of corrosion potential, E_{corr}, and corrosion current density, i_{corr}, are assumed: -350 mV and 1.0 μA/cm^2 for rebar A; -400 mV and 2.0 μA/cm^2 for rebar B; -300 mV and 0.4 μA/cm^2 for rebar C; and -200 mV and 0.1 μA/cm^2 for rebar D. For simplicity the four passive steel bars in the bottom layer are given the same electrochemical parameter values as were used in the previous examples, i.e. $E_{corr} = +100$ mV and $i_{corr} = 2$ nA/cm^2.

Upon galvanic coupling of the eight rebars mentioned, two rebars (A and B) turn into anodic members, whereas six rebars (C, D, E, F, G and H) develop as cathodic components. This result is shown in Figures 7a and 7b which provide an isometric view of the potential changes, both in the concrete and at the steel/concrete interface. The potentials of the cathodic members have shifted towards more negative values, whereas the potentials of the two anodic members have moved in the positive direction. In Figure 7b the vertical scale has been magnified in order to show the different polarisation behaviour of the four corroding bars. For both steel bars C and D, cathodic polarisation results in a decrease in the corrosion rate of 52% (rebar C) and 95% (rebar D). The corrosion rates of rebars A and B are accelerated by 14% for rebar A and 123% for rebar B. The changes in anodic and cathodic current densities for the corroding steel bars are illustrated in Figure 8. Due to their relatively low resistance to polarisation, it appears that corroding steel bars can be very effective cathodes in comparison to passivated steel.

A comparable current distribution was found in experimental testing, which also confirmed that corroding steel can act as a cathodic component due to galvanic coupling.

6 CONCLUDING REMARKS

Galvanic macrocell corrosion in reinforced concrete results in complex electrochemical interactions. These effects can be quantified by a computational analysis. The mathematical representation of a corrosion process involving macrocell interaction is outlined. The galvanic corrosion problem is solved in terms of the equations which characterise electric fields in an ion conductive medium and the boundary conditions which describe the non-linear electrode kinetics of corroding and passive steel embedded in concrete.

The numerical analysis presented permits a quantitative prediction of corrosion current density distributions as affected by the galvanic interaction for arbitrary reinforcement steel geometries, electrochemical characteristics of the steel/concrete interface, and concrete conductivity distributions.

This method is applicable to practical situations involving galvanic corrosion, sacrificial anode and impressed-current cathodic protection designs, and stray current in reinforced concrete structures.

In most practical cases involving active/passive galvanic couples the concrete resistance has a minor direct effect on the magnitude of the galvanic current,[5] the passive steel has a more marked influence. This results in a strong cathodic polarisation of the passive steel corresponding to very negative values of the steady state potential. As a consequence potential mapping may erroneously identify the passive steel as being actively corroding.

For galvanic cells made up of corroding steel bars, the concrete resistance is dominant. When polarised in the cathodic direction, corroding steel is a far more effective cathode than passive steel.

For multi-component galvanic cells, involving actively corroding and passivated steel bars, anodic and cathodic components cannot always be distinguished beforehand. Under these conditions the corroding steel exhibiting the lowest potential values may act as a sacrificial anode, thereby cathodically protecting the corroding steel showing less negative corrosion potentials.

References

1. P. Schießl and M. Schwarzkopf, *Betonwerk + Fertigteil-Technik*, 1986, **52**, 626.
2. M. Raupach, 'Zur chloridinduzierte Makroelementkorrosion von Stahl in Beton', Beuth Verlag, Berlin, 1992, 433.
3. J. Nöggerath, 'Zur Makroelementkorrosion von Stahl in Beton', Dissertation ETH Zürich, Zürich, 1990, 9310.
4. K. Oberle and H. Wheat, 'The effect of repairs on the corrosion of reinforcing steel in concrete slabs', in 'Corrosion90', National Association of Corrosion Engineers, Las Vegas, 1990, Paper number 313.
5. J. Gulikers and J. van Mier, 'Accelerated corrosion by patch repairs of reinforced concrete structures', in 'Rehabilitation of concrete structures' (Eds: D. Ho and F. Collins), RILEM, Melbourne, 1992, 341.
6. N. Berke and M. Hicks, 'Electrochemical methods for determining the corrosivity of steel in concrete', in 'Corrosion testing and evaluation' (Eds: R. Baboian and S. Dean), American Society for Testing and Materials, STP1000, Philadelphia, 1990, 425.
7   C. Andrade, V. Castelo, C. Alonso and J. González, 'The determination of the corrosion rate of steel embedded in concrete by the polarization resistance and ac impedance methods', in 'Corrosion effect of stray currents and the techniques for evaluating corrosion of rebars in concrete' (Ed: V. Chaker), American Society for Testing and Materials, STP906, Philadelphia, 1986, 43.
8   C. Hansson, *Cement and Concrete Research*, 1984, **14**, 574.
9   E. Schlangen and J. van Mier, *Cement and Concrete Composites*, 1992, **14**, 105.
10 E. Schlangen, 'Computational aspects of fracture simulations with lattice models', in 'Fracture mechanics of concrete structures', AEDIFICATIO Publishers, Freiburg, 1995, 913.

CORROSION OF STEEL IN THE AREA OF CRACKS IN CONCRETE - LABORATORY TESTS AND CALCULATIONS USING A TRANSMISSION-LINE-MODEL

M. Raupach
Bauingenieur Sozietät Aachen
Sasse-Schießl-Fiebrich-Raupach
Steppenbergallee 226
D-52074 Aachen/Germany

1 INTRODUCTION

Although there are numerous recent reports and studies describing reinforcement corrosion in crack zones,[1-3] only a few results are available from practice-oriented tests describing the dominant mechanism for the case of chloride-induced reinforcement corrosion in the crack zone.[4]

In order to clarify the corrosion mechanism and the dominant influencing variables, especially the influence of crack width, laboratory tests were first performed on cracked reinforced concrete beams. Test results and a mathematical model[5] were then used to calculate the effect of crack distances and the effect of crack width limitation by using smaller rod diameters, on the predicted rates of steel removal due to chloride-induced corrosion.

2 ELECTROCHEMICAL PRINCIPLES

The electrochemical principles of steel corrosion in concrete have been described in detail.[6] The high alkalinity of the pore solution in concrete fundamentally provides durable protection for steel in concrete. A passive layer of ferrous oxides is formed on the steel surface, virtually excluding any possibility of iron dissolution.

This protection can be removed only through carbonation of the concrete or if a critical chloride content at the steel surface is exceeded (depassivation), allowing the reinforcement to corrode where a sufficient supply of moisture and oxygen at the steel surface are available.

The resulting corrosion is an electrochemical process, taking place in two substeps, as in a battery (Figure 1).

corrosion process

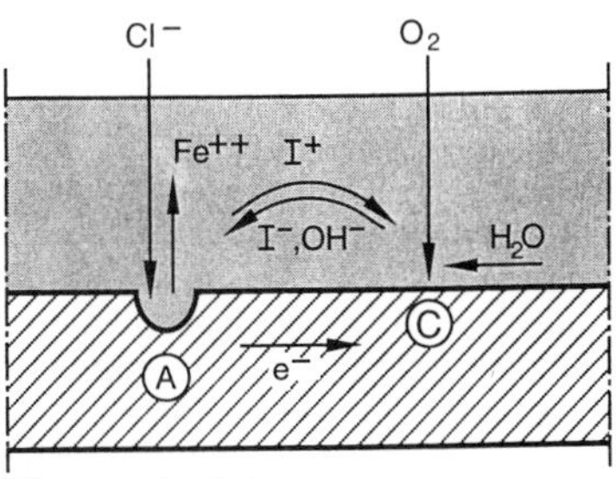

reactions

anode : $Fe \longrightarrow Fe^{++} + 2e^-$

cathode : $\frac{1}{2}O_2 + H_2O + 2e^- \longrightarrow 2(OH)^-$

$Fe(OH)_2 + X \cdot O_2 + Y \cdot H_2O \longrightarrow$ rust

Figure 1 *Schematic representation of the corrosion process*

• The anodic subreaction occurs at the actual corroding pits, where iron ions enter into solution, releasing two electrons per ion.

• The cathodic subreaction is not harmful to the steel. It usually occurs alongside the anodically-acting regions of the steel surface, where the free electrons react with water and oxygen, forming hydroxyl ions. These hydroxyl ions in turn react with the ions of iron in solution, forming the corrosion products, which are generally deposited near the anode.

In reinforced steel structures, anode and cathode may be microscopically adjacent, or, especially in cases of chloride-induced corrosion, up to a few meters apart (Figure 2). Whereas the anodic reaction is confined to steel surface zones where the critical chloride content causing depassivation is exceeded, the cathodic reaction may occur virtually anywhere.

Two different corrosion mechanisms are theoretically possible for steel corrosion in the region of cracks:

• In Mechanism 1, the anodic and cathodic subprocesses take place in the crack zone. Anodes and cathodes are extremely small and locally not separable (microcell corrosion). Oxygen supply to the cathodically-acting surface zones is mainly through the crack.

• In Mechanism 2, the reinforcement in the crack zone acts mainly as an anode, the passive steel surface between the cracks forming the cathode. In this case, oxygen transport to the cathode is chiefly via the uncracked area of the concrete (macrocell corrosion). Much higher corrosion rates are to be expected than in Mechanism 1, since the steel surface involved in the cathodic subprocess is much larger.

The laboratory tests described below were carried out in order to determine whether the steel between the cracks acts cathodically during reinforcement corrosion in the crack zone.

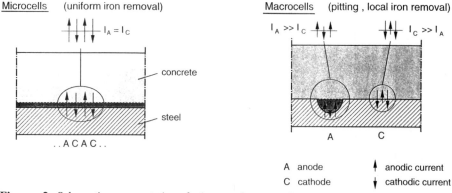

Figure 2 *Schematic representation of micro- and macrocell-corrosion*

3 LABORATORY TESTS

3.1 Test Set-Up

In order to be able to generate cracks with defined crack widths, reinforced concrete beams (70 x 15 x 9,7 cm) were clamped against steel girders (Figure 3). As a further measure to ensure unambiguous test conditions, only a single crack was generated in each beam, on the top side in the centre field.

In order to carry out electrical cell current measurements, it is necessary to interrupt the steel reinforcement at the points where the electric current is to be measured. A low-resistance ammeter can then be used to determine the current between the reinforcing steel zones, via cables leading to an external measuring point.[5] In the present instance, the reinforcement in the immediate vicinity of the crack, which can be depassivated by the action of the chlorides, therefore had been separated from the neighbouring zones, which can act only as cathodes, in order to observe any macrocell formation. The depassivated steel surface is, however, very small and extends about 1 cm inside both crack flanks.

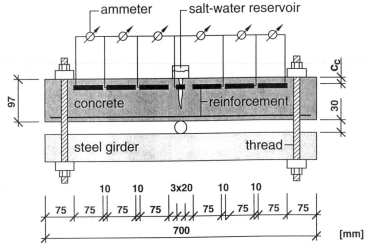

Figure 3 *Test set-up and measuring points for laboratory tests on cracked reinforced concrete beams*

It is not possible to induce a crack through the enveloping surface area of a roughly 2 cm long section of reinforcing steel, since the required anchoring length is substantially greater. This problem was solved by using a reinforcing steel which had been coated with an epoxy resin by a coating company except for a 2 cm central section. This allows the crack to be induced through the enveloping surface area of the uncoated section, while the coated sections are used to apply the force and cannot act significantly as cathodes.

The potentially cathodic reinforcement adjacent to cracks was simulated by adding three reinforcing steel sections on each side of the crack, allowing cathodic action to be determined as a function of distance from the crack used (Figure 3). The rod diameter was 14 mm.

The side faces and undersides of the beams were coated with an epoxy-resin-based coating to prevent most of the effects of rapid drying of the beams via these areas; this simulates the situation in a large component more closely.

3.2 Concrete Composition

The concrete in the tests is composed of 300 kg/m^3 OPC 35 F with a w/c-ratio of 0.6 and an AB aggregate grading curve according to DIN 1045. This concrete has Strength Class B 35 and standard consistency according to DIN 1045.

3.3 Performance of the Tests

3.3.1 Fabrication and Storage of the Specimens. The beams were cast in steel moulds, removed after one day and stored for a total of two days in a humid room.

Three days after concreting, each of the beams was clamped against the steel girders and cracks of the desired width were induced. The required central position of the crack was obtained by inserting a 0.5 mm thick, 7 mm deep plastic strip in the surface of the green concrete as a crack initiator.

The specimens were stored in a climatic chamber at a temperature of 20 ± 1 °C and 80 ± 5 % relative atmospheric humidity.

3.3.2 Wetting. The tests described here relate solely to chloride-induced corrosion and not to corrosion caused by carbonation of the concrete.

In order to establish clear conditions and prevent depassivation elsewhere than near the crack, chloride wetting was confined within 2 cm wide wetting frames placed on the top of the beam in the crack zone (Figure 3). In the tests described here, wetting was started 28 d after casting of the specimens.

On the basis of results from studies[7] in the splash zone of a motorway near Düsseldorf during the harsh winter of 1986/87, a 1 % chloride solution was poured into the wetting frame above the crack once weekly for a period of 24 h. Twelve wetting periods were followed by two periods in which tapwater without added chlorides was introduced. The specimens were then stored without further wetting at 80 % relative atmospheric humidity in a climatic chamber; after 1 year the cycle of 12 wetting periods was repeated in order to simulate real conditions as closely as possible.

3.3.3 Cell Current Measurement. The cell currents were measured under computer control, using a data logging system developed at the Institute for Building Materials Research (ibac) at the Technical University of Aachen, Germany.[5]

The electric current values can be converted directly into corrosion rates: a current of 100 μA is equivalent to an annual mass of iron of 911 mg transformed into rust. Assuming that the entire steel surface of 8.80 cm[2] in the crack zone is corroded, the steel removal rate will be 132 μm/a. Since it is unlikely that the entire steel surface in the crack zone will corrode and since removal of the depassivated steel surface is not uniform, the actual local corrosion rates will, however, be considerably higher.

3.4 Test Results

3.4.1 General. Figure 4 shows the time curves of the measured cell currents between the steel in the crack zone and the six steel sections outside the crack for the basic mix and a crack width of 0.5 mm. An extremely high cell current of some 200 μA is observed immediately after the first chloride wetting, corresponding to a local steel removal rate in the crack zone of more than 250 μm per year. When the salt solution is removed from the wetting frame the next day, however, the corrosion rate declines significantly. The eleven

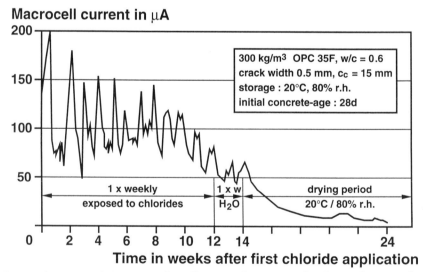

Figure 4 *Measured time curves for cell currents between steel in the crack zone and steel sections outside of the crack*

successive chloride wettings which follow are also characterized by current maxima during the wetting phase and a subsequent decrease in cell current. There are probably several reasons for the fact that the current peaks decrease over time, although the chloride concentration in the crack zone increases with each wetting:

• The permeability of the concrete diminishes with increasing concrete age, increasing the electrolytic resistance and inhibiting all ionic transport processes;

• Chloride binding is progressive over time;

• The formation of corrosion products in the crack zone obstructs material transport to and from the corrosion pits.

It is not at present possible to judge which of these causes is finally decisive. It may be anticipated that all three play a significant role.[5]

Wetting with tapwater containing no added chloride in the 13th and 14th weeks after initial chloride wetting produces a cell current curve similar to that in the preceding periods. After wetting ends, the concrete in the crack zone dries out and cell currents fall to values below 10 µA during the succeeding 10 weeks. This behaviour documents the known decisive influence of humidity on the corrosion rate of the reinforcement.

In some specimens, macro-corrosion commenced only after a number of wetting cycles, or did not occur at all.

The following section will discuss first the test results indicating macrocell formation and then the influence of concrete cover, concrete composition and crack width.

3.4.2 Corrosion Mechanism. By measuring the electric current between the reinforcing steel electrodes (Figure 3) it is possible to determine for each region of the reinforcing steel whether more electrons have been received than sacrificed (cathode) or vice versa (anode).

Figure 5 shows the current balance of each of the reinforcing steel sections for the basic mix on the first day after chloride wetting, with a crack width of 0.5 mm. In order to show the effect of the reinforcement/crack distance, the widths of the bars correspond to the lengths of the reinforcing steel sections.

It is apparent that regions of the reinforcement outside the crack behave cathodically up to a distance of more than 20 cm from the crack. The current density of the cathodes falls with increasing distance from the crack, because electrolytic resistance rises with crack distance.

The evaluation of the current balances for the specimens aged up to four months and for the other concrete mixes and covers investigated in the tests showed that the current

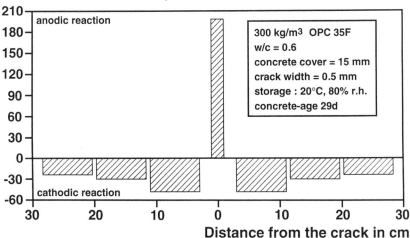

Figure 5 *Current balances of individual reinforcing steel sections during initial chloride wetting*

distribution deviates only slightly from that depicted in Figure 5.

The conductivity of the concrete is accordingly high enough to allow formation of quite large cathodes at the prevailing relative atmospheric humidity of 80 %, roughly representing the annual average for Germany.

In general, it may be concluded from the results of cell current measurements that Corrosion Mechanism 2 in Figure 2 is usually the dominant corrosion mechanism in crack zones.

3.4.3 Influence of Concrete Cover on the Corrosion Rate. Figure 6 plots the total mass of steel removed during the 24-week test period, as calculated from the measured cell current curves,[5] against the tested concrete covers, w/c-ratios and crack widths. Figure 7 shows the total mass losses after a period of two years, i.e. after a total of 2 x 14 chloride or water wetting periods. As expected, it is evident that mass losses increase with time mass losses from specimens depassivated during the first wetting period are roughly twice as high after 2 years as after 24 weeks. Particularly high growth rates for the mass losses are observed in specimens which were not depassivated until the second year. One specimen with a concrete cover of 35 mm, w/c = 0.5 and a crack width of 0.1 mm suffered virtually no corrosion throughout the whole test period.

In terms of concrete cover, it may be inferred from Figures 6 and 7 that increasing the concrete cover from 15 to 35 mm leads to much reduced removal rates. This relationship is already known from other research work.[1-3]

3.4.4 Influence of the Water/Cement-Ratio. A reduction in the w/c-ratio from 0.6 to 0.5 yielded a further reduction in steel mass loss in the crack zone (Figures 6 and 7). The influence of w/c ratio is especially pronounced after 24 weeks, while the influence of the w/c-ratio becomes much smaller after one year.

These effects may be explained by the fact that the period up to depassivation is prolonged by a reduction of the w/c-ratio, but that after the onset of corrosion the w/c-ratio has only a negligible influence. This result applies, however, only to the tested w/c-ratios between 0.5 and 0.6. With larger differences (e.g. between 0.4 and 0.7), a significant influence may be anticipated even after the onset of corrosion; this factor was not investigated in the present study.

3.4.5 Influence of Crack Width. The results in Figures 6 and 7 indicate that corrosion currents usually increase with increasing crack width given the selected test conditions and observation periods, but it is also evident that concrete cover and composition have a much greater influence than crack width.

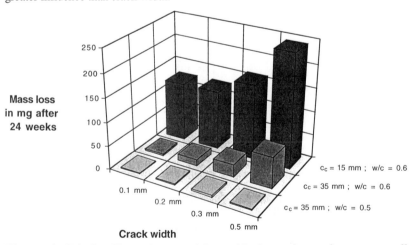

Figure 6 *Calculated losses in mass of the steel in the crack zone due to macrocell corrosion after a test period of 24 weeks (cf. Figure 3)*

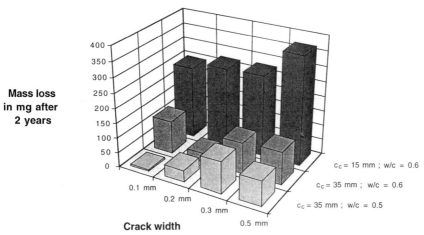

Figure 7 *Calculated losses in mass of the steel in the crack zone due to macrocell corrosion after a test period of 2 years (cf. Figure 6)*

In general, the influence of crack width decreased with time, as illustrated by the specimen with a concrete cover of 35 mm and a w/c-ratio of 0.6. While mass loss was still increasing with crack width after 24 weeks, no continuing influence of crack width was detectable after two years. This result is again due to the fact that the time to depassivation increases as crack width decreases; on the other hand, there is no significant relationship between crack width and corrosion rate after depassivation under the conditions of the present study.

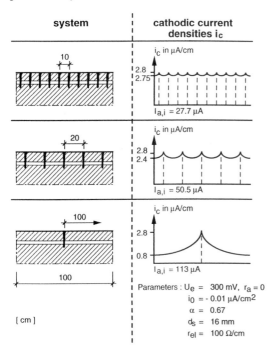

Figure 8 *Simplified model used for the calculations of the influence of geometry on macrocell-corrosion (explanations given in the text and literature[5])*

4   CALCULATIONS OF THE CORROSION RATE OF REINFORCEMENT IN THE CRACK ZONE THROUGH THE FORMATION OF MACROCELLS

4.1 General

As clearly demonstrated by the cell current measurements described above, macro-corrosion cells with local anodes in crack zones and large cathodes between the cracks, effective up to several decimetres from the crack, occur in chloride-induced reinforcement corrosion in the crack zone. The corrosion rate of such macro-corrosion cells can be calculated on the basis of simplified assumptions.[5]

Figure 8 shows the simplified corrosion model of a macrocell consisting of one anode and different arrays of cathode-elements. The electrical currents between the anode and the cathode-elements cause voltage drops, IR, within the concrete and polarisations, ΔU, at the surfaces of the cathode-elements. The cathodic polarisation-curve of passive steel in concrete shown in the Evans-Diagram (Figure 8) has been determined for 60 specimens with different concrete-mixes.[5] As the anodic polarisation depends on various factors, it has been assumed, that the anodes are not polarisable ($U_{R,a} = U_{C,a}$) or that the corrosion potential at the anode $U_{C,a}$ is known. The concrete resistance, r_{el}, is related to the distance between the steel elements.

The following section will first discuss the influence of crack distance on the reinforcement corrosion rate in the crack zone,[5] and then go on to calculate and explain the effects of a crack width limitation through reducing the steel diameter.

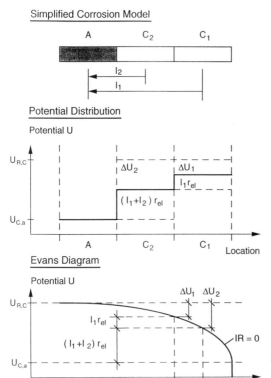

Figure 9 *Influence of crack distance on reinforcement corrosion rates in the crack zone due to macrocell formation*

4.2 Influence of Crack Distance

Figure 9 presents calculated results showing the effect of crack distance on current distribution and steel corrosion rate in the crack zone.[5] The following assumptions are made:
* unidimensional electrical field;
* the distance to the crack is assumed to be the maximum cathode size for each anode, i.e. overlapping electrical fields for several anodes are not taken into account;
* steel diameter: 16 mm;
* voltage between active anode and cathode: 300 mV;
* polarization of the anode is negligible, i.e. as could occur with high chloride and moisture contents;
* electrical resistance of the concrete per unit length: 100 Ω/cm, i.e. incompletely dried (partially wet) concrete;
* the polarization behaviour of the cathode is described by $i_0 = 0.01$ μA/cm^2 and $\alpha = 0.67$, i.e. there is no oxygen deficiency at the cathode, i.e., the concrete is not permanently water-saturated.[5]

Figure 9 shows that the cathodic currents decrease significantly with increasing distance from the crack, agreeing with the results of laboratory tests presented in Figure 5.

Since the cathodically-acting regions of the steel surface may extend a distance of several decimetres from the crack, the potential corrosion rate is reduced where distance to the crack is small. It is, for example, apparent from Figure 5 that raising the crack distance from 10 cm to 20 cm results in an approximate doubling of the steel corrosion rate in the crack zone, whereas an additional fivefold increase in crack distance from 20 cm to 1 m leads only to a further doubling of the corrosion rate.

These calculations indicate that a significant reduction in overall corrosion rate occurs when a number of anodes are close to one another.

4.3 Influence of Crack Width Limitation Through Reduction of Rod Diameter

It is known that crack widths can be reduced by limiting the rod diameter. The effects of such a diameter limitation on the steel corrosion rate in the crack zone are calculated below.

It is assumed that the number of reinforcing rods is selected to ensure that the overall steel cross-section remains constant. If it is likewise assumed that the length of depassivated steel in the crack zone increases in direct proportion to the steel diameter, the depassivated surface area of steel in the crack zone will be four times the rod cross-section. Given a constant overall steel cross-section, it follows that the total depassivated area will remain the same irrespective of rod diameter.

In terms of the cathodically-acting regions of the steel surface, a reduction in rod diamter has two main effects:
• the total surface area of a single steel bar between the cracks increases in inverse proportion to the ratio of the diameter and
• the crack distance decreases less than proportionately to the ratio of the diameters.[8]

Whereas increasing the total surface area of the steel fundamentally tends to accelerate the reinforcement corrosion rate in the crack zone, reduction of the inter-crack distance has the opposite effect, as already indicated in Section 4.2. Since the steel diameter has a greater influence on the total surface area of the steel than on the crack to crack distance, however, a reduction in rod diameter will generally lead to higher corrosion rates in the crack zone, as shown below on the basis of a comparative calculation. This applies only to a case in which the influence of the cathodic reaction is not negligibly small. However, if the concrete has not dried out to too great an extent, and if the chloride content above the critical limit value is present in the crack zone, the cathodic reaction always plays a significant role.[5]

The following calculation was made in order to quantify the effect of a diameter reduction:

In a flexurally-stressed beam with a width of 40 cm and an effective height related to crack-width limitation of 8.75 cm, a reinforcement cross-section of 6 cm^2 is required.[8] In the calculation, the expected corrosion rates for a reinforcement configuration with two 20 mm diameter-rods (Case 1) and twelve Ø 8 mm diameter-rods (Case 2) are compared. All other assumptions correspond to those given in Section 4.2.

Table 1: *Results of comparative calculations on the effect of a crack width limitation by restricting the rod diameters on the loss of cross-section of the steel in the crack zone.*

Parameters	Dimensions	Case 1	Case 2	Case 2 / Case 1
Reinforcement	mm	2 Ø 20	12 Ø 8	-
Cross section of the steel	cm^2	6.28	6.03	0.96
Depassivated steel surface area	cm^2	25.1	24.1	0.96
Total perimeter of the steel bars	cm	12.6	30.1	2.4
Distance from crack to crack	cm	16.2	9.5	0.6
Total macrocell current	μA	78	135	1.7
Mean anodic current density	μA/cm^2	3.11	5.60	1.8
Mean rate of steel removal	μm/a	36.1	65.0	1.8
Mean annual loss in cross section of the steel bars	%/a	0.36	1.63	4.5

Table 1 summarizes the various parameters and the calculated results for the two cases. As already noted, given the above assumptions, the depassivated area in the crack zone amounts to four times the steel cross-sectional area .

In Case 2, the total steel-surface area increases in inverse proportion to the ratio of the respective rod-diameters, i.e. by a factor of 20/8 = 2.5. According to Equation (5) in reference 8, the mean crack distances are 16.2 cm (Case 1) or 9.5 cm (Case 2) respectively. It may be inferred from the influences of the steel-surface area and inter-crack distance that the total corrosion rate in Case 2 will be 1.7 times higher than in Case 1. The mean metal removal rates can be determined from the current densities,[5] which rise by a factor of 1.8 due to the reduction in steel diameter.

However, it is not the removal rate but the loss in cross-section which is decisive for assessment of the effective strength of reinforced concrete structures. The loss in cross-section is obtained by dividing the mean removal rates by the radius of the rod. This shows that the losses in cross-section are inversely proportional to the ratio of the diameters. In the present example, a reduction in rod diameter from 20 mm to 8 mm results in an increase in annual loss in cross-section by a factor of 1.8 x 20/8 = 4.5.

The same calculation was performed for the assumption that only every second or fourth crack is depassivated. Results showed that the losses in cross-section increase by a factor of 5.75 if every second crack or by a factor of 7.25 if every fourth crack is depassivated. The results can be explained by the fact that the increased circumference of the steel has a greater effect with small diameters, if the distance between the depassivated cracks increases.

In general, it may be stated that the losses in cross-section of the steel are greatly increased by reducing the rod diameter. The factor for increased loss of cross-section is roughly 1 to 3 times the reciprocal of the rod diameters.

5 CONCLUSIONS

Cell current measurements on cracked reinforced concrete specimens show that chloride-induced steel corrosion in the crack zone involves the formation of macro-corrosion cells. The steel in the crack zone acts as the anode and the steel between the cracks, up to a distance of several decimetres from the cracks, acts as the cathode. The steel corrosion rate in the crack zone is therefore influenced by the conditions between the cracks.

It was also established that the steel removal rates in the crack zone fall significantly with increasing concrete cover and are slightly lower for a concrete with a w/c-ratio of 0.5 than for one with a w/c-ratio of 0.6. The influence of crack width declined significantly with test duration, until after a period of two years no clear relationship between crack width and steel removal rates in the crack zone was observable.

Using the method of calculation previously described in the literature,[5] it was shown that the corrosion rate in the crack zone also decreases with shorter crack distance, because the cathodically-acting steel surface area for each crack is smaller.

Finally, it was shown that limiting the size of reinforcement in order to restrict the crack widths results in higher losses in cross-section despite the reduced crack distance and crack width. Calculations indicated that the increase is usually about 1 to 3 times the reciprocal of the ratio of the rod diameters.

These calculations and the results of the laboratory tests clearly indicate that the problem of reinforcement corrosion in crack zones cannot solely be solved by crack width limitation in the range between roughly 0.3 and 0.5 mm; corrosion protection must be assured primarily through adequate concrete quality and cover.

References

1. G. Hartl and W. Lukas, *Betonwerk und Fertigteil-Technik* , 1987, **7**, 497.
2. G. Rehm, U. Nuernberger and B. Neubert, *Schriftenreihe des Deutschen Ausschusses für Stahlbeton* , 1988, **390**, 43.
3. P. Schießl, *Schriftenreihe des Deutschen Ausschusses für Stahlbeton*, 1986, **370**, 10.
4. K. Okada and T. Miyagawa, Detroit Michigan: American Concrete Institute, *Performance of Concrete in Marine Environment*, 1980 (ACI SP-65), pp. 237-254.
5. M. Raupach, Berlin: Beuth, *Schriftenreihe des Deutschen Ausschusses für Stahlbeton*, 1992, **433**, Thesis.
6. P. Schießl, RILEM: Technical Committee 60-CSC, New York: *Chapman and Hall*, 1988 (RILEM Report).
7. J. Weber, Karlsruhe: Institut für Massivbau und Baustofftechnologie. 1988, *Abschlußkolloquium des Forschungsschwerpunktprogramm der DFG*, Karlsruhe 4./5. Okt. 1988, 55.
8. P. Schießl, *Schriftenreihe des Deutschen Ausschusses für Stahlbeton,* 1989, **400**, 157.

INFLUENCE OF CRACKING AND WATER CEMENT RATIO ON MACROCELL CORROSION OF STEEL IN CONCRETE

Y. Ohno*, S. Praparntanatorn** and K. Suzuki*

*Department of Architectural Engineering
Faculty of Engineering
Osaka University, Japan
**Department of Civil Engineering
Faculty of Engineering
Thammasat University, Thailand

1 INTRODUCTION

Almost all international codes of reinforced concrete design relate permissible crack widths to anticipated exposure conditions in order to avoid crack induced corrosion. Narrow cracks can be obtained by reduction of the concrete cover. However, the reduction of the cover shall facilitate the ingress of chloride ions.

A large number of researchers have tried to clarify the corrosion problems. Under natural conditions, corrosion is a slow process. Some studies took as long as 10-25 years to be completed. [1,2] One of the facts found from these studies is that the location of rust and cracks is usually coincident. Reference 3 concluded that there is no relationships between crack width and degree of corrosion. Reference 4 reported an interesting result, among the 136 cracks that induced more than 1% of chloride ions by cement weight, only 58 initiated corrosion. This result certainly suggests that there are both active and passive cracks in multi-crack specimens. The previous works of the authors showed that the largest crack of a multi-crack beam tend to initiate corrosion first.[5] It was also found that macrocell corrosion within multi-crack specimens should result in passive cracks. Therefore, the study of the relationship between cracking and corrosion using multi-crack specimens may not be appropriate.

This paper presents a study of crack induced corrosion in single-crack specimens subjected to accelerated exposure. The parameters were crack degree, water to cement ratio, and cover depth. The corrosion characteristics are interpreted through galvanic currents and electrochemical characteristics of anodes.

2 EXPERIMENTAL PROCEDURES

2.1 Test Specimens

Concrete was made from portland cement, river sand, 10 mm maximum size aggregate, tap water and a water reducing agent. The mix proportions are shown in Table 1. The configuration of the beam specimens is shown in Figure 1.

The beams were reinforced with a 13 mm diameter steel bar coated with epoxy resin. The 14 mm diameter and 48 mm long mild steel rod of which the ends were

Table 1 *Mix Proportions (kg/m³) and Compressive Strength (MPa)*

W/C	Water	Cement	Sand	Gravel	σ_c
70%	190	271	834	953	27.5
50%	175	350	736	1030	45.3
30%	160	533	577	1082	83.0

epoxy coated was embedded in the specimens. Two electrically connected 14 mm diameter and 150 mm long stainless steel rods were also embedded in the specimens in order to facilitate cathodic reactions. In order to yield only one crack, the specimens were precut 10 mm in depth at the corner of the midsection of the tensile face. The single flexural cracks were introduced and sustained by back-to-back loading as shown in Figure 1. The variables were crack width (zero (N), small (S), and large (L) according to the stress levels), water to cement ratio, and cover depth. The specimen designation, stress levels, and ranges of cracks are given in Table 2. Half of the uncracked specimens in Series (2) were made of chloride contaminated concrete (0.1 % chloride weight of concrete).

2.2 Exposure Conditions

The galvanic cells were set up by electrical connection between the embedded steels. The cells were subjected to an accelerated test comprising automatically repeated cycles of one day wetting in 65℃, 3.1% NaCl solution and one day drying in the empty tank. The schematic system is shown in Figure 2. The total period of the

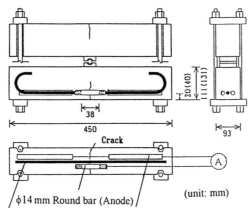

Crack

φ14 mm Round bar (Anode)

(unit: mm)

D13 mm Reinforcing steel φ 14 mm Stainless steel (Cathode)
coated with epoxy resin

Figure 1 *Details of specimens*

Table 2 *Parameters of the Experiment*

Series (1)

Water Cement Ratio (%)	Level of Crack Width* (mm)	Cover Thickness (mm)
70	N(0), S(0.1)	40
50	N(0), S(0.1), L(0.2)	20, 40
50	S(0.1), L(0.2)	20, 40

Series (2)

Water Cement Ratio (%)	Level of Crack Width* (mm)	Cover Thickness (mm)
70	N(0), L(0.2)	40
50	N(0), L(0.2)	40
50	N(0), L(0.2)	40

*S (L): Steel stress is 150 (250) MPa at the cracked section.
Specimen designation (Ex. L 40·50 (1))
    First letter: Level of crack width, First number: Cover thickness,
    Second number: Water cement ratio, Last number: Series number

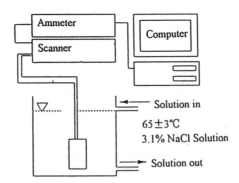

Repeated Cycles of One Day Wetting in 65°C, 3.1% NaCl Solution
and One Day Drying in Empty Tank

Figure 2 *Schematic accelerated test set up*

exposure tests was 35 cycles (70 days) and 70 cycles (140 days) for Series 1 and 2, respectively. For half of the specimens in Series (2), bare stainless steel rods exposed directly to the solution were used as the cathodes from the 39th cycle.

2.3 Examination and Evaluation

The macrocell currents were automatically detected and recorded during the exposure test. The measuring system is shown in Figure 2. The electrochemical characteristics of the half cell potential, Ec, and polarization resistance, Rp, were periodically measured against an Ag/AgCl reference electrode and a platinum counter electrode during the drying periods. At the end of the exposure test (Series 1), the specimens were split and the weight loss and corroded areas of the anode steels were measured.

3 RESULTS AND DISCUSSION

3.1 Corroded Area

Corrosion was found in all cracked specimens in Series(1). It was also found in some uncracked specimens with a 0.70 w/c ratio. For uncracked specimens made of 0.30 w/c concrete, almost all of the embedded steels were free of rust. Undoubtedly, flexural cracks induce corrosion more easily than perfect concrete cover. The difference is more pronounced in lower w/c concrete.

3.2 Macrocell Currents

The current-time relationships observed are presented in Figure 3. The corrosion cells, except for the case of the uncracked or 0.3 w/c concrete specimens, generated the expected macrocell current (mild steel acting as an anode and stainless steel acting as a cathode) during the early wetting cycles of the accelerated exposure. The galvanic currents between the mild steel anodes and the stainless steel cathodes flowed all the time, even during the drying periods. All the current-time curves show similar patterns, like mountain shapes, with single or multiple peaks. Initially, the currents rose to their first peak and then dropped to lower levels. Some specimens showed a continuous decline in current peaks until the end of the exposure test. The continuous reduction is similar to that found in galvanic couples, between precracked specimens and bare stainless steel, under synthetic sea-water.[6] The reduction and its magnitude reflect the accumulation of corrosion products at the steel-concrete interface. The characteristics of the current-time relationships indicate the substantial role played by anodic dissolution of steel in concrete.

For the corrosion cells where the cathodes were embedded in the crack specimens, the maximum current during the wetting cycles was 200μA which was nearly twice that found in the uncracked specimens under this accelerated test[7] and a 400 mV corrosion cell.[8] The current during the drying cycle was less than 50μA. This reduction in current can be attributed to the decreases in temperature and restricted water supply, and to the increased resistivity. Compared with the uncracked specimens in reference 7, the cracked specimens exhibit higher currents suggesting an

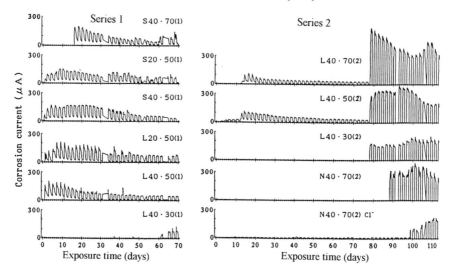

Figure 3 *Current time curves*

easier degree of anodic dissolution in the cracked sections. In other words, the cracked area reduces the accumulation of corrosion products around the steel surface which would otherwise inhibit anodic dissolution. The anodic dissolution in concrete is an important controlling factor in macrocell corrosion.

When the anode steel was connected to the external cathode steel, a large macrocell current flowed immediately. This result confirms that the properties of the concrete surrounding the cathode have a significant effect on the cathodic reactions. For the cracked specimens made with 0.7, 0.5 and 0.3 w/c ratio concrete, after coupling with the external cathode, the maximum current was 600, 400 and 250µA, respectively. This suggests that w/c ratio i.e. the properties of the concrete surrounding the anodes also has an effect on macrocell corrosion in the cracked specimens.

The macrocell current in the uncracked specimens in series 2 commenced some days after external cathode coupling. Comparing uncracked specimens made with chloride contaminated concrete and these made with chloride free concrete, it appears that the 0.1% chloride (by concrete weight) contamination may not initiate significant corrosion prior to the corrosion activated by chloride penetration. The characteristics of chloride contaminated concrete and chloride penetration effects on macrocell corrosion need further study. The proposed accelerated test may be an appropriate methodology.

3.3 Degree of Macrocell Corrosion

Figure 4 shows the relationship between the summation of the macrocell current and the weight loss of the anode steel. It is clear that the summation of the macrocell current is nearly proportional to the weight loss. The degree of macrocell corrosion may be represented by the summation of current over time (Sum. I · Δt). Figure 5 shows

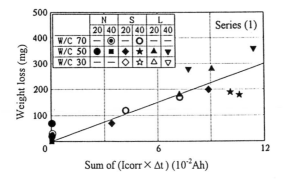

Figure 4 *Relationships between summation of corrosion current and weight loss*

the comparisons of macrocell corrosion against w/c ratio, crack width, and cover thickness.

As shown in Figure 5 (a), the extent of macrocell corrosion depends on crack width. There is a trend for the degree of corrosion to increase as crack width increases. The relationship is particularly marked for the 0.30 w/c specimens. The principal factor affecting macrocell corrosion in cracked concrete appears to be w/c ratio. It can be seen in Figure 5 (c) that the degree of macrocell corrosion is considerably higher for cracked specimens made with 0.50 w/c than those made with 0.30 w/c. The same result was observed in reference 7.

The effect of cover thickness on macrocell corrosion is shown in Figure 5 (b). For 0.30 w/c cracked specimens, there is a moderate trend for the degree of corrosion to decrease as the cover increases. The tendency is not apparent for cracked specimens made with higher w/c concrete. This may be due to the gap and interface between the steel and the concrete being affected by bleeding of the fresh concrete. The results need further investigation.

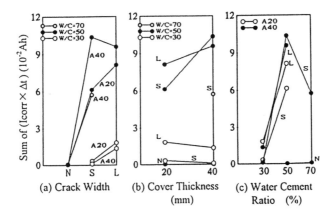

Figure 5 *Summation of corrosion current   (Series(1))*

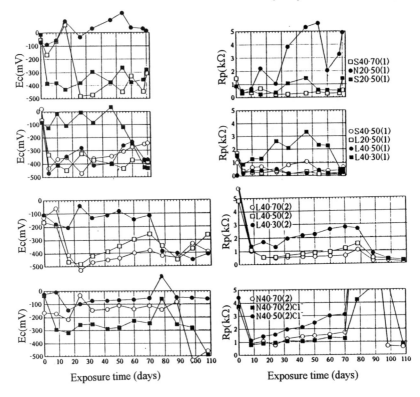

Figure 6 *Typical electrochemical characteristics*

3.4 Electrochemical Characteristics

Figure 6 shows the typical change with time in the half cell potential Ec and the polarization resistance Rp taken at the center of the specimens. The electrochemical characteristics of the specimens (N20 · 50(1), N40 · 70(2), and N40 · 50(2)Cl⁻) in which corrosion current did not flow were markedly different from the others. In these cases, the Ec and Rp increased over time. For specimens where the current flowed during the first wetting cycle, the initial Ec and Rp values observed after exposure were relatively low. It was noticeable that the specimens with higher magnitudes current had lower values of Rp. The results show that the electrochemical characteristics over time corelate the initiation time and the magnitude of corrosion current.

3.5 Electrochemical Characteristics and Degree of Macrocell Corrosion

Figure 7 shows the relationships between Ec and the total current (sum. I · Δt) at the 35th cycle. The -76 and -226 mV horizontal lines in the figure indicate the probability of corrosion according to ASTM. The electrode potentials given in the standard are corrected for the different electrode used. The probability of corrosion is

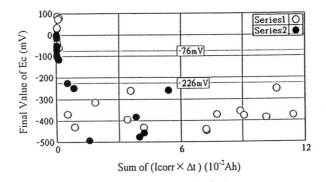

Figure 7 *Relationships between half cell potential (Ec) and
summation of corrosion current*

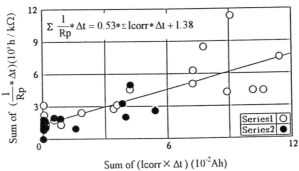

Figure 8 *Relationships between sum of 1/Rp ×Δ t and
summation of corrosion current*

95 %, when Ec is less than -226 mV. When Ec is more than -76 mV, the corrosion
probability is 5 %.   The results of the present study conform well to the criteria of
ASTM.   Thus, the risk corrosion of steel in concrete can be non-destructively
evaluated by Ec measurement.   The extent and rate of corrosion, however, cannot be
evaluated from the value of Ec.

The reciprocal of Rp is theoretically proportional to the rate of corrosion.
Therefore, the integration value of the reciprocal of Rp with time should correspond to
the corrosion degree.   The relationship between the sum of $1/Rp \times \Delta t$ and the sum of
corrosion current shown in Figure 8 appears to confirm this assumption.

4 CONCLUSIONS

Based on the results, it is concluded that:
(1) The proposed accelerated test can be applied effectively to study reinforcement
    corrosion within a short period of time.   The total exposure of time in the study

was only 70 days or 140 days.

(2) Flexural cracks in concrete cause corrosion to occur more readily than uncracked concrete. Crack width has a significant effect on the degree of corrosion in good concrete.

(3) Compared with the uncracked specimens, the cracked specimens had a flatter current-time relationship and a higher current magnitude. Anodic dissolution is facilitated in cracked concrete.

(4) The most important factor influencing the degree of macrocell corrosion in cracked-concrete is the w/c ratio. The total macrocell current increased as the w/c ratio increased.

(5) Change in the electrochemical characteristics with time corelate well with the initiation time and the magnitude of the corrosion current.

Acknowledgments

This study was performed in the laboratory of the Department of Architectural Engineering, Osaka University. The assistance given by Yuji Yamashita, a graduate student of Osaka University during the experimental program is sincerely appreciated.

References

1. P. Schiessl, 'Admissible crack width in reinforced concrete structure, Contribution II 3-17. Preliminary Reports Vol. II', Inter-Association Colloquium on the Behavior of in Service of Concrete Structures, 1975

2. E. F. O'Neil, 'Study of reinforced concrete beams exposed to marine   environment, Performance of Concrete in Marine Environment', ACI SP-65, American concrete Institute, Detroit, 1980, 113.

3. A. W. Beeby, 'Concrete in the ocean: Cracking and corrosion', Technical report No.1, C&CA, Department of Energy, 1978.

4. G. Hartl and W. Lukas, 'Investigations on the penetration of chloride into concrete and on the effect of cracks on chloride-induced corrosion of reinforcement', Betonwerk + Fertigteil-Technik, 1987, 497.

5. K. Suzuki, Y. Ohno and S. Praparntanatorn, 'Mechanism of steel in cracked concrete, Corrosion of reinforcement in concrete, 3rd International symposium', Society of Chemical Industry, 1990, 19.

6. O. Vennesland and O. E. Gjorv, 'Effect of cracks in submerged concrete sea structures on steel corrosion', *Material Performance*, 1981, 49.

7. Y. Ohno K. Suzuki and S. Praparntanatorn, 'Macrocell corrosion of steel in uncracked concrete', 'Corrosion and Corrosion Protection of Steel in Concrete, International Conference', Sheffield Academic Press, 1991, 1, 224.

8. H. T. Cao. and V. Sirivivatnanon, 'Corrosion of steel in concrete with and without silica fume', *Cement and Concrete Research*, 1991, 21, 316.

EFFECTIVE CATHODE TO ANODE RATIO AND REINFORCEMENT CORROSION IN CONCRETE

C. Arya[a] and P.R.W. Vassie[b]

[a]South Bank University
Wandsworth Road
London SW8 2JZ, U.K.

[b]Transport Research Laboratory
Crowthorne
Berkshire RG11 6AU, U.K.

1 ABSTRACT

Corrosion of steel in concrete was studied using three models: (1) a 4 m long beam containing a bar in seven segments, one mild steel and the remaining stainless steel, (2) a 5.25 m long beam in six sections, five 1 m long and one 250 mm long, containing mild steel bars at 250 mm centres and (3) a slab, 300 mm wide x 150 mm deep x 1.2 m long, containing mild steel electrodes generally at 35 mm centres. In all cases corrosion was initiated by using concrete dosed with 3% Cl from NaCl. Corrosion activity was monitored by half cell potential, linear polarisation resistance and galvanic current measurements using a zero resistance ammeter.

The results show that both the cathode to anode area ratio and anode to cathode separation are important factors influencing localised corrosion. A possible interpretation of the LPR and galvanic current results is provided.

2 INTRODUCTION

Reinforcement corrosion remains the most serious cause of the premature deterioration of concrete structures worldwide. Many methods have been proposed to combat this problem, the most practical being to delay the onset of corrosion either by treating the surface of the concrete with waterproofing membranes, carbonation coatings, silanes/siloxanes,[1] or by reducing the permeability of concrete by lowering the w/c ratio and incorporating cement replacement materials.[2,3] Recent evidence suggests, however, that such measures alone may not be sufficient in all circumstances.[4,5] In order to limit the effect of reinforcement corrosion during the life of the structure it may also be necessary to reduce the rate of corrosion following initiation.

An examination of the localised corrosion process suggests that the rate of corrosion is related to the effective cathode to anode ratio.[6] One method of limiting reinforcement corrosion would perhaps be to restrict the effective cathode to anode ratio in structures.

The principal aim of this paper is to present results of a preliminary investigation of how (i) cathode to anode area ratio (ii) anode to cathode separation distance and (iii) concrete quality affect the corrosion of steel in concrete. The corrosion process was studied using three models, the construction details of which are given below.

3 EXPERIMENTAL PROCEDURE

3.1 Specimen Design

Model A was made of six segments of AISI type 316 stainless steel reinforcement, 25 mm in diameter and 600 mm-665 mm in length (Figure 1).[7] The ends of the segments were tapped and connected to adjacent sections using short lengths of threaded nylon rod. The nylon also served to insulate adjacent sections of reinforcement. The centre bar of the assembly was a segment of mild steel reinforcement, 25 mm in diameter and 20 mm in length. Each segment was electrically connected at one point using stainless steel type 316 wire, sleeved with PTFE. This was to allow any or all of the stainless steel segments to be electrically connected to the mild steel section via an external connection.

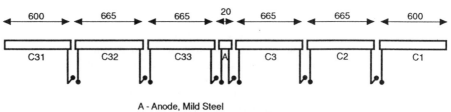

A - Anode, Mild Steel
C1, C2, C3, C31, C32, C33 - Cathodes, Stainless Steel
Dimensions in mm

Figure 1 *Layout of segmented bar with all switches in the open circuit position*

The segmented bar was placed centrally in a steel mould, 80 mm x 150 mm x 4000 mm long. Silver-silver chloride, half-cells (P_o, P1, P2, P31, P32) were attached to the segmented bars at the positions shown in Figure 2 using nylon ties.

Two beams (A1 and A2) containing the above assembly were cast. The proportions of the two mixes used are given in Table 1. The chloride content in both mixes was 3% by weight of cement. The chloride ions were introduced by dissolving appropriate quantities of sodium chloride into the mix water.

Table 1 *Mix proportions*

	Mix 1	Mix 2
OPC	1	1
Zone 2 Sand	2.4	1.5
Thames Valley Flint 10 mm	2.9	3
Water	0.65	0.45
Superplasticiser	-	1.7%
NaCl	3% Cl*	3% Cl*

* by weight of cement

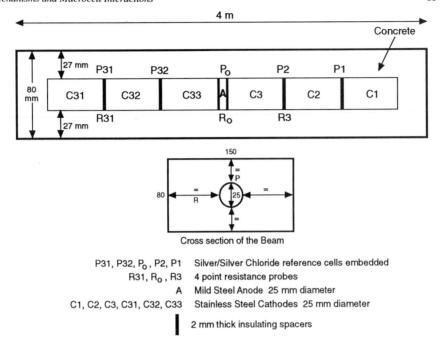

Figure 2 *The test beam showing the position of the anode, cathodes, reference electrodes and resistance probes*

Resistance probes (R_o, R3, R31) each consisting of four stainless steel type 316 plates of dimensions, 10 mm x 10 mm, and stainless steel connecting wire were cast into the beam at the locations shown in Figure 2.

Model B was a 150 mm wide x 170 mm high x 5.25 m long beam in six sections, five 1 m long cathode beams and one 250 mm long anode beam (Figure 3). Each cathode beam contained four segments of mild steel, 16 mm diameter x 120 mm long at 250 mm centres. The anode beam contained a 16 mm diameter mild steel bar coated in a polymer modified grout and sealed with heat shrink tubing. The central 30 mm length of bar was left bare, thereby allowing it to act as an anode. The anode and cathode beams were laid end to end and connected electrolytically using capillary matting which was kept permanently moist by immersing the bottom end of the matting in a tray of water.

The mix proportions used for the anode beam were the same as mix 1 (Table 1). The cathode beams were also cast from concrete with the same mix proportions but excluding the chloride.

Model C was a slab 300 mm wide x 150 mm high x 1200 mm long, containing twenty one mild steel bars 16 mm diameter and 240 mm in length, generally placed at 35 mm centres (Figure 4). One bar was made to act anodically with respect to the others by casting a 50 mm square block of concrete, dosed with 3% Cl by weight of cement, around the bar. The length of the bar in contact with the chloride-bearing concrete was actually only 30 mm. The remainder of the bar was masked from the concrete by coating the surface in a polymer modified grout and sealing with heat shrink tubing as per the anode beam in model B. The mix proportions used for the slab were the same as mix 1 (Table 1) but excluding the chlorides.

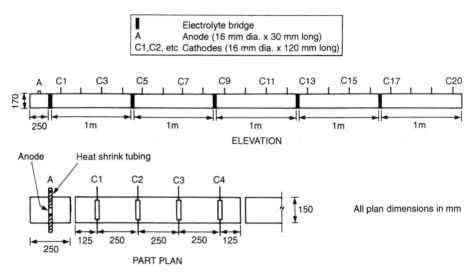

Figure 3 *Beam B showing the position of the anode, cathodes and electrolyte bridges*

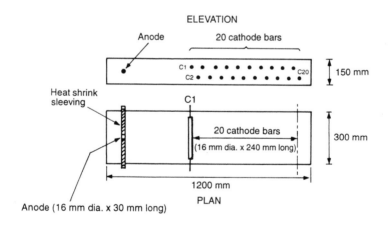

Figure 4 *Slab C showing the position of anode and cathodes*

3.2 Testing Procedure

All the test specimens were cured for a period of at least 56 days in a fog room at a relative humidity of 95% and temperature of 21° C prior to testing.

In all cases, the galvanic current flowing in the specimens for various cathode to anode connections was measured using a zero resistance ammeter (ZRA). The corrosion current density of beam B and slab C was also determined by linear polarisation resistance (LPR) measurements. The LPR current density was determined using an automatic potentiostat linked to a ramped generator. The ramp applied was from -10 mV to +10 mV from the rest potential at a sweep rate of 0.1 mV/sec using an external counter/reference electrode system.

The resistances between the electrodes were measured using an AC bridge to give a measure of the concrete resistances and, in the case of beam B, to ensure that the anode and cathode beams remained electrochemically connected during the test period. All readings were taken once steady state conditions had been attained.

4 RESULTS AND DISCUSSION

4.1 Compressive Strengths

Table 2 shows the average twenty eight day compressive strengths of the concretes in mixes 1 and 2.

Table 2 *Average 28 day compressive strengths*

	Mix 1		Mix 2
	with 3% Cl	0% Cl	
Compressive strength (N/mm^2)	31	30	44

4.2 Cathode to Anode Ratios

Figure 5 shows the steady state current densities in beams A1 and A2 and slab C as a function of the cathode/anode area ratio. In beams A1 and A2 the stainless steel segments were electrically connected to the mild steel anode in the order: C3, C33, C2, C32, C1 and C31. In slab C the cathodes were connected to the anode in the order C1 to C20.

The current densities of beams A1 and A2 increased as the cathode/anode area ratio increased up to at least 190, due to an increase in the cathodic surface random reactions. The current density in slab C reached a maximum value of 3.5 μA/cm^2 at a cathode/anode area ratio of approximately 40. Presumably the galvanic current did not increase further as more cathodes were added was because the anodic reactions were at their maximum level.

The rate of increase in current density was higher for slab C than beam A1 despite the fact that both specimens were made from concrete with essentially the same mix composition. This was probably attributable to differences in the location of anodic and cathodic areas in the two specimens.

The current densities were considerably higher in beam A1 than beam A2, presumably due to the fact that the concrete in beam A2 had a lower water/cement ratio and hence lower permeability than in beam A1. The lower current obtained from the low permeability mix can probably be explained by:

    1)   the higher resistances measured in beam B (Table 3), even though both beams contained 3% Cl by weight of cement

2)   lower rates of transport of oxygen and ferrous ions producing restrictions to the cathode and anode reaction kinetics.

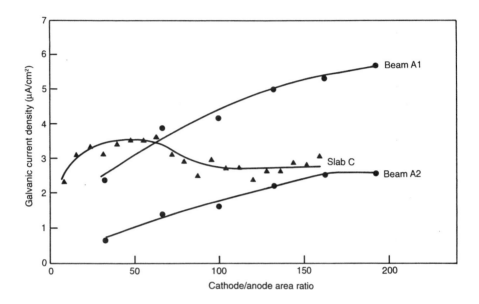

Figure 5 *Effect of cathode/anode area ratio on the galvanic current density*

Table 3 *Average resistances in beams A1 and A2*

Resistance Probe	Resistance (Ohms)	
	Beam A1	Beam A2
R31	200	250
R_O	215	260
R3	190	250

4.3 Distance Between Anode and Cathode

Figure 6 shows the steady state current densities for beams A1 and A2 and beam B for increasing distance between the anode and cathode. It can be seen that generally the greater the distance between the anode and cathode, the smaller the current density.

Comparing the performance of beams A1 and A2, it can be seen that with the better quality mix in beam A2, the reduction in current density is more pronounced.

The only effect of increasing the anode-cathode separation is to increase the length of the return path for negative ions transferring charge from the cathode to the anode through the concrete. This ohmic effect substantially reduces the current although it is worth noting that the most distant cathodes cause significant corrosion currents to flow. The large throwing power of the anode, particularly in the poorer quality concrete, explains why large reductions in bar cross-section can occur over very short time spans in concrete structures.

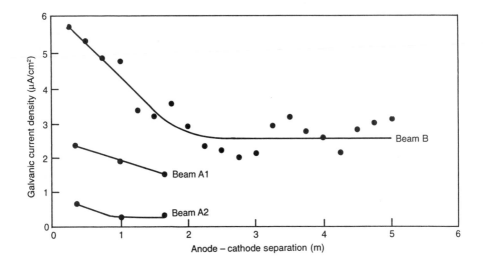

Figure 6 *Effect of distance between anode and cathode on galvanic corrosion current density*

The development of a concrete mix that limits the throwing power of the anode would produce a marked reduction in the area of the cathode and hence also in the corrosion current in concrete structures. The test specimens used in this study had the reinforcement in one dimension only, therefore a reduction in the throwing power of the spheroidal anode field in a structure would result in a much greater reduction in cathode area than was achieved in the test beams.

4.4 Comparison of LPR and Galvanic Currents

Figures 7 and 8 show the LPR and galvanic current densities for, respectively, beam B and slab C. It can be seen that in almost all cases, the LPR current density exceeds the corresponding galvanic measurement. It is suggested that this is due to the fact that the galvanic technique only measures the current flowing between the electrodes whereas LPR measures the total corrosion activity.[8,9]

More significant perhaps was the fact that the LPR current densities hardly changed as either the cathode/anode area ratio or the distance between anode and cathode increased.

4.5 Anode and cathode potentials

Tables 4 and 5, for beams A1 and A2 respectively, compare the potentials of each section in the connected and isolated modes. These results clearly show that even cathode sections more than a metre distant from the anode are being significantly polarised.

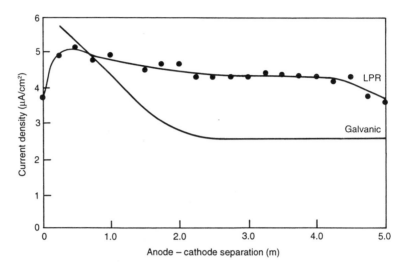

Figure 7   *Effect of distance between anode and cathode for beam B on current measurements using LPR and galvanic couples*

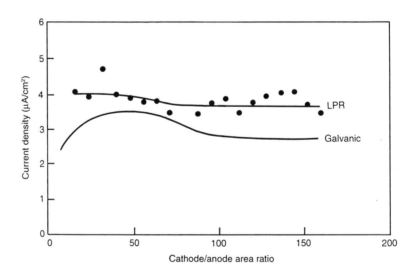

Figure 8   *Effect of cathode to anode area ratio for slab C on current measurements using LPR and galvanic couples*

Table 4   *Potentials of cathode elements for beam A1 in the isolated and connected modes, versus the Ag/AgCl reference electrodes*

Cathode	Potential (mV)	
	Isolated	Connected
C31	-35	-260
C32	-30	-300
C33	-70	-370
C1	-40	-270
C2	-50	-320
C3	-100	-360

Table 5   *Potentials of cathode elements for beam A2 in the isolated and connected modes, versus the Ag/AgCl reference electrodes*

Cathode	Potential (mV)	
	Isolated	Connected
C31	-160	-280
C32	-120	-260
C33	-100	-340
C1	-100	-300
C2	-95	-270
C3	-85	-330

4.6 Discussion

The results can be summarised as follows:

a. The anode current density as measured by the LPR method increased when the first cathode was connected but subsequently remained approximately constant as additional cathodic bars were connected.

b. The galvanic current increased as additional cathodic bars were added up to about three bars and then remained approximately constant.

c. The potential of the cathodic bars changed significantly when they were connected to the anode even when there was no change in the LPR or galvanic currents.

d. The connection of a single cathode bar to the anode bar has the effect of increasing the LPR current and generating a significant galvanic current even when the cathode is more than 1 m distant from the anode.

Possible explanations for and significance of these results are:

1. The maximum anodic current density occurs at relatively low cathode/anode area ratios.

2. Although further increases in the number of cathode bars connected does not increase the anodic current density the fact that the galvanic current does increase implies that the cathodic current is redistributed over the available cathodic area. The regions of the anode bar that were acting cathodically in the disconnected state

probably become anodic because the incipient anode[10] will develop when an alternative source of cathode in chloride free concrete is provided. The state at which the anode bar is behaving entirely as an anode may correspond to the maximum anodic current since the anodic area is then maximised.

3.   Eventually the connection of additional cathode bars no longer produces an increase of the galvanic current, however the cathode bars are significantly polarised. This suggests that these bars form part of the macrocell, but that the current drawn from them is very small. Ultimately at large cathode-anode separations the cathode will not be polarised and will not form part of the macrocell due to ohmic effects.

4.   The presence of a large cathode/anode ratio in a macrocell does not seem to have much influence on the rate of anode dissolution; however it could be very significant in terms of concrete repairs since in a reinforced concrete structure incipient anodes could be present anywhere within the boundary of the macrocell. Thus replacement of the anodic and all the cathodic regions of the macrocell should be sufficient to ensure an effective repair.

5.   Even when the distance of the nearest cathode to the anode is more than a metre, significant galvanic currents are generated supporting the argument that if epoxy coated steel is used as reinforcement then it should be used throughout to prevent the possibility of uncoated steel acting as a cathode and forming a macrocell with anodic sites caused by defects in or damage to the epoxy coating.

5 REFERENCES

1.   Department of Transport, Impregnation of Concrete Highway Structure, Department Advice Note BA33/90, Her Majesty's Stationery Office, London, 1990.

2.   R. K. Dhir, M. R. Jones and M. J. McCarthy, *Mag. Concr. Research*, 1994, **46**, No.169, 269.

3.   P. B. Bamforth and W. F. Price, 'Proc. Int. Conference: Concrete 2000', Univ. Dundee, Scotland, ed. R. K. Dhir and M. R. Jones, 1993, **2**, 1105.

4.   E. J. Wallbank, 'The Performance of Concrete Bridges: A 200 Bridge Survey', Department of Transport, Her Majesties Stationery Office, London, 1989.

5.   G. Osborne, 'Proc. P.K. Mehta symposium on Durability of Concrete', Nice, France, 1994, 119.

6.   N. J. M. Wilkins and P. F. Lawrence, Proc. Conference: 'Corrosion of reinforcement in concrete construction' ed. A. P. Crane, Ellis Horwood Ltd, 1983, 119.

7.   C. C. Naish, A. Harker and R. F. A. Carney, Proc. Third Int. Symposium on 'Corrosion of Reinforcement in Concrete', ed. C. L. Page, K. W. J. Treadaway and P. B. Bamforth, Elsevier, London, 1990, 314.

8.   C. Andrade, I. Rz-Maribona, S. Feliu and J. A. Gonzalez, Proc. NACE 'Corrosion 92', Paper 194, 1992.

9.   N. S. Berke, D. F. Shen and K. M. Sundberg, STP 1065, eds. N. S. Berke, V. Chaker and D. Whiting, ASTM, Philadelphia, PA, 1990, 38.

10.   P. R. Vassie, *Proc. Inst. Civ. Engrs.*, Part 1, 1984, **76**, 713.

DETERIORATION OF CONCRETE DUE TO CORROSION OF REINFORCEMENT IN IMMERSED CONCRETE AND PREVENTIVE MEASURES

R. O. Müller

Helbling
Ingenieurunternehmung AG
Hohlstr. 610
CH - 8048 Zürich
Switzerland

1 INTRODUCTION

Water service reservoirs are important buildings in our infrastructure. They are often built from concrete and are ideally suited to store and preserve drinking water, as their mineral composition is similar to the mineral composition of the ground. Therefore, concrete can conserve drinking water without contaminating it with organic compounds which might impede its cleanliness and its taste.

A high life expectancy of these reservoirs, of 50 and more years, is important in order to keep the expenses of our infrastructure at an affordable level.

Sewage plants are also built predominantly from reinforced concrete. They too should attain a reasonable life time to keep the cost of purification of waste water at a reasonable level.

In some middle-aged drinking water reservoirs deterioration and patches of soft concrete at the bottom of the reservoirs have been discovered. Many different theories have been put forward to explain these attacks on the concrete. It is possible that drinking water has become more aggressive towards calcium silicate hydrate and thus is attacking the cementitious part of the concrete. However, this does not explain why in a typical drinking water pool, not all regions of the container are attacked to the same extent.

Other theories suggest that the action of microorganisms could attack the hydrated calcium silicate layer at the surface of the concrete and thus enhance local attacks.[1] However it is not understood why these attacks should form locally at the bottom of the containers, which is the case, and not on the wetted concrete surfaces as well.

A further explanation is, that upon erection of reservoirs not all the surfaces will be in exactly the same condition. Bleeding and settlement will cause irregularities. Differential evaporation of water may also cause differences in the water/cement ratio. But similar patterns of attack have also been observed in reservoirs with walls made of sprayed concrete which should be more homogenous or, at least, show other inhomogeneities than cast concrete reservoirs.

The attacks reflected water levels of operation, not differences in the local quality of the concrete due to manufacturing.

Rendell[2] reported results obtained from concrete slabs which were partially submerged in sea water. The reinforcing bars at different levels were insulated from each other and connected outside the slabs by electrically insulated cables. In this way galvanic cell currents could be measured and it was shown that the partially submerged zone became cathodic while the deeply submerged zone became anodic. Current densities and corrosion rates in the anodic zones were described, but no information was given on the effects on the concrete.

Gerdes and Wittmann[3] published experimental evidence showing that concrete exposed to a direct current deteriorated due to transport of ionic species through the concrete. Damaged areas at the concrete surface in the region of the anode were reported. The same characteristic damage, with spots of weak concrete, could be produced in the laboratory when direct current was forced through immersed concrete slabs. The deterioration of the concrete under the influence of an electric field was explained using a model based on migration processes and subsequent reaction mechanisms. Similar results were observed by Menzel.[4]

2 EXPERIMENTAL

A considerable number of empty water reservoirs were inspected (Figures 1 and 2). The investigations were performed visually and electrochemically by measuring the potential of the rebars by contacting the concrete surface with a wet sponge and a copper/copper sulfate reference electrode. The rebars which were generally in good electrical contact with each other (this was checked in several cases) were connected to the positive input of a high impedance millivoltmeter. The negative input was connected to the copper/copper sulfate reference electrode and touched the concrete surface with a wet sponge. This method is described in ASTM C 876 : Half-cell potential measurement of uncoated reinforcing steel in concrete.[5-8] The method enables the measurement of the rebar potential and assessment of the condition of the rebars without the need to remove the concrete covering.

Figure 1 *50 year old water tank in apparently good condition*

The method is non-destructive and produces information about corrosion reactions in which non-expansive corrosion products are formed. In this case the visual method fails as there are no cracks formed (Figure 3).

In a similar measuring circuit it is also possible to exchange the high impedance (> 10 Mohm) millivoltmeter for a low impedance (100 Ohm) microammeter. In this way it is possible to measure the short circuit direct current. By dividing the open circuit potential (in millivolts) by the short circuit current (in microamperes) it is possible to calculate the direct current resistance of the described electrical circuit.

Figure 2   *Attacks and weak spots on the bottom of the tank shown in Figure 1*

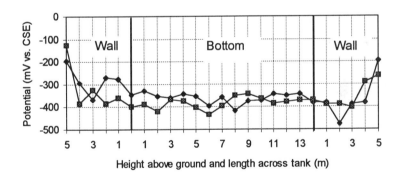

Figure 3 *Potentials measured on two axes in a water tank in Hombrechtikon, Switzerland*

Control measures with alternating current resistance meters showed a good correlation.

The resistance of wet concrete is directly related to the volume of open pores and the conductivity of the pore water. The measuring device is influenced by the quality of the concrete at the contact location of the wet sponge, which has an area of about 5 cm in diameter.

If the resistance values, measured at different points on the concrete surface are recorded, a resistance-diagram can be generated (Figure 4).

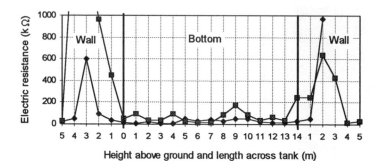

Figure 4 *Electric resistance measured on two axes in a water tank in Hombrechtikon, Switzerland*

3   RESULTS

The potential plots measured in different reservoirs are shown in Figure 3. The walls showed quite high potentials whereas at the bottom of the tanks relatively low potentials of between - 300 mV and - 400 mV were measured. Thus it is clear that a driving force of 100 mV to 300 mV is acting, causing corrosion in the non-aerated zones when the reservoir is filled with conducting water.

At potentials of - 300 to - 400 mV one would not expect corrosion attack to occur in alkaline concrete. There are two reasons which explain this contradictory fact:

1. The concrete covering of the reinforcement causes a 'potential drop'. The measured values of - 300 to - 400 mV correspond to much lower potentials at the interface of the steel and concrete.

2. The measured values of - 300 to - 400 mV are mixed potentials between spots of actively corroding iron with potentials of - 600 mV and a pH of ~ 7 - 8 and passive iron with higher potentials with a pH of ~ 13.

The resistivity diagrams show that the walls in the oxygenated region have a resistance which is 4 to 400 times higher than the resistance at the bottom of the container (Figure 4).

At the base of a column the resistance was measured on a weak spot. The result was 18kΩ. About 15 cm away from the weak spot, the measured resistance was 350 kΩ which was 20 times higher than at the weak spot. This difference was not directly related to the different thicknesses of the covering of the rebars.

At the bottom of empty reservoirs resistance values of 3 kΩ to 20 kΩ are generally recorded. In the same reservoirs, higher up the walls, where higher potentials were measured, the resistance varied from around 50 to 500 kΩ or even more (Figure 4).

In another reservoir, after eight years of service, many brown spots of 1 to 3 cm diameter were observed. The concrete below the spots was weak. As a repair measure the surface was sand-blasted and new layers of white cement mortar with a total thickness of about 3 to 5 mm were applied. After only four years, similar brown spots appeared again, as shown in Figure 5. At such locations the potential was 50 to 100 mV lower than 20 cm away and the resistance was only 3 to 5 kΩ instead of 15 to 25 kΩ.

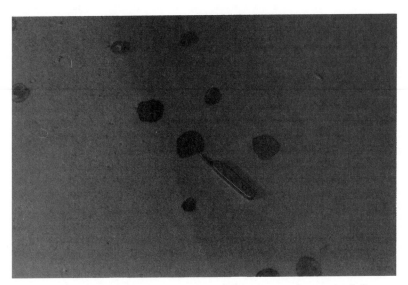

Figure 5 *Attacks and weak spots at the bottom of a tank, repaired four years ago*

4 DISCUSSION

The environment of rebars in alkaline concrete can be described thermodynamically by Pourbaix diagrams.[6] Investigations have shown that the main corrosion product under weakly oxidizing conditions is iron-2-hydroxide. Therefore the diagram with $Fe(OH)_2$ will be used (Figure 6).

The Pourbaix diagram shows that between the region of stable iron (mild steel) at low potentials, and passive steel, which is formed in alkaline and chloride free concrete under highly oxidizing conditions, there is a region of iron-2-hydroxide which is less protective.

In isolated systems this does not lead to enhanced corrosion, as iron and iron-2-hydroxide are stable at low potentials due to the lack of an electromotive force. This means that as long as there is no oxidizing agent, the system is stable. However in aerated concrete the iron will form a passive layer showing higher potentials.

In water tanks and reservoirs between the region of high potentials (e.g. 0 mV) at the air-water interface and the region of low potentials due to oxygen depletion (e.g. - 500 mV) there is a potential difference of a few hundred millivolts. In a filled tank this causes a macro cell to form and a current flows from the anode to the aerated cathode. Due to this current the corrosion rate of the oxygen depleted rebars is greatly enhanced (Figure 7).

In the present paper the term 'anode' will always be used to mean the electrode or metal-electrolyte interface where the oxidation reaction takes place. The cathode refers to the metal-electrolyte interface where the reduction, for example of oxygen, takes place, regardless of the potential at this interface.

Figure 7 shows the anodic reaction whereby steel is oxidized to form positive iron ions. This causes a positive space charge which attracts ions of opposite charge out of the water: firstly bicarbonate and eventually other anions, such as chlorides.

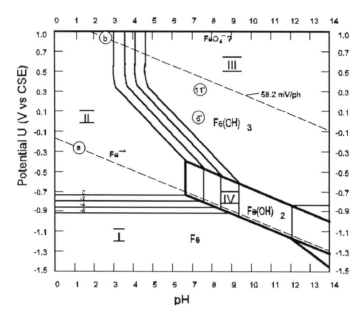

Figure 6 *Potential-pH Diagram for the system iron-water at 25°C. Considered species:*
Fe, Fe(OH)$_2$, Fe(OH)$_3$. Legend: I: Stability of iron, II: Soluble iron salts,
III: passivity, IV: non-protective iron-2-hydroxide.

In the electric field, the ionic current is transported through the cement pores by a
flow of positively charged sodium, potassium, calcium and iron ions into the water and a
flow of negatively charged ions into the cement pores.[3]

In this way the concrete surface is depleted of calcium resulting in a local reduction
in mechanical strength. At the same time a much higher carbonation rate is observed
than should be the case for immersed concrete.[3,4,7]

At the same time the porosity of the concrete is increased near the local anodes. As a
result, a decrease in resistance values has been recorded. The increase in the concentration
of anions such as bicarbonate, chloride and nitrate around the anode stabilizes the pitting
reaction and increases the corrosion rate.

In the cathodic regions the formation of alkaline reaction products at the steel-
concrete interface increases the passivity of the steel surface. In addition the depletion of
ions around the cathodic zone increases the resistance of the concrete.

The observed patchy distribution of damaged areas may be explained by the fact that
the corrosion takes the form of localized shallow pits, which are widely separated from
each other, and that the current flow generates paths of low resistance, as has been
observed for other breakdown mechanisms, such as with high voltage electricity.

The proposed mechanism needs to be elaborated further but does explain why
concrete at the base of the wall can be softened while similar concrete in the same wall and
exposed to identical water composition in the aerated zone shows no deterioration.

As these attacks are quite common (Figure 5) and can be easily explained by the
proposed model it is surprising that there is relatively little evidence in the technical

literature to support the idea of corrosion attack by macro cell currents. There is also no reference to this reaction in the existing technical standards and building codes.

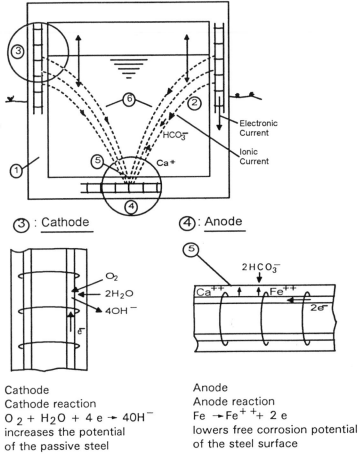

③ : Cathode   ④ : Anode

Cathode
Cathode reaction
$O_2 + H_2O + 4 e \rightarrow 4OH^-$
increases the potential
of the passive steel

Anode
Anode reaction
$Fe \rightarrow Fe^{++} + 2 e$
lowers free corrosion potential
of the steel surface

Figure 7 *Water tank with electrochemical macro cell between steel rebars in the aerated zone and the anaerobically corroding rebars. Legend: 1) Water tank made of concrete, 2) drinking water, 3) cathode reaction, 4) anode reaction (corrosion), 5) brown spots, 6) ionic currents in water and concrete.*

5 PROTECTION

5.1 Conventional Repair Method

The conventional remedial measure for macrocell corrosion consists of exchanging the soft attacked concrete with new layers of high density concrete. However after a few years the weak spots tend to reappear as illustrated in Figure 5.

This may be due to the fact that the rebars on the opposite side of the wall are also corroding. The surrounding concrete, being already weakened, will generally not be exchanged and thus heavy macro cell currents will continue to flow after the repairs have been completed.

5.2 New Method of Repair

As macro cell currents are the cause of the pronounced attacks on the rebars and concrete it is clear that the elimination of the anodic current by lowering the potential of all the embedded rebar surfaces could help to stabilise the concrete. This can be achieved by installing an impressed current anode in the water and thus providing cathodic corrosion protection for the rebars (Figure 8).

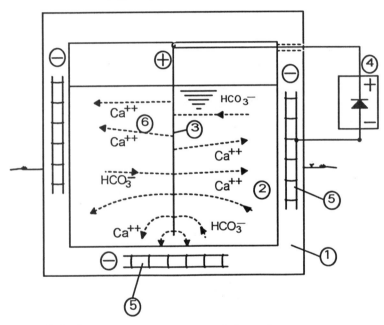

Figure 8 *Cathodic corrosion protection of the rebars in concrete cancels the electrochemical macro cells. Legend: 1) Water tank made of concrete, 2) drinking water (electrolyte), 3) impressed current anode, 4) rectifier, 5) cathodically protected rebars, 6) protective current between anodes and rebars.*

If the weak spots are replaced with new mortar the cementitious coating will show long term stability.

In several water reservoirs anodes have been installed and the rebars have been protected from further attacks. In general the current is adjusted to a level such that steel probes placed in the water at several locations are protected cathodically. This can be assessed by observing the direction of current flow to the probes.

The current consumption is usually between 2 and 5 mA/m² concrete surface. Initially the currents will be higher than this and then diminish with time.

The measurement of off-potentials with a copper/copper sulfate electrode does not reveal the degree of protection because steel surfaces in concrete can be protected at far higher potentials than are necessary to protect steel in water. Thus off-potentials measured in the reservoir at different locations may be higher than would be necessary to protect steel in water.

Several positive examples have shown that this remedial action is an economic and effective way to prevent attacks to our infrastructure. It is hoped that this method will be applied more often in the future as it has many advantages over conventional methods of preservation and restoration of attacked water tanks.

6 REFERENCES

1. D. Schoenen, Fleckartige Farbveränderungen und Zerstörungen von weissen Zementmörtelauskleidungen in Trinkwasserbehältern, *GWF, Wasser - Abwasser*, 1994, **135**, 667.

2. A. Rendell and M. P. Mice, 'Macro Cell Corrosion of Reinforcement in Concrete', UK Corrosion 1991, 3, The Institute of Corrosion, Manchester.

3. A. Gerdes and F. H. Wittmann, 'Electrochemical Degradation of Cementious Materials', 'International Congress on the Chemistry of Cement', 5, 1992, National Council for Cement and Building Materials, New Dehli.

4. K. Menzel and M. Aktas, 'The effects of galvanic current on concrete', *Otto Graf Journal*, 1991, **2**, 217.

5. British Standard BS 1361, Cathodic protection, Part 1, 1991.

6. M. Pourbaix, 'Atlas of Elektrochemical Equilibria in Aqueous Solutions', NACE, Houston, Texas, 1974.

7. R. O. Müller and F. E. Tanner, 'Betonschäden in Trinkwasserreservoirs', *GWA,*.**10**, 1993, SVGW, Zürich.

8. ASTM C 876, 'Half-cell potential measurement of uncoated reinforcing steel in concrete', 1987.

CHLORIDE-INDUCED CORROSION

CORROSION PERFORMANCE OF CHLORIDE BEARING AGGREGATES IN CONCRETE

D. G. Manning and A. K. C. Ip

Ontario Ministry of Transportation
St. Catharines, Ontario L2R 7R4
Canada

1 INTRODUCTION

It has long been recognized that chloride ions play a major role in the corrosion of steel in concrete. The chloride ions may originate from the ingredients of the concrete mixture, or may diffuse through the hardened concrete.

Most building codes attempt to control corrosion-induced deterioration by placing limits on the chloride ion content of the mix ingredients and by specifying the quality of the concrete and thickness of concrete cover appropriate to the exposure conditions.

A number of the aggregates used in concrete in North America contain significant concentrations of chloride ions. This is particularly true in the case of the carbonate aggregates used in Southern Ontario. The chloride ion contents from some sources are sufficiently high to exceed the limits specified in the Canadian standard 'Concrete Materials and Methods of Concrete Construction' for prestressed concrete and, sometimes, for reinforced concrete.

This paper expands upon an earlier paper[1] and describes a series of laboratory investigations undertaken to investigate whether the chloride ions in the aggregate depassivate embedded reinforcement, the development of a test method to discriminate between 'available' and 'unavailable' chloride ions, and the results of a field survey of the corrosion performance of bridge structures containing chloride-bearing aggregates.

2 CHLORIDE LIMITS IN CONCRETE

Establishing limits for the chloride ion content of mix ingredients in concrete is very difficult for three reasons:

i)   Chloride ions are present naturally in many concrete ingredients and specifying a zero chloride ion content is therefore unrealistic.

ii)   There is no agreement on a test procedure which can be used to determine the chloride ions available to depassivate the steel.

iii) The service environment often cannot be described with sufficient precision to determine the risk of corrosion.

2.1 Corrosion Threshold Concept

Although a zero limit for chloride ions is impractical, neither is it necessary because it has been known for a long time that small concentrations of chloride ions in the concrete do not result in an unacceptable level of corrosion of embedded steel. This observation gave rise to the concept of the chloride corrosion threshold.[2] The concept of a threshold implies that there exists a chloride ion concentration below which steel is 'safe' from corrosion and above which, an unacceptable level of corrosion may occur if other necessary conditions, principally the availability of oxygen and moisture, exist to support the corrosion reactions. The concept leads to the initiation-propagation model for corrosion damage[3] but it must be remembered that the corrosion threshold value is not a constant value for all concrete mixtures. The quantity of chloride ions required to depassivate the steel depends on several factors, including the chemical composition of the cement[2] the ratio of the concentration of hydroxyl ions to chloride ions,[4,5] and the type of cation.[6] The corrosion threshold value is also dependent on the method of testing for chloride ion content.

2.2 Test Methods

Test methods for measuring the chloride ion concentration in concrete or concrete ingredients consist of two parts, extracting the chloride ion into solution and measuring the concentration of chloride ions in the solution. Once in solution, the chloride ion concentration is measured by a conventional titration using a standard silver nitrate reagent or by a suitable, calibrated ion-selective electrode.

Test methods therefore differ principally in the method used to extract the chloride ions from the concrete and variations on two basic methods are in common use: digesting in nitric acid and dissolving in water. An alternative approach is to extract the pore water by applying high pressure to the concrete, though this method requires sophisticated equipment, and there is concern as to whether the fraction of pore water extracted is representative of the pore water as a whole.

In North America, the standard acid-soluble test method is AASHTO T 260 'Sampling and Testing for Total Chloride Ion in Concrete and Concrete Raw Materials'.[7] A representative sample is ground to pass through a 300 μm sieve. The chloride ions are extracted by digestion in nitric acid and then measured by titration.

A water-soluble test method, also specified in AASHTO T 260, is very similar to the acid-soluble method. However, the rate of dissolution in water is affected by the extraction time, the temperature of the water and the particle size, and these must be controlled carefully. As a consequence, the measured value is a function of the test method. The chloride ions are extracted by boiling in water for 5 minutes and are then allowed to stand at room temperature for 24 hours prior to analysis of the aqueous phase. The water-soluble test came into general use because the procedure was quicker than the acid extraction and it was thought (erroneously) that the result of the test gave the concentration of chloride ions available to support corrosion.

Based on the work of Lewis,[8] further studies at the Federal Highway Administration Laboratories in the early 1970s[9] showed that for hardened concrete subject to externally applied chlorides, the corrosion threshold was 0.20% acid-soluble chloride ions by mass of cement. The average ratio of water-soluble to acid-soluble chloride ions was reported to be 75 to 80%,[10] although the original work[11] indicates an average of 77% for chloride ions added at the time of mixing and 87% for chloride ions which permeated the hardened cement paste specimens. This work led to the introduction of a limit of 0.15% water-soluble chlorides for reinforced concrete in many North American codes.

2.3 Service Environment

In many situations, both indoors and outdoors, the service environment cannot be described with sufficient precision to determine the risk of corrosion. If the concrete will be exposed to dry conditions, controlling the chloride ion content of the concrete may be unnecessary. At the other extreme, if the concrete will be exposed to chlorides during service, it is logical to limit the chloride contents of the ingredients to the practical minimum. However, there is a wide range of service conditions between these extremes where the risk of corrosion is not only uncertain at the time of construction, but may change during the life of the structure.

3 AGGREGATES IN SOUTHERN ONTARIO

The most common high-quality aggregates used in Southern Ontario are crushed dolomites from the Amabel and Lockport formations which form the cap rocks of the Niagara Escarpment. These rocks contain unusually large concentrations of chloride ions as a result of their deposition in a saline environment on the fringes of the Michigan Basin. The acid-soluble chloride contents of aggregates supplied to the Toronto area in the 1970s from the Amabel formation were found[12] to average 0.125% (average of 25 samples) and those from the Lockport formation averaged 0.080% (average of 5 samples). The corresponding water-soluble chloride contents were 0.041% and 0.026% respectively. The chloride contents of aggregates quarried from the Amabel formation in the Collingwood area were substantially higher. For many concrete mixtures, aggregates from both formations would contribute chloride ions at a concentration above the threshold value for corrosion if all the chloride ions were available to depassivate the reinforcement.

Figure 1, which has been compiled from data in a 1977 report,[12] compares the acid-soluble and water-soluble chloride contents of aggregates from 24 quarries and one pit in Ontario, of which 23 sources were either dolostones or limestones. The water-soluble chlorides were extracted by grinding the rock samples to pass through a 150 μm screen and then boiling the rock powders in water for five minutes. The ratio of water-soluble to acid-soluble chloride ions ranged from 0 for a number of limestones to 75% for a beach gravel. These data illustrate clearly that the ratio of water-soluble to acid-soluble chloride contents varies considerably for individual mix ingredients.

4 LABORATORY INVESTIGATIONS

4.1 Early Screening Tests

Routine testing of concrete for chloride ion content was not undertaken until

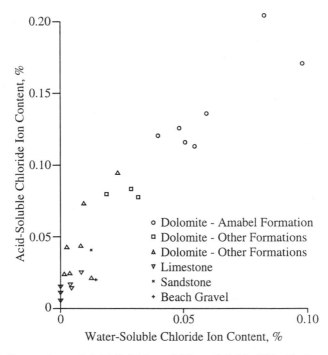

Figure 1 *Comparison of Acid-Soluble and Water-Soluble Chloride Ion Contents of Commercial Aggregates from Southern Ontario*

the early to mid 1970s when the problems of corrosion-induced deterioration of highway structures were first recognized. It was at this time that the high chloride content of the dolomitic limestone aggregates was first documented.

A simple test was conducted in the ministry's laboratories in 1975 on a sample of aggregate from the Amabel formation using particles in the 19 to 26 mm range. The stone was immersed in distilled water for a period of 90 days and the chloride ion content of the water was measured periodically. The measured chloride content was found to be 0.003% after one day and this did not increase over the 90 day period, indicating that the likely source of the chloride ions was the exposed particle surfaces and that chloride contained within the aggregate did not enter into solution. The rock in question was porous (about 2% absorption) and the pores were of a very large size but poorly interconnected. These findings diminished concerns that the aggregate might be contributing chloride ions to the concrete pore water.

In a 1982 experiment, samples of different particle sizes of the same aggregate as in the 1975 test, ranging from retained on 19 mm to passing 53 μm, were prepared and immersed in water for one week. The samples were agitated periodically. After seven days, the chloride ion content of the water was determined and the results are shown in Figure 2. As the particle size decreased, the chloride ion content increased and the relationship was almost linear when plotted on a log/log graph. In other words, as the rock was crushed to expose a greater surface area, the measured chloride ion content increased. This means that the measured chloride ion content

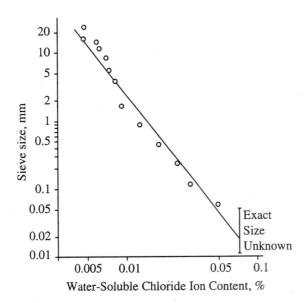

Figure 2 *Effect of Particle Size on Measured Water-Soluble Chloride Ion Content*

is a function of the test method, specifically the preparation of a pulverized sample, and not a measure of the quantity of chloride ions available to depassivate the steel.

In a further test series conducted in 1985,[13] reinforced concrete slabs were cast which contained two aggregates with chloride ion contents of 0.136% and 0.197%. Some slabs also contained various dosages of admixed calcium chloride. Microcell and macrocell corrosion rates were measured under indoor and outdoor conditions. Only the slabs containing calcium chloride at higher levels than the corrosion threshold limit exhibited corrosion. These findings were confirmed by a visual examination of the reinforcement.

It was recognized that, when interpreting chloride data to assess the risk of corrosion, it was necessary to correct for the background chloride content of the concrete when using aggregates with high, but 'unavailable' chloride ion contents. A practice was established by the Ontario Ministry of Transportation of taking cores from sufficiently deep within the concrete to eliminate the effect of external sources of chloride.[14] The chloride content of the concrete most distant from the exposed surface is taken as the 'background' value, provided it is in agreement with the historical value, and is then subtracted from all the measured values before making comparisons with the corrosion threshold value. This approach is not valid if chloride-based admixtures have been used in the concrete. However, the presence of such admixtures is usually evident from the abnormally high chloride values which do not diminish with distance from the exposed surface.

The subtraction of the background value is a pragmatic approach which has been found satisfactory over a number of years. However, a test method which measures only those chloride ions available to initiate corrosion would be preferable, and research to develop such a test method was initiated in the early 1980s. The research was conducted at Queen's University and funded by the Ontario Ministry of Transportation.

4.2 The Soxhlet Test Method

The research which led to use of the Soxhlet method for extracting chloride ions from concrete has been described in detail elsewhere.[1,15] The initial work involved the investigation of several hot and cold water regimes for extracting chloride ions from concrete samples and constituent materials. Tests were conducted on concretes which contained chloride-bearing aggregates, admixed calcium chloride, and chlorides resulting from vacuum saturation of hardened concrete with a 3.5% solution of sodium chloride. The results showed that the Soxhlet apparatus was capable of discriminating between chloride ions available for depassivation of the steel, and those which were unavailable by virtue of being contained in the aggregates or chemically bound in the cement paste.

The Soxhlet extractor consists of a heater, a lower flask, the sample compartment, and a condenser. The extractor contains approximately 100 ml of water in the lower flask. Heat is applied to this flask and vapour from the boiling water passes to the condenser, and the condensate collects in the sample compartment. The sample, consisting of pieces of concrete not larger than 13 mm in diameter, is contained in a porous holder. When the condensate reaches a critical height, the liquid is syphoned into the lower flask and the process repeats. The non-volatile components extracted from the sample accumulate in the lower flask, while each extraction involves fresh, hot distillate. The heat input should be sufficient to give about three extraction cycles per hour. Twenty four hours is normally sufficient to remove all the extractable chloride ions.

As noted elsewhere,[1] the Soxhlet method of extraction was not intended to replace the acid and water extraction methods because, where chloride ion contents are low, or background levels are small, the conventional methods are quicker, less expensive, and adequate. Even where high levels are recorded, engineering approximations, combined with local knowledge, often suffice. However, where uncertainty exists, or there is a need for a discriminating procedure, the Soxhlet method is very useful.

Tests conducted in connection with the development of the Soxhlet method, using water extraction on pulverized samples of concrete, confirmed the effect of particle size on the measured value of chloride ion concentration illustrated in Figure 2. Larger differences in measured values were recorded as the particles were ground finer and finer. This explains why large differences in results can be obtained from duplicate tests, where in one case the powder may be finer than the other.

5 CODE REQUIREMENTS

The 1990 edition of the National Standard of Canada Can/CSA-A23.1-M90 'Concrete Materials and Methods of Concrete Construction' introduced limits for the chloride ion content of concrete ingredients, through the following clause:

'The water-soluble chloride ion content by mass of the cementing material in the concrete before exposure shall not exceed the following values for the indicated applications:

(a)   for prestressed concrete, 0.06%;

(b)   for reinforced concrete exposed to a moist environment or chlorides or both, 0.15%;

(c)   for reinforced concrete exposed to neither a moist environment nor chlorides, 1.0%.

Note: These limits may be exceeded if the Owner can be satisfied that no corrosion problems have occurred in the past in concrete structures made with similar materials and exposed to similar conditions'.

The clause was not changed in the 1994 edition of the standard.

The test method specified (CSA Test Method A23.2-4B) involves pulverizing the material to finer than a 315 μm sieve, boiling in water for 5 minutes, followed by standing for 24 hours. As an alternative, the code permits the measurement of the acid-soluble chloride content, although the same limits apply.

The note to the clause was included in recognition of the long history of the use of chloride-bearing dolomitic limestone aggregates. Unless the note were invoked by the Owner, these aggregates could not be used in prestressed concrete. Aggregates from the Amabel formation could not be used in reinforced concrete exposed to a moist environment, or chlorides, in service, unless the proportion of cementing material were higher than that used in many commercial concretes. The introduction of the clause caused controversy in the aggregate and concrete industries. While casual observation suggested that there was no difference in the long-term corrosion performance of concrete structures containing chloride-bearing and chloride-free aggregates, it was decided that the time was appropriate for a systematic investigation of field structures.

6 FIELD INVESTIGATION

6.1 Experimental Design

A detailed investigation was performed on a small number of bridge structures which had been in service for at least 25 years. For each structure, the corrosion performance of components which had been exposed to deicing salts during service was compared with that of components not exposed to deicing salts. In this way, the service environments of the components were as similar as possible, except for the effect of an external source of chlorides.

The condition surveys of the structures included a visual inspection of the concrete surface, a delamination survey by sounding, corrosion potential measurements (according to ASTM C-876-87), rate-of-corrosion measurements (using a commercial three-electrode, linear polarization instrument), and removal of cores for determination of chloride ion content and examination of the condition of the bars.

6.2 Test Sites

The field investigations were conducted on the components of three bridges built between 1959 and 1968: Sixteen-Mile Creek Bridge, Credit River Bridge, and a structure in the Highway 403/ Highway 2 Interchange. The concretes contained dolomitic limestone coarse aggregate with a high chloride content. The 'background' levels of chloride ions in the concretes varied from 0.046 to 0.074% by mass, as measured by the acid-soluble extraction procedure.

6.2.1 *Highway 403/Highway 2 Interchange Structure.* The Highway 403/Highway 2 Interchange structure is a four-span, voided, post-tensioned slab bridge built in 1968. The 'background' concentration of chloride ions in the concrete was found to be 0.074% by mass of the concrete.

Cracking and rust staining were observed on the barrier walls and sidewalks, and also on those areas of the abutments exposed to surface run-off through leaking expansion joints. All the twelve spirally reinforced columns (each with a surface area of about 10 m^2) were in good condition and there was no cracking, rust staining or delamination. The columns were not exposed to surface run-off but were exposed to light splashing from the traffic.

Corrosion potential measurements were taken on a 10 m^2 section on the west side of the north abutment and on the six columns located at the east and west sides of the bridge. The percentages of readings falling within each of the three ranges normally associated with passive, uncertain, and active corrosion activity are given in Table 1. The potential measurements indicated that the reinforcement in the columns was passive, whereas the steel in the abutment was corroding actively.

Rate-of-corrosion measurements were performed at selected locations on two of the six columns. Table 2 shows the percentages of the measurements falling within the ranges associated with insignificant, little and moderate corrosion activity.[16] The readings showed that there was insignificant or little corrosion activity in the columns.

Severe corrosion was observed on the bars in the two cores removed from the north abutment. The average chloride content at the level of the steel (with concrete cover of 45 mm) was found to be 0.252% by mass of concrete. The anticipated

Table 1 *Corrosion Potential Measurements - Hwy. 403/Hwy. 2 Interchange Structure*

Test Location	Corrosion Potentials (mV vs. CSE)		
	>-200	-200 to -350	<-350
Abutment	0	4	96
Column 1	100	0	0
Column 2	100	0	0
Column 3	100	0	0
Column 4	100	0	0
Column 5	96	4	0
Column 6	98	2	0

Table 2 *Rate of Corrosion Measurements - Hwy. 403/Hwy. 2 Interchange Structure*

Test Location	Corrosion Current (μA/sq. cm)		
	<0.20	0.20 to 1.0	>1.0
Column 1	6	94	0
Column 4	0	100	0

corrosion threshold level for the concrete was approximately 0.030% plus the background contribution of 0.074% by mass of concrete.

There was no corrosion on the bars in the three cores removed from the columns. The chloride content of the concrete in these cores was similar to the background chloride level of the concrete, indicating that chloride ions from de-icing salts had not penetrated the concrete to any significant degree. It is recognized that the absence of corrosion on the reinforcing bars is not sufficient proof that there is no corrosion activity because the cores could have been taken from the cathodes on the bars. The lack of corrosion is presented as corroborating the data from the corrosion potential, rate-of-corrosion, and chloride measurements.

6.2.2 Credit River Bridge. The Credit River Bridge is a seven-span reinforced concrete arch bridge built in 1961. The 'background' concentration of chloride ions in the concrete was found to be 0.046% by mass.

Concrete patches and rust staining were visible at several locations on the deck soffit and piers as a result of corrosion caused by extensive leakage through the deck joints. The portions of the piers not exposed to surface run-off were generally in good condition.

Corrosion potential readings taken from 10 m^2 sections on two piers and arch beams are given in Table 3. The potential readings indicated that while approximately 30 to 75% of the readings taken from the piers were within the range of active corrosion, most readings on the arches were within the passive range, and none were in the range of active corrosion.

Rate-of-corrosion measurements performed at selected locations on the piers and arches are shown in Table 4. The results show that the readings taken on the arch beams were associated with insignificant or little corrosion, whereas areas on all the other faces indicated moderate corrosion.

Corrosion was observed on one of the two bars removed from the piers, and the chloride content at the level of the steel (with concrete cover of 65 mm) was found to be 0.110% by mass, which exceeds the anticipated corrosion threshold value of 0.076% by mass of concrete. The bars in the two cores removed from the arches were in good condition. The chloride ion content at the level of the steel was the same as the background level.

Table 3 *Corrosion Potential Measurements - Credit River Bridge*

Test Location	Corrosion Potentials (mV vs. CSE)		
	>-200	-200 to -350	<-350
West Pier, North Face	3	69	28
West Pier, East Face	3	64	33
West Pier, West Face	0	25	75
West Pier, Arch Face	93	7	0
East Pier, North Face	0	58	42
East Pier, East Face	0	58	42
East Pier, West Face	0	54	46
East Pier, Arch Face	82	18	0

Table 4 *Rate of Corrosion Measurements - Credit River Bridge*

Test Location	Corrosion Current (μA/sq. cm)		
	<0.20	0.20 to 1.0	>1.0
West Pier, North Face	0	87	13
West Pier, West Face	0	90	10
West Pier, East Face	0	60	40
West Pier, Arch Face	19	81	0
East Pier, North Face	0	37	63
East Pier, West Face	0	38	62
East Pier, East Face	0	0	100
East Pier, Arch Face	62	38	0

6.2.3 Sixteen-Mile Creek Bridge. The Sixteen-Mile Creek Bridge is a five-span composite section steel and concrete bridge built in 1959. Each of the four piers consists of three rectangular reinforced columns. The 'background' chloride concentration in the concrete was found to be 0.072% by mass.

There was spalling and rust staining on the curbs of the deck. The south face of the south column of each pier exhibited spalling, cracking, and rust staining, as a result of exposure to surface run-off through open drainage. The other columns, which were not exposed to the surface run-off, were in good condition.

Corrosion potential measurements, taken from different sections of one pier, are summarized in Table 5. The area of each test section was about 13 m^2, and the two test areas on the base of the pier were directly below the south face and east face of the south column. The potential measurements showed that the steel in areas exposed to the surface run-off was corroding actively, whereas none of the readings in those areas located away from the surface run-off indicated active corrosion.

Rate-of-corrosion measurements from the test sections on the north column and south column are given in Table 6. The results show that the readings taken from the north column were associated with little corrosion, but almost 90% of the readings on the south column indicated moderate corrosion.

Table 5 *Corrosion Potential Measurements - Sixteen-Mile Creek Bridge*

Test Location	Corrosion Potentials (mV vs.CSE)		
	>-200	-200 to -350	<-350
North Column, South Face	91	9	0
Central Column, South Face	88	12	0
South Column, South Face	0	0	100
South Column, East Face	0	36	64
South Column, West Face	0	0	100
Base, South Face	0	7	93
Base, East Face	0	13	87

Table 6 *Rate of Corrosion Measurements - Sixteen-Mile Creek Bridge*

Test Location	Corrosion Current ($\mu A/sq.\ cm$)		
	<0.20	0.20 to 1.0	>1.0
North Column, South Face	0	100	0
South Column, South Face	0	12	88

Corrosion was observed on the bar in the core removed from the south column, and the chloride content at the level of the reinforcement (with concrete cover of 40 mm) was found to be 0.215% by mass of concrete. There was no corrosion on the bars in the two cores removed from the north and central columns. The concentration of the chloride ions from de-icing salts at the level of the steel was well below the anticipated corrosion threshold level.

7 CONCLUSIONS

i)   The high concentration of water-soluble chloride ions measured in the dolomitic limestone concrete aggregates used in Southern Ontario is a function of crushing of the aggregates as required by standard test procedures. A direct relationship exists between the measured chloride content and the fineness of the aggregate particles, because of the greater surface area exposed.

ii)   The Soxhlet procedure was identified as an effective method of discriminating between those aggregates which will contribute chloride ions to the concrete and depassivate embedded steel, and those in which the chloride ions remain bound within the aggregate particles.

iii) A field investigation of bridge structures, which had been in service for over 25 years, revealed corrosion only in components where deicing salts had penetrated to the reinforcement. In no case had chloride ions contained in the aggregates initiated corrosion of the reinforcement.

8 REFERENCES

1.   D. G. Manning, 'Reflections on Steel Corrosion in Concrete', American Concrete Institute, Detroit MI, 1992, SP-131, 321.

2.   ACI Committee 222, 'Corrosion of Metals in Concrete', *J. American Concrete Institute*, 1985, **82**, 3.

3.   K. Tuutti, 'Corrosion of Steel in Concrete', Swedish Cement and Concrete Institute, Stockholm, 1982.

4.   D. A. Hausmann, 'Steel Corrosion in Concrete', *Materials Protection*, November 1967, 19.

5.   V. K. Gouda, 'Corrosion and Corrosion Inhibition of Reinforcing Steel I Immersed in Alkaline Solutions', *Br. Corros. J.*, 1970, **5**, (9), 198.

6.   C. M. Hansson, Th. Frolund and J. B. Markussen, 'The Effect of Chloride Cation Type on the Corrosion of Steel in Concrete by Chloride Salts', *Cement and Concrete Research*, 1985, **15**, 65.

7.   'Standard Method of Sampling and Testing for Total Chloride Ion in Concrete and Concrete Raw Materials', Test Method T 260-84, AASHTO, Washington, D.C., 1984.

8.   D. A. Lewis, 'Some Aspects of the Corrosion of Steel in Concrete', 'Proc. of First International Congress on Metallic Corrosion', Butterworths, London, 1962, 547.

9.   K. C. Clear and R. E. Hay, 'Time-to-Corrosion of Reinforcing Steel in Concrete Slabs, Vol 1, Effect of Mix Design and Construction Parameters', Report No. FHWA-RD-73-32, Federal Highway Administration, Washington D.C., 1973.

10.   K. C. Clear, 'Evaluation of Portland Cement Concrete for Permanent Bridge Deck Repair', Interim Report No. FHWA-RD-74-5, Federal Highway Administration, Washington D.C., 1974.

11.   H. A. Berman, 'Determination of Chloride in Hardened Portland Cement Paste, Mortar and Concrete', Report No. FHWA-RD-72-12, Federal Highway Administration, Washington D.C., 1972.

12.   C. Rogers and G. Woda, 'The Chloride Ion Content of Concrete Aggregates from Southern Ontario', Report EM-17, Ministry of Transportation and Communications, Downsview ON, 1977.

13.   B. B. Hope and A. K. C. Ip, 'Chloride Corrosion Threshold in Concrete', Report ME-87-02, Ontario Ministry of Transportation, Downsview ON, 1987.

14.   D. G. Manning and D. H. Bye, 'Bridge Deck Rehabilitation Manual, Part One: Condition Surveys', Report SP-016, Ontario Ministry of Transportation, Downsview ON, 1984.

15.   B. B. Hope, J. A. Page and J. S. Poland, 'The Determination of the Chloride Content of Concrete', *Cement and Concrete Research*, 1985, **15**, 863.

16.   K. C. Clear, 'Measuring Rate of Corrosion of Steel in Field Concrete Structures', Transportation Research Record No. 1211, Transportation Research Board, Washington D.C., 1989, 28.

EFFECT OF TEMPERATURE ON PORE SOLUTION CHEMISTRY AND REINFORCEMENT CORROSION IN CONTAMINATED CONCRETE

M. Maslehuddin*
C. L. Page**
Rasheeduzzafar*
A. I. Al-Mana*

* King Fahd University of Petroleum and Minerals
Dhahran 31261, Saudi Arabia

**Aston University
Aston Triangle, Birmingham B4 7ET, U.K.

1 INTRODUCTION

The hot and humid environment in many arid and semi-arid countries around the world contributes to the poor durability performance of concrete. Reinforcement corrosion is the predominant form of deterioration of concrete in these regions. While the accelerating effect of temperature on any chemical reaction is well known, meagre data are available on the role of temperature on reinforcement corrosion. This lack of knowledge is particularly apparent when one is faced with establishing a threshold temperature value for the acceleration of reinforcement corrosion.

Along the Arabian Gulf coast, concrete structures are also prone to temperature related deterioration, as the ambient temperature in this region is as high as 45 to 50 °C. At this temperature, the concrete surface temperature may be as high as 70 °C, due to solar radiation.[1] High temperature may also compromise the chloride binding capacity of cements. The conjoint effect of an increase in the chemical reaction and increased chloride concentration, due to elevated temperature, is deleterious in structures in which chlorides are inadvertently admixed as part of the mix constituents or they penetrate the hardened concrete from the external environment.

In this study, the effect of temperature on the pore solution composition and reinforcement corrosion was investigated.

2 EXPERIMENTAL WORK

The experimental work was divided in two series. In *Series I*, the effect of temperature and chloride contamination on the pore solution composition in two ordinary Portland cements, OPC-A (C_3A: 8.5%), OPC-B (C_3A: 14.5%) and one SRPC (C_3A: 3.5%) was investigated. Cylindrical mortar specimens, 49 mm Ø and 75 mm height, cast in plastic vials, were used for extracting the pore fluid. The sample containers were sealed on all sides to eliminate evaporation of water. Dune sand of specific gravity 2.62, absorption 0.57% and fineness modulus 1.3 was used in the cement mortar specimens. A sand to cementitious materials ratio of 2.0, and effective water to cementitious materials ratio of 0.50 was used in all the mortar mixes. The specimens were contaminated with 0.8% Cl⁻, by weight of cement. Chloride additions were made by dissolving the required quantities of analar grade sodium chloride. The cement and sand were mixed in a mortar mixer until a uniform mix was obtained and then placed in the plastic vials. These were then tamped lightly to eliminate entrapped air. After 24 hours of laboratory conditioning, the plastic vials were placed in an oven maintained at 25, 40, 55, and 70 °C. The pore solution was extruded after 7, 14, 28, 60 and 90 days of exposure.

After the designated exposure periods, mortar specimens were retrieved from the oven and directly placed on the base of a high pressure pore press and the pore solution was extruded. The pore solution was analyzed to determine the OH⁻ and Cl⁻ concentrations.

In *Series II*, prismatic concrete specimens, 200 x 190 x 75 mm with three 12 mm diameter steel bars, were cast for evaluating the effect of temperature and chloride contamination on reinforcement corrosion. An effective cover of 30 mm was provided both at the top and bottom. The cover was obtained by using plastic spacers. Two bars were used as working electrodes, while the third which was placed in the centre of the specimen was used as a counter electrode in the linear polarization resistance measurements. A small diameter plastic tube was inserted into the concrete specimen for making connection to the reference electrode. The three bars were provided with electrical connection for monitoring the corrosion activity. The ends of the bars and the electrical wire-steel junction were coated with a cement paste overlaid by an epoxy-coating to reduce the risk of crevice corrosion.

After casting, the specimens were covered with plastic sheet and kept at laboratory conditions for 24 hours after which time they were demoulded, covered with wet burlap and kept moist for 13 days. They were then placed in a controlled humidity chamber in which the relative humidity was maintained at 75%, and the desired temperature was adjusted.

Reinforcement corrosion was monitored by measuring corrosion current density using the linear polarization resistance method at periodic intervals. The potentials were measured against a saturated calomel electrode (SCE). To determine the corrosion current density, the electrical leads from the concrete specimens were connected to a EG&G PAR Model 350 A Potentiostat/Galvanostat. The polarization resistance (Rp) was determined by conducting a linear polarization scan in the range of $\pm$ 10 mV of the corrosion potential. A scan rate of 0.1 mV/s was used. The corrosion current density (I_{corr}) was determined using Stern Geary formula.[2] Anodic and cathodic Tafel constants of 120 mV were used.

3 RESULTS AND DISCUSSION

3.1 Pore Solution Composition

The OH⁻ concentration in the uncontaminated and contaminated SRPC mortar specimens are plotted against exposure temperature in Figure 1, while those for OPC-A and OPC-B are plotted in Figures 2 and 3, respectively. The OH⁻ concentration in the contaminated specimens was more than that in the uncontaminated specimens. This increase may be attributed to the cation type associated with the chloride ions. When sodium chloride is added to cement, the chloride ions react with C_3A leading to the formation of calcium chloroaluminate, while Na^+ cations are released into the pore solution as shown below:

$$2NaCl + Ca(OH)_2 \rightarrow CaCl_2 + 2Na^+ + 2OH^- \tag{1}$$

$$3CaO.Al_2O_3.10H_2O + CaCl_2 \rightarrow 3CaO.Al_2O_3.CaCl_2.10H_2O \tag{2}$$

The OH⁻ concentration in both the contaminated and uncontaminated specimens exposed at 25 and 40 °C was more or less similar. However, when the exposure temperature was raised to 55 °C, the alkalinity was significantly reduced. At 70 °C, the OH⁻ concentration in the contaminated and uncontaminated specimens was more or less similar. The reduction in the alkalinity of the pore solution with increasing temperature, particularly above 40 °C, indicates the instability of calcium chloro aluminate hydrate (Friedel's salt) at elevated temperatures.

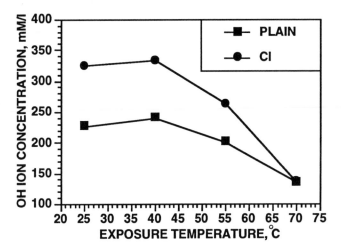

Figure 1 *Variation of OH⁻ with exposure temperature in SRPC mortar specimens*

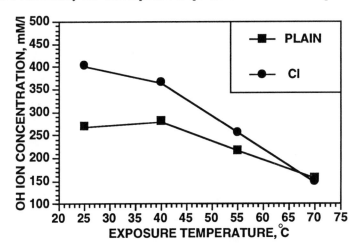

Figure 2 *Variation of OH⁻ with exposure temperature in OPC-A cement mortar specimens*

The decrease in the OH⁻ concentration in the uncontaminated specimens due to elevated temperature may be attributed to the instability of calcium sulpho-aluminate hydrate, formed due to the reaction of gypsum, normally added to regulate the setting time of cement with C_3A. This also results in an increase in the sulfate concentration in the pore solution. When sulfate ions are released into the aqueous phase, electrical neutrality necessitates either the removal of an equivalent quantity of cations or fixation of other anions, amongst which hydroxyl ions constitute the main available species. This effect has been shown by Herr and Wieker [3], who observed that when temperature was increased from 20 to 60 °C, the OH⁻ concentration decreased from 600 mM/l to about 400 mM/l, after 180 days of hydration.

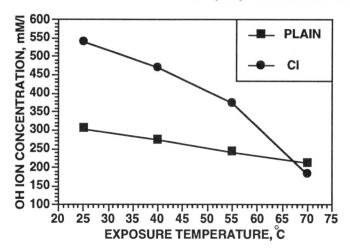

Figure 3 *Variation of OH⁻ with exposure temperature in OPC-B cement mortar specimens*

The chloride concentration in the contaminated mortar specimens, expressed as a proportion of its original concentration, is shown in Table 1. The chloride concentration in the three cements investigated was observed to increase with increasing exposure temperature. An exposure temperature of 55 °C and above was observed to be detrimental from the chloride binding standpoint. Further, the data in Table 1 indicated a 130 to 270% increase in the chloride concentration as the temperature was increased from 25 to 70 °C. These results indicating an increase in the chloride concentration due to a rise in the exposure temperature are in agreement with those reported by Roberts,[4] who evaluated the effect of temperature on the solubility of a pure calcium chloro-aluminate compound and observed that its solubility in water and in solutions of calcium sulfate and calcium hydroxide increased with increasing temperature. Results of studies conducted by Arya *et al.*,[5] on the factors influencing the chloride binding in Portland cements, however, contradict Roberts's findings. According to them, exposure temperature has an insignificant influence on the chloride concentration in the pore solution. Hussain and Rasheeduzzafar,[6] however, reported higher chloride concentration in the specimens exposed at 70 °C compared to those cured at 25 °C. The contradiction between the results of the present study, those reported by Roberts[4] and Hussain and Rasheeduzzafar[6] on one hand, and by Arya *et al.*[5] on the other may be attributed to the fact that Arya *et al.*[5] exposed the specimens at 38 °C. As is manifested in the results of this study, the decrease in chloride binding was only appreciable for exposure temperatures of more than 40 °C. In a later study, however, Arya *et al.*[7] while working on the methods of determining the free chlorides in chloride-contaminated OPC mortars also studied the effect of temperature of the solvent. They observed that for exposure temperatures of more than 45 °C the total chlorides passing into solution were more than those dissolving at lower temperatures. They indicated that bound chlorides are released at temperatures higher than 45 °C. These results[7] are in agreement with the data obtained in this study, which indicated that considerable chloride ions are released in the pore solution at exposure temperature of more than 55 °C. It should be noted that a concrete surface temperature of 55 °C corresponds to an ambient temperature of 35 °C, due to the solar radiation effect. This threshold temperature of 35 °C, detrimental for chloride binding, is close to a value of 30 °C suggested by ACI 305 to define hot-weather conditions.

Table 1 *Chloride Concentration in the Pore Solution*

Temperature (°C)	Chloride Concentration (% of original concentration in the mix water)		
	SRPC	OPC-A	OPC-B
25	47.5	41.0	23.5
40	49.3	54.8	33.8
55	61.9	62.7	48.3
70	70.8	64.8	62.2

The variation in the chloride ion concentration with C_3A for the four temperatures investigated and for the contaminated specimens is shown in Figure 4. These curves indicated a decrease in the chloride concentration with increasing C_3A content of cement. This trend was observed at all temperatures. The beneficial role of C_3A, in binding chlorides, by forming insoluble calcium chloro-aluminate hydrate, $3CaO.Al_2O_3.CaCl_2.10H_2O$, at lower temperatures of 20 to 25 °C is well established.[8-10] It was, however, affected by the exposure temperature. The chloride concentration, at 25 °C, in SRPC (C_3A: 3.5%) was twice that in OPC-B (C_3A: 14.5%). This ratio, however, was reduced to 1.12 in the specimens exposed at 70 °C.

The variation in the Cl^-/OH^- with exposure temperature in the three cements investigated is shown in Table 2. These values increased with the exposure temperature. Again, the increase in Cl^-/OH^- for a change in temperature from 25 to 40 °C was less significant than that for a change from 40 to 70 °C.

Thus, the concomitant effect of an increase in the chloride concentration and decrease in the alkalinity, at elevated temperature, 55 °C and above, increases the Cl^-/OH^- leading to depassivation of steel. Also, if moisture is available, the rate of steel corrosion is expected to increase due to an increase in the temperature.

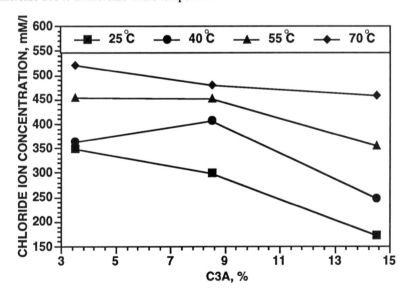

Figure 4 *Effect of C_3A content of cement on chloride binding*

Table 2 *Chloride to Hydroxyl Ion Ratio in the Contaminated Specimens*

Temperature (°C)	Chloride to Hydroxyl Ion Ratio		
	SRPC	OPC-A	OPC-B
25	1.08	0.75	0.32
40	1.08	1.11	0.53
55	1.72	1.81	0.95
70	3.81	3.24	2.52

3.2 Reinforcement Corrosion

The influence of temperature on I_{corr} on steel in the SRPC concrete specimens is plotted in Figure 5. The I_{corr} was observed to increase almost linearly with temperature. In the chloride-contaminated concrete specimens, exposed at 25 °C, the I_{corr} was 1.05 $\mu A/cm^2$, compared to a value of 2.08 $\mu A/cm^2$ measured in specimens exposed at 70 °C, indicating a 98% increase in the corrosion activity. The I_{corr} on steel in the uncontaminated specimens, however, did not exhibit a significant variation with the temperature. Figure 6 shows the influence of temperature on the I_{corr} on steel in the OPC-A concrete specimens. An increase in the I_{corr} with temperature was noted in these specimens also. The I_{corr} on steel in the chloride-contaminated specimens exposed at 25 and 40 °C was more or less similar. However, these values increased almost linearly with temperature in specimens exposed at more than 40 °C. The I_{corr} on steel in the concrete specimens exposed at 25 °C was 0.4 $\mu A/cm^2$, whereas it was 1.8 $\mu A/cm^2$ in the specimens exposed at 70 °C. The effect of temperature on I_{corr} on steel in the OPC-B concrete specimens is plotted in Figure 7. These data also indicated an increase in the reinforcement corrosion activity with exposure temperature. The I_{corr} on steel in the chloride-contaminated concrete was 0.23 and 1.4 $\mu A/cm^2$ for exposure temperatures of 25 and 70 °C, respectively.

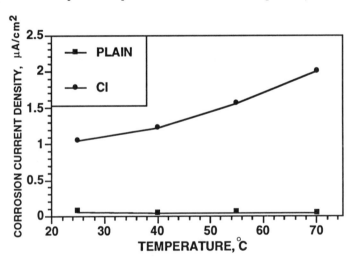

Figure 5 *Effect of temperature on corrosion current density on steel in SRPC concrete specimens*

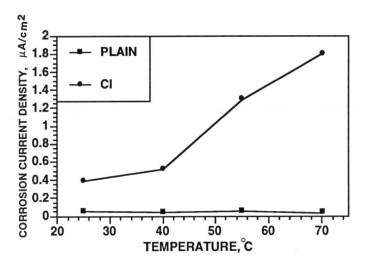

Figure 6 *Effect of temperature on corrosion current density on steel in OPC-A concrete specimens*

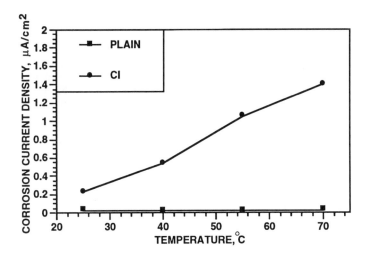

Figure 7 *Effect of temperature on corrosion current density on steel in OPC-B concrete specimens*

The data in Figures 5 to 7 indicate two to six times increase in the I_{corr} when the exposure temperature was raised from 25 to 70 °C. This indicates that temperature significantly affects the kinetics of reinforcement corrosion. The increase in the corrosion activity may be attributed to the acceleration of the electrochemical reactions. In studies conducted by Benjamin and Sykes[11] pitting potential was observed to decrease with the exposure temperature. Schiessl and Raupach[12] also reported an increase in the corrosion current density when the temperature was increased from 15 to 20 °C. The mean value of the acceleration factor was reported to be 1.4. In studies conducted by Maslehuddin *et al.*[13] pitting potentials of -225, -264, -368 and -386 mV SCE were noted for exposure temperatures of 25, 40, 55 and 70 °C, respectively, in the chloride-contaminated electrolyte representing the concrete pore solution. Investigations conducted by Henriksen[14] on the electrochemical behaviour of steel in $Ca(OH)_2$ solution containing 0.1 N NaCl indicated a linear variation in the pitting potential with a change of 25 mV for every 10 °C. Similarly, the corrosion rate, measured as the current required to maintain a passive potential on the specimen, increased linearly up to 50 °C, with the rate doubling for every 10 °C.

The increase in the corrosion activity due to temperature is of particular concern in the hot and arid environments where the concomitant presence of chloride contamination and elevated temperatures may drastically reduce the useful service life of reinforced concrete structures.

The data developed in this investigation pertains to situations where the chloride salts may be inducted into concrete through mixture ingredients, such as unwashed aggregates or usage of brackish water for mixing or curing of concrete. This is a common construction practice in the Arabian Gulf where desalinated or potable water is relatively scarce. However, another situation which is commonly observed in the Arabian Gulf region, is related to permeation of chloride ions to the steel-concrete interface through extraneous sources during the service-life of concrete structures. An increase in the reinforcement corrosion activity may also be envisaged in such situations. At elevated temperatures, the diffusion of the aggressive species to the steel-concrete interface, will be accelerated[15] in addition to a decrease in the chloride binding capacity of cement.

4 CONCLUSIONS

In this investigation, the effect of temperature on the pore solution chemistry and reinforcement corrosion in the contaminated SRPC and OPC was evaluated.

The results indicated that temperature significantly influences the chloride binding capacity of cements. In the chloride-contaminated specimens, the chloride binding was influenced by the exposure temperature. The chloride concentration in the specimens exposed at 25 and 40 °C was not significantly different. However, an appreciable increase in the chloride concentration was noticed for exposure temperatures of 55 °C and above. The chloride concentration at 70 °C was 1.5 to 2.7 times that at 25 °C. The increase in the chloride concentration in all the specimens exposed at temperatures of 55 °C and above is attributable to the decomposition of the Friedel's salt.

The chloride binding capacity of cements increased with the C_3A content. It was, however, affected by the exposure temperature. The chloride concentration at 25 °C in SRPC (C_3A: 3.5%) was twice that in OPC-B (C_3A: 14.5%). This ratio, however, reduced to 1.12 in the specimens exposed at 70 °C.

At temperatures of 25 and 40 °C, the OH^- concentration in the contaminated specimens was more than that in the uncontaminated specimens. The increase in the alkalinity of the contaminated specimens, compared to the uncontaminated specimens, is attributable to the reaction of chloride ions with the C_3A content of cement leading to the formation of sodium hydroxide which considerably increases the alkalinity of the pore solution. At higher temperatures, 55 °C and above, the alkalinity is reduced, so much so that at 70 °C, the alkalinity of the contaminated specimens is nearly equal to that of the uncontaminated specimens. This reduction in the OH^- also indicates the instability of calcium chloro-aluminate at higher temperatures. The OH^- concentration in the uncontaminated specimens also decreased with increasing temperature, particularly above 55 °C. This reduction may be attributed to the instability of calcium sulpho-aluminate hydrate formed due to the reaction of gypsum with C_3A.

The Cl^-/OH^- also increased with temperature. This is attributed to the combined effect of increase in the chloride concentration and decrease in the alkalinity.

The corrosion current density on steel in the chloride-contaminated SRPC and OPC concrete specimens was also affected by the exposure temperature. The corrosion activity increased by 2 to 6 times due to an increase in the temperature from 25 to 70 °C, the acceleration factor for every 15 °C being in the range of 1.26 to 1.87. This may be attributed to an increase in the rate of chemical reactions.

Acknowledgements

The authors acknowledge the Research Institute and Department of Civil Engineering, King Fahd University of Petroleum and Minerals for support of this research.

References

1. C. H. Jaegermann, D. Ravina and B. Pundak, *Concrete and Reinforced Concrete in Hot Countries, Proceedings, International RILEM Symposium*, Haifa, August 1971, Vol. II, 339.
2. M. Stern and A. L. Geary, *Journal of Electrochemical Society*, 1957, **104**, 56.
3. R Herr and W. Weiker, *Proc. 9th International Conference on Alkali-Aggregate Reaction in Concrete*, The Concrete Society, London, 1992, 440.
4. M. H. Roberts, *Magazine of Concrete Research*, 1962, **14**, No. 42, 143.
5. C. Arya, N. R. Buenfeld and J. B. Newman, *Cement and Concrete Research*, 1990, **20**, No. 2, 291.
6. S. E. Hussain and Rasheeduzzafar, *Cement and Concrete Research*, 1993, **23**, No. 6, 1357.
7. C. Arya, N. R. Buenfeld and J. B. Newman, *Cement and Concrete Research*, 1987, **17**, 907.
8. Rasheeduzzafar, S. E. Hussain and S. S. Al-Saadoun, *Cement and Concrete Research*, 1991, **21**, No. 5, 777.
9. M. Maslehuddin, Rasheeduzzafar, C. L. Page, A. I. Al-Mana and A. J. Al-Tayyib, *Proc., 4th International Conference on Deterioration and Repair of Reinforced Concrete in the Arabian Gulf,* (Ed: G. L. Macmillan), Bahrain Society of Engineers, Manama, 1993, 735.
10. W. R. Holden, C L Page and N. R. Short, *Corrosion of Reinforcement in Concrete Construction*, (Ed: A. P. Crane), Ellis-Horwood, Chichester, 1983, 143.
11. S. E. Benjamin and J. M. Sykes, *Corrosion of Reinforcement in Concrete*, (Ed: C. L. Page, K.W.J. Treadaway and P. B. Bamforth), Elsevier Applied Science, London, 1990, 59.
12. P. Schiessl and M. Raupach, *Corrosion of Reinforcement in Concrete*, (Ed: C.L. Page, K.W.J. Treadaway and P. B. Bamforth), Elsevier Applied Science, London, 1990, 59.
13. M. Maslehuddin, N. R. Jarrah, O. A. Ashiru and A. I. Al-Mana, *Proc. Sixth Middle East Corrosion Conference*, Bahrain, 1994, 597.
14. J. F. Henriksen, *Corrosion Science,* 1980, **20**, 1241.
15. C. L. Page, N. R. Short and A. El-Tarras, *Cement and Concrete Research*, 1981, **11**, 395.

INFLUENCE OF EXTERNAL CONCENTRATION AND TESTING TIME ON CHLORIDE DIFFUSION COEFFICIENT VALUES OF STEADY AND NON-STEADY STATE MIGRATION EXPERIMENTS

C. Andrade, M. Castellote, D. Cervigón and C. Alonso

Institute of Construction Sciences 'Eduardo Torroja'.CSIC.
C/ Serrano Galvache, s/n.
Apdo. 19002.
28033 Madrid. Spain.

1 INTRODUCTION

Cathodic protection was one of the earliest techniques to be derived from the application of electrical fields to reinforced concrete[1]. Higher currents and voltages were then used for concrete desalinization[2] or for testing chloride permeability[3]. However, it is only recently that a sound theoretical background has been[4,5] developed to describe the processes when an electrical field is applied to the concrete.

This theoretical support has been based on the application of the Nernst-Plank and Nernst-Einstein equations. Both equations model mass transport in electrolytes. The Nernst-Plank equation is formulated as follows:

$$-J(x) = D_j \frac{\partial C_j(x)}{\partial x} + \frac{z_j F}{RT} D_j C_j \frac{\partial E(x)}{\partial x} + C_j V(x) \tag{1}$$

where:

$J(x)$	=	unidirectional flux of species j (mol/cm²s)
D_j	=	diffusion coefficient of species j (cm²/s)
∂C	=	variation of concentration (mol/cm³)
∂x	=	variation of distance (cm)
z_j	=	electrical charge of species j
F	=	Faraday's number (cal/volt.eq⁻¹)
R	=	gas constant (cal/mol K)
T	=	absolute temperature (K)
C_j	=	bulk concentration of the species j (mol/cm³)
∂E	=	variation of potential (V)
V	=	artificial or forced velocity of ion (cm/s)

Einstein considered a particular case in which either the ions move only due to the electrical field (no diffusion or convection takes place) or the migration movement is equal and opposed to the diffusion movement. Equation 1 can then be solved as follows:

$$D_{eff} = \frac{RT}{nF^2} \Lambda_{Cl} \qquad (2)$$

where:

D_{eff} = Effected diffusion coefficient (cm²/s)

Λ_{cl} = Chloride conductivity (ohm⁻¹.cm².eq⁻¹)

The quantification of ion motion under the influence of an electrical field has multiple applications, may not even imagined yet. Among them, attracting the interest of several researchers, is the possibility of calculating chloride diffusion coefficients by means of an accelerated test[6]. Previous trials have been published by the authors[5,7].

In the present paper, the influence of the chloride concentration of the external solution, and the testing time, are analyzed. Both steady and non-steady state testing conditions have been considered.

2 EXPERIMENTAL

A type I-OPC was used for fabricating the concrete. The chemical analysis is given in Table 1. The mix was prepared with 380 kg/m³ of cement and a w/c ratio of 0.4.

Table 1 *Chemical analysis of the cement used*

| | | | | | | | | Parameter % | |
L.I.	I.R.	SiO$_2$	Al$_2$O$_3$	Fe$_2$O$_3$	CaO	MgO	SO$_3$	Free CaO	Cl⁻
3.45	1.97	19.37	6.12	3.13	62.86	1.78	3.23	1.28	0.013

Cylindrical specimens of 7.5 cm diameter and 15 cm height were prepared and cured for 28 days (under water). After curing, some of the specimens were cut into sections 1 cm thick for tests under steady-state conditions. The arrangement of the cell used is shown in Figure 1.

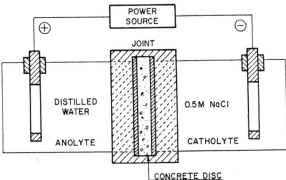

Figure 1 *Cell arrangement used for the steady-state measurements*

For the non-steady state experiments, a reservoir was glued to the top of the cylindrical specimens as shown in Figure 2. After completion of the test period these specimens were cut into sections 1 cm thick. The surface was ground in 2 mm increments in order to establish the chloride profile. The electrodes were made of steel (either bars sheets or nets were used).

Four chloride concentrations were tested: 0.05 M, 0.1 M, 0.5 M and 1 M. The migration cell for steady-state tests contains 360 ml of these solutions as catholyte and the reservoir of the non-steady-state test cells contained 300 ml of the solutions. The anolyte chamber of the migration cell is filled with distilled water. At the bottom of the specimen in the non-steady-state tests, the anode is placed in contact with the concrete through a wet sponge.

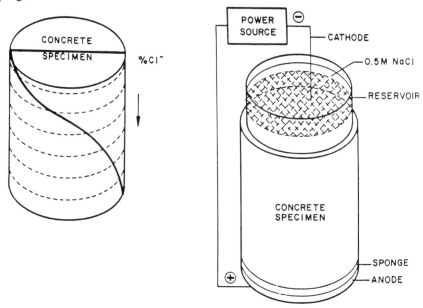

Figure 2 *Arrangement used for the non-steady-state measurements*

A voltage of 12 volts was applied between the electrodes and maintained through out the test. The variations in chloride concentration and pH value were recorded. The experimental details have been described elsewhere[8].

The solution to the steady state regime can be obtained by means of the Nernst-Plank equation expressed as[4,5]:

$$D_{eff} = \frac{J_{Cl}RTl}{z_{Cl}FC_{Co}\gamma\Delta E}$$ (3)

where:

$\gamma =$ activity coefficient
$l =$ concrete disc thickness (cm)
$J_{Cl} =$ chloride flow (mol/cm²s)

The solution for the Nernst-Einstein equation is expressed as:

$$D_{eff} = \frac{RT}{nF^2} \frac{it_{Cl}}{\Delta E} \frac{1}{A} \frac{1}{C_{Co}\gamma z_{Cl}}$$
(4)

In the case of non-steady-state conditions, the expression used was:

$$C_x = C_s \, erfc \, \frac{x}{2\sqrt{\frac{Fz\Delta E}{RT} D_{app} t}}$$
(5)

where:

C_x = chloride concentration at depth x.
C_s = chloride concentration at the specimen surface.

3 RESULTS

3.1 Steady-state Tests

The pH values recorded in both the anolyte and catholyte are given in Figure 3. As expected the catholyte becomes alkaline very rapidly and the anolyte acidifies reaching very low pH values.

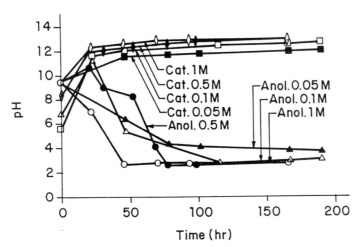

Figure 3 *Evolution of pH values during the tests in the two chambers (catholyte and anolyte) of the cell of figure 1 for the four chloride concentrations.*

The chloride evolution in the anolyte and catholyte is shown in Figure 4 for the four solutions tested. Figures 5 and 6 show the diffusion coefficients, D_{eff}, calculated by means of equations (3) and (4) respectively. In both cases the values become the lower as the chloride concentration of the catholyte increases, except in the case of the most dilute solutions, whose D_{eff} presented a value too low when calculated by means of equation (3). This anomaly can be attributed to the fact that due to the dilute concentration of chloride, the chloride is wasted and therefore, the concept of constant initial chloride concentration can no longer be considered. In consequence, the calculations were repeated for all cases using, not the initial, Co, concentrations, but those which exist in the catholyte when the steady-state regime is reached. The corrected values which were higher than those originally obtained are also shown on Figures 5 and 6.

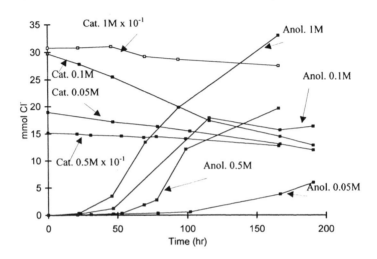

Figure 4 *Evolution of chloride concentration in the catholyte and anolyte for the four chloride concentrations in the steady-state experiments.*

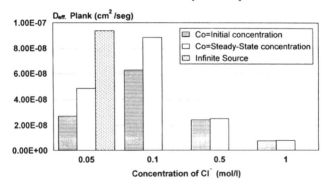

Figure 5 *Diffusion coefficient values, D_{eff}, calculated from the Nernst-Plank equation. The correction for the 0.05 M Cl⁻ solution derived from the actual concentration during the steady-state regime, is also shown, as well as a case where a very big reservoir is used as catholyte (infinite source)*

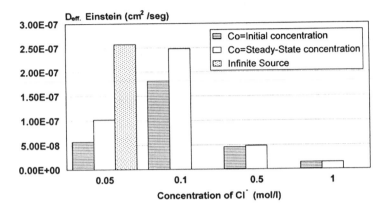

Figure 6 *Diffusion coefficient values, D_{eff}, calculated from the Nernst-Einstein equation. The correction for the most dilute solution is also shown as well as that for the case of an infinite source.*

Further experiments investigated the effect of changing the arrangement of the migration cell and using a very high reservoir as catholyte, (Figure 7) in order to attain conditions of constant initial concentration or of effectively infinite source. The evolution of chloride concentration for this case is shown in Figure 8. The chloride concentration in the anolyte is found to be much higher than in the case of a finite source (small reservoir). The D_{eff} value obtained is also higher as expected. Table 2 gives all the values obtained.

Table 2 *Diffusion coefficients obtained under different testing conditions*

		Chloride Concentration (M)			
Diff. Coeff. x $10^{-8}cm^2/s$		0.05	0.1	0.5	1
Deff Nernst-Plank	IC	2.7	6.2	2.4	0.7
	AC	4.8	8.8	2.5	0.7
	IS	9.4	-	-	-
Deff Nernst-Einstein	IC	5.7	1.8	4.6	1.4
	AC	10.2	24.8	4.8	1.4
	IS	25.7	-	-	-
Dapp	TL	0.1	0.2	0.2	0.2
	N-S-S	0.1	0.4	0.0	0.1

D_{eff}: *Effective diffusion coefficient* D_{app}: *Apparent diffusion coefficient*
IC: *Initial Concentration* TL: *Time lag*
AC: *Averaged steady-state concentration* N-S-S: *Non steady-state tests*
IS: *Infinite source*

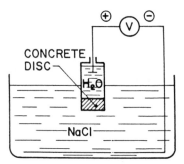

Figure 7 *Arrangement for steady-state measurements which reproduces the case of an infinite source*

Finally, it should be noted that the values of D_{eff} obtained through equation (4) are always higher than those obtained through equation (3).

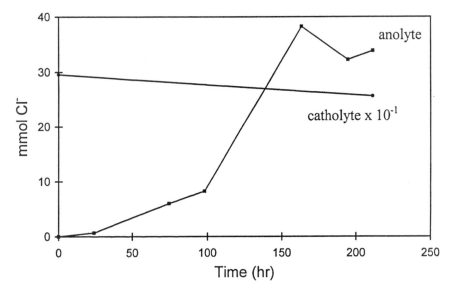

Figure 8 *Evolution of chloride concentration in the catholyte and anolyte for the arrangement shown in Figure 7*

In addition to these calculations, another attempt was made using the concept of time-lag described previously[9]. The diffusion coefficient so obtained is not D_{eff} but the apparent coefficient, D_{app}, that is, that obtained in non-steady state conditions. The values of D_{app} calculated from the delay time (time-lag) between the initiation of the experiment and the establishment of a steady-state regime, are also shown in Table 2 and Figure 9. The results are almost independent of the initial concentration, Co, perhaps due the small thickness (1 cm) of the concrete disc.

3.2 Non-steady-state Conditions

The experiments had two aims: 1) to obtain D_{app} values as a function of the external chloride concentration and 2) to find out whether the duration of the test may influence the D_{app} value.

Figure 9 shows the D_{app} values obtained for the four chloride concentrations tested. The duration of the experiments was different for each test as follows: for 0.005 M, 472 h., for 0.1 M, 189 h., for 0.5 M, 167 h. and for 1 M, 472 h. In this case no test was repeated to check the effect of an infinite source on the most dilute solution. The dependence of D_{app} on concentration is similar to that obtained under steady-state conditions.

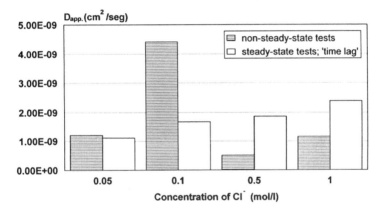

Figure 9 *Diffusion coefficient values, D_{app}, calculated from the non-steady-state experiments with the four initial chloride concentrations.*

The chloride profiles after three different testing periods, for the 0.5 M solution, are given in Figure 10. The results are surprising since they show that C_s decreases with time instead of increasing or remaining constant. It appears that a fixed amount of chloride entered the specimen at the beginning of the test and penetration then stopped. The electrical field "pushes" the chloride concentration during the rest of the testing period.

In view of this surprising evolution of the profile, the total amount of chloride which entered was calculated by integration of the chloride-time area. The total amount of chloride remaining in the external reservoir was also analyzed. The results are shown in Table 3, and confirm the earlier assumption.

The chlorides seem to penetrate during the first few days until a maximum is reached after 7 and 15 days of testing. After this time, no more chlorides penetrate even though chloride remains in the external reservoir. This behaviour can be attributed to the alkalinization produced within the catholyte which induces a reduction in the transport number of the chlorides and therefore, restricts their movement within the electrical field. This effect "blocks" the chloride penetration. This behaviour also indicates the need to use large external reservoirs in order to minimize the effect of cathodic alkalinization. It may also explain the abnormally low value found in Figure 8 in the case of the 0.5 M solution, as the testing time was shorter.

Table 3 *Chloride content in solution and concrete over test time*

		Time (days)	
mmol Cl	*6*	*14*	*22*
Specimens	5.68	8.85	8.46
Reservoir	132.5	124.13	124.8

The relatively high chloride concentration found in the specimen "skin" is attributed to the higher cement content (due to bending) in the top of the specimen.

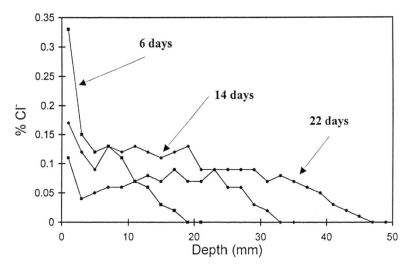

Figure 10   *Chloride profiles after three different periods of testing for the case of the 0.5M Cl⁻ solution*

4 DISCUSSION

4.1 Dependence of D Value on Chloride Concentration

A basic statement in analytical chemistry says that higher ion concentrations lead to an increasing number of ion-ion interactions and therefore, a lower ionic activity. Thus, in dilute solutions the ions move restricted only by the nature of the solvent, while in more concentrated solutions, the ion-ion interactions result in a lower activity. This is why reference parameters, such as ionic mobility, are always standarized to infinite dilution.

In consequence, regarding the chloride penetration of concrete, an increase in the external chloride concentration does not always lead to an acceleration of the test or an increase of the diffusional driven force. As the chloride concentration increases, the chloride activity decreases, until a certain limit of efficiency is reached.

This is supported by the present results were the D values show a dependence on the external chloride concentration. Higher D values are found for the more dilute external concentrations.

Although this basic principle is well known, it is surprising that this is not taken into consideration when testing chloride diffusion in concrete. Not only is it not considered, but higher and higher external chloride concentrations are proposed with the misleading goal of accelerating the chloride penetration.

The Diffusion Coefficient, D, is a function of the concentration. That is, D is not a constant. Further research is needed in order to derive a mathematical expression for this dependence in the form $D = f(C)$. Furthermore, the function may be not simple but may depend also on the concrete porosity and cement type (binding)[6], which may make modelling of the situation very complex. In any case, many more experimental systematic studies are needed, particularly in order to ascertain whether the changes in D with chloride concentration are due to the ion-ion interactions, or to a non-linear binding.

The independence of D_{app} obtained from time-lag experiments may indicate that under the present conditions, the binding is less significant than the ion-ion interactions.

All of these influences on D make it difficult to define a standard procedures which can be used for predictive purposes.

4.2 Theoretical Modelling and Experimental Procedures

In the steady-state situation, either equation (3) or (4) was used. The D values obtained are termed D_{eff} since they represent the diffusion when the concrete disc is already saturated with chlorides and no more binding is taking place. These D_{eff} values are about one order of magnitude higher than the D_{app} values (non steady-state). This difference is due to the binding ability and the different mathematical expression used. For predictive purposes in real conditions, it would appear that D_{app} gave better results.

The similarity of the values of D_{app} obtained from time-lag experiments and those obtained in non-steady-state experiments suggests that this seems to be a promising short method of obtaining D_{app} values.

D_{app} values were obtained through equation (4) which applies to semiinfinite media and constant surface concentration (infinite source). Since the chlorides do not reach the bottom of the specimens during the test time, it is possible to apply the semiinfinite condition. However, the condition of constant surface concentration does not apply and it is even, doubtful from Figure 10 whether an instantaneous source condition would model the process much better.

Finally, it has to be stressed that those last comments apply to migration experiments and care has to be taken in extrapolating them to natural diffusion experiments. Only by comparing parallel migration and natural diffusion experiments can it be seen whether migration tests model real diffusion conditions[8]. Tests are developing. The results will be reported further.

5 CONCLUSIONS

The conclusions that may be drawn up from present results are:
1) The external chloride concentration influences the D value measured either in steady-

state or in non-steady-state conditions. D values increase with decreases in the chloride proportion.

2) From this it can be deduced that $D = f(C)$ and therefore, D_{eff} obtained from steady-state conditions does not model real behaviour. In principle, D_{app} values may be more accurate.

3) The dependence of $D = f(C)$ makes it difficult to standardize a test for predictive purposes.

4) In the case of migration experiments under non steady-state conditions in a finite specimen, the solution of Fick's second law based on considering semiinfinite media, infinite source and constant surface concentration does not always hold. Either the external reservoir has to be maintained at constant concentration or other theoretical solutions have to be applied.

5) The alkalinization produced at the cathodic chamber during the treatment induces a stopping of Cl⁻ migration, which may serve to explain the need to use intermitent treatments during chloride desalinization.

Acknowledgement

The authors express their gratitude to ENRESA, the Spanish Agency for Radioactive Waste Storage, for the funding provided to develop the present research. They also grateful the grant to M. Castellote provided by the Ministry of Education and Science of Spain.

References

1. D. A. Hausmann, 'Criteria for cathodic protection of steel concrete structures', *Materials Protection*, 1969 October, 23.
2. D. R. Lankard, 'Neutralization of chloride in concrete', Batelle Columbus Laboratories, U.S.D.O.T.-FHWA, National Technical Information Service, Washington D.C., 1975, FHWA-RD-76-6-P.B-255309.
3. D. Whiting, 'Rapid determination of the Chloride Permeability of Concrete', Report no. FHWA-RD-81-119, August 1991, NTIS DB no. 82140724.
4. C. Andrade, 'Calculation of Chloride Diffusion Coefficients in Concrete from Ionic Migration Measurements', *Cement & Concrete Research*, 1993, **23**,724.
5. C. Andrade, M.A. Sanjuán, A. Recuero and O. Río, 'Calculation of Chloride Diffusivity in Concrete from Migration Experiments in Non-steady-state Conditions', *Cement and Concrete Research*, 1994, **24**, no 7, 1214.
6. Sergi, G., Yu, S.W. and Page, C.L., 'Diffusion of Chloride and Hydroxile Ions in Cementitious Materials Exposed to a Saline Environment', *Magazine of Concrete Research*, **44**, 1992, no.158,63.
7. C. Andrade, C. Alonso and M. Acha, 'Chloride Diffusion Coefficient of Concrete Containing Fly-ash Calculate from Migration Experiments', Proc. of the Int. Conference on Corrosion and Corrosion Protection of Steel in Concrete, (R.N. Swamy), Sheffield, July 1994, II, 783.
8. C. Andrade and M.A. Sanjuán, 'Experimental Procedure for the Calculation of Chloride Diffusion Coefficients in Concrete from Migration tests', *Advances in Cement*

Research, 1994, **6**, 23, 127.

9. J. Cranck, 'The Mathematics of diffusion', Ed. Oxford University, Press New York,1975.

10. C. Andrade, J.M. Díez, A. Alamán, C. Alonso, 'Mathematical Modelling of Electrochemical Chloride Extraction from Concrete', *Cement and Concrete Research*, 1995, **25**, no.4, 727.

TIME TO CRACKING FOR CHLORIDE-INDUCED CORROSION IN REINFORCED CONCRETE

Youping Liu & Richard E. Weyers
Department of Civil Engineering
Center for Infrastructure Assessment and Management
Virginia Polytechnic Institute and State University
Blacksburg, Virginia 24061-0105, USA

1 INTRODUCTION

Chloride-induced steel corrosion is one of the major deterioration mechanisms for steel reinforced concrete structures. The high alkaline environment of good quality concrete forms a passive film on the surface of the embedded steel which normally prevents the steel from corroding.[1] However, under chloride attack, the passive film is disrupted or destroyed and the steel spontaneously corrodes.[2] Since the volume of the corrosion products is about four to six times larger than that of iron, the surrounding concrete cracks when the expansive pressure exceeds the tensile strength of concrete.

Since the late 60s, chloride de-icing salts used on roadways in the United States have increased greatly, currently about 10 million tons of salts are used annually.[3] It has been reported that more than 40 percent of the highway bridges in the United States are structurally deficient or functionally obsolete. The cost of repairing and replacing these bridges is estimated at $70 billion.[4] Approximately 20 percent of the total estimated cost is due to the corrosion deterioration of concrete bridges. There are insufficient funds to address all the present repair, rehabilitation, and replacement needs. However, through the use of deterioration models, cost-effective decisions concerning the right time to repair, rehabilitate or replace existing structures can be made as well as predicting future maintenance and replacement needs and selection of the most effective corrosion abatement systems.

2 BACKGROUND

A deterioration model for predicting the remaining life of a typical bridge deck developed by Cady and Weyers has three phases:[5] diffusion, corrosion to visible damage and deterioration to end of functional service life. The corrosion phase is the time period from initiation of corrosion to cracking of the cover concrete. Bazant developed a physical-mathematical model to determine the time to cracking for chloride-induced corrosion based on a steady-state corrosion rate.[6,7] In Bazant's model, the time to cracking is a function of corrosion rate, cover depth, spacing, and mechanical properties of concrete, tensile strength, modulus of elasticity, Possion's ratio and creep coefficient. A sensitivity analysis of Bazant's theoretical equations demonstrates that for the parameters of concrete strength, cover depth, reinforcing size, horizontal spacing and corrosion rate, corrosion rate is the most significant parameter in determining the time to cracking of the cover concrete. Unfortunately, Bazant's model has never been validated experimentally.[7]

Two devices which are used for measuring the corrosion rate (corrosion current density) both in the laboratory and in the field are K. C. Clear's 3LP and the Geocisa Geocor devices, both use the theories of polarization resistance to obtain evidence of corrosion activity. The corrosion current density is a function of the polarization resistance through the Stern-Geary equation:[8]

$$I_{corr} = \frac{(\beta_a \beta_c)}{2.3R_p(\beta_a + \beta_c)} \tag{1}$$

where $\beta_a =$ the anodic Tafel slope
$\beta_c =$ the cathodic Tafel slope
$R_p =$ the polarization resistance

A difference between the 3LP and Geocor devices is that the Geocor device has a guard ring electrode which is used to confine the influence area of the counter electrode by actively confining the polarization current during the measurement process. Another significant difference is the polarization rate. The Geocor polarization rate is device controlled based on the rate of corrosion, whereas the 3LP device is operator dependent within a set of guidelines. In addition, the devices use different Tafel slope values in calculating the corrosion current density. The result of these differences is approximately an order of magnitude difference in the measured corrosion current density between the instruments. The general guidelines for interpreting the results of the 3LP and Geocor as supplied by the instrument manufacturers are as follows:[9,10]

Table 1   3LP and Geocor Guidelines

I_{corr} ($\mu A/cm^2$)	Geocor Device Corrosion State	I_{corr} ($\mu A/cm^2$)	3LP Device Expectation
<0.1	Passive	< 0.2	No damage expected
0.1 - 0.5	Low corrosion	0.2 - 1.0	Damage possible 10-15 years
0.5 - 1.0	Moderate	1.0 - 10.0	Damage possible 2-10 years
> 1.0	High corrosion	> 10.0	Damage possible < 2 years

This paper reports on a study which was initiated to validate or modify a set of theoretical equations for field linear polarization measurements using the 3LP and Geocor devices. A partial factorial experiment was designed to simulate a typical bridge deck based on sensitivity analysis of Bazant's model.

2.1 Experimental Design

A partial factorial experiment was designed to achieve five different corrosion rates, two concrete cover depths, reinforcing sizes and spacings, and exposure conditions. A total of 56 specimens were constructed with six admixed chloride contents, 0.0, 0.36, 0.71, 1.42, 2.85 and 5.69 kg/m^3; concrete cover depths of 5.08 and 7.62 cm; reinforcing diameters of 15.87 and 18.75 mm and spacings of 20.32 and 15.24 cm. The concrete mixture proportions are presented in Table 2.

Table 2 Mixture proportions

Series (chloride) kg/m^3	0.0	0.36	0.71	1.42	2.85	5.69
Cement kg/m^3	381	381	379	379	379	337
Water kg/m^3	173	160	159	155	167	162
W/C Ratio	0.45	0.42	0.42	0.41	0.44	0.43
Coarse Agg. kg/m^3	1068	1065	1077	1084	1067	1078
Fine Agg. kg/m^3	718	718	715	706	712	713
Daravair g/m^3	367	319	319	367	363	416
Salt, NaCl kg/m^3	0	0.6	1.2	2.4	4.8	9.6
Chloride kg/m^3	0	0.36	0.71	1.42	2.85	5.69
Slump cm	15	10	17	8	10	10
Unit Wt. kg/m^3	2232	2196	2148	2232	2236	2139
Air Content, %	3.2	5	5.4	4.2	4.7	6.7

Each specimen is a 118 cm square slab, 20 cm thick and contains five electrically isolated steel reinforcing bars. Forty specimens are outdoors and the others indoors in order to maintain a near constant concrete moisture content and temperature. Indoor specimen surfaces are wetted with water once a week and covered with a loose fitting sheet of clear plastic. The indoor exposure temperature varied from a low of 20°C in the winter to a high of 32°C in the summer.

The 3LP and Geocor devices were used to measure the corrosion current density. Measurements were performed once a month, and concrete temperatures at the bar depth were also recorded. Metal loss measurements were performed in accordance with ASTM G1-90, Method C.3.5 and compared with the measured corrosion rates for specimens which cracked. Statistical analysis of the initial corrosion measurements demonstrated that the variability for each experimental cell along a reinforcing bar and within a specimen is small with the largest variability being between specimens within corrosion cells. Thus, the number of monthly measurements within an experimental cell was reduced to seven or eight for each device.

3 RESULTS AND DISCUSSIONS

3.1 Measured Corrosion Rates

During the three and one half years of observations reported on here, only one outdoor test series has cracked. The highest chloride content, 5.69 kg/m³, outdoor specimens with 5.08 cm cover depth cracked after 1.84 years. No cracking has been visually observed in other outdoor specimens and indoor specimens. The monthly corrosion rates for the zero and the 5.69 kg/m³ admixed chloride specimens are presented in Figures 1 through 8. Presented are the monthly averages and 95% confidence limits for seven or eight measurements, two or three measurements for each of three slabs in a test series. Note that F, W, S, S refer to fall, winter, spring, and summer months in Blacksburg, VA, USA. Figures 1 through 4 present the zero admixed chloride measurements, Figures 1 and 2 are the 3LP indoor/outdoor specimens and Figures 3 and 4 represent the Geocor measurements for the same test specimens. Likewise, Figures 5 and 6 are the 3LP, 5.69 kg/m³ admixed chloride specimens and Figures 7 and 8 represent the Geocor readings for the same admixed chloride series.

As shown, the corrosion rates appear to decrease with time for both the zero and admixed chloride series. The decrease in measured corrosion rates is greater for the 3LP measurements than the Geocor measurements. For the zero admixed case, the reduction in corrosion rate may be related to the formation of the passive film. Whereas, for the admixed chloride case, the decrease may be related to a decrease in the cathode to anode area ratio as the amount of corroded surface area increases and an increase in the time taken for the corrosion reactants to diffuse through the corrosion products layer.

Figures 5 through 8 show the influence of cyclic weather conditions and the interaction between temperature and concrete moisture content on the measured corrosion rates. The measured corrosion rate is lowest in the winter (W), increases in the Spring (S), decreases slightly in the Summer (S), increases again in the Fall (F) and then decreases to its lowest rate in the Winter. Thus, the measured corrosion rates are sensitive to changes in temperature and moisture.

The coefficient of variation for three admixed chloride series, indoor specimens, are presented in Figures 9 and 10. The same chloride series, outdoor specimens, are presented in Figures 11 and 12. The variability, illustrated by the larger coefficient of variations, is greater for the Geocor device than the 3LP device. The higher variability for the Geocor device is most likely to be related to the increased accuracy of an imprecise corrosion process afforded by the guard ring. Whereas, without the guard ring the polarization current tends to spread out and creates a less precise effect.

3.2 Metal Loss and Cracking

According to Faraday's Law, the loss of iron over time can be obtained from the area under the curve of corrosion current versus time. The calculated annual metal loss for the six test series for the 3LP and Geocor devices are presented in Figures 13 and 14. There is a significant difference, generally a factor of 10, between the results of the two measurement methods. The metal loss increases with increasing chloride content greater than 1.42 kg/m³. There is little difference between outdoor and indoor specimens within the same amount of admixed chloride. However, cracking has not occurred in indoor specimens with the highest

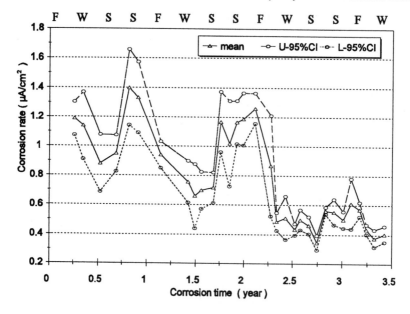

Figure 1.   *Corrosion rate vs time for 3LP, zero admixed chloride series, indoor slabs.*

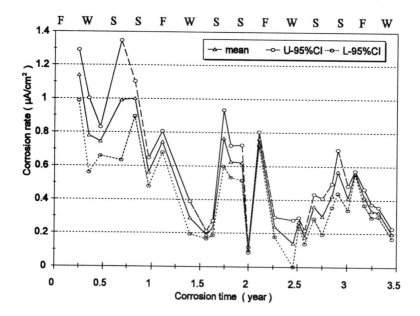

Figure 2.   *Corrosion rate vs time for 3LP, zero admixed chloride series, outdoor slabs.*

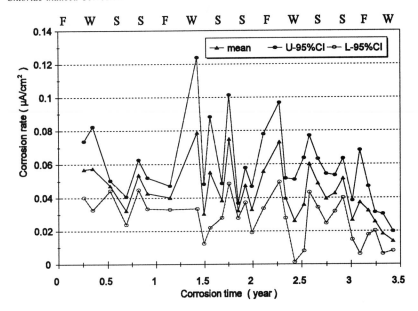

Figure 3. *Corrosion rate vs time for Geocor, zero admixed chloride series, indoor slabs.*

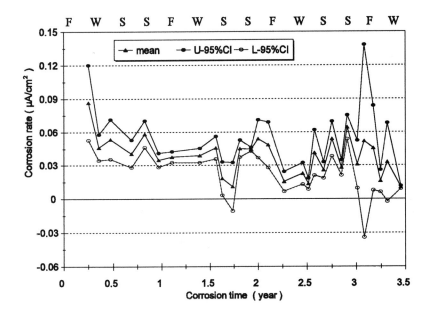

Figure 4. *Corrosion rate vs time for Geocor, zero admixed chloride series, outdoor slabs.*

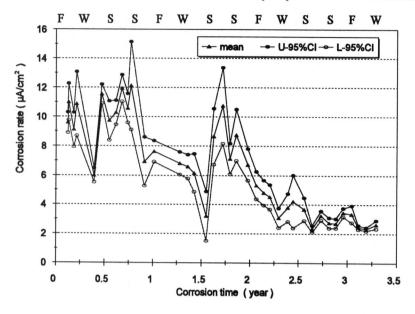

Figure 5.   *Corrosion rate vs time for 3LP, 5.69 kg/m³ admixed chloride series,*
                 indoor slabs.

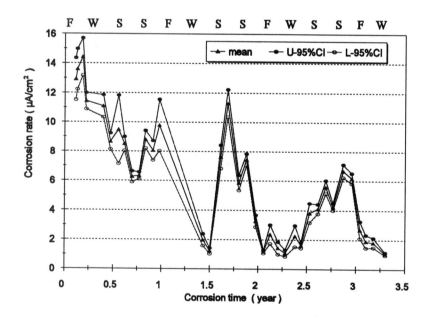

Figure 6.   *Corrosion rate vs time for 3LP, 5.69 kg/m³ admixed chloride series,*
                 outdoor slabs.

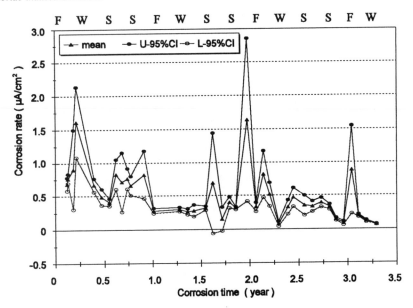

Figure 7. *Corrosion rate vs time for Geocor, 5.69 kg/m³ admixed chloride series, indoor slabs.*

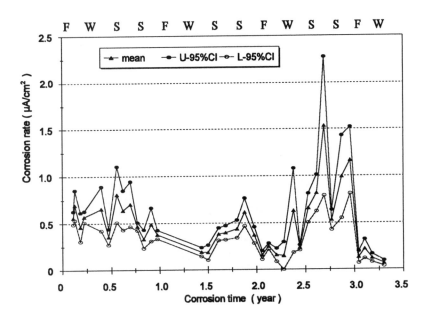

Figure 8. *Corrosion rate vs time for Geocor, 5.69 kg/m³ admixed chloride series, outdoor slabs.*

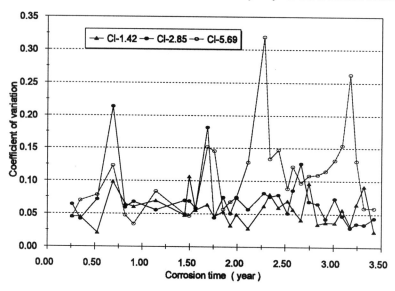

Figure 9.   *Coefficient of variation for 3LP measurements of three admixed chloride series, indoor slabs.*

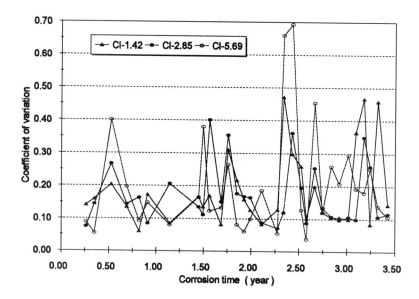

Figure 10.   *Coefficient of variation for Geocor measurements of three admixed chloride series, indoor slabs.*

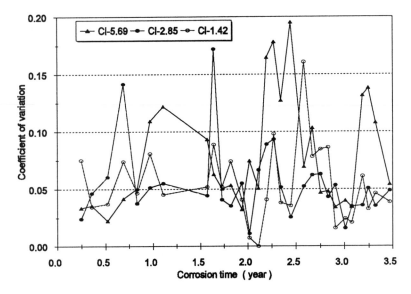

Figure 11. *Coefficient of variation for 3LP measurements of three admixed chloride series, outdoor slabs.*

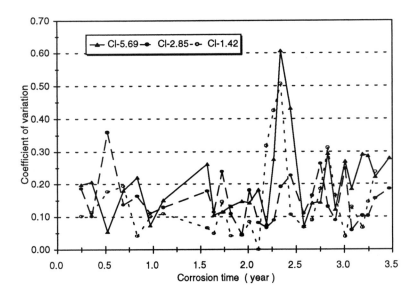

Figure 12. *Coefficient of variation for Geocor measurements of three admixed chloride series, outdoor slabs.*

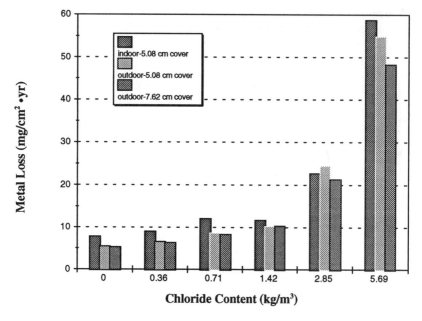

Figure 13. *Calculated metal loss from 3LP measurements.*

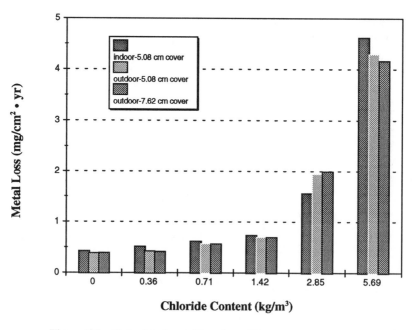

Figure 14. *Calculated metal loss from Geocor measurements.*

chloride content, whereas cracking has already occurred in outdoor specimens for the same test series. Since the indoor specimens are kept moist the corrosion products could possibly be migrating away from the reinforcing steel and thus not creating such high internal pressure.

The mean corrosion rate can also be calculated by the metal loss method, the results of which are presented in Table 3. As shown in Table 3, the mean calculated corrosion rates for the specimens with admixed chloride content more than 1.42 kg/m³ are relatively high with possible damage occurring in 2 to 10 years based on the general guidelines presented with the 3LP.

Using the corrosion rates from Table 3, the corrosion time to cracking can be calculated from Bazant's equation. However, comparing the observed cracking time of 1.84 years for the outdoor, 5.08 cm cover depth, 5.69 kg/m³ admixed chloride series with the calculated time to cracking for these specimens predicted by Bazant's equation (93 days) using the 3LP corrosion current shows that the predicted time is much shorter than the observed one. Thus, Bazant's equation underestimates the time to cracking using the measured corrosion rates by 1.5 years. This is due to limitations in the model; for example, it assumes that all corrosion products create pressure and uses a calculated corrosion product density in determining the internal pressure produced due to corrosion.

Table 3   Mean calculated corrosion rates from 3LP and Geocor calculated metal loss over 3½ years

Chloride kg/m³	Indoor		Outdoor (5.08 cm cover)		Outdoor (7.62 cm cover)	
	3LP µA/cm²	Geocor µA/cm²	3LP µA/cm²	Geocor µA/cm²	3LP µA/cm²	Geocor µA/cm²
0.0	0.85	0.046	0.60	0.042	0.58	0.043
0.36	0.99	0.056	0.72	0.046	0.70	0.045
0.71	1.32	0.067	0.95	0.062	0.91	0.061
1.42	1.29	0.081	1.13	0.077	1.13	0.076
2.49	2.50	0.17	2.69	0.21	2.23	0.22
5.69	6.40	0.51	6.00	0.47	5.29	0.46

Another possible reason that Bazant's model underestimated the time to cracking is that the rate of rust production in concrete may not be a linear relationship of the product of the corrosion rate and time as stated by Faraday's Law because the iron ions have to diffuse through the rust layer before oxidation can take place.

3.3 Modelling the Time to Cracking

3.3.1 Equivalent Corrosion Rate. Corrosion of iron in concrete is a dynamic process. As shown in Figures 15 and 16, temperature significantly influenced the measured corrosion rate in the outdoor specimens, the corrosion rates increased with increasing

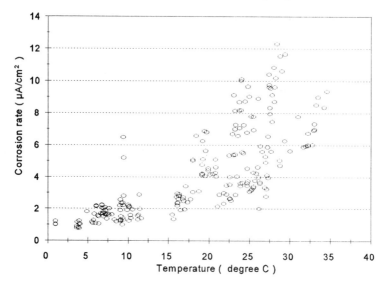

Figure 15. *Influence of temperature at reinforcing steel depth on 3LP corrosion rates, 5.69 kg/m³ admixed chloride series, outdoor slabs.*

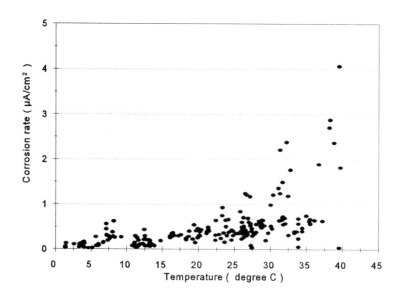

Figure 16. *Influence of temperature at reinforcing steel depth on Geocor corrosion rates, 5.69 kg/m³ admixed chloride series, outdoor slabs.*

temperature at the depth of the steel. However, the moisture content also influences corrosion rate. For field structures, moisture content at the reinforcing bar depth is a function of temperature, humidity and time, as previously shown. Thus measured corrosion rates at a point in time cannot be directly used for predicting the time to cracking. To simplify, an equivalent corrosion rate (annual mean corrosion rate) is suggested for use in predicting the metal loss over time.

Generally, as temperature increases by about 10°C, the electrochemical reaction rate doubles.[11] Based on this approximation, an exponential equation for the influence of temperature on the corrosion rate is suggested to estimate the equivalent corrosion rate measured by the 3LP:

$$i_{corr} = i_m \times 2^{(T_{corr}-T_m)/10} \tag{2}$$

where: i_{corr} equivalent corrosion rate ($\mu A/cm^2$)
      i_m measured corrosion rate at temperature T_m by 3LP ($\mu A/cm^2$)
      T_{corr} annual average temperature (°C)
      T_m temperature at the depth of the reinforcing steel (°C)

The annual average temperature in 1991 to 1994 was about 11.7°C in Blacksburg, Virginia, USA (see Figure 17). The equivalent uniform corrosion rates for the 3LP device calculated from equation 2 and by the weight loss method (ASTM G 1-90, Method C.3.5) are summarized in Table 4. As shown, results from the two methods are in good agreement with each other. However, since the Geocor corrosion current density measurement is about a factor of 10 less than the 3LP device, the Geocor measurements would not correlate with the weight loss measurements. Reasons for this lack of correlation may be that Geocor device measurements are more accurate at the instant the measurement is taken. Whereas, the 3LP device being an unguarded device, the spreading of the polarization current presents a more averaging effect of an imprecise corrosion process.

Table 4 Estimated equivalent corrosion rate and measured rate by ASTM G1-90

Cracked Specimens	Equivalent Corrosion Rate ($\mu A/cm^2$)	ASTM G1-90 ($\mu A/cm^2$)
O28596-1	2.10 - 2.60	
O28596-2	2.30 - 2.60	2.35
O28596-3	2.30 - 3.00	

3.3.2 Critical Weight of Rust Products. The limiting amount of the corrosion products that causes chloride-induced cracking of steel reinforced concrete structures depends on the cover depth, reinforcing bar diameter and spacing, and the mechanical properties of the concrete. The critical weight of rust products may be expressed as,

$$W_{cr-rust} = k\Delta D \tag{3}$$

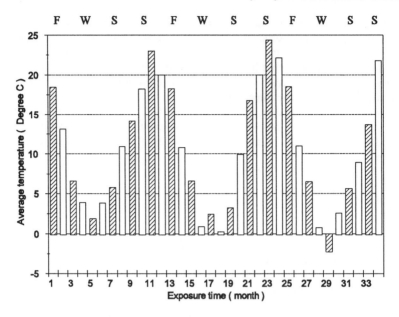

Figure 17. *Monthly mean temperature in Blacksburg, Viginia, USA.*

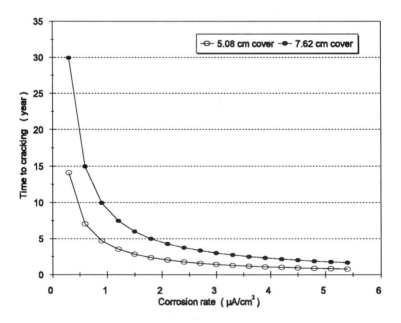

Figure 18. *Time to cracking vs corrosion rate.*

where

$$k = (½)(1/\rho_{rust} - \alpha/\rho_{steel})^{-1}, \text{ for } \rho_{rust} = 3000 \text{ kg/m}^3, \rho_{steel} = 7850 \text{ kg/cm}^3 \tag{4}$$

(α equals molecular weight of Fe divided by the molecular weight of rust products. For example, α is 0.523 for Fe → Fe(OH)$_3$, 0.622 for Fe→Fe(OH)$_2$.)

ΔD is the increase in rebar diameter (cm), which can be estimated from Bazant's equation:[7]

$$\Delta D = 2f_1(L/D)\delta_{pp} \text{ (inclined cracking or spalling)} \tag{5}$$

$$\text{or } \Delta D = 2f_1(S/D-1)\delta_{pp} \text{ (cover peeling or delamination)} \tag{6}$$

where: f_1 tensile strength of concrete
        L   cover depth (cm)
        S   rebar spacing (cm)
        D   rebar diameter (cm)
        δ_{pp} bar hole flexibility

3.3.3 Growth of Rust Products. As the rust layer grows thicker, the ionic diffusion distance increases, and the rate of rust formation decreases because the diffusion is inversely proportional to the oxide thickness.[12]

$$dW_{rust}/dt = k_p/W_{rust} \tag{7}$$

integrating:

$$W_{rust}^2 = 2k_p t_{corr} \tag{8}$$

where: $k_p = (1/\alpha) 9.11 i_{corr}$, i_{corr} corrosion rate (μA/cm^2)
        α related to rust products
        t_{corr} corrosion time (year)

From equation 8, it can be shown that the rust growth mainly depends on the corrosion rate and exposure time.

3.3.4 Time to Cracking. According to equation 8, the time to cracking can be given as follows:

$$t_{corr} = W_{corr-rust}^2/2k_p \tag{9}$$

For a typical concrete structure, the time to cracking is mainly determined by the corrosion rate and the concrete cover depth (see Figure 18).

The predicted time to cracking for specimens with 5.69 kg/m^3 of admixed chloride and 5.08 cm cover depth from the above equations is 1.62 to 2.11 years depending on the density of the different rust products. The observed time to cracking was 1.84 years, thus the observed value is within the predicted values.

Since only one series of specimens has cracked, the above model still needs further verification or modification, which will be based on the observed time to cracking for the other series.

4 CONCLUSIONS

1. Chloride-induced corrosion rates in concrete vary with changes in environmental exposure conditions such as temperature, moisture, chloride content at the depth of the reinforcing steel and corrosion time.
2. Temperature has a significant effect on the field measurement of corrosion rates using the 3LP and Geocor devices. However, the equivalent corrosion rate corrected for temperature for the 3LP is in good agreement with that of the measured weight loss, ASTM G1--90, Method C.3.5.
3. The critical weight of rust which causes concrete cracking mainly depends on the cover depth and mechanical properties of concrete.
4. The model proposed here to predict the time to cracking using the critical weight of rust products and equivalent corrosion rate gives a good result to an actual observed time to cracking.

References

1. C. L. Page and K.W.J. Treadaway, *Nature*, 1982, **297**, 109.
2. G. J. Verbeck, 'Mechanism of Corrosion in Concrete', *Corrosion of Metals in Concrete*, ACI SP-49, 1975.
3. R. Baboian, 'Synergistic Effects of Acid Deposition and Road Salts on Corrosion', *Corrosion Forms and Control for Infrastructures*, ASTM STP 1137, 1992.
4. E. J. Fasullo, 'Infrastructure: The Battlefield of Corrosion', *Corrosion Forms and Control for Infrastructures*, ASTM STP 1137, 1992.
5. P. D. Cady and R. E. Weyers, *Cement, Concrete & Aggregate*, 1983, **5**, 81.
6. Z. P. Bazant, 'Physical Model For Corrosion in Concrete Sea Structures - Theory', *Journal of the Structural Division*, ASCE, 1979, ST6, 1137.
7. Z. P. Bazant, 'Physical Model for Steel Corrosion in Sea Structures - Applications', *Journal of the Structural Division*, ASCE, 1979, ST6, 1155.
8. M. Stern and A. L. Geary, *Journal of Electrochemical Society*, 1957, **104**, 56.
9. K. C. Clear, 'Measuring Rate of Corrosion of Steel in Field Concrete Structure', *Transportation Research Record*, 1992, **1211**, 28.
10. J. P. Broomfield, J. Rodriguez, L. M. Ortega and A. M. Garcia, 'Corrosion Rate Measurements in Concrete Bridges by Means of the Linear Polarization Technique Implemented in a Field Device', Paper presented at the ACI Fall Convention, Minneapolis, MN, USA, November, 1993.
11. P. Schiessl and M. Raupach, 'Corrosion of Reinforcement in Concrete', (C. L. Page, K. W. J. Treadaway, P. B. Bamforth eds.), Society of Chemical Industry, Elsevier Applied Science, London, 1990, 49.
12. S. A. Bradford, 'Corrosion Control', New York, 1992.

EFFECT OF STRUCTURAL SHAPE AND CHLORIDE BINDING ON TIME TO CORROSION OF STEEL IN CONCRETE IN MARINE SERVICE.

A. A. Sagüés and S. C. Kranc

Department of Civil and Environmental Engineering
University of South Florida
Tampa, FL 33620
U.S.A.

1 INTRODUCTION

Corrosion of reinforcing steel can seriously limit the service life of concrete bridge substructure components in marine applications. The history of the corrosion process may be divided into two general phases: an initiation period, in which the reinforcing steel surface is still passive as a result of contact with the high pH concrete pore water, and a propagation period that starts when chloride ion contamination at the steel surface exceeds a critical threshold level for passivity breakdown. Corrosion rates in the propagation period are relatively fast, and externally visible cracks and spalls in the concrete cover are often observed only a few years after corrosion initiation. As a result, present design trends for long term durability emphasize prolonging the initiation period as far as possible. Concrete-based strategies to attain this goal include improving the concrete quality to reduce the rate of chloride ion penetration, and maintaining an adequate concrete cover thickness. Approaches oriented to the reinforcing steel include increasing the critical chloride concentration threshold, either by the addition of corrosion inhibitors to the concrete mix, or by using alternative rebar materials.

Quantitative prediction of the service life of a structure is a crucial step in determining how appropriate or cost-effective the various approaches for durability may be. Any prediction is based on the speed of chloride accumulation at the rebar surface, and computation of the time necessary to achieve the critical concentration threshold. A simplified method of prediction consists of assuming a planar geometry with a concrete surface that has a constant level of chloride ion contamination. Chloride ion transport through the concrete cover is then assumed to take place by homogeneous diffusion with time-invariant diffusivity. Application of standard solutions of the one-dimensional diffusional equation then yield a prediction of the time needed to attain the threshold concentration level at the rebar cover depth. This simplified approach is only useful for very rough or comparative estimates, because of the numerous complications in the actual systems. A partial list of these complications includes variation of the surface level of chloride contamination with time, both short- and long-range heterogeneity introduced by the presence of aggregates, aging of the concrete, non-diffusional transport of both water and chloride ions, the presence of cracks in the concrete, effect of gradients of porosity, temperature and electric potential, local changes in the critical threshold due to polarization of the steel in corrosion macrocells, deviations from planar geometry, and the effect of chloride trapping by mechanisms such as chemical binding.

Inclusive treatment of all the effects mentioned above in a durability prediction model is difficult at best.[1-6] However, valuable insight can be achieved when some of the factors are isolated for examination. In this study, combined effects of structural geometry variations and chloride binding on the outcome of durability predictions have been investigated with mathematical modeling for conditions relevant to marine bridge substructure components.

The following modeling constraints and assumptions were used:

1. An individual structural element was considered, at a fixed elevation and not coupled electrochemically to other portions of the structure. The concrete was assumed to be fully mature, water-saturated and with a constant level of porosity. The heterogenous nature of the concrete was recognized but no special property variations near the surface were assumed. The effects of stress and cracking were ignored.

2. Three system geometries were considered: a) flat wall with reinforcing steel placed below a constant concrete cover depth; b) a convex square corner corresponding to the intersection of two walls, with a reinforcing steel bar placed at equal distance from both surfaces and c) a circular reinforced concrete column section, with reinforcing steel bars placed at a fixed distance from the outer perimeter.

The chloride ion activity at the concrete surface was assumed to be constant with time. Temperature was considered to be constant. Electromigration and synergistic effects with carbonation reactions were ignored.

3. The chloride ions were assumed to be present in either of two conditions: 'free chlorides', dissolved in the pore water, and 'bound chlorides' which were chemically or physically absorbed by the concrete. In the following, the concentration of free chlorides per unit volume of pore water is designated as C_f and the concentration of free chlorides per unit volume of concrete is designated as C_F. The concentrations of bound chlorides and total (free plus bound) chlorides, which can only be defined per unit volume of concrete, are designated by C_B and C_T respectively. Lower case subscripts are used to express magnitudes for the pore water volume only and upper case subscripts for the entire concrete volume.

4. It was assumed that only the free chlorides were able to move through the concrete. Chloride transport was assumed to take place only by diffusion through the pore water. The flux j_x of chloride ions diffusing through the pore water can be written as:

$$j_x = -D\nabla C_f \tag{1}$$

The diffusion coefficient, D, is not the diffusion coefficient D_o which would relate concentration gradients to flux in an unconstrained electrolyte, but rather a parameter modified by the tortuosity and constrictivity of the pore structure.[7] For simplicity, Equation (1) assumed that the flow is driven by concentration rather than chemical activity gradients, but activity coefficients could be incorporated in the treatment, if desired.

5. Chloride transport was assumed to occur only through open pores with a microstructure such that the effective pore area for transport could be equated with the total porosity, ϵ, so that:

$$\frac{\partial C_T}{\partial t} = \nabla \cdot (\epsilon j_x) \tag{2}$$

2 DIFFUSION COMBINING BINDING OF CHLORIDES

Converting the free chloride concentration in the pore water to a total volume basis:

$$C_F = \epsilon C_f \tag{3}$$

and by the assumption of spatially constant porosity, a cancellation of terms results in Equation (2) and the diffusion equation becomes:

$$\frac{\partial C_T}{\partial t} = \nabla \cdot (D \nabla C_F) \tag{4}$$

Following the treatment of Nilsson et al,[4] since $C_T = C_B + C_F$

$$dC_T = dC_F \left(1 + \frac{\partial C_B}{\partial C_F}\right) \tag{5}$$

and as a result, Equation (4) becomes:

$$\frac{\partial C_T}{\partial t} = \nabla \cdot \left[\frac{D}{1 + \dfrac{\partial C_B}{\partial C_F}} \nabla C_T \right] \tag{6}$$

The value of the term $\partial C_B / \partial C_F$ at a given time and position is determined by the mechanism responsible for the binding of chlorides. In this paper, the modeling calculations have been limited to cases in which bound chlorides result from an instantaneous equilibrium reaction between free chlorides and the concrete surrounding the pore water,[2] so that C_B is a function of C_F only. That function is known as the binding isotherm for the system. If the isotherm consists only of the direct proportionality $C_B = k_o C_F$, the system is said to display linear binding. Calling $D' = D/[1 + \partial C_B / \partial C_F]$, then for the linear case $D'_{linear} = D/(1 + k_o)$ and if D is constant Equation (6) reduces to

$$\frac{\partial C_T}{\partial t} = D'_{linear} \nabla^2 C_T \tag{7}$$

This equation could be viewed as describing the behavior of a system in which the total chloride moves as if the binding were not taking place, but with a diffusion coefficient smaller than D by a factor $[1 + k_o]$. Thus, the result of linear binding is to uniformly slow down the overall transport of chloride ions compared to a case without binding.

Actual near-equilibrium isotherms exhibit distinctly non-linear behavior.[4,6-10] Since in an equilibrium isotherm C_B, C_F and C_T are uniquely related, then in Equation (6) one can write $D/[1 + \partial C_B / \partial C_F] = D'(C_T)$. Equation (6) thus modified may be viewed as describing a system in which overall chloride motion is determined by a diffusion coefficient $D'(C_T)$ which varies with concentration even if D itself is constant.

3 CASES STUDIED

The test geometries used were as indicated in Section 1. The surface condition of the concrete was considered to fall in one of two extreme cases: complete submersion, in which the surface of the concrete is always in contact with seawater, and an evaporative regime position a short elevation above the high tide region, in which the seawater splashing the concrete surface evaporates sufficiently to reach the solubility limit of chloride ions (but with the concrete pores still filled with solution). In the first case the concentration of chloride ions in the pore water near the surface was assumed to be constant with time and equal to $C_{sfw} = 25$ kg/m^3 (typical of seawater). In the evaporative regime the surface pore concentration was considered to become stable early enough in the life of the structure to also be treated as a constant value, chosen to be $C_{sfe} = 250$ kg/m^3 (a round figure bracketing the nominal solubility limit of chloride salts in water). The concrete porosity was assumed to be $\epsilon = 0.1$ in all cases. Therefore at the surface $C_{SFW} = 2.5$ kg/m^3 and $C_{SFE} = 25$ kg/m^3.

Three chloride binding regimes were considered: negligible binding, and two idealized isotherms. Both isotherms involved a maximum concentration of bound chlorides $C_C = 5$ kg/m^3, typical of observed values[3,9] and for simplicity were considered to follow Langmuir behavior[2] such that:

$$C_B^{-1} = (C_F k C_C)^{-1} + C_C^{-1} \tag{8}$$

At small values of C_F, $C_B \approx k\, C_C\, C_F$, and the behavior approaches linear binding with a coefficient $k_o = k\, C_C$. At large values of C_F, $C_B \approx C_C$ and the behavior resembles unbound chloride diffusion since the binding effect has reached saturation. One of the isotherms had $k\, C_C = 3$ ('soft' binding case, representative of values often reported in the literature[2]). The other isotherm, with $k\, C_C = 300$, represented almost immediate saturation of the binding reaction with very low free chloride concentrations ('hard' binding isotherm, as an idealized extreme case).

4 MATHEMATICAL SOLUTION

Numerical solutions of Equation (6) for constant D were obtained by means of a conventional finite difference approach utilizing a forward time step-central space explicit scheme. The equation was formulated, using the non-dimensional variables $x = x/L$, $t = t/(L^2/D)$, and $C_T = C_T/C_o$ where L, D and C_o are scaling parameters that can be used to fit the general solutions to individual cases. Thus formulated, Equation (6) becomes non-dimensional and D disappears from the equation (tortuosity and constrictivity adjustments that refer directly to D_o, not addressed here, would require special treatment). The term $\partial C_B / \partial C_F$ must be evaluated for the value of those magnitudes when $C_T = C_T\, C_o$, so even though the equation is non-dimensional the result depends on the choice of isotherm parameters. Although the isotherms studied here obey Equation (8), the numerical method permits using any other $C_B = f(C_F)$ relationship desired, either in analytical form or as a tabulation of experimental data. The problem was set up to correspond to a finite x domain with $C_T = 0$ at $t = 0$ within the domain and $C_T = 1$ for all values of t at both domain ends. The value of C_o was chosen so that the value of C_T at the domain ends corresponded through Eq.(8) to $C_F = C_{SFW}$ or $C_F = C_{SFE}$. Typically, 100 non-dimensional x intervals were used for the one-dimensional and cylindrical cases. The two-dimensional corner case was solved with a two-dimensional grid typically 100 non-dimensional x

intervals wide in each direction, with boundary conditions arranged similarly to the other cases. For this study, the non-dimensional time progression was limited in the one-dimensional and corner cases to domains in which semi-infinite behavior was prevalent. However, since the problem layout was finite the same computations can be easily extended to examine the behavior of finite thickness slabs or narrow square/rectangular columns if desired. The finite problem was necessarily addressed in the cylindrical case, for which Equation (6) was formulated in radial coordinates with L=r (the column radius).

5 RESULTS AND DISCUSSION

For each of the semi-infinite one-dimensional and corner cases the form of the equation and the boundary conditions permit collapsing the solutions for a given isotherm as a single function of the time variable, $T = 4Dt/x^2$, where x is the distance into the concrete from the flat concrete surface (for the corner case, the distance from either face but at points equidistant from both faces). The relationships between C_T, C_B and C_F in one-dimensional cases have been discussed elsewhere[2,4]. Since the initiation of corrosion is expected to concern primarily the values of C_f (and C_F when ϵ is constant), the following analysis will address the latter variable.

Figure 1 shows the calculated values of C_F as a function of T for each of the six semi-infinite flat wall cases, coded as per Table 1. The no-binding curves (NEF, NWF) are identical to the usual error-function solution $C_F = C_{SF}(1-\mathrm{erf}(T^{-1/2}))$ with C_{SF} equal to C_{SFE} or C_{SFW} respectively. The graphic representation permits easy evaluation of the effect of binding type and exposure regime on the time to corrosion initiation given a certain chloride concentration initiation threshold, $C_{fi} = C_{FI}/\epsilon$ for a given combination of rebar cover, x, and diffusion coefficient, D. For plain reinforcing steel, C_{fi} values in the absence of passivating inhibitors are often said to approximate a [Cl⁻]/[OH⁻] ratio ≈ 0.6,[11] which corresponds roughly to 0.07 kg/m³ $\leq C_{FI} \leq 0.7$ kg/m³ for $\epsilon=0.1$ and pore solution $12.5 < \mathrm{pH} < 13.5$.[12] Thus, for the submerged concrete cases ('WF' family of curves) the change from no binding to 'soft' binding represents an improvement by a factor of ≈ 3 in the time to corrosion initiation if C_{FI} is in the range considered. The 'hard' binding case represents roughly an additional factor of 3 improvement. This additional improvement takes place even though the assumed saturation concentration of bound chlorides

Table 1 *Cases computed and nomenclature*

BINDING	Regime	Semi-infinite Flat wall	Semi-infinite Corner	Cylindrical Column
NONE	Submerged	NWF	NWC	NWR
	Evaporative	NEF	NEC	NER
'SOFT'	Submerged	SWF	SWC	SWR
	Evaporative	SEF	SEC	SER
'HARD'	Submerged	HWF	HWC	HWR
	Evaporative	HEF	HEC	HER

(5 kg/m³) was the same in both cases. The additional improvement is due to the fact that at the relatively low concentrations relevant to the initiation threshold the effective diffusion coefficient approaches the limit $D'_{linear} = D/(1+ko) = D/301$ for the 'hard' binding case as compared with $D/4$ for the 'soft' binding case.

A sensitivity factor, $S = dlog(T)/dlog(C_F)$ (the inverse of the slope of the curves), can be defined to describe the sensitivity of the time for corrosion initiation to variations in C_{FI}. The value of S, which is usually about 1/2 to 1/3 for the no-binding cases, becomes smaller the 'harder' the binding (which causes the increase in concentration with time at a given x to be more abrupt). This observation may be important in assessing the suitability of corrosion control measures based on material modifications or surface treatments aimed at increasing the value of C_{fi}.[5,13] The nature of the binding isotherm of the concrete used may in those cases affect the relative increase in time to corrosion initiation (and cost effectiveness) associated with the material alternative considered.

For the evaporative exposure cases ('EF' family) the times to corrosion initiation were, as expected from the larger surface chloride concentration, smaller than for the corresponding submerged exposure cases. In the 0.07 kg/m³ $\leq C_{FI} \leq 0.7$ kg/m³ range considered earlier, the relative reduction in time to corrosion initiation due to binding (and to the change from 'soft' to 'hard' isotherm) is less than for the submerged case. This results from the smaller ratio of bound to total chloride (compared to the submerged case) encountered for the evaporative case in the high concentration portion of the profile. At

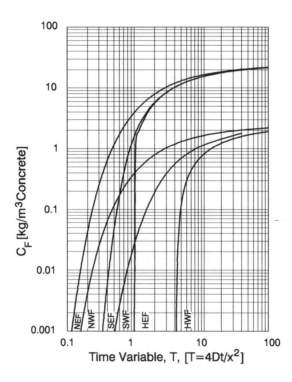

Figure 1   *Free chloride as function of the time variable for the semi-infinite one-dimensional cases (see Table 1).*

lower chloride concentrations, the effect of introducing binding becomes more pronounced in both the evaporative and submerged cases. However, for a given binding condition the sensitivity factor, S, is smaller in the evaporative case than in the submerged case. This reflects the more abrupt rise in concentration with time at a given location that can be expected as the driving gradients become greater. Otherwise, the same general observations given for the submerged case also apply to the evaporative case.

Figure 2 shows the calculation results for the semi-infinite corner geometry. The distance x corresponds to the concrete cover as measured from either surface, but the calculations apply only to bars placed equidistant from either surface (bars far away enough from the diagonal would be in conditions approaching the flat cases discussed earlier). The general shapes of the curves resemble those in Figure 1, but are somewhat distorted and shifted toward lower time values. The shift to lower times reflects the simultaneous entry of chlorides from both faces, resulting in a faster buildup along the diagonal than away from the corners. The same overall observations concerning binding and surface concentration effects can be made as for the flat geometry cases. However, it is important for design purposes to explore the differences in time to corrosion initiation of the corner (T_C) and flat wall (T_F) cases.

A derating factor, $F_D = T_C/T_F$ was calculated for each binding and surface concentration case, and the results are presented in Figure 3 ('CORNER'). The solid curve corresponds to the no-binding cases, for which the results are a function only of C_{FI}/C_{SF}

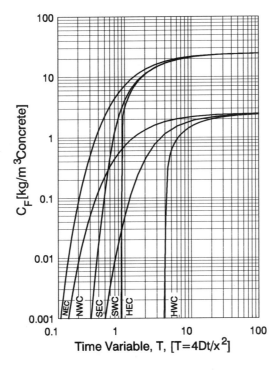

Figure 2   *Free chloride as function of time variable for the semi-infinite two-dimensional cases (see Table 1)*

since the plain diffusion equation (as in Equation(7)) can be scaled directly. Computation of F_D for the no-binding cases could be made with high precision by contrasting with standard solutions to the plain diffusion equation. Resolution in the computation of F_D for the binding cases was limited, but within that resolution, the results for the binding cases (denoted by * in Figure 3) were found to approximate the no-binding behavior when plotted as a function of C_{FI}/C_{SF}. In all cases the derating became more severe as C_{FI} increased. The likely C_{FI} range for corrosion initiation without inhibitors mentioned earlier (0.07 kg/m^3 $\le C_{FI} \le$ 0.7 kg/m^3) merits examination. This range corresponds to $0.003 \le C_{FI}/C_{SF} \le 0.3$ when both submerged and evaporative regimes are considered. Within that range, the maximum derating was $\approx 50\%$ in the submerged regime but typically less than $\approx 20\%$ in the more severe evaporative regime.

The cylindrical cases were solved by formulating Equation(6) in radial coordinates. In these inherently finite-dimension cases the solution cannot be expressed in terms of a single variable to cover all space-time combinations. Thus, instead of a single $C_F(T)$ curve for each surface concentration and binding case (Figures 1 and 2), a family of curves is obtained with x/r as a parameter, where r is the radius of the column. This is illustrated in Figure 4 for the SER case. As before, a derating factor $F_D = T_R/T_F$ was calculated to compare the time to corrosion initiation in the round geometry cases (T_R) to that of the flat wall cases. The results are presented in Figure 3 ('ROUND') for values of x/r=0.1, 0.2 and 0.3. The solid curves corresponds to the no-binding cases, for which the results are a function only of C_{FI}/C_{SF} for the same reasons indicated earlier. The no-binding derating curves were calculated with high precision using standard solutions for diffusion in

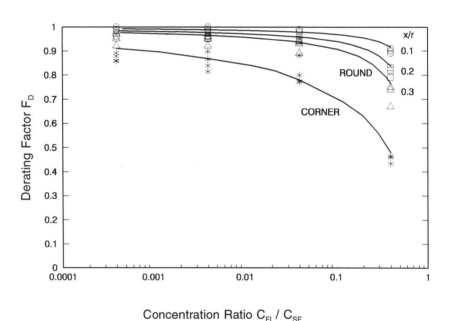

Figure 3   *Reduction in time to corrosion initiation in semi-infinite corner cases and in round cases compared with the semi-infinite flat wall.*

cylinders. As in the corner geometry problem the computation of F_D for the round geometry binding cases was limited in resolution, but the results (denoted by $\circ$, $\square$ and $\triangle$ for $x/r = 0.1$, 0.2 and 0.3 respectively)) were found to approximate the no-binding behavior. For structural columns in bridges the concrete cover, x, is usually a small fraction (≤ 0.2) of the radius. The calculations showed that the derating factor in the time to corrosion initiation under these circumstances is smaller than in the flat-vs-corner case and does not exceed 10% over much of the range of C_{FI} values of interest.

The above analyses can be easily tailored to other geometries (such as three-sided corners and finite thickness slabs) and other regimes of C_{FI} of interest, such as the higher levels that would result from the application of passivating corrosion inhibitors[2,5,13] or corrosion resistant alloys and other treatments.[14,15] As shown in Figure 3, the geometric derating factor could in those cases become more severe and have a more important impact on design. Computations with finer x spacing are in progress for both the corner and round cases to allow for more precise definition of the effects of binding on F_D and results will be presented in the future. It should be kept in mind that the results presented here addressed only selected combinations of geometric and binding effects. While the findings can serve to assist in guiding design decisions, the simultaneous effect of other critical factors mentioned in Section 1 such as the effect of aging of the concrete and partial water saturation[1,3] must be carefully examined.

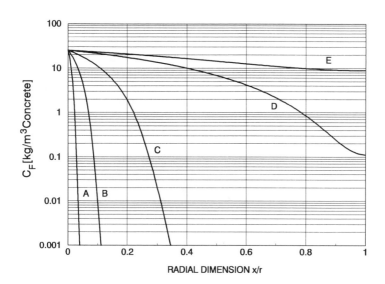

Figure 4   *Concentration profiles for the column case SER for increasing dimensionless time $t=0.0001$ (A); 0.001 (B); 0.01 (D) and 0.2 (E).*

6 CONCLUSIONS

The results of the investigation showed that:

1. The time to corrosion initiation in corners or round columns was reduced compared to a flat wall configuration by a factor that became less important as the surface free chloride concentration increased. For the conditions examined and within the resolution of the numerical solutions obtained, the effect of system geometry on time to corrosion initiation was comparable in the binding and no-binding cases.

2. The relative reduction in the time to corrosion initiation in two-sided corners and columns with small cover/radius ratios typical of marine bridge substructures was moderate for the concentration thresholds usually associated with plain steel without inhibitors. More severe derating factors are expected if protection measures are adopted that elevate the initiation threshold.

7 ACKNOWLEDGMENT

This investigation was funded by the Florida Department of Transportation and the Federal Highway Administration. The opinions, findings and conclusions presented here are those of the authors and not necessarily those of the funding agencies.

8 REFERENCES

1.   A. V. Saetta, R. V. Scotta, and R. V. Vitaliani, *ACI Materials Journal*, 1993, **90** (5), 441.
2.   C. J. Pereira and L. L. Hegedus, Eighth International Symposium on Chemical Reaction Engineering, Institution of Chemical Engineers Symposium Series No. 87, 1984, 427.
3.   P. S. Mangat and K. Gurusamy, *Cement and Concrete Research*, 1987, **17** (4), 640.
4.   L. O. Nilsson, M. Massat and L. Tang, SP 145-24., American Concrete Institute, Detroit, 1994, 469.
5.   N. S. Berke, in 'Corrosion Effects of Stray Currents and the Techniques for Evaluating Corrosion of Rebars in Concrete', (Ed: V. Chaker), ASTM STP 906, American Society for Testing Materials, Philadelphia, 1986, 78.
6.   C. L. Page, P. Lambert and P. Vassie, *Materials and Structures*, 1991, **24**, 243.
7.   A. Atkinson and A. K. Nickerson, *Journal of Materials Science*,1984, **19**, 3068.
8.   L. Tang and L. O. Nilsson, *Cement and Concrete Research*, 1993, **23**, 247.
9.   Rasheeduzzafar, S. Hussain and S. Al-Saadoun, *ACI Materials Journal*, 1992, **89**, 3.
10.   J. Tritthart, *Cement and Concrete Research*, 1989, **19**, 586.
11.   D. Hausmann, *Materials Protection*, 1967, **6**, 19.
12.   S. Goñi and C. Andrade, *Cement and Concrete Research*, 1990, **20**, 525.
13.   N. S. Berke and M. C. Hicks, in 'Corrosion Forms and Control for Infrastructure', ASTM STP 1137, (Ed: V. Chacker), ASTM STP 1137, American Society for Testing and Materials, Philadelphia, 1992, 207.
14.   Treadaway, K., Cox, R. and Brown, B., *Proc. Instn. Civ. Engrs*, 1989, **86**, 305.
15.   Yeomans, S., *Corrosion*, 1994, **50**, 72.

LABORATORY TESTING OF FIVE CONCRETE TYPES FOR DURABILITY IN A MARINE ENVIRONMENT

Rob B. Polder

TNO Building and Construction Research,
P.O. Box 49, 2600 AA, Delft, Netherlands

1 INTRODUCTION

Penetration of chloride from sea water is the main cause of depassivation of steel reinforcement and subsequent damage to marine concrete structures, particularly in the splash zone. High chloride penetration resistance may be obtained by adding hydraulic or pozzolanic minerals to the cement. Examples are blast furnace slag (BFS) which has been used in concrete in the Netherlands at high replacement levels for many decades[1] and silica fume and fly ash which are being used in other countries. In order to evaluate the potential durability of a specific mix in the marine splash zone environment during a test period of up to two years, a set of laboratory tests was proposed. Subsequently, five concrete mixes were tested in a study ordered by SMOZ and CUR.[2]

2 DURABILITY OF CONCRETE IN MARINE ENVIRONMENT

The service life is the sum of the corrosion initiation period and the corrosion propagation period.[3,4] In a marine environment, the initiation period is determined by the time taken for chloride penetration through the concrete cover to the reinforcement, until the steel loses its passivation and corrosion starts. The propagation period is characterised by the rate of corrosion and the amount of corrosion that is necessary before concrete cover will crack. Major cracking of the concrete cover is considered to indicate the end of the service life. Without repair measures, structural safety may become compromised in a relatively short length of time.

Within a marine structure, the risk of corrosion is generally highest in the splash zone, because both chloride penetration and corrosion rate are relatively rapid. Chloride penetration of the surface layers may be accelerated by a partial drying out of the concrete and subsequent capillary absorption of sea water. The corrosion rate after depassivation is high because the concrete remains wet due to the frequent exposure to splash, so the concrete resistivity is relatively low. Furthermore, because the concrete is not completely saturated with water, sufficient oxygen can reach the steel for corrosion to occur.

3 LABORATORY TESTING FOR DURABILITY

The durability prediction of the investigated concretes was found by adding the the time-to-depassivation, calculated from the rate of chloride diffusion and the projected cover depth, and the the time-to-cracking, calculated from the concrete resistivity and the tensile strength of each mix, using a modification of a life time model from literature.[3]

4 CORROSION INITIATION

The effective chloride diffusion coefficients, D_{eff}, were calculated from diffusion profiles that were fitted to the observed chloride profiles determined after 1.5 years of immersion in 3.5% NaCl solution. It was assumed that the transport of chloride through the concrete may be modelled as a diffusion process, which is justified for the following reasons.

Firstly, the effect of capillary absorption is only significant for low cover depths and in the short term. In the long term and for higher cover depths, chloride transport will be dominated by diffusion. This was confirmed from experiments involving concrete exposure to marine splash for 6 years,[5] where the penetration was modelled as diffusion and coefficients were obtained which were similar (for similar concrete mixes) to those from 16 years submersion in the North Sea.[6]

Secondly, it was shown experimentally that the influence of chloride binding by cement components can be included in the effective diffusion coefficient.[7,8]

For the input values of the calculation the following data were used. The surface chloride concentration, C_s, was estimated from measurements on marine structures to be about 5% chloride by mass of cement.[9] Based on splash zone exposure tests, a value of 4.5% is suggested for service life design calculations.[10] From the present laboratory tests slightly lower C_s values were found, namely between 3 and 4%. Consequently, 4% is taken as the value for C_s in these calculations.

Regarding the critical chloride content or threshold value, C_{cr}, relatively low values of about 0.4% by mass of cement have been adopted in many national standards. Recently it has been established that significant corrosion rates in alkaline (i.e. non-carbonated) concrete occur only at chloride contents of at least 1%.[11] It appears that 1.0% chloride by mass of cement is a reasonable, possibly even conservative, value for good quality concrete such as that used in most marine structures (and for the mixes investigated here).

If the D_{eff} of a particular concrete mix is known from experiments, then by using Fick's second law of diffusion with the specified values for C_s and C_{cr}, for a given cover depth, L, the time-to-initiation, t_i, can be calculated with:

$$t_i = \frac{L^2}{2.65 * D_{eff}}$$

(1)

5 CORROSION PROPAGATION

After depassivation, the corrosion rate is assumed to be determined by the resistivity of the concrete. This is a simplification, for which there is experimental and theoretical support. It may be assumed that the anodic process is not rate-controlling at chloride contents over 1%.[11] The cathodic process is not rate-controlling because the concrete is not completely water saturated, which allows sufficient oxygen to penetrate to the steel. From macrocell experiments, the resistivity has been found to be the major factor controlling the corrosion rate for concretes with a relatively high resistivity.[12] In low resistivity concretes, other factors have appeared to be dominating the corrosion rate. Pitting corrosion may be regarded a special case of macrocell corrosion, where the anode-to-cathode distance is smaller than in a true macrocell. A linear relationship between the inverse of concrete resistivity and steel corrosion rate has been established from steel polarisation resistance experiments in mortars involving various cement types in the laboratory.[13-15] From experiments on concrete blocks exposed to marine splash for 6 years, a similar relationship between resistivity and corrosion rate was found,[16] suggesting resistive control for resistivities greater than about 70 Ωm. Using these data, Bazant's theoretical model was calibrated. It appeared to produce very high corrosion rates, probably due to an overestimated maximum potential difference between the anodic and cathodic sites of 1 V. A more realistic value would be about 100 mV. Using the calibrated model, the corrosion rate, CR, is given by:

$$CR = \frac{1000}{\rho_{concrete}}$$

(2)

with $\rho_{concrete}$ the concrete resistivity in Ωm, giving CR in μm Fe/year.

To calculate the length of the propagation period from a given corrosion rate, the amount of corrosion required to cause cracking must be known. From practical observations, it has been found that the critical amount of corrosion for bars with 10 mm cover depth was 200 μm.[4] Experiments and calculations have shown that very fine cracks are caused by as little as 20 μm corrosion, and cracks wider than 0.3 mm originate from 100 to 200 μm of steel thickness loss.[17,18] Considering the stresses caused by the expansion and the elasticity of concrete (taking creep into account, it has been suggested that the critical amount of corrosion is proportional to the tensile strength of the concrete.[3] These results are in general agreement with the other data.

The length of the propagation period, t_p, can be calculated from the amount of corrosion that causes cracking divided by the corrosion rate. Assuming that the corrosion rate is determined by the resistivity according to equation 2, this means that:

$$t_p = 12.5 * 10^{-3} * f * \rho_{concrete}$$

(3)

with t_p in year, and f, the tensile strength, in N/mm^2.

6 INVESTIGATED CONCRETE MIXES

The five concrete mixes investigated included three mixes with additional cementing materials, one with lightweight aggregate and one plain portland cement and gravel mix, all with a low water-to-cement ratio (wcr).

Table 1   *Concrete compositions and fresh concrete slump*

Concrete Type	mix S	mix L	mix F	mix P	mix B
Cement Type	OPC	OPC	OPC	OPC	BFSC
Specific Features	5% silica fume	Lytag coarse aggregate	5% silica fume & 10% fly ash	-	70% slag
Cement Content (kg/m^3)	340	353	337	339	338
Water Content (kg/m^3)	146	152^1	145	146	145
Water-to-Cement-Ratio	0.43	0.43^1	0.43	0.43	0.43
Superplasticiser (% by Cement)	2.65	2.65	3.06	2.47	2.02
Slump (mm)	100	80^2	100	120	125

1 exclusive of 10% (m/m) absorbed water with respect to Lytag mass
2 slump was kept relatively low in order to prevent floating of aggregate

The main compositional features and the mix codes were (see also Table 1):
- OPC and (river) gravel, mix P,
- BFSC (70% slag) and gravel, mix B,
- 5% silica fume, OPC and gravel, mix S,
- 5% silica fume, 10% fly ash, OPC and gravel, mix F,
- OPC and sintered fly ash (Lytag) aggregate, mix L,
all with river sand, about 340 kg cement/m^3 (class A), a wcr of 0.43 and 2 to 3% superplasticiser, of which the dosage was adjusted to obtain a slump of about 100 mm.

7 TEST METHODS AND BASIC RESULTS

The tests involved determining:
- compressive and tensile strength of 150 mm cubes (in triplicate) and density after 28 days in a fog room,
- chloride diffusion coefficients after 1.5 years submersion of 100 mm x 100 mm x 300 mm prisms in 3.5% NaCl solutions at three temperatures (20, 30 and 40°C) from analysis for total chloride content of thin slices and fitting diffusion profiles;
- electrical resistivity (108 Hz AC) of 100 mm x 100 mm x 50 mm prisms with two embedded bars (in 4-fold) during 1.5 years exposure to a fog room and in air at 20°C and 80% relative humidity, respectively; after that time the two sets of specimens were exchanged between the two climates and measurements were carried on until 884 days; the cell constant (0.063 m) was obtained from calibration using solutions of known conductivity.

Table 2   *Mechanical properties and densities of concrete after 28 days curing in a fog room*

Concrete Type	mix S	mix L	mix F	mix P	mix B
Compressive Cube Strength (N/mm^2)	64.1	49.1	68.2	50.4	51.2
% of mix P	127	97	135	100	102
Splitting Tensile Strength (N/mm^2)	5.19	3.52	4.93	3.87	3.85
% of mix P	134	91	127	100	99
Density (kg/m^3)	2412	2037	2413	2431	2415
% of mix P	99	84	99	100	99

The results of the mechanical testing and the densities after 28 days are given in Table 2. The mixes S and F have higher strengths (by about 30%) than the mix P due to the addition of silica fume. In terms of strength, mix B is completely equivalent to mix P. Mix L has a normal compressive strength but a slightly low tensile strength.

Chloride penetration profiles fitted to the experimental profiles after 1.5 years of exposure to salt solution at 30°C are given in Figure 1. The calculated effective diffusion coefficients are presented in Table 3. The effect of increased temperature is not shown for individual mixes. The differences between the five mixes are important. The OPC concretes, mix P and mix L, (made with gravel or Lytag, respectively) have high diffusion coefficients. Of these, the Lytag concrete has a higher diffusivity than the gravel/OPC mix. On the other hand, two of the composite (blended) cement concretes, mixes B (blast furnace slag) and F (silica fume and fly ash), have a diffusion coefficient which is about three times lower than mix P. The third blended mix, mix S (silica fume), does not fit into this pattern. It would have been expected to have low chloride diffusion, but in fact was similar to the plain OPC mix. Possibly a higher level of addition would have resulted in increased chloride penetration resistance. It is clear from the increased strength (compared to mix P) that the silica fume has had some effect on the microstructure of the concrete.

The diffusion coefficient for mix P is similar to values found after 16 years submersion in the North Sea for OPC concrete.[6] Blast furnace slag diffusion coefficients after 16 years were 0.3 x 10^{-12}, which is about 2.5 times lower than the present mix B values (0.8 x 10^{-12}). Probably the value decreases due to further hydration of the slag during the long wet exposure.[6] Considering this, the results of both mix P and mix B are in general agreement with data from long term natural sub-sea exposure and may be regarded reliable values for marine (diffusion) conditions. Apparently the lower temperature in sea exposure (approximately 10°C) than in the laboratory (20 to 40 °C) has no strong influence on the diffusion coefficient.

A final remark must be made regarding diffusion in mix L. Diffusion in saturated mix L concrete is faster than in mix P concrete, probably due to additional transport of chloride through channels in the porous aggregate (it has the same matrix). In real splash zone conditions, the concrete and in particular the aggregate would dry out to a

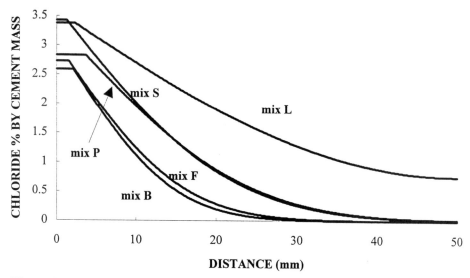

Figure 1   *Fitted chloride diffusion profiles for the five mixes after 1.5 years submersion in 3.5% NaCl solution at 30°C*

certain extent. Consequently the transport would not be able to proceed through the aggregate at the same rate and the effective diffusion would reduce, possibly approaching the value of mix P.

The resistivities as a function of exposure time are shown in Figure 2 (for specimens that started in the fog room) and values after 1.5 years are given in Table 4. Most results were fairly stable for the period between 3 months and 1.5 years. The resistivities in 80% RH (not shown) were much higher than the values in the fog room. Upon transfer of the 80% RH samples to the fog room after about 1.5 years, the resistivities decreased to roughly their fog room values. Only the value of mix L was significantly higher when transferred to the fog room after exposure in 80% RH when

Table 3   *Effective chloride diffusion coefficients after 1.5 years exposure to 3.5% NaCl solution for various exposure temperatures (20°C only mix P) and overall result for each mix; - not tested*

Concrete Type	mix S	mix L	mix F	mix P	mix B
Temperature		diffusion coefficient (* 10^{-12} m^2/s)			
40°C	3.2	7.7	1.0	2.5	0.7
30°C	2.4	8.4	1.2	2.5	1.0
20°C	-	-	-	2.5	-
Overall	3	8	1.1	2.5	0.8

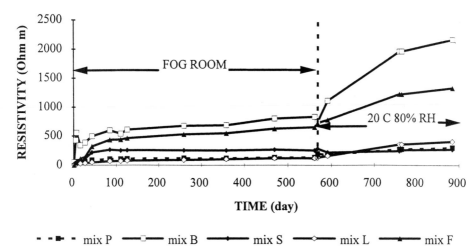

Figure 2 *Concrete resistivity for five mixes during exposure in a fog room for 560 days and another 320 days in air of 20°C and 80% RH*

compared to the constantly wet specimens (+35%). This is probably due to drying out of the pores in the aggregate which are apparently not fully rewetted. In the splash zone at least partial drying out will occur, therefore both values for the resistivity are used in the service life calculation.

For all concretes in the fog room, the resistivity was greater than 100 Ωm from an age of one year onwards, fulfilling the condition that the resistivity must be greater than 70 Ωm, which permits the calculation of the corrosion rate from the resistivity. The observed resistivities correspond well to laboratory and field values for concretes of similar composition (cement type, water-to-cement ratio).[16,19]

8 SERVICE LIFE CALCULATION

The basic transport properties used for the service life calculation and the results in terms of initiation period and propagation period are presented in Table 4. As was mentioned before, for mix L two resistivities (constantly wet and dry → wet) are given and two service life values are calculated. The difference between the two values is small, however.

Table 4 *Service life calculation results for a cover depth of 50 mm*

Concrete Type	mix S	mix L	mix F	mix P	mix B
D_{eff} (x 10^{-12} m²/s)	3	8	1	2.5	0.8
t_i (y)	10	3.7	30	12	37
f (N/mm²)	5.2	3.5	4.9	3.9	3.9
$\rho_{concrete}$ (Ωm)	250	120..160	650	135	830
t_p (y)	16	5..7	40	7	40
t_l (y)	26	9..11	70	19	77

9 CONCLUSIONS

9.1 Concrete Mixes

The following conclusions can be drawn regarding the durability of the five investigated concrete compositions for use in the marine splash zone.

Blended (composite) cements containing blast furnace slag or fly ash and silica fume have improved durability properties in a marine environment, with service lives that are several times longer than the service life of plain portland cement concrete. According to the test results, adding only 5% silica fume to portland cement does not significantly improve the durability. Using porous lightweight aggregate (and portland cement) appears to result in earlier depassivation of the reinforcement than using dense gravel aggregate. This may be an underestimate due to the simplification of the test method.

The calculated service lives in the splash zone range from about 10 to about 80 years, with increasing service life in the order: lightweight aggregate/OPC (mix L) < gravel/OPC (mix P) < gravel/silica fume/OPC (mix S) < gravel/silica fume/ fly ash/OPC (mix F) < gravel/blast furnace slag (mix B). This ranking is in general agreement with the expectations from concrete exposure research and with practical experience. Quantitatively, the service life results are within the range observed in practice (as far as experience exists), so the prediction is reasonable.

9.2 Test Setup

Regarding the test setup, the following conclusions may be drawn.

The service life was calculated by adding the time-to-depassivation (calculated from the chloride diffusion rate and the projected cover depth) and the time-to-cracking (calculated from the concrete resistivity and the tensile strength). This procedure was concluded to produce useful and realistic durability predictions for concrete in the marine splash zone. Simple laboratory experiments with a duration of between one and two years appear to produce suitable results for this calculation.

Chloride penetration was determined satisfactorily from constant submersion tests and subsequent determination of the chloride profile by grinding thin layers and analysing the dust. A suitable exposure temperature was 20 to 30°C. The effect of a higher temperature is negligible.

Measuring the electrical resistivity of the concrete in wet conditions between a half and two years age gives information useful for estimating the corrosion rate. The model used for chloride penetration is more firmly based and calibrated in the field than the model used for the corrosion rate. Therefore the calculated time-to-initiation results are probably more accurate than the time-to-cracking results.

10 REFERENCES

1.   J.G. Wiebenga, 'Durability of concrete structures along the North Sea coast of the Netherlands', in 'Performance of concrete in marine environment', ASTM special publication SP-65, paper 24, 1980, 437-452.
2.   R.B. Polder, 'Durability of new concrete types for marine environment', TNO Building and Construction Research report 94-BT-R0459-02, 1995, to be published,

CUR, Gouda.
3. Z. Bazant, 'Physical model for steel corrosion in concrete sea structures, -theory, - application', *J. Struct Div. Am. Soc. Civ. Eng.*, 1979, **105**, ST6, 1137-1166.
4. K. Tuutti, 'Corrosion of steel in concrete', CBI Stockholm, 1982.
5. P.B. Bamforth, and J. Chapman-Andrews, 'Long term performance of RC elements under UK coastal conditions', 'Proc. International Conference on Corrosion and Corrosion Protection of Steel in Concrete', University of Sheffield, Sheffield, 24-29 July, 1994, 139-156.
6. R.B. Polder and J.A. Larbi, 'Investigation of concrete exposed to North Sea water submersion for 16 years', TNO Building and Construction Research report 94-BT-R1548, 1995, to be published, CUR, Gouda.
7. G. Sergi, S.W. Yu and C.L. Page, 'Diffusion of chloride and hydroxyl ions in cementitious materials exposed to a saline environment', *Magazine of Concrete Research*, 1992, **44**, no.158, 63-69.
8. S.W. Yu, G. Sergi and C.L. Page, 'Ionic diffusion across an interface between chloride-free and chloride-containing cementitious materials', *Magazine of Concrete Research*, **45**, no.165, 1993, 257-261.
9. B. Sørensen, 'Penetration of chloride in old marine structures', *Steel in concrete* 1979 **4**, 6-7.
10. P.B. Bamforth and W.F. Price, 'Factors influencing chloride ingress into marine structures', Concrete 2000, Dundee, vol. 2, theme 4, 1993, 1105-1118.
11. R.F.M. Bakker, G. van der Wegen and J. Bijen, 'Reinforced concrete: an assessment of the allowable chloride content', CANMET, Nice, 1994.
12. M. Raupach, 'Zur chloridinduzierten Makroelementkorrosion von Stahl in Beton', Deutscher Ausschuss für Stahlbeton, **433**, Beuth Verlag, Berlin, 1992, in German.
13. C. Alonso, C. Andrade and J. Gonzales, *Cement and Concrete Research*, 1988, **8**, 687-698.
14. W. Lopez and J.A. Gonzalez, 'Influence of the degree of pore saturation on the resistivity of concrete and the corrosion rate of steel reinforcement', *Cement and Concrete Research*, 1993, **23**, 368-376.
15. G.K. Glass, C.L. Page and N.R. Short, 'Factors affecting the corrosion rate of steel in carbonated mortars', *Corrosion Science*, 1991, **32**, 1283-1294.
16. R.B. Polder, P.B. Bamforth, M. Basheer, J. Chapman-Andrews, R. Cigna, M.I. Jafar, A. Mazzoni, E. Nolan and H. Wojtas, 'Reinforcement Corrosion and Concrete Resistivity - state of the art, laboratory and field results', 'Proc. Int. Conf. Corrosion and Corrosion Protection of Steel in Concrete', University of Sheffield, Sheffield, 24-29 July, 1994, 571-580.
17. C. Andrade, C. Alonso, F.J., Molina, 'Cover cracking as a function of bar corrosion: part 1-Experimental test', *Materials and Structures*, 1993, **26**, 453-464.
18. F.J. Molina, C. Alonso and C. Andrade, 'part 2-Numerical model', *Materials and Structures*, 1993, **26**, 532-548.

CHLORIDE INGRESS AND CHLORIDE-INDUCED CORROSION IN REINFORCED CONCRETE MEMBERS

D. W. Hobbs

British Cement Association
Century House
Telford Avenue
Crowthorne RG45 6YS

1 INTRODUCTION

Chloride ions can be present in a concrete as a result of the application of de-icing salts, exposure to a marine environment, airborne salt or from the concrete constituents. Under field conditions chloride ions can enter the concrete by a number of mechanisms, (i) by diffusion, (ii) by capillary suction under wetting and drying conditions, (iii) under a hydrostatic head and (iv) through cracks or defective joints. In concrete the chloride ions are present both as part of the hydrate structure and as free ions in the pore solution.

The age at which a critical free chloride level is reached at the steel-concrete interface depends upon several factors including - degree of exposure to chlorides, pH of the pore solution, water-binder ratio, binder type, curing time, temperature, moisture state and concrete cover.[1-8] In many parts of the world, damage to concrete due to chloride-induced corrosion is the greatest threat to the durability and integrity of concrete structures. It has been estimated that the annual cost of repairs to UK concrete structures due to steel corrosion[9] is about £5 x 10[8]. In the USA, the associated repair bill for concrete bridges alone has been estimated to be $5 x 10[10]/annum.[10] In parts of the Northern Hemisphere, many of the corrosion problems noted in the past 20 to 30 years have resulted from an increasing application of de-icing salts to bridges and roads which has occurred since the 1950s.

There is a widely held belief that most chloride-induced corrosion problems are due to inadequate concrete specification and that specifying lower water-cement ratios, higher depths of cover or alternative binders would greatly reduce the risk of chloride-induced corrosion problems. As a consequence, durability provisions in codes over the past 60 years have become more onerous with minimum cover increasing from 12 mm to 40 mm or more, and with minimum concrete quality by grade increasing, since the 1960s, and with maximum water-cement ratios reducing.[11-14] However, the more onerous durability provisions have not been successful in minimizing chloride-induced corrosion problems.

In the current paper the main causes of chloride-induced corrosion problems are briefly discussed together with comments on the quality of Portland cement concretes and cover to reinforcement which are probably necessary to give a satisfactory design life in structures, subject to exposure similar to that in the UK, where problems due to design and poor construction practices are absent. Because both Portland cement and concrete practices have changed over the years, the paper concentrates primarily on the reported performance of concretes placed since 1960.

2 QUALITY OF CONCRETE CONSTRUCTION

Two major faults in concrete construction which can lead to problems in concrete members are inadequate cover to reinforcement and inadequate compaction of cover concrete.[15-18] In situations where chloride ingress into concrete occurs, each of these causes can lead to premature deterioration of the affected members.

In a survey of 200 UK concrete bridges, it was found that low cover to reinforcement resulted in spalling of concrete in 77 of the bridges examined.[15] A recent example is the Chiswick flyover in London (author's judgement). Much of this spalling was probably due to chloride-induced corrosion. Surveys of concrete structures elsewhere have also led to the same conclusion that where deterioration occurs, the concrete cover is often low indicating poor cover control. For example, a survey of buildings in Australia[16] showed that at 227 areas of damage, the average cover was only 5 to 6 mm, with a maximum cover of 18 mm. This compared with specified cover in the range 25-40 mm. In a follow up investigation, a detailed survey was made of buildings and bridges under construction.[17] The results are summarized in Figure 1, where the percentage of observed covers exceeding various proportions of the nominal cover is shown. It can be seen that the quality achieved on bridge sites is superior to that obtained on building sites but even so, 4% of the observed covers in the bridges under construction were less than 0.6 of the nominal cover.

The second major cause of premature chloride-induced corrosion problems is inadequate compaction of the cover concrete. This most commonly occurs at the soffit of beams or where the cover to reinforcement is below specification (Figure 2). In a recent survey of 320 bridges along the coast of Norway, it was found that the underside of beams often had quite severe deterioration due to chloride-induced corrosion.[18] A high proportion of this deterioration was attributed to inadequately compacted concrete and insufficient cover.

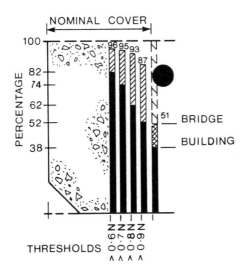

Figure 1 *Percentage of observed covers exceeding various thresholds*[17]

3 ROLE OF DESIGN

Investigation of chloride-induced corrosion problems in bridges has shown that poor performance can often be attributed to inadequacies in design. In the survey of UK bridges referred to earlier, 25 of the bridges had spalling or rust in areas subject to chloride contamination.[15] Typical defects identified were spalling of the top corners of piers, spalling of the abutments just below the bearing shelf, or horizontal cracks with rust staining in similar locations. These defects occurred because failure of the expansion joints above the piers had allowed salt-laden water to run onto, and pond, on the tops of the piers and abutments. In these locations the concrete is particularly vulnerable to chloride-ingress, firstly, due to bleeding and settlement, secondly, due to incomplete compaction in the chamfered regions, thirdly, due to the likely presence of plastic settlement cracks induced by the reinforcement located close to the top of the piers and abutments and fourthly, if the detailing is inadequate, due to the presence of structural cracks in the bearing area (Figure 2). Chloride-ingress and the resulting deterioration could have been avoided by placing drainage channels beneath the joint or by ensuring that joints were not present above piers and abutments.

Another defect in some of the bridges surveyed was spalling near the base of piers exposed to salt spray.[15] There are a number of possible factors which could have contributed

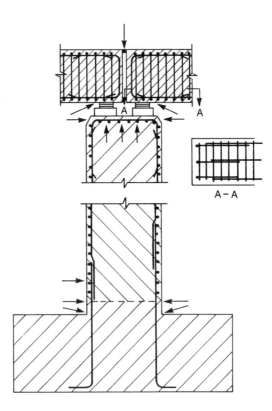

Figure 2 *A bridge pier showing with arrows possible easy access routes for chlorides which can result from inadequacies in both or either design and construction.*

to this poor performance - cold joints, low quality kickers and lapped joints reducing bar spacing, resulting in incomplete compaction of the concrete (Figure 2). The latter is possibly a common fault. For example, in Copenhagen a high proportion of bridge columns have corrosion problems.[19] A major factor associated with this poor performance is considered to be the presence of lapped joints at ground levels, resulting in reduced spacing between bars and increased difficulties in achieving adequate compaction.[19] The fact that many of the columns were at low points of the road profile also aggravated the chloride ingress problem at ground level.

4 QUALITY OF CONCRETE

4.1 Corrosion

Tables 1 and 2 summarize the observations made in a number of field studies on reinforced concretes subject to chloride ingress where it has been concluded that the concrete performed either poorly or well. All of the concretes referred to are Portland cement (PC) concretes. Examination of the information in these tables clearly illustrates that quality of concrete as given by water-cement ratio is a major parameter influencing the risk of chloride induced corrosion. It is deduced from the observations that, for concrete exposed to a marine environment or de-icing salts, water-cement ratios of up to 0.45 and minimum covers greater than 30 to 40 mm can result in a 50 year working life. This is in agreement with ENV 206: 1992[30] and ENV 1992-1-1: 1991,[31] wherein a maximum water-cement ratio of 0.45 and a

Table 1 *Performance of concrete structures subject to marine exposure*

Investigator	Structures	When built	Concrete Strength (N/mm^2)	Intended cover (mm)	Water-cement ratio	Condition
Bamforth et al[20]	Concrete blocks. UK	1987	39.4	40	0.65	Half cell potential measurements showed corrosion commenced at 2 years
Oshiro et al[21]	Building. Japan	1984	21	30 & 40	0.63	Half cell potential measurements showed corrosion after 3 years.
Liam et al[22]	Jetty. Singapore	1968	-	70	0.52(e)*	At 24 years cracking and spalling due to chloride ingress.
Gautefall[23,24]	Concrete blocks. Norway	1983	-	30 (facing sea), 50 (facing land)	0.54	Surface potentials indicated no corrosion at 9 years.
Sandvik et al[25]	Concrete platforms. North Sea	1971 onwards	-	>50	<0.45 <0.40 splash	Excellent
Somerville[26]	Precast pretensioned members. UK	Early 1940s onwards	60-80	>30 to 40	0.35 to 0.42	A history of good performance.

* estimated

Table 2 *Performance of concrete structures subject to de-icing salt or airborne salt*

Investigator	Structures	When built	Concrete strength (N/mm²)	Intended cover (mm)	Water-cement ratio	Condition
Bashenini et al[27]	Desalination plant (SRPC) Middle East	1982	21	50 25 where spalling	0.7(e)*	Concentrated brine attack. Cracking and spalling at 10 years.
Radain et al[28]	Coastal building Middle East	~1982	-	10-15	-	Cracking of columns attacked by water leaking from pipes.
Stolzner[29]	Bridge piers Denmark	1970	-	30-40	0.40-0.45	At 21 years negligible chlorides below 10 mm.
Henriksen et al[19]	Columns of 20 bridges Denmark	1940-90	-	30-40	0.40	'Predicts' 50 year life.
Author's examination 1995	Central bridge piers, M1. UK	About 1960	C45?	25-28	0.45 (2:1:1 mix)	At 35 years condition good apart from some minor spalling where cover low (probably <10 mm).
Somerville[26]	Precast pretensioned members. UK	Early 1940s onwards	60-80	>30 to 40	0.35 to 0.42	A history of good performance.

* estimated

minimum cover of 40 mm are recommended. However, ACI 201.2R-92[31] recommends that the w/c ratio 'be as low as possible, preferably less than 0.40' and for severe marine exposure, that the minimum cover to steel be 75 mm and 50 mm where salt is applied to the concrete.

It follows from this sub-section and the previous two sections that nearly all chloride induced corrosion problems are probably due to inadequate design, construction faults, poor maintenance and the occasional use of concretes of high w/c ratio. In the author's experience, inadequate cover is the most common cause of premature deterioration in reinforced concrete.

4.2 Chloride Ingress: Marine Exposure

Measurements of total chloride ingress into concrete in the tidal and splash zone have been made by a number of investigators.[20-25,33-38] It is these regions of reinforced concrete elements which are most susceptible to chloride induced corrosion. For such concretes it is often assumed that the chloride ingress into concrete can be described by Fick's second law of diffusion:

$$C_x = C_s [1 - \text{erf}(x)/2(D_{ce}t)^{1/2}] \tag{1}$$

where C_x is the total chloride level at depth x at age t, and C_s and D_{ce} are the best fitting values for the surface chloride level and effective diffusion coefficient respectively.

Figure 3 shows how well equation 1 fits the chloride profile measured on PC concretes of differing water-cement ratio subject to splash or tidal zone exposure.[20,25,33] It is clear from the figure that w/c ratio is a major parameter influencing chloride ingress into concrete. This is illustrated further in Figure 4 where the effective diffusion coefficient determined from equation 1 is shown plotted against water-cement ratio for a number of structures subject to splash or tidal zone exposure. Figure 5 is a similar plot for concretes containing slag, fly ash and silica fume. The ages of the concretes range from 5 to 30 years. In the investigations referred to where the effective diffusion coefficient was not reported by the author(s), a best fitting value for D_{ce} has been determined for the purposes of this paper using the procedure proposed by Nagana and Naito.[35] The high coefficients observed by Liam et al[22] are probably

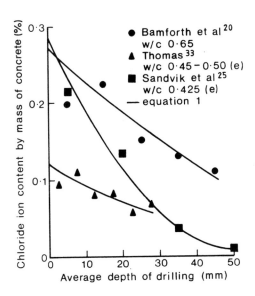

Figure 3 *Chloride concentration profiles in concretes subject to marine exposure. 5 years,[20] about 11 years,[25] 30 years.[33]*

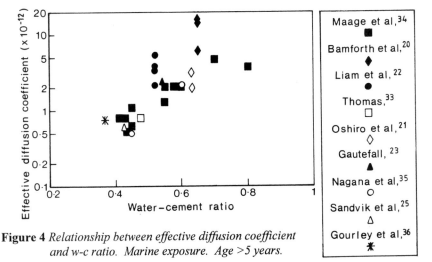

Figure 4 *Relationship between effective diffusion coefficient and w-c ratio. Marine exposure. Age >5 years.*

due to the high exposure temperatures (~30°C). The D_{ce} for the PC concrete with a w/c ratio of 0.37 was for a concrete pile which had micro-cracked during driving and bending.[36] At water-binder ratios in the range 0.4 to 0.5, D_{ce} ranges from 5 to 11 x 10^{-13} m^2/s and C_s from about 0.1 to about 0.4 per cent by mass of concrete for PC concretes[33,34]. For concretes containing fly ash, slag and silica fume[22,23] D_{ce} ranges from 2 to 14 x 10^{-13} m^2/s, but with a lower mean D_{ce} than that deduced for PC concretes. Higher values for D_{ce} are applicable to concretes subject to tropical marine exposure[22,37] or a hydrostatic head.[36,37] For high quality PC off-shore structures Browne[38] quotes a D_{ce} of 1 x 10^{-13} m^2/s which is lower than any of the values plotted in Figures 4 and 5.

In Figure 6, D_{ce} is shown plotted against age for PC concretes with water-cement ratios ranging from 0.40 to 0.50. For ages in the range 6 to 30 years, no clear dependence of D_{ce} upon age is apparent.

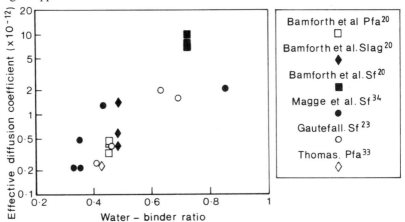

Figure 5 *Relationship between effective diffusion coefficient and w-b ratio. Marine exposure. Age >5 years.*

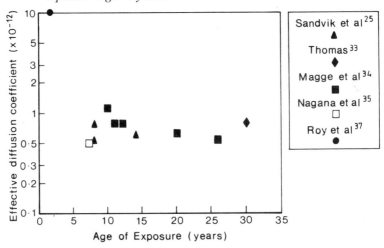

Figure 6 *Relationship between effective diffusion coefficient and exposure age. Marine exposure. PC concretes 0.4 ≤ w/c ≤ 0.5.*

Service life predictions are sometimes based upon limiting total chloride ion levels at the reinforcement. Such an approach is unsatisfactory because there is no unique total (or free) chloride ion concentration which initiates steel corrosion. The critical total chloride ion concentration increases as the binding of chlorides increases, as the hydroxyl ion concentration increases, as the degree of water saturation increases and as the water-binder ratio decreases.[1-8,22,36,37] Replacing part of a PC by fly ash and slag can, depending upon the alkali levels of the binder components, increase or reduce the hydroxyl ion concentration.[38] Replacing part of a PC by silica fume will reduce the hydroxyl ion concentration.[38]

In specifications a maximum allowable total chloride level of 0.4% by mass of binder is sometimes specified for reinforced concrete.[14] This chloride level is often taken to be the level at which corrosion commences,. However, for low w/c ratio concretes, subject to frequent wetting and drying, total chloride levels close to the steel in excess of 1.0% by mass of PC may be required.[39] This translates into an allowable total chloride level of about 0.17 to 0.20% by mass of concrete. Assuming (i) that this value is the total chloride level at which corrosion commences, (ii) w/c 0.45, (iii) minimum cover 40 mm, (iv) UK type exposure conditions, (v) D_{ce} 7 x 10^{-13} m^2/s, (vi) C_s 0.4% by mass of concrete and (vii) that equation 1 is valid, gives 50 to 60 years before corrosion commences and a service life of possibly 60 to 85 years.[36]

4.3 Chloride Ingress: De-icing Salt Exposure

Chloride ingress into the surface layers of concrete exposed to intermittent salt-laden water (or airborne salt) can rarely be controlled by a simple diffusion process. Chloride ingress into the cover concrete will often occur by capillary suction and wash-out of some of the chlorides can occur when the concrete is exposed to rain or non-salt-laden vehicle spray. Ingress of chloride ions into the concrete will also be complicated by the process of carbonation which will change the bound chloride level and the permeability of the affected concrete. Carbonation can reduce or increase the permeability of the surface layers.[40] The influence of carbonation will be most marked in parts of structures protected from direct rain. For example, in the case of bridge piers, it may be a year or two before they are subject to their first significant vehicle spray. For such elements, significant drying and carbonation of the cover concrete is possible.

Measurements of total chloride ingress into concrete subject to intermittent exposure to salt-laden water or vehicle spray or airborne salt have been made by a number of investigators.[41-43] Some of the results obtained are shown plotted in Figures 7 to 9. Comparison of these figures with Figure 3 indicates, in the case of concretes containing fly ash or slag, a marked departure from the chloride profile predicted by simple diffusion. The effect is less marked for PC concretes. At a depth of about 30 mm the limited data indicate that the total chloride level in a non-PC concrete is probably similar or lower than for a PC concrete.

Fitting Fick's second law of diffusion to the reported chloride levels in PC bridge piers gives the following values for the 'apparent D_{ce}' and C_s:

- Henriksen,[19] 20 undamaged bridge piers constructed between 1940 and 1990, 'D_{ce}' from 0.1 to 1.8 x 10^{-12} m^2/s with a mean of about 1.2 x 10^{-12} m^2/s, C_s from 0.02 to 0.17% by mass of concrete with a mean of about 0.09%. Henriksen[19] attributed this variability to 'varying concrete qualities from bridge to bridge'.

- Stolzner,[29] a bridge built in 1970 'D_{ce}' of 1.4 x 10^{-13} m^2/s with C_s of 0.03% respectively by mass of concrete.

- Thomas,[33] a 10 year old bridge, 'D_{ce}' 3.1 x 10^{-12} and C_s 0.06% by mass of concrete.
- TRL,[42] 2 bridges age 5 years, D_{ce} of 1.7 x 10^{-12} and 2.7 x 10^{-12} m^2/s and C_s of 0.09 and 0.045% by mass of concrete respectively.
- In the case of car park T - beams, Funahashi[44] quotes 'D_{ce}' from 1.3 to 1.8 x 10^{-12} m^2/s and C_s from 0.08 to 0.26% by mass of concrete. In this particular case, wash-out of chlorides was unlikely.

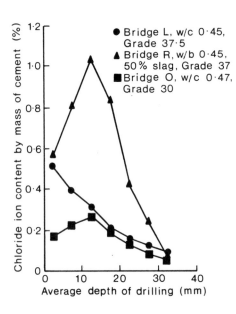 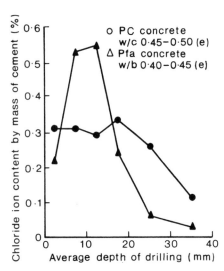

Figure 7 *Chloride profiles for bridge piers.[42]*   **Figure 8** *Chloride concentration profiles*
        Height 1 m. Distance to slow traffic   *for bridge piers exposed to*
        lane 4.75 to 5.0 m. Age 5 years.   *de-icing salts.[33] Age 10 years.*

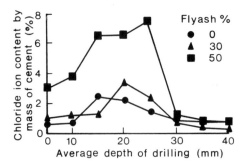

Figure 9 *Chloride concentration profiles for cubes stored 50 m from sea.[43] w/b 0.6. Age 3 years*

Assuming that (i) the total chloride level at which corrosion commences is 0.06% by mass of concrete,[19] (ii) w/c 0.45, (iii) minimum cover 40 mm, (iv) D_{ce} 2 x 10^{-12} m^2/s, (v) C_s 0.10% by mass of concrete and (vi) equation 1 is applicable to PC bridge piers subject to de-icing salts gives approximately 50 years before corrosion commences and a service life of possibly 60 to 75 years. In the author's view, the assumption of a propagation period prior to cracking and spalling of between 10 to 25 years[19] may be an underestimate. The performance of low cover bridge pier concrete indicates a possible propagation period of 20 to 30 years.

5 CONCLUSIONS

The following conclusions are based on the reported literature dealing with chloride ingress into field concretes and their effects and the author's experience:

1. Most chloride-induced corrosion problems in concrete are due to inadequate design, construction faults, poor maintenance or the use of concretes of high w/c ratio.
2. Inadequate cover to reinforcement is the most common cause of premature deterioration in concrete subject to chloride ingress.
3. Much more attention needs to be given to reducing chloride ingress into concrete by improved design, in particular, constructability and better construction practices.
4. Provided reinforced concrete structures are properly designed, placed and maintained, a maximum water-Portland cement ratio of 0.45 and a minimum cover of 40 mm should result in a design life greater than 50 years for concrete exposed to a marine environment or de-icing salts. This conclusion is broadly supported by field performance.

6 ACKNOWLEDGEMENT

The author is grateful to his colleague, Mr M Webster, for helpful discussion and for carrying out a number of the calculations of the effective diffusion coefficient and surface chloride levels.

7 REFERENCES

1. K. Petterson, Swedish Cement and Concrete Research Institute, CBI Report 2:92, 1992.
2. D. A. Hausman, *Materials Protection*, 1967, November, 19.
3. P. S. Mangat and B. T. Molloy, *Mat. Struct.*, 1992, **25**, 404.
4. K. Petterson, 'Proc. Int. Conf. on Corrosion and Corrosion Protection of Steel in Concrete, Sheffield, July 1994', (Editor: R. N. Swamy), Sheffield Academic Press, Sheffield, 1994, 461.
5. S. Guiguis, H. T. Cao and D. Baweja, 'CANMET/ACI Int. Conf. on Durability of Concrete, Nice, 1994', (Editor: V. M. Malhotra), ACI, SP-145, 263.
6. R. H. Dhir, M. R. Jones and M. J. McCarthy, *Proc. Instn. Civ. Engrs., Structs. and Bldgs.*, 1993, **99**, 167.
7. D. F. Rasheeduzzafar and K. Mukarram, 'Proc. Katherine and Bryant Mather Int. Conf. on Concrete Durability, Atlanta, 1987', (Editor: J. M. Scanlon), ACI, SP-100, 1987, **2**, 1477.

8. A. Rosenberg, C. M. Hausson and C. Andrade, 'Materials Science of Concrete 1', (Editor J. P. Skalny), A. Cer. Soc. 1989, 285.

9. National Materials Advisory Board Committee, 'Concrete Durability: a multi billion dollar problem', NMAB-437, National Academy Press, Washington, DC, 1987.

10. Building Research Establishment, Garston, UK, Private Communication, 1994.

11. A. W. Beeby, 'Concrete 2000, Dundee, 1993', (Editor: R. K. Dhir), E. and F. N. Spon, 1993, 1, 37.

12. D.S.I.R., Report of the Reinforced Concrete Structures Committee of the Buildings Research Board, HMSO, London, 1934.

13. BSI, CP114: Part 2: Metric Units: 1969.

14. BSI, BS 8110: Part 1: 1985.

15. G. Maunsell and Partners, 'The performance of concrete in bridges: A survey of 200 bridges', Dept. Trans., 1989.

16. D. Griffiths, M. Marosszeky and D. Sade, 'Proc. Katherine and Bryant Mather Int. Conf. on Concrete Durability, Atlanta, 1987', (Editor: J. M. Scanlon), ACI, SP-100, 1987. 1703.

17. M. Marosszeky and M. Chew, *Concr. Int.*, 1990, 12, 59.

18. T. Ostmoen and A. Aas-Jakobsen, 'Proc. Conf. on Chloride Penetration into Concrete Structures, Chalmers University of Technology, Sweden', (Editor: Lars-Olof Nilsson), Chalmers Tekniska Hogskola, Goteborg, 1993, 108.

19. C. F. Henriksen and E. Stoltzner, 'Proc. Conf. on Chloride Penetration into Concrete Structures, Chalmers University of Technology, Sweden', (Editor: Lars-Olof Nilsson), Chalmers Tekniska Hogskola, Goteborg, 1993, 166.

20. P. B. Bamforth and J. F. Chapman-Andrews, 'Proc. Int. Conf. on Corrosion and Corrosion Protection of Steel in Concrete, Sheffield, July 1994', (Editor: R. N. Swamy), Sheffield Academic Press, Sheffield, 1994, 139.

21. T. Oshiro and S. Tanikawa, 'Proc. Int. Conf. on Protection of Concrete, Dundee, 1990' (Editors: R. K. Dhir and J. W. Green), E and F. N. Spon, 1990, 469.

22. K. C. Liam, S. K. Roy and D. O. Northwood, *Mag. Concr. Res.*, 1992, 44, 205.

23. O. Gautefall, 'Proc. Conf. on Chloride Penetration into Concrete Structures, Chalmers University of Technology, Sweden', (Editor: Lars-Olof Nilsson), Chalmers Tekniska Hogskola, Goteborg, 1993, 148.

24. O. Gautefall, Norwegian Institute of Technology, Private Communication, November 1995.

25. M. Sandvik and S. O. Wick, 'Proc. Conf. on Chloride Penetration into Concrete Structures, Chalmers University of Technology, Sweden', (Editor: Lars-Olof Nilsson), Chalmers Tekniska Hogskola, Goteborg, 1993, 159.

26. G. Somerville, 'Proc, Conf. on Bridge Modification, London 1994', (Editor: B. Pritchard), Thomas Telford Publications, London, 1995, 25.

27. M. S. Bashenini, S. E. Hussain and I. S. Paul, 'Proc. Int. Conf. on Corrosion and Corrosion Protection of Steel in Concrete, Sheffield, July 1994', (Editor: R N Swamy), Sheffield Academic Press, Sheffield, 1994, 61.

28. T. A. Radain, F. F. Wafa and T. A. Samman, 'Proc. Int. Conf. on Corrosion and Corrosion Protection of Steel in Concrete, Sheffield, July 1994', (Editor: R N Swamy), Sheffield Academic Press, Sheffield, 1994, 51.

29. E. Stolzner. 'Proc. Conf. on Chloride Penetration into Concrete Structures, Chalmers University of Technology, Sweden', (Editor: Lars-Olof Nilsson), Chalmers Tekniska Hogskola, Goteborg, 1993, 355.

30. BSI. DD ENV 206: 1992.
31. BSI. DD ENV 1992-1-1:1992.
32. American Concrete Institute, Manual of Concrete Practice, Part 1, 1994, 201. 2R-1.
33. M. A. A. Thomas, 'Proc. 5th. Int. Conf. on Durability of Building Materials and Components, Brighton, 1990', (Editors: J. M. Baker, P. J. Nixon, A. J. Majumdar and H. Davies), E. and F. N. Spon, 1990, 383.
34. M. Maage and S. Helland, 'Proc. Conf. on Chloride Penetration into Concrete Structures, Chalmers University of Technology, Sweden', (Editor: Lars-Olof Nilsson), Chalmers Tekniska Hogskola, Goteborg, 1993, 125.
35. H. Nagana and T. Naito, 'Int. Conf. on Concrete in the Marine Environment, London, 1986', The Concrete Society, Slough, 1986, 211.
36. J. T. Gourley and D. T. Bieniak, 'Proc. Sym. on Concrete 1983. The Material for Tomorrow's Demands, Perth, October, 1983', Institution of Engineers, Australia, 1983, No. 83/2, 41.
37. S. K. Roy, L. K. Chye and D. O. Northwood, *Cem. Concr. Res.*, 1993, **23**, 1289.
38. D. W. Hobbs, 'Alkali-Silica Reaction in Concrete', Thomas Telford Publications, London, 1988.
39. P. Sandberg and K. Petterson, Lund Institute of Technology, Private Communication, 1995.
40. L. J. Parrott, 'A review of carbonation in reinforced concrete', British Cement Association, Crowthorne, 1987, C/1-0987.
41. M. D. A. Thomas, *Proc. Inst. Civ. Eng.*, Part 1, 1989, **86**, 1111.
42. P. Vassie, Transport Research Laboratory, Crowthorne, Private Communication, 1995.
43. M. M. Salta, 'Proc. Int. Conf. on Corrosion and Corrosion Protection of Steel in Concrete, Sheffield, July 1994', (Editor: R. N. Swamy), Sheffield Academic Press, Sheffield, 1994, 794.
44. M. Funahashi, *ACI Mat.J.*, 1990, **87**, 581.

CHLORIDE DIFFUSION MODELLING FOR MARINE EXPOSED CONCRETES

E. C. Bentz, C. M. Evans and M. D. A. Thomas

Department Of Civil Engineering,
University Of Toronto,
Ontario, Canada, M5S 1A4

1 INTRODUCTION

As part of a wider study on the effects of fly ash on the durability of concrete, a substantial number of reinforced concrete specimens, with a wide range of mix proportions and curing conditions, have been placed in the tidal zone of the Building Research Establishment's (BRE) marine exposure site on the Thames Estuary. Early data (up to two years) related to chloride ingress and reinforcement corrosion in these concretes have been reported in previous papers and clearly indicated the efficacy of fly ash in improving the durability of marine concrete.[1,2] Diffusion coefficients were calculated for these concretes by fitting the chloride concentration profiles at various ages to Crank's solution of Fick's second law of diffusion, which is:[3]

$$\frac{C_x}{C_s} = 1 - erf\left(\frac{x}{2\sqrt{D_a t}}\right) \tag{1}$$

where: C_s = chloride concentration at the surface
C_x = chloride concentration at time t at a distance x from the surface
D_a = apparent diffusion coefficient

However, it was acknowledged that such an approach may not be valid for the conditions used in the study for the following reasons:

- Concretes were unsaturated at the time of exposure and significant chloride ingress occurs due to capillary suction during the first tidal immersion
- Tidal conditions produce wetting/drying cycles in the surface zone which result in the surface concentration increasing with exposure (i.e. $C_s = f_1(t)$)
- The diffusion coefficient tends to decrease with time due to continued hydration (i.e. $D_a = f_2(t)$).

In most situations, one or more of the above limitations are likely to apply. For highway structures periodically exposed to chlorides (due to seasonal application of de-icing salt) the effects of capillary suction will be more pronounced and may even be the dominant transport mechanism. Despite these limitations, it has become commonplace to derive diffusion coefficients for structures from concentration profiles using Equation 1, and to use the same to extrapolate to the end of service life (i.e. to predict the time for a critical chloride level to reach the steel).[4-6]

This paper describes the development of a numerical model to determine diffusion coefficients from chloride profiles taking account of sorption effects and changing surface concentrations. The model has been used to determine diffusivity values for concrete specimens after various periods at the BRE marine exposure site.

2 EXPERIMENTAL DETAILS

Concrete prisms (100 x 100 x 300 mm) were cast with a range of fly ash contents (0% to 50% fly ash) and a specified 28-day compressive strength of 35 MPa. After 24 hours curing in the mould, specimens were stripped and stored in air at 20°C and 65% RH until 28 days. At this age, the specimens were transported to the BRE marine exposure site and placed in the tidal zone. Specimens were retrieved at 1, 2 and 4 years for the determination of chloride concentration profiles (and reinforcing steel weight loss). Parallel specimens were placed in seawater in the laboratory at 28 days and chloride profiles determined after 1 and 28 days immersion. This was carried out to assess the depth of chloride penetration due to capillary suction effects in the unsaturated concrete. Full details of the materials, concrete mix proportions and test methods used are given elsewhere.[1,2] Chloride concentration profiles are shown in Figure 1.

3 NUMERICAL MODELLING

3.1 Numerical Methods

It was necessary to assume that the samples were homogeneous and isotropic (i.e. diffusivity constant throughout section). It was assumed that all sorption took place in the first 28 days after exposure and that the sample was fully saturated in the modelling region.

Internal chloride binding values were taken from Arya,[7] who investigated the chloride binding capacity of cement pastes with varying w/c ratios with similar U.K. Class F fly ashes to that used in this study. Binding levels were assumed to vary linearly with w/c and fly ash content from 46% to 68% for 0 to 50 % fly ash.

In general, ionic diffusion in porous media is governed by Fick's laws. Fick's second law, used as the equation of chloride mass conservation, is:

$$J_{Cl} = \frac{\partial C(x,y)}{\partial t} = D_i\left(\frac{\partial^2 C}{\partial x^2} + \frac{\partial^2 C}{\partial y^2}\right) \qquad (2)$$

where: J_{Cl} = the ion flux through a unit area per unit of time
C = is the chloride concentration at time t and co-ordinates (x,y)
D_i = is the intrinsic diffusion coefficient.

This was the governing equation used in the model. The numerical methods used to determine diffusion values in one and two dimensions for the given data were:

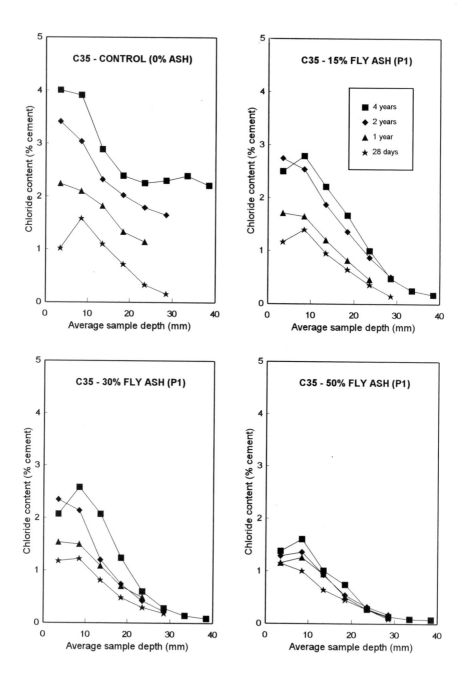

Figure 1 *Measured Chloride Concentration Profiles - from BRE Exposure Site*

- develop a numerical model of the governing equation that could consider binding and any initial profile of concentration;
- use a least squares technique to determine the best fit diffusion constant, given that the diffusion profile begins from a 28 day sorption profile.

Diffusion coefficients were determined for one dimensional penetration by a direct finite difference implementation of Fick's second law in one direction. Assuming central difference form for spatial derivatives, the following approximation is achieved:

$$\frac{\partial C}{\partial t} = D \frac{C_{i+1} - 2C_i + C_{i-1}}{\Delta x^2} \tag{3}$$

The temporal derivatives were taken with a Crank-Nicolson theta differencing scheme (i.e. $\theta = 0.5$). The second spatial derivative is shown here in its non finite-difference form:

$$\frac{C_i^{t+\Delta t} - C_i^t}{\Delta t} = \theta \left[D \frac{\partial^2 C}{\partial x^2} \right]^{t+\Delta t} + (1-\theta) \left[D \frac{\partial^2 C}{\partial x^2} \right]^t \tag{4}$$

From these equations, a linear set of equations can be set up to be solved for each time step. The concentration at a new time step depends only on the previous time step, and can be solved explicitly.

A similar approach was taken for the 2-D case. Fick's second law was developed into a finite difference approximation form using the same assumptions. Solving one of the larger trials would have resulted in 2500 equations in 2500 unknowns with a wide bandwidth. Rather than solving with a linear set of equations, a successive-overrelaxation procedure was used which iteratively converges on the correct concentration profile for each time step.

3.2 Boundary and Initial Conditions

The boundary and initial conditions used were the same for the 1-D and 2-D cases. In the 2-D case, the boundary was the entire edge of the sample, whereas it was a point for the 1-D case. To utilise symmetry, only part of the cross section was analysed. For the 1-D case this part was a line from the outer edge to the centre of the sample. For the 2-D case it was a quarter of the total cross section. To impose this symmetry, the slope of the concentration curve was set to zero at the inside edges, in the direction of symmetry.

For the outer edge boundary condition, it was necessary to examine the experimental data. The data at the outer data point (3.5 mm) showed strong variability, so this section of the sample was ignored. This variability was likely to have resulted from cyclic wetting and drying which may tend to concentrate chlorides; furthermore, attack by sulphate (from seawater) and carbon dioxide (from the atmosphere) will influence chloride binding at the surface. Consequently, the outer edge of the model was assumed to be at the point 8.5 mm from the edge (centre of the second depth interval sampled; 6-11 mm). An examination of the data at this depth showed a fairly linear increase in concentration over time (Figure 2). Thus the boundary condition on the outside was a linearly increasing concentration from the level at 28 days (sorption) to the measured concentration at the target year.

The initial concentration condition was the experimental profile of 28 day sorption, and the measured background level elsewhere (0.1% Cl⁻ by mass of cement).

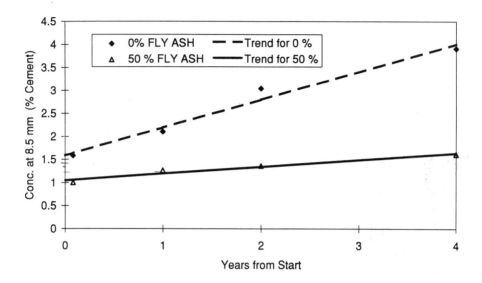

Figure 2 *Change in Surface Chloride Concentration with Time*

3.3 Model Performance

The 2-D model was found to converge most rapidly at an overrelaxation factor of about 1.1. An analysis showed 1 day to be the maximum time step that gave consistent results. The grid size variation was checked as well. It was found that the shape of the curve was basically identical for 24 or 12 blocks per modelled half of the sample. The chosen grid size was 34 because as such the experimental data fell exactly onto nodes. The tolerance value on the successive over-relaxation was found to be optimal at 0.4.

3.4 Solution Procedure

To determine the diffusion coefficients for the given data, the time stepping was started from 28 days using the sorption profiles shown in Figure 1. It was then stepped to the desired set of data, i.e. the 1, 2 and 4 year profiles. This was repeated for a number of different diffusion coefficients, and the one with the least squares of differences was chosen as the best fit. Figure 3 compares the concentration profile from the numerical model with the experimental data. It can be seen from the graphs that the fit of the model to the data is good. The shape of the experimental data trends is the same as defined by the model.

A proper sensitivity analysis was not done. However, it was noted the sensitivity to binding and the time step were low. The model was sensitive to the surface chloride concentration and the initial profile.

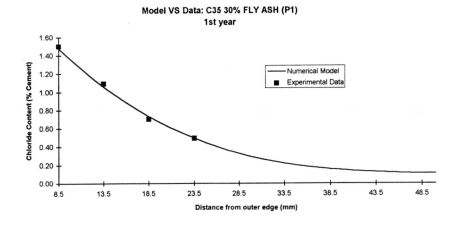

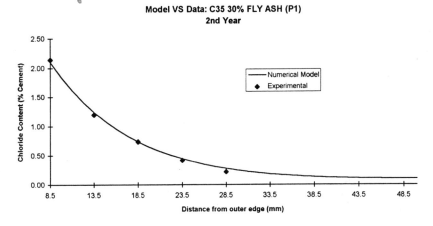

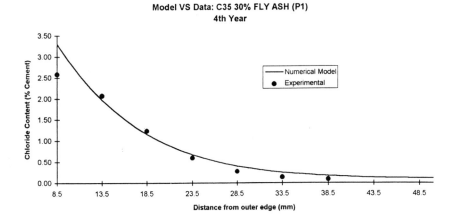

Figure 3 *Comparison of Numerical Model to Experimental Data*

4 DISCUSSION

Table 1 shows the diffusion values for the samples at 1, 2 and 4 years determined using the numerical model. At all ages, there is a general trend of reducing diffusivity with increasing fly ash content. In all cases, the diffusion coefficients decrease with time, presumably due to continued cement hydration or interaction between the diffusing species and the hydrates which may alter the microstructure of the cement paste.

No significant difference was observed between the results from the 1-D and 2-D models for the fly ash concretes. However, the 2-D model gave significantly lower values for the OPC concrete compared to the 1-D results for the same concrete. This is expected since 2-D effects are only important when the chlorides from the orthogonal face reach the profile being considered. Examination of the data in Figure 1 suggests that this effect is only significant in the OPC concrete.

Figure 4 shows the effect of the initial sorption profile on the calculated diffusion coefficient using the numerical model (for concrete with 0% and 50% ash). Ignoring the effect of initial sorption (i.e. assuming initial profile due to diffusion) increases the calculated diffusion coefficient. The error becomes more marked with increasing fly ash content and appears to be negligible in the OPC concrete. The profiles in Figure 1 show that the initial sorption of chlorides, as represented by the 28 day profile, accounts for a significant proportion of the chloride present after further exposure for 4 years, especially for the concrete with 50% ash. Consequently, the assumption that these chlorides penetrated by diffusion rather than sorption introduces considerable error in the diffusion coefficient calculated from concentration profiles. The initial sorption profile becomes less significant on long-term profiles in concrete with higher diffusivity.

The coefficients presented in Table 1 are average coefficients for the entire period of exposure. However, it is possible to calculate diffusion coefficients between other intervals of time for which profiles have been determined. For example, the average diffusion coefficient between 2 and 4 years may be calculated using the profiles obtained at 2 and 4 years as the initial and final states in the numerical model. Coefficients calculated in this manner for the periods 28 days to 1 year, 1 year to 2 years, and 2 years to 4 years, are shown in Figure 5, and are compared with average values calculated for the whole exposure period (i.e. the values in Table 1). Reducing the averaging period accentuates the decrease in diffusivity with time, the effect being particularly marked for the concrete with higher fly ash levels. This is reflected by the data in Figure 1 which shows that there is very little penetration of chlorides in concrete with 30 or 50% ash beyond the first year (i.e. profiles very close together).

Table 1 *Results from Numerical Model (Average values for exposure period)*

Concrete (% ash)	Model	Diffusion Coefficients (10^{-12} m^2/s)		
		1 year	2 year	4 year
0%	1-D	20	15	12
	2-D	15	11	8.0
15%	1-D	3.2	4.5	2.3
30%	1-D	5.0	1.3	2.6
50%	1-D	3.2	1.7	1.0

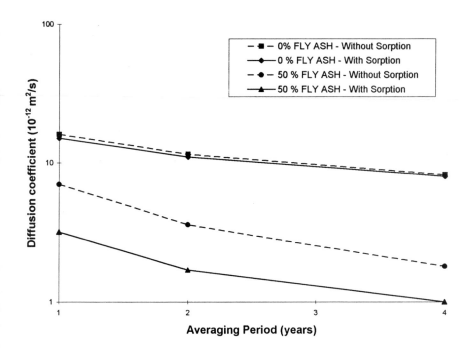

Figure 4 *Effect of Initial Sorption on Diffusion Coefficient Calculated from Profile*

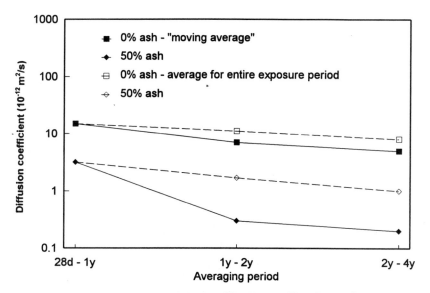

Figure 5 *Diffusion Coefficients Calculated for Various Time Intervals*

The average diffusion coefficients calculated for the time interval from 2 to 4 years were 5×10^{-12} m^2/s and 2×10^{-13} m^2/s for concrete with 0% and 50% fly ash, respectively. This represents a reduction of 25 times in the fly ash concrete compared to OPC concrete of the same specified strength. The implications of the reduced diffusivity on service life are considerable.

Figure 6 shows estimated concentration profiles at 50 years as determined by the numerical model. In this case, the surface concentration was allowed to increase with time as shown in Figure 2 up to an arbitrarily chosen maximum value of 5.0% chloride by mass of cement (for both concretes). The diffusion coefficients used were constant with time and equal to the 4 year values presented in Table 1 (i.e. 8.0 and 1.0×10^{-12} m^2/s, for 0% and 50% ash, respectively). Assuming a threshold chloride concentration for corrosion of 0.40% (by mass of cement) for these concretes,[2] the data in Figure 6 show that steel with 40 mm cover would still be protected by the fly ash concrete after 50 years exposure in a tidal marine environment. However, cover depths in excess of 140 mm may be required to avoid corrosion in plain concrete of the same specified strength.

The model presented here was developed solely for the purpose of calculating diffusion coefficients for the unique set of materials and exposure conditions of the BRE study and is not necessarily applicable to other situations. The results from the model highlight the considerable advantage that can be gained through the incorporation of fly ash in to the concrete. The results also demonstrate the importance of separating the effects of different transport mechanisms (e.g. sorption and diffusion) and this may be more important in highway structures which are affected to a greater depth by cyclic wetting and drying, and, hence, the mechanisms of unsaturated flow. Models should be capable of dealing with changing properties of the concrete and its exposure, especially diffusivity and surface chloride concentration.

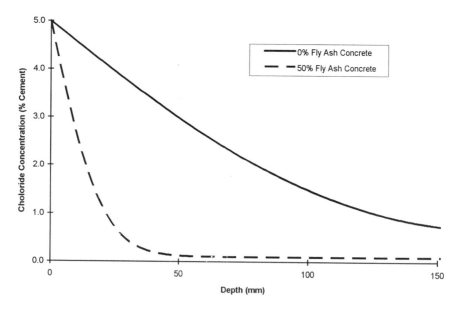

Figure 6   *Predicted Chloride Concentration Profile at 50 Years*

5 CONCLUSIONS

A numerical model was developed to determine chloride diffusion coefficients in concrete from concentration profiles. The model uses a direct finite difference implementation of Fick's second law and corrects for the effects of initial sorption, changing surface concentration with time and chloride binding by the hydrates.

Collectively, the results obtained using the numerical model demonstrate that application of the standard solution (i.e. Equation 1) to concentration profiles may be inappropriate and result in significant error in the calculated diffusion coefficient. These errors appear to be more marked for concretes with lower diffusivity (e.g. higher levels of fly ash) and result in higher calculated coefficients.

The average diffusion coefficient calculated using the numerical model over the time interval 2 to 4 years was reduced by 25 times for concrete with 50% fly ash compared with OPC concrete of the same specified strength. This will have a substantial effect on the service life of concrete structures built with these materials.

References

1. M. D. A. Thomas, 'Corrosion of Reinforcement in Construction', (Ed. C. L. Page et al.), Elsevier Applied Science, London, 1990, 198.
2. M. D. A. Thomas, *Mag. Conc. Res.,* 1991, **43** (156), 171.
3. J. Crank, 'The Mathematics of Diffusion', Oxford University Press, Oxford, 1975.
4. R. D. Browne, 'Performance of Concrete in a Marine Environment', (Ed. V. M. Malhotra), ACI SP-65, American Concrete Institute, Detroit, 1980, 169.
5. P. D. Cady and R. E. Weyers, *Cem. Concr. & Aggreg,* 1983, **5** (2), 81.
6. N .S. Berke and M. C. Hicks, *Corros.,* 1994, **50** (3), 234.
7. C. Arya, N. R. Buenfeld and J. B. Newman, *Cem. Concr. Res.,* 1990, **20** (2), 291.

WATER ABSORPTION, CHLORIDE INGRESS AND REINFORCEMENT CORROSION IN COVER CONCRETE: SOME EFFECTS OF CEMENT AND CURING

L. J. Parrott

British Cement Association,
Telford Ave,
Crowthorne,
Berks RG11 6YS

1 INTRODUCTION

In recent years there has been considerable interest in the rate of water absorption into cover concrete:

a) measurement of the rate of water absorption is rapid and simple; it provides an indication of the transport properties of cover concrete that reflects the effects of cement reactivity, water/cement ratio, compaction and curing.[1-6]

b) the rate of water absorption can be correlated with carbonation and air permeability for many conditions.[4,5]

c) water absorption can rapidly increase the rate of corrosion in depassivated reinforcing steel to levels that will cause cracking and spalling.[7-9]

It is becoming increasingly recognised from field investigations[10-13] and from natural exposure studies of test specimens[14-16] that absorption can be an important transport mechanism for the ingress of chlorides into reinforced and prestressed concrete. In field investigations of highway structures it was noted that reinforcement corrosion tended to be more severe where chlorides penetrated into bridge concrete via faulty drains and leaking joints.[11] The natural exposure studies of Bamforth[14,15] and Salta[16] indicated that, with exposure to sea spray, penetration of chlorides into the outer few centimetres of cover concrete was dependent upon capillary suction initially and upon long-term diffusion at greater depths. These results also indicate that the beneficial chloride resistance often observed with well-cured blended cement concretes is much less evident if chloride ingress involves absorption.

Laboratory studies have often employed short drying periods (14 days or less) between cycles of wetting with chloride solution[17-20]: this limits the penetration of the "drying front" and the extent of subsequent absorption of chloride. MacInnis[21] examined chloride ingress after specimens were dried to equilibrium at 105°C and found that absorption from sodium chloride solution decreased slightly as the concentration increased in the range 2 to 10%. McCarter[22] dried his test specimens for 21 days at 55% relative humidity in order to obtain a more realistic moisture condition prior to testing.

The ease with which chloride in solution can penetrate dry or partially dry concrete suggests that chloride-induced reinforcement corrosion could, under unfavourable exposure conditions, develop shortly after completion of a structure. This is substantiated by the results of Dass[20] for some low strength concretes. Chloride-induced corrosion levels were similar for Portland and Portland-slag concretes. Another study by Otsuki[18] showed that, for a constant water/binder ratio of 0.5, chloride ingress and reinforcement corrosion were reduced when Portland cement was replaced by 30% pulverised fuel ash or 50% ground granulated blastfurnace slag.

The present report investigates the rate of water absorption, chloride diffusion, chloride absorption and associated corrosion in concretes made, at a constant water/cement ratio, with a range of 17 Portland, Portland-limestone and granulated blastfurnace slag cements cured for 1, 3 or 28 days under sealed conditions: specimens were uniaxially dried for 6 or 18 months at 20°C and 60% relative humidity prior to testing. The results are analysed in relation to parallel, reported measurements of strength, carbonation and cover concrete transport properties[23,24] and in relation to reinforced concrete durability and the use of early-age tests for the assessment of concrete quality.

2 EXPERIMENTAL

The characteristics of the cements and concrete mixes are shown in Table 1: further details are given in reference 23. The concretes were all made with a free water/cement ratio of 0.59 and a cement content of 320 kg/m^3. The concrete was mixed for four minutes in a horizontal pan mixer. The slump of the fresh concrete was measured six minutes later and, fifteen minutes after the start of mixing, the concrete was remixed for one minute. The slumps of the fresh concretes were in the range 40 to 175 mm.

The concrete was cast into 100 mm cube moulds and compacted on a vibrating table. One mould was fitted with four 6.4 mm diameter mild steel reinforcing rods at 4, 8, 12 and 20 mm from the vertical face to be exposed. The steel rods were marked, cleaned and weighed prior to fitting in the moulds so that the corrosion loss, after 6 months of dry exposure and 4 weeks of chloride exposure, could be determined by weighing after extraction from the concrete and cleaning.[24-26] Eight cubes were cast for compressive strength measurements at 1 day, 3 days, 28 days and 18 months: further cubes were used for weight loss, absorption, air permeability and carbonation measurements. The strength, air permeability and carbonation tests are reported elsewhere.[23,24] With the exception of the specimens for strength tests all cubes were sealed on five faces and the remaining face was exposed to laboratory air at 60% relative humidity and 20°C after 1, 3 or 28 days of sealed curing. The weights of all exposed cubes were measured at scheduled times after the start of exposure. Uniaxial absorption after 6 or 18 months exposure was achieved by placing the exposed face of the test specimen on a wire grid that was 1 mm below a water surface. The gain of weight was measured after 1, 2, 4 and 6 hours of wetting. The rate of absorption in cover concrete was characterised by the weight of water absorbed in 4 hours.[5]

Table 1 *Slump and strength of concretes made with different cements (Free Water/Cement = 0.59, cement content 320 kg/m^3)*

Name	Cement		Cube strength (MPa)	
		ENV197 type & class	28d	18m
D54		CEMI 52.5R	45.5	53.8
U1		CEMI 32.5R	35.0	49.9
U2	85% U1+15% limestone	CEMII/A-L 42.5R	37.1	47.5
U3	75% U1+25% limestone	CEMII/B-L 32.5R	33.5	39.9
U4		CEMI 52.5R	45.7	52.7
U5	85% U4+15% limestone	CEMII/A-L 42.5R	36.0	41.0
U6	75% U4+25% limestone	CEMII/B-L 32.5R	30.0	34.1
U7	50% U4+50% GGBS	CEMIII/A 42.5	38.8	51.9
U8	25% U4+75% GGBS	CEMIII/B 32.5	29.0	42.6
U9	75% U4+25% GGBS	CEMII/B-S 52.5	44.4	53.1
F1		CEMI 42.5R	39.5	47.1
F2		CEMI 42.5	41.0	50.0
F3		CEMI 42.5R	39.7	48.6
F4	75% F2+25% limestone	CEMII/B-L 32.5	24.7	39.3
F5	79% F1+19% limestone	CEII/A-L 32.5R	30.4	39.2
F6	80% F3+5%PFA+5%GGBS+10%LS	CEII/A-M 32.5R	33.2	42.9
F7		CEIII/C 32.5	25.5	38.8

In order to stimulate reinforcement corrosion after 6 months of drying the cover concrete was wetted by immersing the exposed face to a depth of 1 mm in a 10% sodium chloride solution for 6 hours and then supporting the face above a water surface to maintain a relative humidity close to 100% for 28 days. The initial 6 month period of exposure at 60% relative humidity was intended to approximate the most severe seasonal drying conditions that might arise in marine or highway structures. The use of a 10% sodium chloride solution was intended to produce a condition intermediate between sea water and deicing salt exposure. A gravimetric method was used to determine reinforcement corrosion.[24-26] A corrosion loss of 1 g/m^2 in 28 days corresponds to a steel corrosion depth of 1.65 μm/year.

Some limited tests were undertaken to determine the depth of chloride penetration resulting from diffusion in resaturated concrete. After 3 days curing and exposure for 6 months at 60% relative humidity, the test cube was immersed to a depth of 1 mm in water for 7 days and then in 10% sodium chloride solution for 21 days. The cube was then split and dried at 105°C for 30 minutes before spraying with fluorescein solution (1 g/l of 70% ethyl alcohol) and silver nitrate solution (0.1 molar). A dark region develops where chloride ions are present and the depth of chloride penetration can then be measured.[27]

3 RESULTS AND DISCUSSION

3.1 Water Absorption Rate

A plot of absorption after 18 months versus that after 6 months was approximately linear. Absorption increased by about 25% between 6 and 18 months presumably because of continued drying. A slightly greater increase was observed for concretes containing the greater quantities of granulated blastfurnace slag. Figure 1 illustrates the wide range of water absorption rates that were obtained by the use of different cements in concretes with a constant water/cement ratio. There was no clear

4 hr water absorption after 6 months (kg/m²)

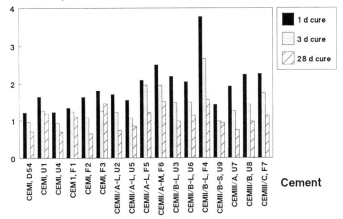

Figure 1 *Effect of cement type and curing upon water absorption after 6 months exposure*

Power exponent

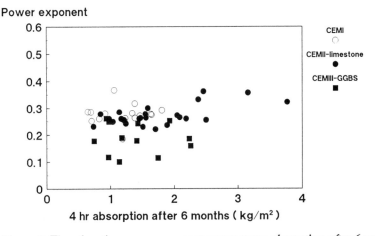

4 hr absorption after 6 months (kg/m²)

Figure 2 *Time function power exponent versus water absorption after 6 months exposure*

correlation with cement type although the absorption rate broadly increased as Portland cement clinker was replaced with limestone or granulated blastfurnace slag. In common with earlier studies[5,6] the weight of water absorbed did not increase linearly with the square root of wetting time: instead of a power exponent of 0.5, values in the range 0.1 to 0.4 were obtained, Figure 2. The results indicated that the power exponent increased slightly with exposure time for CEMI and CEMII cements, consistent with a closer approach to moisture equilibrium. However the results obtained with CEMIII cements suggest that low values of the power exponent may be associated with the pore structure gradients that would be expected with these slowly reacting slag cements.

4 hr water absorption, 1 or 28 d cure (kg/m²)

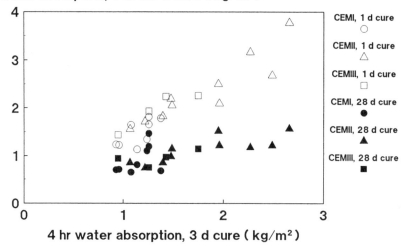

4 hr water absorption, 3 d cure (kg/m²)

Figure 3 *Water absorption rate after 6 months exposure for 1 and 28 day curing versus that for 3 day curing*

The effects of curing upon the rate of water absorption after 6 months exposure at 60% relative humidity are illustrated in Figures 1 and 3. It is evident from Figure 1 that prolonged curing was especially beneficial for the concretes with the higher values of water absorption rate. The results showed that the 4 hour absorption for 1 day curing was about 135% of that for 3 day curing after 6 and 18 month periods of exposure. The 4 hour absorption for 28 day curing averaged about 75% of that for 3 day curing.

The loss of weight during the initial four days of drying can be regarded as a simple indicator of cover concrete transport properties: it reflects the porosity and rate of moisture diffusion in a surface zone approximately 15 to 35 mm deep.[28] Figure 4 indicates that there is a close relationship between the 4 hour water absorption, measured after 6 months exposure, and the initial 4 day weight loss.

The results indicated that there was not a close correlation between the water absorption rate and previously reported[23] values for air permeability of the cover concrete. A closer correlation was observed where the 4 hour absorption, measured after 18 months exposure, was plotted against the cube strength at 8 days after the end of curing. The use of this strength parameter accounted well for the combined effects of cement type and curing although only small reductions of absorption rate are observed as the strength rises above 30 MPa. An earlier investigation indicated that a similar strength-based approach could also account for the effect of water/cement ratio upon water absorption rate.[5]

Carbonation is controlled by the diffusion of carbon dioxide to a penetrating reaction front.[29,30] If the rate of water absorption is indicative of the gas transport properties of the carbonated cover concrete then a correlation with carbonation depth might be expected, at least for concretes made with cements from a particular chemical group; indeed this has been observed in other studies.[4,5,31] The results in Figure 5 exhibit a correlation between carbonation depth and 4 hour water absorption for Portland, Portland-limestone and some Portland-slag cements but greater depths of carbonation are observed when cements contain more than 50% slag. Such behaviour has been attributed to the greater amount of calcium silicate hydrate and the smaller amount of calcium hydroxide with hydrated slag cements; carbonation of calcium silicate hydrate leads to an increase of gas permeability and carbonation of calcium hydroxide leads to a decrease of gas permeability.[32,33]

4 hr water absorption after 6 months (kg/m²)

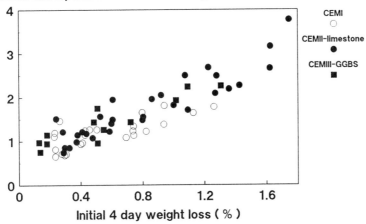

Figure 4 *Water absorption rate after 6 months exposure versus initial 4 day weight loss*

Carbonation depth after 18 months (mm)

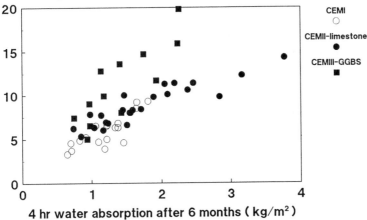

Figure 5 *18 month carbonation depth versus water absorption rate measured after 6 months.*

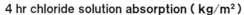

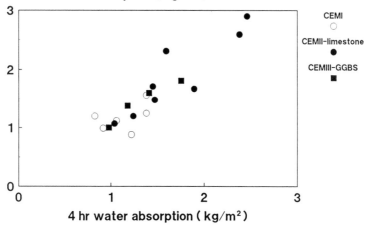

Figure 6 *Absorption of 10% sodium chloride solution versus absorption of water*

The absorption rate results of MacInnis[21] indicated that, on average, absorption of 10% sodium chloride solution was about 4% faster than absorption of water. The present results, shown in Figure 6, indicate that the absorption rate of 10% sodium chloride solution is about 8% greater than the absorption rate of water: however it can be concluded that water absorption results are broadly indicative of chloride absorption. Furthermore the relationship in Figure 6 is not dependent upon cement type.

The corrosion levels measured after 6 hours of 10% sodium chloride absorption followed by 28 days over water are shown in Figure 7. A corrosion rate of 30 g/m^2 in 28 days, if sustained, might be expected to cause visible cracking of cover concrete in about 2 years.[34] The highest rates of corrosion were observed for concretes containing 75% or more of granulated blastfurnace slag (cements U8 and F7) although high corrosion rates were also observed when Portland-composite cement F6 and Portland-limestone cements F4 and F5 were used. High rates of corrosion were not observed when plain Portland cements were used. It is evident from Figure 7 that where the rate of corrosion is greatest it can be moderated by an increased depth of cover to the reinforcement; where it is least it is not greatly affected by the depth of cover. The 6 hour period of absorption would cause the ingress of a limited quantity of sodium chloride solution such that chemical or physical binding might be expected to reduce the chloride ion concentration in the concrete pore solution. This might be a factor for those concretes where corrosion rates are low and independent of the depth of cover. If the cover is 20 mm or more then, apart from concretes made with cements F6 and F7, the time for visible cracking of the cover is likely to be 10 years or more. However even with the most favourable cements and 20 mm cover the corrosion propagation period for visible cracking is unlikely to exceed about 30 years and this, if coupled with a short initiation period, would be unlikely to provide an acceptable service life for a reinforced concrete element. It would be of practical interest to extend the present study to examine the effects upon corrosion of cover

Corrosion (g/m²)

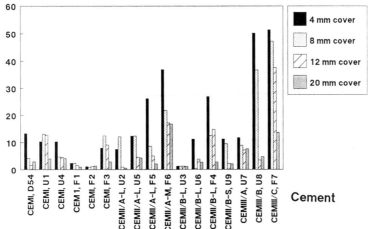

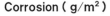

Figure 7 *Effect of cement and depth of reinforcement cover upon corrosion induced by absorption of sodium chloride solution*

Depth of chloride diffusion (mm)

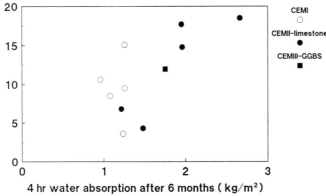

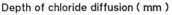

4 hr water absorption after 6 months (kg/m²)

Figure 8 *Depth of chloride diffusion in rewetted concrete versus 4 hour absorption, after drying for 6 months*

depths greater than 20 mm and periods of chloride absorption in excess of 6 hours. There was no direct correlation between the level of chloride-induced corrosion for a particular depth of cover and the 4 hour absorption of water or of sodium chloride solution.

The measured depths of chloride diffusion in cover concrete, after resaturation with water, are plotted in Figure 8 against the 4 hour water absorption measured during resaturation. There is a broad increase in the depth of chloride diffusion as the absorption rate increases and there is no obvious effect of cement type upon the relationship. Dhir[31] also reported a broad relationship between chloride diffusion and water absorption but he found that, for a given absorption rate, concretes containing granulated blastfurnace slag gave lower depths of chloride diffusion than those containing Portland cement. These beneficial effects of granulated blastfurnace slag were observed by Dhir after about one month of dry exposure whereas the present results involved a dry exposure period of 6 months; possibly the benefits of granulated blastfurnace slag become moderated by prolonged drying and by the changes of cover concrete microstructure associated with drying and carbonation.[32,33]

4 CONCLUSIONS

Experimental data were presented to illustrate the effects of cement type and curing upon water absorption rate, chloride ingress and chloride-induced reinforcement corrosion in cover concrete after exposure for 6 or 18 months at 20°C and 60% relative humidity. Three curing periods (1, 3 and 28 days) and 17 cements, with various proportions of granulated blastfurnace slag or limestone, were used to make concretes with a constant water/cement ratio of 0.59 and 28 day strengths in the range 26 to 46 MPa. There was an increase of about 25% in the rate of water absorption as the period of initial dry exposure was increased from 6 to 18 months. The time function for the rate of water absorption followed a power law with an exponent in the range 0.1 to 0.4. Water absorption rates for 1 and 28 days curing were about 135 and 75% of those for 3 day curing regardless of cement type. The water absorption rate broadly increased in the cement sequence CEMI, CEMII, CEMIII but the performances of these cements overlapped.

The relationships between water absorption rate, water loss during initial drying, cube strength and depth of carbonation were examined. The results for a wide range of concretes indicated that early-age measurements of cube strength and weight loss during drying could be used to assess water absorption rates after 18 months of drying. The depth of carbonation was correlated with the water absorption rate for many concretes but greater depths of carbonation were observed for a given water absorption rate if the cement contained 70% or more of granulated blastfurnace slag.

The rate of absorption of a 10% chloride solution was similar to that of water. Corrosion induced by absorption of chloride solution broadly reduced with increasing cover and increased in the cement sequence CEMI, CEMII, CEMIII, but the performances of these cements overlapped. Cements with slag contents of 70% or more led to the greatest levels of corrosion. Limited measurements of chloride diffusion in cover concrete, after resaturation with water, indicated that chloride penetration broadly increased with an increase of water absorption rate measured during resaturation.

REFERENCES

1. C. Hall, Mag. Concrete Res., 1989, **41**, 51.
2. E. Senbetta and C. Scholer, ACI Jour., Jan 1984, 82.
3. S. Kelham, Mag. Concrete Res., 1988, **40**, 106.

4. F. Rostasy and D. Bunte, 'Proc IABSE Symp Durability of Structures', Lisbon, 1989, 145.

5. L. Parrott, Materials & Structures., 1992, **25**, 284.

6. L. Parrott, Materials & Structures., 1994, **27**, 460.

7. K. Tuutti, 'Corrosion of steel in concrete', Swedish Cement and Concrete Institute, 1982, Report F04, 468 pages.

8. K. Muller, 'Proc RILEM Symp Long-term observation of concrete structures', Budapest 1984, **1**, 9.

9. N. Saeki, N. Takada and Y. Fujita, Trans. Jap. Concrete Inst., 1984, **6**, 155.

10. ACI Committee 362. 'Guide for the design of durable parking structures'. ACI 362.1R-94, 1995, 40 pages.

11. V. Novokshchenov, Concrete Internat., 1991, **13**, 43.

12. E. Schjolberg, 'Proc Chloride penetration into concrete structures', Goteborg Univ., P-93:1 N535, Jan 1993, 274.

13. M. Bashenini, S. Hussain and I. Paul, 'Corrosion and corrosion protection of steel in concrete', Sheffield Academic Press, 1994, **1**, 61.

14. P. Bamforth and D. Pocock, 'Proc Third Symp Corrosion of Reinforcement in Concrete Construction', Elsevier Applied Science, 1990, 119.

15. P. Bamforth and B. Chapman-Andrews, 'Corrosion and corrosion protection of steel in concrete', Sheffield Academic Press, 1994, **1**, 139.

16. M. Salta, 'Corrosion and corrosion protection of steel in concrete', Sheffield Academic Press, 1994, **2**, 794.

17. J. Aldred, 'Pacific Corrosion '87', Melbourne, Nov. 1987, 8 pages.

18. N. Otsuki, S. Nagataki and K. Nakashita, 4th CANMET/ACI Conf Fly Ash, Silica Fume, Slag and Natural Pozzolans in Concrete, Istanbul 1992, Suppl Papers, 1055.

19. L. Johansson, 'Proc Chloride penetration into concrete structures, Goteborg Univ., P-93:1 No 535, Jan 1993, 244.

20. K. Dass, T. Raj and P. Gupta, 'Proc Corrosion and Corrosion Protection of Steel in Concrete', Sheffield Academic Press, 1994, 387.

21. C. MacInnis and Y. Nathawad, 'Durability of building materials and components', ASTM STP 691, 1980, 485.

22. W. McCarter, H. Ezirim and M. Emerson, Mag. Concrete Res., 1992, **44**, 31.

23. L. Parrott, Mag. Concrete Res., 1995, **47**, 103.

24. L. Parrott, Materials & Structures, to be published.

25. L. Parrott, 'Proc Protection of Concrete', Dundee, Sept 1990, 1009.

26. L. Parrott, Mag. Concrete Res., 1994, **46**, 23.

27. Unicemento, Italian Standard UNI 7928, December 1978.

28. L. Parrott, British Cement Association Report C/9, March 1991, 20 pages.

29. RILEM Tech. Comm. 116-PCD, 'State of art report on permeability and concrete durability', Chapman & Hall, 1995.

30. L. Parrott, BRE/BCA Report C/1, July 1987, 126 pages.

31. R. Dhir, P. Hewlett, E. Byars and J. Bai, Concrete, 1994, **28**, 25.

32. J. Kropp and H. Hilsdorf, 'Proc Materials Science and Restoration', Esslingen Univ., 1983, 153.

33. T. Bier, MRS Symp Proc, **85**, 'Microstructural development during hydration of cement', 1987, 123.

34. L. Parrott, '3rd CANMET/ACI Conference on Durability of Concrete', ACI SP-145, Nice 1994, 283.

A MODEL FOR EVALUATING THE WHOLE LIFE COST OF CONCRETE BRIDGES

P. R. Vassie and R. S. Rubakantha

Transport Research Laboratory
Old Wokingham Road
Crowthorne
Berkshire RG45 6AU

1. INTRODUCTION

Reinforced concrete bridges are deteriorating largely due to the effects of corrosion of the reinforcing steel initiated by chlorides from road de-icing salt. This deterioration has resulted in the need for major repair work typically when the bridge is only about 20 years old[1]. Bridges are designed for a life of 120 years and although this includes provision for periodic maintenance, a maintenance free life (MFL) of only 20 years is not satisfactory, particularly as the methods for repairing corroded reinforced concrete are not well established[2,3].

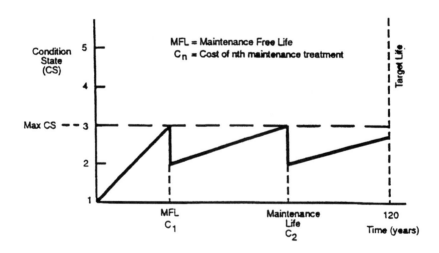

Figure 1 *Schematic diagram the Whole Life Costing Process*

In order to overcome these difficulties the industry has developed a number of design details and materials for new construction aimed at preventing or delaying the onset of corrosion (Table 1).

Table 1 *Durability Options for Concrete Bridges*

Durability Option	Examples	Mechanism
Concrete coating	a. waterproofing membrane b. permeable formwork	Prevent contact between concrete and chloride ions
Concrete impregnant	a. silane b. siloxane	Make concrete surface hydrophobic to repel aqueous solutions
Increased cover		Increase the time for chloride ions to diffuse to the reinforcement
Concrete modifications	a. low water/cement ratio b. cement replacement materials	To reduce the rate of diffusion of chlorides
Steel coatings	a. epoxy coating b. galvanising	To prevent corrosion when chlorides reach the reinforcement
Non corrodible reinforcement	a. glass fibre b. carbon fibre c. stainless steel	Not susceptible to corrosion in saline media
Modified design details	a. continuous decks b. integral abutments	Eliminate the need for expansion joints and prevent chloride leakage

The impact of these durability options on the cost of construction and the MFL is varied. The concept of Whole Life Cost (WLC) has been used to appraise the economics of alternative durability options taking account quantitatively of the advantage of increased MFL and disadvantage of increased construction cost[4].

This paper describes a model for the calculation of WLC that includes consideration of factors such as the varying rates of deterioration encountered on different parts of a bridge, the impact of maintenance on the user of the bridge and the use of discounted cash flow.

2. WHOLE LIFE COSTING

The principle of whole life costing is explained schematically in Figure 1. The objective is to keep the condition of the bridge above some set level throughout its required life. To achieve this objective maintenance will be required at certain times to counteract the effects of deterioration. Maintenance work will improve the condition and reduce the rate of subsequent deterioration. From Figure 1 it can be seen that the following items of information are needed to calculate the WLC:

- The cost of construction at time, t = 0

- The MFL (the age of the bridge when major maintenance is first required)
- The lifetime of the maintenance treatment (the time before deterioration necessitates further maintenance work)
- The estimated cost of the maintenance work at the times when it is required.

On this basis it would appear that the calculation of WLC is reasonably straightforward. However, in practice there are considerable difficulties in estimating the MFL, the life of maintenance treatments and the maintenance costs. Whole life costing consists of a projection of future maintenance costs to the present time. Future projections of this type are open to abuse unless steps are taken to ensure that all the costs associated with future maintenance are included and an analysis is made of the effect of the inherent uncertainty in the value of the input parameters. The remainder of this paper describes the procedures that have been adopted to ensure that the calculated value of WLC is reliable.

3. FACTORS TO BE CONSIDERED TO ENSURE THAT THE CALCULATED WLC IS RELIABLE

3.1 The Bridge

Construction and maintenance costs and MFL may vary according to bridge type for a particular durability option. Thus to appraise the economics of alternative durability options it is necessary to eliminate the variability arising from different bridge types. For our model the bridge consisted of seven elements all of which are commonly used in trunk road and motorway construction; a drawing of the bridge is given in Figure 2 and the elements are listed in Table 2.

Table 2 *Elements used in the test bridge for whole life costing*

End supports	Intermediate supports	Deck spans
Wall abutment	Leaf pier	Voided slab
Bank seat abutment	Columns and crossbeam	Beam and slab
		Steel beam and slab

3.2 When to carry out first maintenance and repeat maintenance treatments

Different elements and parts of elements of a bridge deteriorate at different rates and hence will require maintenance at different times[1]. For example the top surface of a waterproofed bridge deck deteriorates very slowly because chloride ions do not come into contact with the concrete. Vertical faces of piers and abutments deteriorate at an intermediate rate because chloride ions only come into contact with the concrete during damp weather in the winter. Deck ends and abutment shelves under leaking expansion joints deteriorate most quickly because the concrete is exposed to chlorides for a high proportion of the time. The MFL values for bridge elements, designed to current specifications and exposed to salt spray and leaking salt, are listed in Table 3. These values are based on current experience of deterioration of concrete bridges.

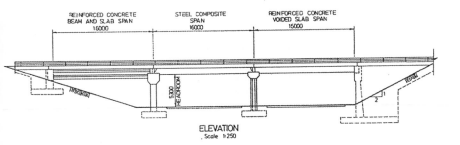

Figure 2 *General arrangement of hybrid bridge*

Table 3 *Maintenance free lives for the existing specification (years)*

	End support	Intermediate support	Deck
Leakage Area	20	20	20
Spray Area	60	60	45

The time for first maintenance depends on the MFL of the selected durability option and the time for repeat maintenance depends on the life of the maintenance method that is used. Since most durability options and maintenance methods have been used only to a limited extent and for a short period they have yet to develop a track record and therefore uncertainty exists about the duration of their effective lives.

3.3 Maintenance Costs

The full maintenance cost depends on a number of factors that include:
- gaining access
- equipment
- engineering and labour
- materials
- traffic management
- user delays.

Most of these costs have been obtained from Bills of Quantity for maintenance schemes. Traffic management and user delay costs are harder to determine since they depend on the maintenance method, element to be maintained, the density and mix of traffic, and the extent, duration and length of the lane closures required. Nevertheless it is essential to take account of traffic management and user delay costs because in some circumstances these are the major constituents of the maintenance cost[5].

3.4 The Contribution of Traffic Management and User Delay Costs to the Maintenance Cost

Some types of maintenance on some elements do not necessitate lane closures and hence do not give rise to user delays or require traffic management. When the maintenance work does involve interference with the traffic it is sometimes possible to mitigate the effect by:

- combining the bridge maintenance with a road maintenance scheme where lanes are already closed.
- carrying out the maintenance at night when traffic volumes are much smaller
- using narrow lanes
- providing incentives to minimise the duration of the disruption.

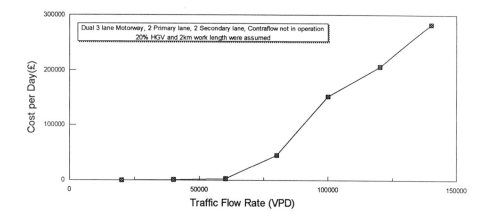

Figure 3 *Cost vs Traffic Flow Rate (for Post-Tensioned concrete bridge)*

User delay costs are calculated using the DOT program QUADRO or from the QUADRO ready reckoner tables in the Trunk Road Maintenance Manual[6].User delay costs depend mainly on:

- traffic flow rate (vehicles per day)
- percentage of heavy goods vehicles
- type of road and number of lanes/carriageways closed
- length (km) and duration (days) of closure.

They are particularly sensitive to the traffic flow rate and for a given type of road there is a critical flow rate below which delay costs are small and above which the delay cost increases very quickly with flow rate (Figure 3).

The implications of the above discussion are that in many cases delay costs can be avoided or minimised but that on heavily used roads they can be more than a factor of ten times the other maintenance costs. All of these factors are taken into account by the model to ensure that a realistic maintenance cost is calculated.

3.5 Discounted Cash Flow

This is a technique for taking into account the time value of money[7]. In very simple terms economic growth leads to an increase in the value of money with time. Thus a certain amount of maintenance will cost more, in real terms, today than it will in the future. The model takes account of this effect by discounting at the current test discount rate for highway infrastructure, 8%, using the standard formula

$$NPV = C_o/(1+i)^n \qquad (1)$$

where NPV is the net present value of maintenance work carried out in year n.
     C_o is the maintenance cost in year, $t = 0$.
     i is the discount rate.

In general NPV calculations are carried out for cost and benefits. However, in this case where economics of durability options are being appraised, it is only necessary to do the calculation for costs because it is assumed that the bridge is kept open throughout the period, regardless of which durability option is selected so the benefit will be the same for all options. Lane closures arising from maintenance work are accounted for as a cost rather than a reduced benefit.

To find the discounted WLC for a bridge the discounting formula is summed over all the elements and for all the maintenance treatments. The implications of discounting at 8% are shown in Figure 4. After 30 years the NPV is only one tenth of the maintenance cost in year 0, hence it could be argued that there is no point in extending the WLC period beyond 30 years. However in certain circumstances user delay costs can be so large that the NPV is significant even after as much as 50 years particularly if the expected growth in traffic is taken into account. In our model a period of 60 years is normally used.

Comparing the simple model for calculating WLC in section 2 and the additional factors described in this section it is clear that the complexity has increased because the WLC depends on many factors some of which are interdependent. To take account of this increased complexity and reduce the calculation time, computer modelling is appropriate and justified.

4. SENSITIVITY ANALYSIS

It is essential to test all models for the sensitivity of the result to uncertainty in the values of the input parameters. This is particularly important for models that involve projections of future events where uncertainty is inherent. Our model tests for sensitivity in two ways:
i.   Deterministic method
ii. Probabilistic method

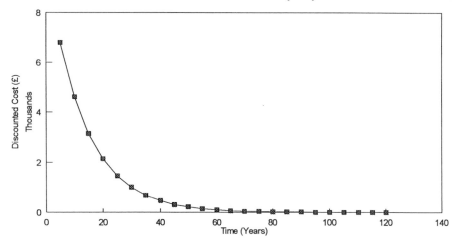

Figure 4 *Variation of Discounted Maintenance Cost with Time*

4.1 Deterministic Method

This is a very simple method that considers just one uncertain parameter at a time. The range of the uncertainty in the input parameter is assessed and the WLC is calculated at a number of discrete points within the range. For example if the best estimate of MFL for a particular durability option, element and type of exposure to chlorides is 50 years and the range of uncertainty is ± 25 years then WLC calculations may be repeated for MFL = 25, 35, 45, 50, 55, 65, and 75 years. Although this method is easy to carry out it has two important limitations that can make the results misleading:

i.   Only one uncertain parameter can be considered at a time. This means it is difficult to assess the effect of uncertainty in several input parameters acting simultaneously, particularly if they are interdependent.

ii. Each value of an input parameter that is tested is assumed to have the same probability of occurrence. This is unrealistic and in general, values close to the best estimate would be expected to have a much greater probability of occurrence than values distant from the best estimate.

4.2 Probabilistic Method

The two limitations of the deterministic method can be overcome by using a probabilistic analysis. A Monte-Carlo type method is used to carry out typically several hundred 'what-if' analyses[8]. For example "what is the WLC when the MFL is 20 years?" The procedure is explained in Figure 5; it consists of the following steps:

i.   produce cumulative statistical distributions to describe the uncertainty of each input parameter

ii. generate random numbers in the range 0 to 1 and for each number read off the corresponding value of the uncertain parameter from the cumulative distribution

iii. calculate the WLC using this input value

iv. repeat with sufficient random numbers until the cumulative statistical distribution of WLC values produced has converged to within a suitable bound.

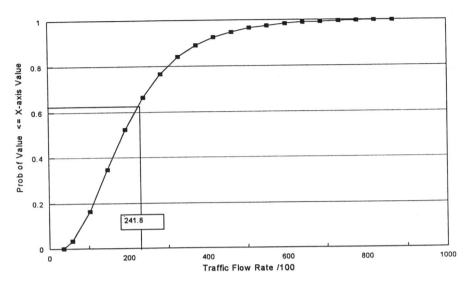

Figure 5 *Cumulative Distribution for Traffic Flow Rate/100*

The large number of calculations involved can be easily carried out on modern personal computers. The disadvantage of the probabilistic method is that it is more difficult to interpret the result because instead of producing a single value of WLC a distribution of values is produced that has to be interpreted in terms of probabilities.

5. APPLICATIONS OF THE WLC MODEL

There are many ways in which the model can be applied and two examples are given in this section.

5.1 Comparing the Cost Effectiveness of Durability Options

The comparison can be made graphically by plotting the WLC cost distributions for a number of durability options together. Figure 6 shows the distributions for three types of durability option. Durability option C has a high initial cost and a long life; its WLC is constant over almost the entire probability range and is higher than the WLC of options A and B over almost the entire probability range. Thus in general option C is less cost effective than options A or B. Durability options A and B may be compared in a similar way.

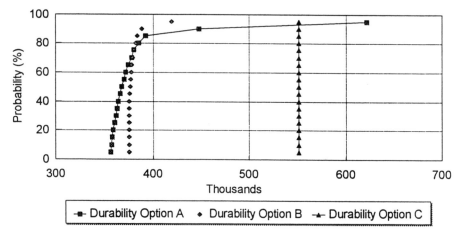

Figure 6 *Comparison of Durability Options*

5.2 Determining the Value of an Input Parameter when the WLC for Two Options are Equal

We may have the situation where the WLC for durability option D is less than the WLC of durability option E at a particular traffic flow rate. A method of back analysis is used to find by how much the traffic flow rate needs to increase in order for the WLC of the two options to become equal. If the flow rate exceeds this value the WLC of option E will be lower than the WLC of option D. A typical set of results is shown in Figure 7.

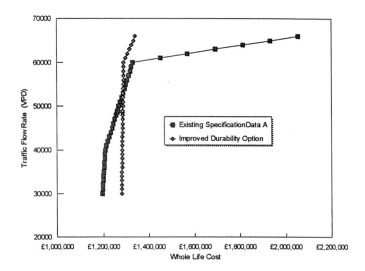

Figure 7 *Flow Rate at which Durability Option Becomes Economical*

6. CONCLUSIONS

The WLC model described in this paper provides an appropriate and reliable method for calculating the WLC of concrete bridges, appraising the economics of a range of durability options and investigating the effect of variation of input parameters.

The model takes account of the different deterioration rates that occur in different parts of a bridge, user delay costs arising from repair work, traffic growth and cost discounting. The model incorporates deterministic and probabilistic sensitivity analysis to test the reliability of the results.

7. REFERENCES

1. E. J. Wallbank, "The Performance of Concrete Bridges," A Report for the Department of Transport, HMSO, 1989.
2. P. R. Vassie, "Reinforcement Corrosion and the Durability of Concrete Bridges," *Proc. Inst. Civ. Engrs Part 1*, 1984,**76**, 713.
3. P. R. Vassie, "Durability of Concrete Repairs: the effect of steel cleaning procedures," Transport Research Laboratory, Research Report 109, Department of Transport, Crowthorne, 1987.
4. D. J. O. Ferry and R. Flanagan, "Life Cycle Costing," CIRIA Report No 122, 1991.
5. D. R. B. Jones, "The appraisal of traffic delay costs," Concrete Bridge Development Group Seminar, London, 1995.
6. Department of Transport, Trunk Road Maintenance Manual, 1993
7. M. Spackman, "Discount Rates and Rates of Return in the Public Sector," Govt. Econ Service Working Paper No 113, H M Treasury, 1991.
8. A. M. Law and W. D. Kelton, "Simulation Modelling and Analysis," McGraw-Hill, New York, 1982.

ENVIRONMENTAL EFFECTS ON REINFORCEMENT CORROSION RATES

C. Hawkins and M. McKenzie

Transport Research Laboratory
Crowthorne
Berkshire RG11 6AU

1 INTRODUCTION

Over the last 20 years an increasing number of corrosion related problems have arisen affecting reinforced concrete highway structures. The main cause has been chlorides from de-icing salts penetrating to the level of the steel reinforcement. The factors that influence the time taken to initiate corrosion are fairly well understood and procedures have been implemented to extend this period - for example the use of lower water-cement ratios and increased concrete cover. It has also been established that, once corrosion starts, the rate of reinforcement corrosion is influenced by air temperature and humidity, the chloride content of the concrete, the water-cement ratio of the concrete, and the concrete cover depth to the reinforcement.[1,2] However the relative importance of each of these factors and, in particular, their interaction are less well understood. A knowledge of the relative importance of concrete properties and the exposure environment on corrosion rates would assist in defining measures that could be taken to mitigate the effects of corrosion both at the design stage and during the life of the structure.

The experimental work described in this report studied the combined effects of the following factors on the corrosion rate of steel in concrete: temperature and humidity varying over a range appropriate to highway structures, (including conditions in which the temperature and humidity were cycled to simulate typical daily variations); surface moisture; concrete cover typical of good and poor quality construction; a range of water-cement ratios representative of good and poor quality concrete; chloride content; and conditions of cure.

2 EXPERIMENTAL INVESTIGATION

2.1 Specimen Design and Construction

Eighteen concrete specimens with dimensions of 460 mm by 135 mm by 210 mm (Figure 1) were cast; each contained corrosion macrocells to measure corrosion rates. Each macrocell consisted of a 16 mm diameter bar of mild steel reinforcement adjacent to a 16 mm diameter bar of austenitic stainless steel. The bars protruded from the side of the test specimen and were connected through a zero resistance ammeter (ZRA). In a corrosive environment the mild steel bar acted as an anode and the stainless steel bar

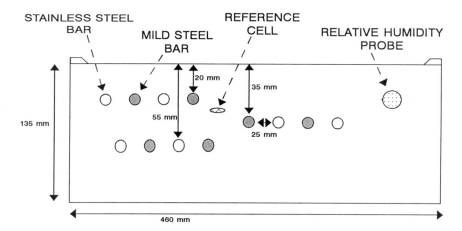

Figure 1   *Cross section of test specimen*

as a cathode; the corrosion current flowing between them was measured using the ZRA and gave an estimate of the corrosion rate. When measurements were not being taken the two bars remained connected to prevent the cell from depolarising.

To limit edge effects equal lengths of the ends of each bar were coated with a plasticised cement mix then sealed in heat shrink sleeving. This end protection extended 50 mm into the specimen leaving a central uncoated bar length of 100 mm inside the specimen.

Each specimen contained two macrocells at 20 mm cover, two at 35 mm cover and two at 55 mm cover. This replication provided information on the variability of current measurements under the same conditions. Each specimen also contained a Ag/AgCl reference cell. Selected specimens also contained an embedded relative humidity (RH) probe at the upper level of macrocells. The top of each specimen was recessed to allow water to be ponded on the surface.

The specimens were cast in nine pairs to cover all combinations of 0.5%, 1% and 3% chloride (added as sodium chloride) by weight of cement, and water-cement ratios of 0.45, 0.55 and 0.65. The concrete mixes were made with Ordinary Portland Cement and Thames Valley Aggregate. Mix designs are shown in Table 1. Specimens with 0.5% chloride and water-cement ratio of 0.45, and specimens with 3.0% chloride and 0.65 water-cement ratio had the embedded RH probes. One of each pair of specimens was demoulded after 24 hours and allowed to cure in air for seven days; the other was also demoulded after 24 hours but was then cured under damp hessian and polythene for seven days. After curing, the sides and base of each specimen were sealed so that any effects of humidity or surface moisture would be due to action on a single face. This was intended to simulate the conditions in a large mass of concrete. The electrical connections necessary to carry out the corrosion current measurements were then made to each specimen.

It took about six weeks after casting to complete the sealing and electrical connections of the specimens. During this time the specimens were stored in a laboratory where the controlled temperature was about 20 °C and the RH was about 50%.

Table 1   *Concrete Mix Designs*

Water-Cement Ratio	Cement /kg m⁻³	20 mm Aggregate /kg m⁻³	10 mm Aggregate /kg m⁻³	Sand /kg m⁻³	Mean 28 Day Strength /N mm⁻²
0.45	400	509	500	745	57
0.55	345	516	498	765	40
0.65	300	462	448	910	33

2.2 Testing

2.2.1 Tests at Constant Temperature and RH. The specimens were placed inside a walk-in environmental cabinet in which the temperature and RH of the atmosphere could be adjusted to set levels, or cycled. Initially, conditions of temperature and RH were maintained at a constant set level of 25 °C and 50% RH; of the constant conditions to be used this was closest to the conditions prevailing during the preparation of the specimens after casting. Corrosion currents were measured on a daily basis, along with the potential of each macrocell in relation to the embedded Ag/AgCl reference cell. Where present, the embedded probe was used to measure the specimen RH. The aim was to wait until the corrosion currents stabilised before the set levels were changed but this was not always possible in the time available. Changes were therefore made when the currents measured on the upper bars became fairly constant - i.e. showing only small, random, day to day variations - and currents on the lower bars were changing only slowly.

The set conditions and order in which they were applied were: 25 °C and 50%RH; 25 °C and 90%RH; 5 °C and 90%RH; 5 °C and 50%RH.

2.2.2 Ponding Trials. The effect of surface moisture on the corrosion rate was investigated by ponding the specimens with tap water. The temperature in the environmental cabinet was adjusted to 25 °C and the humidity to 50%RH. Some preliminary experiments were carried out in which some specimens were ponded with tap water for two days and the corrosion currents in one of the upper macrocells recorded on an hourly basis. This showed that a significant response was occurring within hours of ponding but that the currents continued to rise for several days. Specimens were then allowed to dry out until the currents fell back to values approaching the pre-ponding levels. Specimens were then ponded with tap water for six days. The water was then removed and the specimens allowed to dry out for a further ten days. Throughout the ponding and drying out, daily measurements of corrosion current were taken.

2.2.3 Tests under Cyclic Temperature and RH. Three series of tests were carried out to assess the effect of variations in temperature and RH approximating diurnal cycles. First the temperature was set at 25 °C and the RH was cycled over a range of 50% to 90% to 50% over 24 hours for a period of about 20 days. The RH was then set to 70% and the temperature was cycled from 5 °C to 25 °C to 5 °C over a 24 hour period for about ten days. Finally temperature and humidity were varied together with low temperature corresponding to high RH. To follow changes over the diurnal cycle, corrosion currents were logged automatically. However only a limited number of currents could be logged at any one time; this was controlled by the number of measurement channels in the data loggers used. Therefore the corrosion current in one macrocell from the upper level in each specimen was logged on an hourly basis; and in addition measurements of one macrocell from the middle and lower levels were logged for air cured specimens with 3% chloride and water-cement ratios of 0.55 and 0.65.

2.2.4 Tests under Exterior Exposure. Following tests in the environmental cabinet, specimens which had shown appreciable corrosion currents were exposed outside at TRL. This involved all specimens with 3% chloride and specimens with 1% chloride and a water-cement ratio of 0.65. Specimens were initially placed horizontally so that any rain would collect in the ponding recess and remain there until it evaporated. Currents from a single macrocell in the upper layer of each specimen were logged on an hourly basis over a 25 day period. Air temperature was logged on an hourly basis along with specimen temperature for the air cured specimen with 3% chloride and a water-cement ratio of 0.65.

A second trial was carried out in which the specimens were tilted so that water would wet the surface but would not pond. This trial ran for about 30 days during which the currents were also logged from single macrocells in the middle and lower positions in air cured samples with 3% chloride and water-cement ratios of 0.55 and 0.65.

3 RESULTS

3.1 Tests at Constant Temperature and RH

3.1.1 Corrosion Currents. During the initial set conditions of 25 °C and 50%RH the corrosion currents in all specimens showed a decrease with time after their introduction into the environmental cabinet. This is illustrated in Figure 2 which shows the average value of the current in each pair of macrocells for the air cured specimen with 3% chloride and a water-cement ratio of 0.65. It is likely that the corrosion currents were altering as a result of the continuing cure of the specimens as well as adjusting to

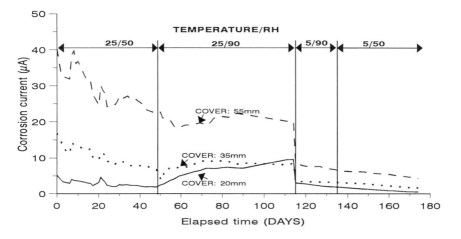

Figure 2   *Change in corrosion currents with time in environmental cabinet: air cured specimen; 3% chloride ion by weight of cement; water-cement ratio of 0.65*

the external humidity conditions. In general the bars with 20 mm cover had the lowest corrosion currents and were stabilising faster than bars with deeper cover. Changing to an RH of 90% resulted in a gradual increase in the corrosion currents over several weeks for bars with 20 mm cover so that the currents eventually became significantly higher than previously. Response for bars with a greater cover was much less pronounced. Once currents in bars with 20 mm cover were fairly stable the temperature was reduced to 5 °C; this led to a rapid fall in the corrosion currents in all bars followed by further, slower changes. The relatively small size of the specimens and the fact that the steel bars protruded from the samples could explain the initial rapid response. In a large structure the response is likely to be slower. The final change in RH back to 50% resulted in further reductions in the corrosion current.

Chloride content had a dominant effect on the corrosion currents. The maximum current measured for 0.5% chloride samples was less than 0.5 μA, for 1% chloride samples it was less than 2 μA, but for the 3% chloride samples maximum current was greater than 20 μA. For the specimens with 3% chloride, the water-cement ratio had a clear influence with the corrosion currents in specimens with a water-cement ratio of 0.45 being markedly lower than in specimens with higher water-cement ratios (Figure 3).

An analysis of variance was carried out on the results from samples with 3% chloride to assess the significance of the factors in the experimental design. This showed that, overall, corrosion currents were significantly higher for higher water-cement ratios, higher temperatures, increased RH and increased cover; there was no overall significant effect of cure. However there were also significant interactions whereby the effect of one factor depended on the level of another. In particular it can be seen from Figure 3 that changes in corrosion current as the RH was changed from 50% to 90% are much greater for bars with 20 mm cover than for bars with greater depths of cover, and in individual cases low cover combined with high RH led to the highest corrosion currents.

3.1.2 Half Cell Potentials. Half cell potentials gave a general indication of the degree of corrosion in that for specimens with 0.5% chloride all potentials were more positive than -100 mV; for specimens with 1% chloride the readings tended to be between -100 mV and -200 mV; for specimens with 3% chloride the readings were generally more negative than -200 mV. As environmental conditions changed half cell readings also changed as expected (i.e., becoming more negative as conditions became more corrosive). However these changes were relatively small. The limitations of using half cell potentials as indicators of corrosion rate were apparent in that there was often a considerable spread in the corrosion current values for the same half cell potential.

3.1.3 Relative Humidity in Specimens. When first introduced into the environmental cabinet at 50% RH, the humidity probes in the specimens showed a range of values but all were above 50%. As time passed the RH in each specimen showed little tendency to decline. When the external RH was altered, the RH in the specimens followed the direction of change but the rate of change was quite slow. Humidity also fell when the temperature was reduced.

3.2   Ponding Trials

During ponding, specimens with 3% chloride and water cement ratio of 0.55 showed dramatic increases in corrosion currents for the bars with 20 mm cover. Once the water was removed currents declined at a slower rate and had not returned to pre-ponding levels after one week (Figure 4). Similar results were observed in specimens with 3% chloride and a water-cement ratio of 0.65 except that the bar with 35 mm cover

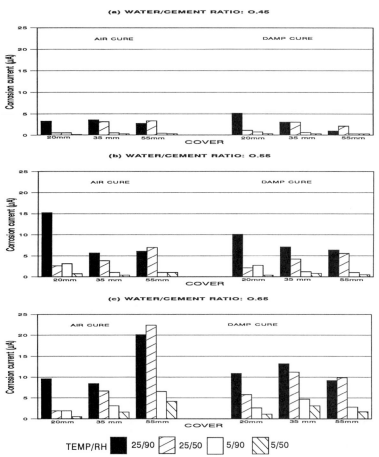

Figure 3 *Average corrosion currents under constant conditions at the end of each test period: Specimens with 3% chloride ion by weight of cement*

in the damp cured specimen also showed a significant increase in corrosion current during ponding. Of the remaining specimens only the specimens with 1% chloride and a water-cement ratio of 0.65 showed any increases in current and these tended to be rather less than the increases in the other specimens particularly for bars with 20 mm cover. The extent of the effect of ponding is shown in Figure 5 where the maximum current measured during the ponding trials is compared with the currents recorded at constant temperature and humidity.

3.3 Test under Cycling Temperature and RH

Corrosion currents showed no significant response to typical diurnal variations in humidity at constant temperature. However the corrosion currents did show a close correlation with temperature variations at constant humidity although there was a lag of about two hours between the two cycles. A similar correlation was observed when both temperature and humidity were cycled together due to the dominant effect of temperature.

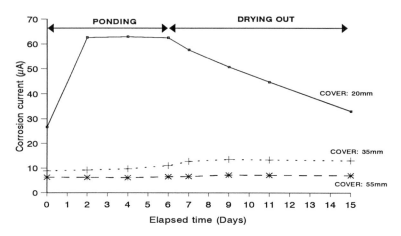

Figure 4   *Corrosion currents during ponding/drying out at 25°C and 50%RH:*
           air cured specimen; 3% chloride ion by weight of cement; water-
           cement ratio of 0.55

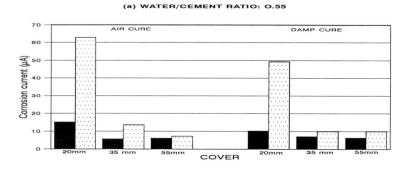

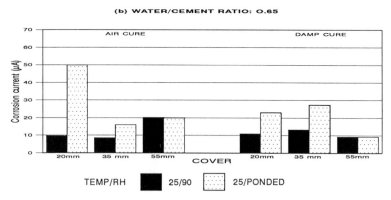

Figure 5   *Maximum corrosion currents under ponding and under constant*
           conditions of temperature and humidity for specimens with 3% chloride
           ion by weight of cement

3.4   Tests under Exterior Exposure

The corrosion currents logged for an upper macrocell in each specimen showed large hourly variations which closely followed the temperature variations. To smooth out some of this variation, average daily corrosion currents and temperatures were calculated from the hourly results. Average daily specimen temperatures were very similar to the average daily air temperatures. Only the specimens with water-cement ratios of 0.55 and 0.65 with 3% chloride content showed significant corrosion currents. At the start of the exposure trial the specimens all had surface water in the ponding recess. Weather conditions then stayed dry for the next 16 days at which time it rained and all ponding recesses once again became full of water. There was sufficient rain during the rest of the period to keep the surface visibly wet on all specimens. Corrosion currents gradually fell during the dry period presumably as a result of the specimens drying out. This effect was sufficient to override the effects of temperature which showed an increase over the same period. However once the specimens became wet there were dramatic increases in the corrosion currents (Figure 6). This was certainly not a result of temperature change as this stayed fairly constant during the wet period.

In the second exposure trial the specimens were tilted so that they would become wet but water would not pond. Corrosion currents showed day to day variations that corresponded to the daily temperature variations but otherwise stayed at roughly the same level except for a downward trend observed in the air cured specimen with a water cement ratio of 0.65. The weather conditions were fairly wet throughout this trial so that even though ponding did not occur the specimens would be expected to be moist. In this second trial the corrosion currents lay between those measured in the dry and ponded stages of the first trial at roughly equivalent temperatures. Currents measured in bars with 35 or 55 mm cover were similar to or lower than those measured in bars with 20 mm cover (Figure 7).

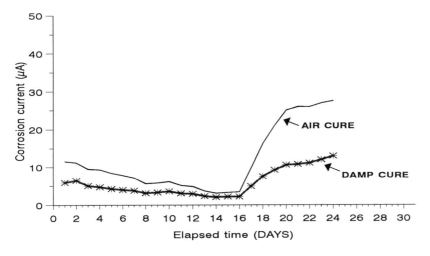

Figure 6   *Effect of ponding on corrosion currents during exterior exposure: specimens with 3% chloride ion by weight of cement and water-cement ratio of 0.65*

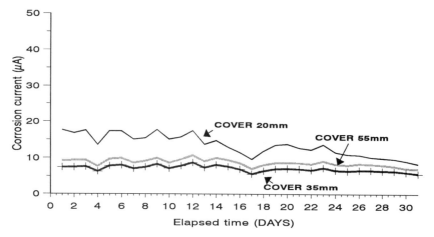

Figure 7   *Corrosion currents for different covers during exterior exposure of specimen with 3% chloride ion by weight of cement and water-cement ratio of 0.65*

Measurements under drier conditions would be necessary to confirm the observation made from laboratory ponding trials that bars with greater cover are less influenced by changes in surface moisture.

4 DISCUSSION

The results from this series of experiments have confirmed that corrosion currents for steel in concrete increase with increasing chloride content and for concretes with higher water-cement ratios. The types of cure used in these trials had no consistent effect on corrosion current. Because the chloride was cast in, the chloride levels used do not yield useful information on critical chloride levels for structures where chlorides penetrate in from outside. These results only show the general effects of increasing chloride levels.

Of the direct environmental effects, corrosion currents increase with increasing temperature, increasing humidity and when the concrete is wet. Response to temperature changes was very rapid in these trials partly because of the design and small size of the test specimens. Response to temperature changes in a bridge is likely to be much slower. Response to humidity was much slower and the currents did not respond to typical diurnal changes in RH. The extent of the effects of humidity and surface moisture were particularly dependent on the concrete cover. For concretes with higher water-cement ratios, bars with low cover could eventually show large differences in the corrosion rate as humidity changed from low to high values, and rates could become particularly high when the concrete became saturated under ponding conditions. It was not possible in the time available to test whether ponding for extended periods would also increase the corrosion rates of bars with greater covers.

These results suggest that certain actions should be taken to reduce the corrosion rate of steel in concrete structures. At the construction stage the use of a water-cement ratio of 0.45 is clearly beneficial in reducing the rate of corrosion. Specifications for higher concrete covers of around 50 mm, aimed at delaying the onset of chloride induced corrosion, will also offer benefits when corrosion does start in that the large corrosion currents associated with surface moisture and low cover will be avoided. Under conditions where atmospheric humidity remains low for long periods, a low cover of around 20 mm could lead to lower corrosion rates. However such conditions are likely to occur only rarely in the UK and any such benefits are outweighed by the benefits of high cover already outlined. Designs should avoid features which would allow water to pond.

During the life of a structure, exposure to chlorides should be restricted and exposure to moisture limited. Silanes are already used to prevent chloride ingress; such coatings might also reduce corrosion rates by reducing the moisture content of concrete. It would, however, be necessary to verify the practical effectiveness of such measures by laboratory and full scale trials.

5   CONCLUSIONS

Corrosion currents were greater for higher chloride contents and for higher water-cement ratios. The type of cure had no consistent significant effect.

Higher temperatures led to increased corrosion currents. In these tests response to temperature was rapid, partly as a result of the specimen design and size, and corrosion currents followed diurnal changes in temperature.

Response to humidity depended on concrete cover. High humidity led to increased corrosion currents for bars with low cover; there was less effect as the cover increased. Corrosion currents did not follow diurnal changes in humidity.

Surface moisture, particularly where it lay on the surface for extended periods, could lead to large increases in corrosion currents for bars with low cover in concrete with high water-cement ratios.

To reduce corrosion rates, structures should be designed with low water-cement ratios and high cover. Designs should avoid features which would allow the surface to remain wet. During the life of a structure, exposure to chlorides and moisture should be restricted as far as possible.

References

1.   C. L. Page and P. Lambert, 'Analytical and electrochemical investigation of reinforcement corrosion', Department of Transport TRRL Contractor Report 30, Transport Research Laboratory, Crowthorne, Berks, 1986.

2.   P. Schiessl and M. Raupach, 'Macrocell steel corrosion in concrete caused by chlorides', Proceedings of the Second Canmet/ACI International Conference on Durability of Concrete, August 4-9, Montreal, Canada, 1991.

DEFINITION OF EXPOSURE CLASSES AND CONCRETE MIX REQUIREMENTS FOR CHLORIDE CONTAMINATED ENVIRONMENTS

P B Bamforth

Taywood Engineering Limited
345 Ruislip Road
Southall
Middlesex UB1 2QX

1 INTRODUCTION

The corrosion of reinforcement in concrete is a world-wide problem. Its extent is evident by the size of the concrete repair market and the attention given to it by the international research community. This paper deals with corrosion caused by chloride ingress and, in particular, with defining the surface chloride level in relation to the exposure conditions.

Concrete is porous and when it is exposed to chlorides in solution a number of mechanisms operate. If the concrete is not fully saturated, chlorides are absorbed into the unfilled spaces by capillary action.[1,2] When the surface dries, the chloride remains and, with successive cycles, builds up in the surface layer until a limiting value is reached.[3] The time taken to achieve a stable surface level depends not only on the severity of chloride deposition but also on the prevailing climatic conditions. For example, wash-out can occur from surfaces exposed to rainfall and run off.

As chlorides migrate into the concrete, some are absorbed onto, and react with, the cement hydrates.[4] This slows down the rate of chloride migration.[5] In saturated concrete, diffusion is the principal transport process, but even in this case there may be a concentration of chlorides reaching levels which are two to four times that of the surrounding saline solution. This is due to a process defined as chloride 'condensation'[6] which has been attributed to physical adsorption of chlorides on to the pore surfaces by the electric double layer in the cement. Thus, to define a surface chloride level is, in itself, a complicated process. However there are four essential requirements of any values which are selected.

i)   They must be measurable with an acceptable level of reliability.
ii)   They must be in a form which is consistent with the assumed process of chloride migration and the predictive models which are used.
iii) They must be representative of values which are observed in practice.
iv)   For design purposes they should err on the side of caution and represent characteristically high values (e.g. upper 95% confidence limits).

An ideal model for chloride ingress would include the effects of absorption and evaporation, surface washout, chloride binding and diffusion, and be able to predict the free chloride in the pore solution at the depth of the reinforcement. While such a model is theoretically achievable and models currently exist which include some of these features,[2,7] engineers have tended to adopt a much more simplistic approach using the diffusion model empirically, with predictions based on the total chloride content. One advantage of this approach is the abundance of field data against which to validate the predictive process. It also provides a basis for sensitivity studies and for defining concrete performance requirements and cover in relation to design life.

2 THE CHLORIDE DIFFUSION MODEL

In structures, chloride ingress occurs under transient conditions. For one dimensional flow into a semi-infinite medium the error function solution to Fick's Second Law of Diffusion[8] is used as follows:

$$C_x = C_s \left[1 - \mathrm{erf} \frac{x}{2\sqrt{D_c t}} \right] \tag{1}$$

where D_c is the diffusion coefficient
          t is the time of exposure
          C_x is the concentration at depth, x, after time, t
          C_s is the surface concentration
          erf   is the error-function

If there are significant quantities of cast in chlorides then equation 1 can be modified to take this into account.[9] Using equation 1 a true value of D_c is only achieved if the concrete is homogeneous, D_c is constant, C_s is constant and the chlorides either do not react with the concrete or if the ratio of bound to fixed chlorides is constant. These conditions do not in apply in practice, but the author[10] has used the model empirically in order to utilise the enormous quantity of published data. Empirically derived values of (notional) surface chloride, C_{sn}, age dependent (effective) diffusion coefficient, D_{ce}, and threshold level, C_t, are used and it is within the context of such a model that surface chloride limits are discussed.

3 TYPICAL VALUES OF SURFACE CHLORIDE LEVEL

Substantial published data are available from laboratory tests, from natural exposure trials and from structures up to 60 years old and results for PC concretes are summarised in Table 1.[11-28] Sometimes the chloride profiles close to the surface are not entirely consistent with diffusion theory. For example, on exposed structures which are not permanently saturated, the profiles may be distinctly different after winter and summer exposure, as rainfall causes washout.[29] Thus, caution must be exercised in deriving values of C_{sn} from a single profile, at one age, only. Values should be obtained over a long period in order to define a reliable value.

Results obtained by the author from exposure trials on a UK splash zone site at Folkestone[30,31] are shown in Figure 1(a) for PC concretes. These are consistent with published data,[11-28] falling typically within the range 0.3 to 0.7% by concrete weight. However, two important features of the author's results should be highlighted.

Table 1   *A Review of Measured Values of Surface Chloride*

Ref no.	Location	Age (years)	No.of results	Surface Chloride (% conc) Range	Mean
	Structures				
11	Bridge deck in US	13	4	0.56 - 0.65	**0.60**
12	Tidal/splash zone, Singapore	24	5	0.18 - 0.43	**0.29**
13	Submerged, tidal and splash zones, Australia	14	9	0.08 - 0.36	**0.24**
14	Tidal, splash and spray zones, Denmark				
	Langeland bridge	15	12	0.01 - 0.37	**0.22**
	Stignaes harbour	11	1	0.24	**0.24**
		16	2	0.65 - 0.77	**0.71**
	Holsskov harbour	20	4	0.20 - 0.46	**0.32**
		16	2	0.15 - 0.17	**0.16**
15	Seawall,UK	30	1	0.11	**0.11**
16	Marine fort, UK, tidal and splash zones	34	5	0.18 - 0.71	**0.41**
17	Coastal bridge piers, Norway, splash zone, leeward surfaces	15	24	0.18 - 0.61	**0.32**
18	Coastal structures in Japan, splash zone	7.5	1	0.76	**0.76**
		17	6	0.08 - 0.25	**0.15**
		23	10	0.31 - 0.80	**0.44**
		24	1	0.15	**0.15**
		32	4	0.23 - 0.52	**0.38**
		32	4	0.21 - 0.27	**0.24**
		55	1	0.58	**0.58**
		58	5	0.13 - 0.37	**0.26**
		58	3	0.70 - 0.98	**0.82**
19	Marine discharge structure	30	10	0.04 - 0.6	**0.19**
	in Japan, splash and tidal	30	10	0.03 - 0.74	**0.31**
20	Coastal structures in Japan			0.20 - 1.00	**0.51**
	Natural exposure trials				
Author	Splash zone, UK	6	54	0.20 - 0.67	**0.42**
21	Tidal zone, UK	2	1	0.44	**0.44**
22	Coastal exposure, Norway	2	7	0.39 - 1.06	**0.62**
23	Tidal exposure in Japan	4	3	0.56 - 0.76	**0.64**
24	Tropical tidal, splash and atmospheric zone	1.5	6	0.20 - 0.50	**0.39**
25	Tidal exposure, UK	5	7	0.17 - 0.96	**0.54**
26	Marine exposure, Japan	6		0.15 - 0.82	**0.42**
	Laboratory trials using seawater				
27	UK laboratory	2	6	0.39 - 0.68	**0.53**
28	Australian laboratory	21 day	12	0.23 - 0.67	**0.42**
		1	12	0.35 - 0.65	**0.44**

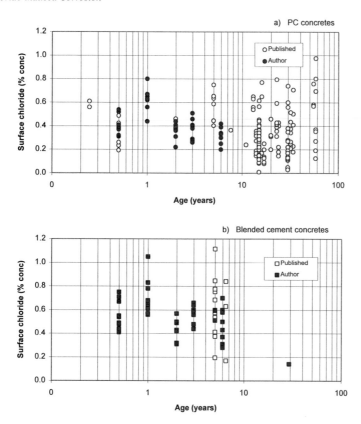

Figure 1 *The Variation of Csn with Time for PC and Blended Cement Concretes*

i)   While Csn varied with time, there was no significant tendency for values either to increase or decrease. Thus, after the first six months exposure it would be reasonable to assume a constant average value of Csn for predictive purposes. But the results also demonstrate the danger in deriving this average value from a single result.

ii)   There was a significant influence of the mix type. Concretes containing blended cements (PC with either 30% pulverised fuel ash, pfa, or 70% ground granulated blastfurnace slag, ggbs) consistently sustained higher values of surface chloride.

For PC concrete under natural exposure conditions the mean value of Csn was 0.36% wt of concrete, with the upper 95% confidence limit being 0.79%. The distribution of results, independent of age, is shown in Figure 2(a). In a survey of Japanese structures, Moringa[20] reported similar findings with 50% of values >0.35% Cl, 25% >0.51% Cl and 10% > 0.73% Cl.

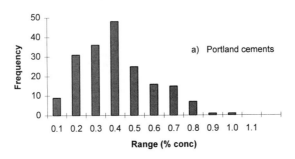

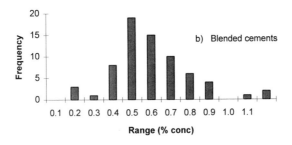

Figure 2   *Histograms Showing the Frequency Distribution of Values of C_{sn} for PC and Blended Cement Concretes*

4 INFLUENCE OF LOCATION

It is self evident that the proximity to the source of chloride must have a significant effect on chloride deposition. Results obtained from surveys of maritime structures are shown in Figure 3.[12-14,16,17] Although the data are scattered they clearly demonstrate that the splash zone is the most severe zone with regard to the accumulation of surface chlorides. The results also demonstrate the widely varying values of C_{sn} and the need to accommodate values up to about 0.8% wt of concrete.

Other factors such as the orientation of a particular surface, the direction of prevailing winds and the degree of exposure to rainfall affect the microclimate around a structure. For example a survey of the Hadsel Bridge in Norway[17] showed very high values of C_{sn} on leeward faces (4% wt of cement), while on the windward faces chloride levels were appreciably lower, (<0.6% wt of cement) due to wash-down by driving rain.

5 MIX DESIGN AND CURING

The influence of mix design on curing is complex. At the most simplistic level, it might be assumed that a surface which is more porous will achieve a higher chloride level and, to a degree, this is correct. The author's results after six months exposure[32] have shown a relationship between C_{sn} and sorptivity and, in tests on surface treatments, Pfiefer and Scali[33] reported proportionality between water absorption and chloride content for cylinders immersed in 15% NaCl solution. Frederiksen[34] has also related surface chloride levels to the open porosity of the cement paste. However, the author's study also revealed that extending curing caused an increase in C_{sn}[30] this being attributed to the increase in cement hydration leading to enhanced binding of chlorides at the surface. This is consistent with the work of Tang and Nilsson[35] who have related the level of chloride binding to the amount of C-S-H gel which has been produced.

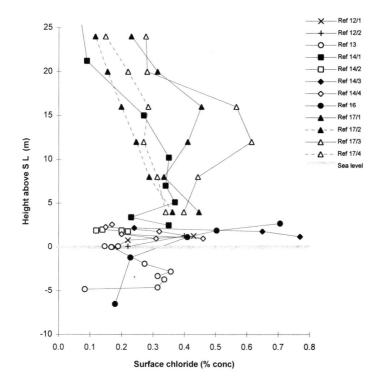

Figure 3 *Variations in Surface Chloride Levels in Relation to Sea Level*

This being the case, concretes with a higher cement content might be expected to exhibit a tendency for higher surface chloride levels. Based on limited published data[16,21,22,23,36] Bamforth and Price[30] suggested a relationship between C_{sn} and cement content. However, the inclusion of additional data[25,28,41,42] indicates that such a simple relationship is unlikely. Results are shown in Figure 4 for PC and blended cement concretes. While the suggested design curve[30] still represents a reasonable upper bound for PC concretes, many of the blended cement mixes achieved higher levels of surface chloride.

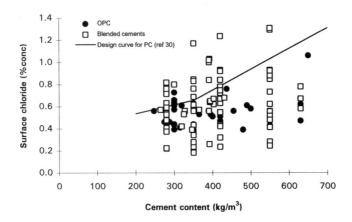

Figure 4   *The relationship between C_{sn} and Binder Content*

6 INFLUENCE OF CEMENT TYPES AND ADMIXTURES

In the author's study of splash zone conditions the higher surface levels observed for blended cement concretes (Figure 1(b)) were related to higher sorptivity at an early age. The average difference was about 15% over six years of exposure.[31] In a separate study, Cook et al[28] found higher values of C_{sn} with the use of GGBS. C_{sn} was dependent on the ratio of PC to GGBS with a peak value at between 40% and 60% GGBS. This was observed for three cement types and four levels of binder content from 280 kg/m³ to 550 kg/m³. In contrast, Thomas *et al*[21] observed a reduction in C_{sn} with increasing proportions of PFA. However, these results were obtained in the tidal zone where the concrete remains wet for long periods and is less prone to salt accumulation by evaporation.

Results for blended cements are shown in Figures 1(b) and 2(b).[15,21,22,25,27,28,29,31,36,38] The mean value for results from natural exposure trials and structures was 0.51% (wt of concrete) with an upper 95% confidence limit of 0.94%. Values are up to 15% higher than the values obtained for PC concretes and this is believed to be due to the enhanced binding characteristics, particularly for PFA and GGBS mixes. It is necessary, therefore, to take account of the cement type when predicting a value of C_{sn}.

The author also investigated the influences of a lignosulphate water reducing admixture and a calcium stearate integral waterproofer on levels of C_{sn}[30]. The former had no significant effect, while the waterproofer caused a reduction in C_{sn} of about 10%, this being consistent with the lower sorptivity of the waterproofed concrete. In a separate study by Price,[39] a high performance integral waterproofer was investigated which reduced the sorptivity by about 75%. However, the influence on C_{sn} was much less marked, being of the order of 20%. This again suggests that there are significant factors, in addition to sorptivity, which influence the surface chloride level.

7 PREDICTING THE LEVEL OF SURFACE CHLORIDE

There are clearly several factors which determine the level of surface chloride. These are related not only to the environment, but also to the concrete itself. It is, therefore, difficult to predict the likely value without specific knowledge of the concrete mix type and the microclimate around the structure. Until further data is available, and a much more thorough analysis is carried out on the results already in the public domain, it is therefore necessary to adopt values which are conservative and which minimise the risk of under design in relation to concrete quality and cover. For design purposes, the following exposure limits (by wt of concrete) are proposed (Table 2):

Table 2 *Suggested Exposure Classes and Surface Chloride Levels*

		PC	Blended Cement
Extreme exposure	C_{sn} =	>0.75%	>0.9%
Severe exposure	C_{sn} =	0.5% to 0.75%	0.6% to 0.9%
Moderate exposure	C_{sn} =	0.25% to 0.5%	0.3% to 0.6%
Mild exposure	C_{sn} =	< 0.25%	<0.30%

8 WHEN IS THE C_s VALUE CRITICAL

The time to corrosion activation is determined by the relationship between C_{sn}, D_{ce}, cover and the chloride threshold level. In order to identify in which combinations C_{sn} is critical in influencing service life a sensitivity study has been carried out. The following ranges of values have been investigated (Table 3):-

Table 3 *Range of Values used to Assess When C_s becomes Critical*

Surface chloride levels, C_{sn}	0.2% to 1.0% by concrete weight
Effective diffusion coefficients, D_{ce}	5 x 10^{-12} m²/sec to 1 x 10^{-13} m²/sec
Threshold levels, C_t	0.4% to 1.0% by cement weight
Cover	50 mm to 75 mm.

Some typical results are illustrated in Figure 5 for 50 mm cover, a threshold level of 0.4% by cement weight and a range of values of D_{ce}. Depending on the required time to activation, the conditions in which C_{sn} is influential may be defined. For example, for a 50 year period, and for concretes with values of D_{ce} greater than about 3 x 10^{-13} m²/sec differences in C_{sn} would demand changes in cover and concrete quality. For lower values of D_{ce}, while C_{sn} still influences time to activation, it will always be greater than the required 50 years. Clearly, as cover and threshold level increase, the quality of concrete for which C_{sn} is influential reduces. In the most optimistic case considered (75 mm cover and 1.0% by cement weight threshold), C_{sn} need only be considered when the value of D_{ce} is greater than about 1×10^{-12} m²/sec, unless a much longer design life is specified.

With many civil engineering structures being specified with design lives in excess of 100 years, the sensitivity study illustrates that C_{sn} will be an influencing factor and if durability is to be predicted based on the approach described herein, then levels of C_{sn} must be quantified.

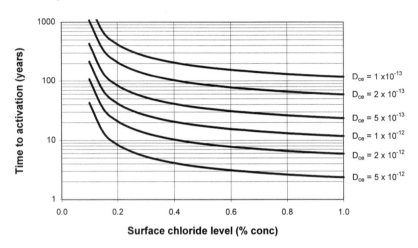

Figure 5   *The Relationship Between C_{sn}, D_{ce} and Time to Activation for 50 mm Cover and 0.4% (Cl by cement wt) Threshold Level*

9 RECOMMENDED MIX CLASSES

Based on the three levels of chloride exposure (0.25%, 0.5% and 0.75% wt of concrete) graphs have been produced to enable acceptable combinations of cover and concrete quality (in terms of D_{ce}) to be defined (Figure 6). Three service life periods have been selected based on the definitions medium, normal and long life in BS7543.[40] It is assumed in this case, that the design life is the time to corrosion activation. In each graph, two threshold levels have been assumed, 0.4% and 1.0% of cement weight. The analysis indicates that with reasonable cover depths (up to 75 mm) for normal and long life structures in severe and extreme exposure conditions, D_{ce} must not exceed 1×10^{-12} m²/sec and in many circumstances must be appreciably lower than this value.

In previous papers, the author has presented values of D_{ce} obtained from his own studies and from data in literature.[30,31,38] These have indicated that normal structural grade PC concretes cannot always be relied upon to achieve acceptably low values of D_{ce}, with values often significantly greater than 1×10^{-12} m²/sec. To achieve lower values very high grade, low w/c mixes are necessary, resulting in strengths which are likely to be much higher than required for structural purposes alone. Blended cement concretes, however, have achieved much lower values of D_{ce} in practice, thus avoiding the need to use excessively high strength grades.

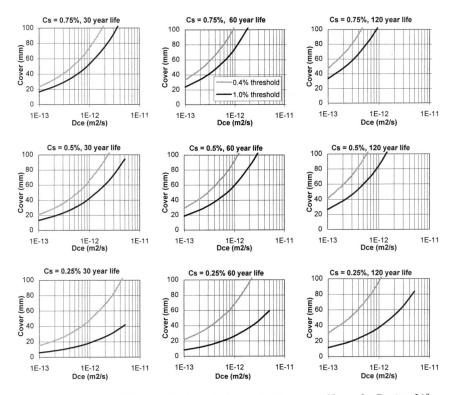

Figure 6 *Limiting Values of D$_{ce}$ in Relation to the Exposure Class, the Design Life, the Chloride Threshold Level and Cover to Reinforcement*

10 CONCLUSIONS

In order to predict service life and to select appropriate mix types it is necessary to be able to predict the exposure conditions in relation to the chloride level in the surface layer of concrete. Substantial data are available from structures which can be used to define appropriately safe levels for design purposes. However, to be more effective in design, further studies are needed to enable surface chloride levels to be defined with greater reliability. In estimating surface chloride levels it has been found that the concrete as well as the environment has an influence and that this is related to both the sorptivity and the binding capacity of the near surface layer. Without data on a specific environment and mix type, a conservative design value of 0.75% wt of concrete is proposed for PC concretes, and for mixes containing blends of PC with either pfa or ggbs this should be increased to 0.90%.

11 ACKNOWLEDGEMENTS

The author wishes to thank the Directors of Taywood Engineering Ltd for permission to publish this paper. The exposure trials were initiated under the EC BRITE programme and monitoring continued within the EC COST 509 project. The UK Department of the Environment are supporting long term measurements, up to ten years, under their Partners in Technology Programme.

12 REFERENCES

1. P. B. Bamforth and D. C. Pocock, 'Minimising the risk of chloride induced corrosion by selection of concreting materials', 'Proc. 3rd Int. Symp. on Corrosion of Reinforcement in Concrete', Society of Chemical Industry, London, 1990, 119.

2. W. R. Grace, 'Chloride penetration in marine concrete - a computer model for design and service life evaluation', Life Prediction of Corrodible Structures, NACE, Ed. R N Parkins, 1994, **1**, 527.

3. T. Oshiro, Y. Horizono, S. Tanigawa and K. Nagai, 'Experimental and analytical studies on penetration of chloride ions in concrete', *Transactions of the Japan Concrete Institute*, 1987, **9**, 155.

4. M. R. Roberts, 'Effects of calcium chloride on the durability of pretensioned of pretensioned wire in concrete', *Magazine of Concrete Research*, 1962, **14(42)**, 143.

5. W. J. McCarter, H. Ezerim and M. Emerson, 'Absorption of water and chloride into concrete', *Magazine of Concrete Research*, 1992, **44***(158)*, 31.

6. S. Nagataki, N. Otsuki, T. H. Wee and K. Nakashita, 'Condensation of chloride ions in hardened cement matrix materials and on embedded steel bars' *ACI Materials Journal*, 1993, **90(4)**, 323.

7. M. Massat, L-O. Nilsson and J-P. Olliver, 'A clarification of the fundamental relationship concerning ion diffusion in porous materials', 'Proc. 3rd Coll. on Concrete'. Technical Academy, Esslingen, 1992, 177.

8. J. Crank, 'The mathematics of diffusion', Second Edition, Oxford Science Publications, Oxford, 1983.

9. S. Chatterji, 'Transport of ions through cement based materials. Part 1 - Fundamental equations and basic measurement techniques', *Cement and Concrete Research*, 1994, **24(4)**, 907.

10. P. B. Bamforth, 'Prediction of the onset of reinforcement corrosion due to chloride ingress', 'Conference on Concrete Across Borders', Danish Concrete Association, Odense, 1994, **1**, 397.

11. M. Funahashi, 'Predicting corrosion-free service life of a concrete structure in a chloride environment', *ACI Materials Journal*, **87** No. 6, 1990, 581.

12. K. C. Liam, S. K. Roy and D. O. Northwood, 'Chloride ingress measurements and corrosion potential mapping of a 24 year old reinforced concrete jetty structure in a tropic marine environment', *Magazine of Concrete Research*, 1992, **44(160)**, 205.

13. J. T. Gourley and D. T. Bieniak, 'Diffusion of chloride into reinforced concrete piles. Proceeding of Symposium on Concrete: The Material for Tomorrow's Demands', The Institution of Civil Engineers, Australia, 1983, 41.

14. B. Sorensen and E. Maahn, 'Penetration of chloride in marine concrete structures', 'Nordic Concrete Research, Publication No. 1', Nordic Concrete Federation, Oslo, 1982, 24.1.

15. M. D. A. Thomas, 'A comparison of the properties of OPC and PFA concrete in 30 year old mass concrete structures', 'Proc. of 5th Int. Conf. on Durability of Building Materials and Components', Brighton, E&F N Spon, 1990, 383.

16. Taylor Woodrow Research Laboratories, 'Concrete with Oceans - Marine durability of the Tongue Sands Tower', CIRIA UEG Technical Report No. 5.

17. B. T. Sand, 'The effect of the environmental load on chloride penetration', 'Nordic Miniseminar on Chloride Penetration into Concrete Structures', Chalmers Uuniversity, Goteberg, Sweden, 1993, 113.

18. K. Uji, Y. Matsuoka and Maruyat, 'Formulation of an equation for surface chloride content of concrete due to permeation of chloride', 'Corrosion of Reinforced Concrete', SCI, Elsevier Applied Science, London, 1990, 258.

19. S. Morinaga, K. Irino, T. Ohta and H. Arai, 'Life prediction of existing reinforced concrete structures determined by corrosion', 'Proc. of Conference on Corrosion Protection of Steel in Concrete', Sheffield University Press, Sheffield, 1994, 603.

20. S. Morinaga, 'Life prediction of reinforced concrete structures in hot and salt-laden environments', 'RILEM Conference on Concrete in Hot Climates', (Ed M. J. Walker), E & F N Spon, London, 1992.

21. M. D. A. Thomas, J. D. Matthews and C. A. Haynes, 'Chloride diffusion and reinforcement corrosion in marine exposed concretes containing PFA', 'Proc. 3rd Int. Symp. on Corrosion of Reinforcement in Concrete', SCI, Elsevier Applied Science, London, 1990, 198.

22. O. E. Gjorv and Θ. Vennesland, 'Diffusion of chloride ions from seawater into concrete', *Cement and Concrete Research*, 1979, **9**, 229.

23. M. Makita, Y. Mori and K. Katawaki, 'Marine corrosion behaviour of reinforced concrete at Tokyo Bay', 'American Concrete Institute SP65', 271.

24. S. K. Roy, L. K. Chye and D. O. Northwood, 'Chloride ingress in concrete as measured by field exposure tests in the atmospheric, tidal and submerged zones of a tropical marine environment', *Cement and Concrete Research*, 1993, **23**, 1289.

25. J. D. Matthews, 'Performance of limestone filler cement concrete', 'Euro-cement: Impact of ENV207 on Concrete Construction', (Ed Dhir and Jones), E&FN Spon, 1994, 113.

26. S. Sakoda, N. Takeda, and S. Sogo, 'Influence of various cement types on concrete durability in marine environment', '9th International Congress on Chemistry of Cement', New Delhi, India, 1992, **VI**, 175.

27. Taylor Woodrow Research Laboratories, 'Concrete in the Oceans - Phase II - Performance of uncracked cover in protecting reinforcement embedded in marine concrete against corrosion', Project P16 Final Report, CIRIA, June 1986.

28. D. J. Cook, I. Hinczak, M. Jedy and H. T. Cao, 'The behaviour of slag cement concretes in marine environment chloride ion penetration' 'SP-114, Fly Ash,

Silica Fume, Slag and Natural Pozzolans in Concrete, Proc. of 3rd Int. Conf', ACI, Detroit, 1989, **2**, 1467.

29. A. V. Saetta, R. V. Scotta and R. V. Vitalian, 'Analysis of chloride diffusion into partially saturated concrete', *ACI Materials Journal*, 1993, 441.

30. P. B. Bamforth and W. F. Price, 'Chloride ingress into marine structures: the effects of location, materials, curing and formwork', 'Concrete 2000, Economic and Durable Construction through Excellence', Dundee, 1993, 1105.

31. P. B. Bamforth and J. Chapman-Andrews, 'Long term performance of r.c. elements under UK coastal exposure conditions', 'Proc. of Conference on Corrosion Protection of Steel in Concrete', Sheffield University Press, Sheffield, 1994, 139.

32. D. C. Pocock and P. B. Bamforth, 'Cost effective cures for reinforcement corrosion - collaboration research brings new developments', *Concrete Maintenance and Repair*, 1991, **5(6)**, 7.

33. F. W. Pfiefer and M. J. Scali, 'Concrete Sealers for Protection of Bridge Structures', NCHRP Report 255, Americal 'Association of State Highway and Transportation Officials, Washington, US, 1991.

34. J. M. Frederiksen, 'Chloride binding in concrete surfaces', 'Nordic Mini Seminar on Chlorides Pentration into Concrete Structures', Chalmers University, Goetberg, Sweden, 1993, 90.

35. L. Tang and L-O. Nilsson, 'Chloride binding capacity and binding isotherms of OPC pastes and mortars', *Cement and Concrete Research*, 1993, **23**, 247.

36. S. L. Marusin, 'Chloride ion penetration in conventional concrete and concrete containing condensed silica fume', 'Proceedings of 2nd Int. Conf. on Fly Ash, Silica Fume, Slag and Natural Pozzolans in Concrete', Madrid, 1986, ACI SP-91, **2**, 1119.

37. J. M. Bijen, 'Durability aspects of the King Fahd Causeway', 'RILEM Conference on Concrete in Hot Climates', (Ed M. J. Walker), E & F N Spon, London, 1992.

38. P. B. Bamforth, 'Concrete classification for r.c. structures exposed to marine and other salt laden environments', 'Structural Faults and Repair 93', Edinburgh, 1993, **II**, 31.

39. W. F. Price, 'A comparison of heat cured PFA concretes with and without the Everdure Caltite System', Taywood Engineering Limited, Research Report No. 014H/88/3705, 1988.

40. British Standards Institution, 'Guide to Durability of buildings and building elements, products and component', BS7543:1992, BSI, 2 Park Street, London, W1A 2BS.

41. R. Polder and L. J. Larbi, 'Investigation of concrete exposed to North Sea water submersion for 16 years', TNO Building and Construction Institute, Report No. 93-BT-R0619-02, 1993.

42. M. A. Mustafa and K. M. Yusok, 'Atmospheric chloride penetration into concrete in semi-tropic marine environment', *Cement and Concrete Research*, 1994, **24(4)**, 661.

MISCELLANEOUS EFFECTS

STUDIES OF CARBONATION AND REINFORCEMENT CORROSION IN HIGH ALUMINA CEMENT CONCRETE

A M Dunster, D J Bigland, K Hollinshead and N J Crammond

Building Research Establishment,
Bucknalls Lane, Garston, Watford,
WD2 7JR, UK

1 INTRODUCTION

High alumina cement (HAC) (also known as calcium aluminate cement) concrete was used extensively in the UK from the 1950s to the early 1970s in the manufacture of pre-cast, pre-stressed concrete beams. After a few well publicised collapses in the early 1970s, HAC has not been recommended for use in structural concrete in the UK. As a consequence, all structural HAC concrete in the UK is now more than 20 years old. With the passage of time longer term durability problems have now begun to affect HAC concrete and the most important of these is reinforcement corrosion due to carbonation of concrete cover.[1]

In HAC concrete, the cement hydration products (principally CAH_{10} and C_3AH_6), react with atmospheric carbon dioxide in a process known as *carbonation*. Once the depth of carbonation exceeds the depth of the reinforcing bars, it appears that the steel is no longer protected from corrosion.[2] HAC concrete is thus beginning to show the same symptoms of reinforcement corrosion already evident in many Portland cement concretes. A programme of research has been initiated at the Building Research Establishment to provide data on the rates and mechanisms which control carbonation and reinforcement corrosion in HAC concrete.

Cubes reinforced with prestressing wire and unreinforced prisms have been cast with two cement contents and two different curing regimes and then exposed to accelerated carbonation or natural carbonation in internal or outdoor environments. Carbonation depth was determined periodically using pH indicators and petrography (optical microscopy) and the condition of the steel was monitored electrochemically.

2 EXPERIMENTAL

2.1 Casting, Curing and Exposure of Concrete Specimens

Details of the mixes are given in Table 1. Mix H400 (cement content 400 kg/m³) was designed to comply with the current guidelines for mix design using HAC.[3] Mix H250 (cement content 250 kg/m³) was formulated to resemble a typical HAC concrete used by the pre-casting industry prior to 1974.

The specimens cast were: prisms (100 mm x 100 mm x 250 mm) for depth of carbonation; discs (50 mm x 150 mm diameter) for oxygen permeability and cubes (100 mm, reinforced with four bars nominally at 20 mm and 30 mm depth) for corrosion rate measurements.

Table 1 *Details of concrete mixes*

Mix No	Curing*	Cement content (kg/m³)	Total w/c	Free w/c	Degree of conversion**
H250	AC	250	0.60	0.55	medium
	WC				high
H400	AC	400	0.40	0.37	medium
	WC				high

* for details of curing see text
** for details, see ref 4.

The specimens were seal cured for four hours. They were then demoulded and further moist cured in a fog room for 20 hours. After 24 hours, the cube and prism specimens were divided into two groups which were given ambient or high temperature water curing respectively. The ambient cured (AC) group were conditioned in the laboratory for seven days before exposure. The high temperature cured (WC) specimens were immersed in water (38°C) for five weeks followed by seven days conditioning in the laboratory as above. Curing of the disc specimens is described in Section 2.3.

Specimens for carbonation and long term corrosion measurements were exposed in one of the following three natural environments: external (exposed); external (sheltered from rain); internal (20°C, 65 % RH). The degree of conversion of companion cubes (unreinforced) was determined at the time of exposure and these values were assigned to one of the categories (high, medium or low conversion) given in BRE current paper CP34/75[4] (Table 1).

In order to obtain corrosion rate data at an early stage in the programme, a number of the reinforced cubes and companion carbonation prisms were carbonated in an accelerated carbonation chamber maintained at 20°C, 65% RH and 10% CO_2 by volume. When carbonation tests indicated that carbonation had penetrated to the depth of the steel, the reinforced cubes were transferred to one of the following environments: 65% RH, 78% RH, 95% RH (all 20°C).

2.2 Carbonation Depth Determinations

The carbonation depth of the prism specimens was determined periodically using the pH indicator phenolphthalein and petrography (optical microscopy). The monitoring intervals for the carbonation prisms were 24 weeks, 52 weeks, 104 weeks and monitoring will be continued to 10 years. Slices (approx 30 mm thick) were broken from the prism specimens perpendicular to their exposed surfaces using a rock splitter. The freshly broken surface was immediately sprayed with a fine mist of indicator solution and the depth of the carbonation front measured one hour after spraying.

There is evidence that the phenolphthalein test for assessing the depth of carbonation in Portland cement concrete[5] can be unreliable for ageing HAC concrete and consequently, optical microscopy was used to confirm the depths of carbonation determined using phenolphthalein. Lump samples were impregnated with Araldite, sliced perpendicular to the exposed surface and made into thin sections 30 microns thick. The thin sections were examined using a petrological (optical) microscope.

2.3 Oxygen Permeability Measurements

Oxygen permeability measurements were carried out using a gas permeability cell developed by Lawrence.[6] After 24 hours moist curing as described in Section 2.1, the specimens were conditioned in a nitrogen atmosphere (65% RH, 20°C) to avoid carbonation for 8 weeks before testing.

2.4 Corrosion Measurements

The steel reinforcement was monitored using the linear polarisation resistance (LPR) technique with IR compensation for the resistance of the concrete. Corrosion current, corrosion potential and concrete surface resistance (from IR compensation) were recorded. The results presented in this paper are for the specimens from accelerated carbonation. Specimens from natural exposure were also studied although there are currently too few data to make meaningful comparisons.

3 RESULTS

3.1 Natural Carbonation and Oxygen Permeability

The natural carbonation and oxygen permeability results presented here are for the ambient cured specimens. Table 2 shows carbonation depth results for concrete specimens exposed internally and externally for a period of two years determined using both phenolphthalein indicator and thin section examination. Although the thin section method indicated a depth of carbonation 1 mm to 2 mm lower than the indicator in some cases, there was a good correspondence between carbonation depths measured using the two methods.

For both mixes, the influence of exposure regime on carbonation depth (in order of increasing depth) was: internal > external (sheltered) > external (fully exposed). For the specimens stored internally, there were not large differences between the carbonation depths of the H250 and H400 mixes (11 mm and 9 mm at 2 years respectively). However, the influence of exposure environment on carbonation depth became more marked at the higher concrete grade. H250 specimens stored externally (sheltered) carbonated approximately 30% less (at 2 years) than specimens of the same concrete stored internally. For specimens fully exposed, the carbonation depth was 50% less than for internally stored specimens. For H400 specimens, the corresponding values were approximately 40% and 83% less than the carbonation depth for internal exposure. For internal exposure, increasing the cement content of the concrete from 250 kg/m^3 to 400 kg/m^3 reduced the carbonation depth at 2 years by only 3 mm (ie. 27%).

Unlike Portland cement concretes,[7] the depth of carbonation for the HAC concretes did not increase proportionally with the square root of time and appeared to level off between 52 weeks and 104 weeks. For example, the internally exposed H400 specimen carbonated to a depth of 7 mm to 8 mm in the first 52 weeks but subsequently carbonated by only a further 1 mm to 2 mm between 52 weeks and 104 weeks. This effect was observed for both mixes and

Table 2 *Carbonation results from indicator and thin section (ambient cured [AC])*

Mix	Storage	Test method (indicator or thin section)	Exposure duration (weeks)		
			24	52	104
H250	Internal		carbonation depth (millimetres)		
		I (TS)	3.0 (3.0)	10.0 (7.5)	11.0 (11.5)
	External (sheltered)	I (TS)	nd (nd)	8.5 (7.0)	9.0 (10.0)
	External (exposed)	I (TS)	nd (nd)	5.0 (5.0)	5.5 (6.0)
H400	Internal	I (TS)	4.5 (0.5)	8.0 (7.0)	9.0 (9.0)
	External (sheltered)	I (TS)	nd (nd)	6.0 (5.0)	5.5 (5.5)
	External (exposed)	I (TS)	nd (nd)	1.5 (1.5)	1.5 (2.0)

nd = not determined

a similar effect was found with the high temperature cured specimens (data not given here). Table 3 shows carbonation data for internal exposure and corresponding oxygen permeability results together with data for identically cured Portland cement concretes of similar mix designs to the HAC concretes (after Thomas).[8] The data show that, for a given carbonation depth, the oxygen permeability of the HAC concretes is approximately an order of magnitude lower than for the Portland cement concretes. Although there are some differences between the mix designs of the HAC and Portland cement concretes shown in Table 2, the carbonation depths at 2 years were broadly similar for both types of concrete with comparable cement contents and w/c ratio.

3.2 Electrochemical Data

The results presented here are from specimens with a cement content of 250 kg/m^3 carbonated at 10% CO_2, as these were the first to carbonate to the depth of the steel. Further data from the study will be published once available.

Figures 1 and 2 show the changes observed in corrosion potential (vs calomel) and corrosion current during carbonation and subsequent exposure of the carbonated specimens to relative humidities of 65%, 78% and 95%. These data are from bars with 20 mm cover in ambient cured concrete and are typical of the other data collected. Figure 3 shows the variation of corrosion current with humidity at the two depths of cover and for both curing regimes after 8 months (AC) and 4 months (WC). Figure 4 shows a similar plot for the

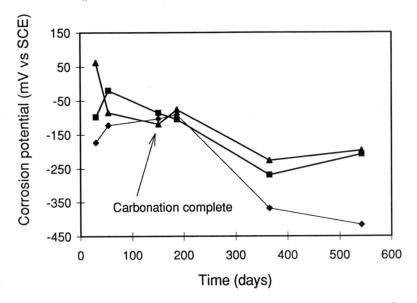

Figure 1 *Corrosion potential in 10% CO_2 carbonated specimens, 250 kg/m³, ambient cure at 20 mm cover. (Relative humidity after carbonation: ♦ = 95%, ■ = 78%, ▲ = 65%).*

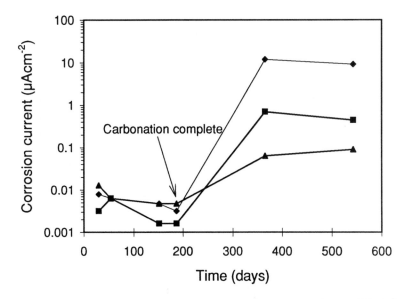

Figure 2 *Corrosion current in 10% CO_2 carbonated specimens, 250 kg/m³, ambient cure at 20 mm cover. (Relative humidity after carbonation: ♦ = 95%, ■ = 78%, ▲ = 65%).*

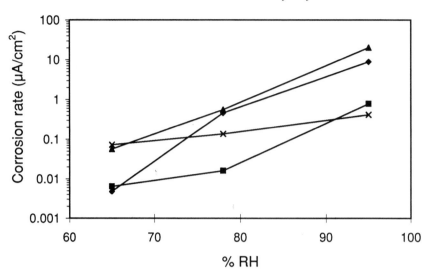

Figure 3 *Relationship between corrosion current and humidity in 10% CO₂ carbonated specimens, 250kg/m³. (Curing and depths of cover: ♦ = AC 20 mm, ■ = AC 30 mm, ▲= WC 20 mm, ✗ = WC 30 mm).*

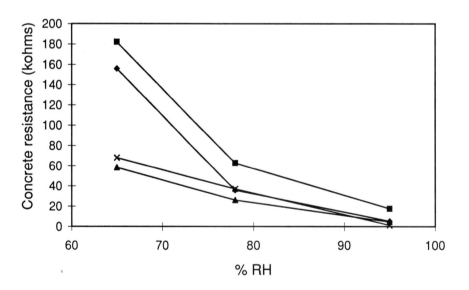

Figure 4 *Concrete surface resistance data for 10 % CO₂ carbonated specimens, 250 kg/m³. (Curing and depths of cover: ♦ = AC 20 mm, ■ = AC 30 mm, ▲= WC 20 mm, ✗ = WC 30 mm).*

Table 3 *Carbonation depth and oxygen permeability results*

Concrete type	Mix no	w/c ratio total and (free)	Internal carbonation depth (mm)		Oxygen permeability (x 10⁻¹⁷ m²)
			1 year	2 years	
HAC	H250	0.60 (0.55)	10.0	11.0	5.84
	H400	0.40 (0.37)	8.0	9.0	1.55
Portland cement*	P350	0.54 (0.49)	6.0	8.0	43.7
	P300	0.63 (0.57)	9.0	11.5	47.5
	P250	0.76 (0.68)	12.0	16.0	65.2

* Data for Portland cements from ref (8).

concrete surface resistance determined during IR compensation.

4 DISCUSSION

4.1 Carbonation and Oxygen Permeability

The data presented here have shown that the depth of carbonation for the natural exposed HAC concretes in three different exposure environments levels off after one year. This may be associated with carbonation reducing the permeability of the concrete surface which may help to protect the interior of what is already a low permeability concrete, from subsequent carbonation. The reduction in porosity and permeability which accompanies the carbonation of Portland cement concretes is well established.[7]

In spite of its low gas permeability, HAC concrete gives similar carbonation depths at 104 weeks to Portland cement concretes of comparable w/c ratio and cement content. Carbonation in HAC involves reaction of carbon dioxide with the calcium aluminate hydrates (which constitute approximately 80% by mass of a mature HAC paste),[9] to form calcium carbonate and aluminium hydroxide. HAC has a lower CaO content than Portland cements (approx 37% compared with approx 64%) and the hydrates have lower CaO contents than the main hydrate phases in Portland cements. The contrast between the CaO content (and hence the carbonate binding ability) of these two types of cement may partly account for the observed influence of permeability on carbonation depth.

Studies by BRE of pre-cast HAC "I" beams in existing buildings (unpublished) have shown that the depth of cover to the steel may be as little as 15 mm. Prediction of long term carbonation from the carbonation results given in this paper is not straightforward and such predictions can be crude. However, if the trend established in the laboratory specimens between 52 weeks and 104 weeks were to continue, carbonation would have reached the steel in such beams within the first ten years.

Phenolphthalein indicator has proved an effective means for determining the carbonation depth in young, ambient cured HAC concretes (up to 2 years age). However, for more highly converted concretes (both young and ageing), the visible indication is often indistinct and fades over time. Phenolphthalein is therefore not recommended for determining carbonation depth in ageing HAC concretes.

4.2 Electrochemical Measurements

The electrochemical data collected from reinforced HAC concrete specimens appear to be similar to those generally observed in Portland cement concrete. Active corrosion is considered to be occurring at corrosion currents greater than 0.1 to 0.2 μAcm^{-2}.[10] Therefore, Figure 3 indicates that the bars at 20 mm cover were actively corroding at 95% and 78% RH and at 95% RH the bars at 30 mm cover were also corroding. At 78% RH the rate of corrosion was an order of magnitude less than at 95% RH. At 65% RH none of the bars showed active corrosion. The currents observed at 95% RH for the ambient cured specimens were an order of magnitude less than for water cured specimens. This behaviour is also reflected in Figure 4 where the surface resistance of the ambient cured specimens at 65% RH is approximately three times that of water cured specimens. This may be due to the water cured specimens needing a longer drying period than the air cured to reach the same moisture condition. However, the resistances of both ambient and water cured specimens had stabilised by the time they were completely carbonated. Figures 3 and 4 also illustrate the strong relationship between the availability of moisture, concrete resistance and the rate of corrosion.

Broomfield et al[11] class rates of corrosion above 1 μAcm^{-2} (equivalent to 11.6 μm per year) as high. The amount of corrosion required to cause cracking depends on a large number of factors and reported values range from 5 to 100 μm. Therefore, the time to cracking at the rate of corrosion observed for 20 mm cover at 95% RH is very uncertain. From these values it could range from under a year to 10 years. At the lower rate of corrosion the time to cracking may not be reduced as it is possible that less corrosion is required to cause cracking at lower rates of corrosion.[12]

5 FURTHER WORK

This paper presents early data from a ten year research programme. Monitoring of corrosion, permeability and carbonation following natural exposure of specimens from mixes H250, H400 and a poor quality mix (206 kg/m^3 cement content, total w/c 0.8) is ongoing. Studies using scanning electron microscopy are planned and these will provide mechanistic data on the carbonation processes.

6 CONCLUSIONS

For young HAC concrete, there is a good correspondence between carbonation depth measured using pH indicator and that determined by thin section.

Carbonation depths at two years for the HAC concretes studied are broadly comparable to those for Portland cement concretes with similar curing, w/c ratio and cement content.

Carbonation rates for the HAC concretes studied were not proportional to the square root of exposure time, levelling off after one year.

For a given carbonation depth, the oxygen permeability of the HAC concretes is approximately an order of magnitude lower than for the Portland cement concretes.

Active corrosion was observed in carbonated HAC concrete for bars with 20 mm cover at both 78% and 95% RH and also for bars with 30 mm cover at 95% RH. At 65% RH (both depths of cover) and at 78% RH (for 30 mm cover) the bars were either not corroding or were borderline between corroding and not corroding.

Clear relationships between corrosion current and RH, and between surface resistance and RH were observed.

High rates of corrosion were observed at 95% RH, which if observed in practice, could cause cracking in under 10 years, although the propagation time could be considerably shorter for the lower corrosion rates which occur at lower relative humidities.

References

1.   N. J. Crammond and R. J. Currie, *Mag. Concr. Res.*, 1993, **45**, 275.
2.   R. J. Currie and N. J. Crammond, *Proc. Instn. Civ. Engnrs. Structs and Bldgs,* 1994, **104**, 83.
3.   Ciment Fondu Lafarge, Guide for use CA4 890, Lafarge Special Cements.
4.   Building Research Establishment, High Alumina Cement concrete in buildings, BRE Current Paper CP34/75, 1975, 13.
5.   Building Research Establishment, Carbonation of concrete and its influence on durability, BRE Digest 405, 1995.
6.   C. D. Lawrence, *Proc. 8th Int. Cong. Chem. Cements, Rio de Janeiro*, 1986, **5**, 29.
7.   L. J. Parrott. A review of carbonation in reinforced concrete. BRE/C&CA report, 1987.
8.   M. D. A. Thomas and J. D. Matthews, Durability of pfa concrete, BRE Report 216, Garston, BRE, 1994.
9.   A. J. Majumdar, B. Singh and R. N. Edmonds, *Cem. Concr. Res.,* 1990, **20**, 197.
10.   C. Andrade, C. Alonso and J. A. Gonzalez, Corrosion Rates of Steel in "Concrete", ASTM STP 1065, (Eds: N. S. Berke, V. Chaker and D. Whiting), ASTM, Philadelphia, 1990, 29.
11.   J. P. Broomfield, J. Rodriguez, L. Mortega and A. M. Garcia, *Proc. Structural Faults and Repair Conference,* 1993, **2**, 155.
12.   C. Andrade, C. Alonso, F. J. Molina, *Materials and Structures*, 1993, **26,** 453.

RISKS OF FAILURE IN PRESTRESSED CONCRETE STRUCTURES DUE TO STRESS CORROSION CRACKING

J. Mietz and B. Isecke

Federal Institute for Materials Research and Testing (BAM)
Unter den Eichen 87
D-12205 Berlin
GERMANY

1 INTRODUCTION

In recent years failures of prestressed concrete structures more than 30 years old with post-tensioned steel members were observed. The fracture appearance shows characteristic signs of hydrogen-induced stress corrosion cracking on the microscopic and macroscopic scale. Figure 1 shows a failed prestressed beam which led to the collapse of the roof of a production hall. Defects related to non-injected ducts or the presence of corrosion inducing substances within the grout which are usually the cause of such failures could not be detected.[1-3] The prestressing wires involved can be classified as 'old type' quenched and tempered steel, strength class St 1420/1570. Steel of this composition had official approval only until the early 1960s. This steel type had shown different problems shortly after or even prior to stressing, and in 1965 the steel composition was modified resulting in an alloy with a substantially lower susceptibility to stress corrosion cracking.

Figure 1   *Failed prestressed beam*

In order to explain the failure mechanism and to estimate future risks of older prestressed concrete structures containing the 'old type' of steel, several research projects were carried out by the Federal Institute for Materials Research and Testing (BAM) in

Berlin and the Research and Materials Testing Institute (FMPA) in Stuttgart.[4,5] The projects deal with investigations on existing structures as well as laboratory work which should yield information as to whether corrosion-induced crack initiation and propagation can take place in quenched and tempered high strength steel of the 'old type' in sufficiently grouted ducts over long service times. Furthermore, our understanding of the corrosion risk to quenched and tempered steel with pre-tensioning under depassivating conditions due to imperfections or carbonation of the concrete should be improved.

2 HYDROGEN INDUCED STRESS CORROSION CRACKING OF PRESTRESSING STEELS

Stress corrosion cracking is the crack initiation and propagation in the presence of certain corrosion media under static tensile stress. In some cases residual stresses within the material are sufficient to initiate this form of corrosion. Contrary to most other manifestations of corrosion, the failure, due to a sudden brittle fracture, often occurs without the formation of detectable corrosion products and is not necessarily characterized by any previously visible damage.

Hydrogen-induced stress corrosion cracking is caused when hydrogen infiltration leads to an embrittlement of the material. For high-strength steels this type of corrosion does not require a specific medium. Only enough atomic hydrogen capable for adsorption is necessary. Atomic hydrogen can be produced by the cathodic reaction of the corrosion process in acid or neutral media. The hydrogen reduction reaction can be split into different steps. In the first step protons are discharged at the phase boundary:

$$H^+ + e^- \rightarrow H_{ad} \tag{1}$$

H-atoms adsorbed at the metal surface recombine in a second step according to the Tafel-equation:

$$H_{ad} + H_{ad} \rightarrow H_2 \tag{2}$$

or the Heyrovsky-reaction

$$H_{ad} + H^+ + e^- \rightarrow H_2 \tag{3}$$

H_2-molecules formed by this reaction can escape to the atmosphere. With kinetic inhibition of the recombination adsorbed hydrogen atoms build up near the surface and can lead to high H-activities resulting in the penetration of hydrogen into the metal:

$$H_{ad} \rightarrow H_{ab} \tag{4}$$

An equilibrium exists between the concentration of adsorbed hydrogen at the surface and dissolved H-atoms in the metal matrix. With an increasing degree of coverage by adsorbed atomic hydrogen the probability of H-absorption increases. The hydrogen uptake will be substantially increased by promoting agents which hinder the recombination to hydrogen molecules. Hydrogen induced stress corrosion cracking of high-strength steels of the quenched and tempered type is characterized by cracks with intercrystalline corrosion along the former austenite grain boundaries.

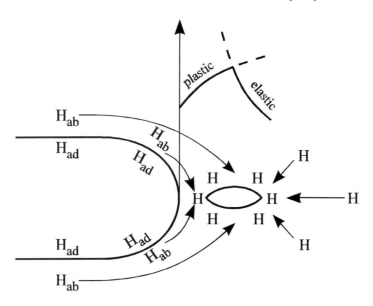

Figure 2   *Mechanism of crack propagation according to decohesion theory* [8]

Among the different hypotheses put forward to explain hydrogen-induced embrittlement the decohesion theory (Figure 2) is likely to be the most relevant for high-strength prestressing steels.[6,7]

According to this theory crack propagation results from an interaction of absorbed hydrogen with the base metal in areas in front of the crack tip where plastically deformed zones can be generated due to the triaxial state of stress. The crack propagation develops stepwise. For each propagation step enough hydrogen must be accumulated through diffusion into the region of the highest triaxial state of stress which occurs at the transition from the plastically deformed zone at the crack tip to the elastically stressed center. This forms an additional crack which then grows and coalesces with the initial crack.

3   RISK ASSESSMENT OF EXISTING STRUCTURES

Prior to the examination of different structures existing literature and reports were investigated and relevant data were analysed and evaluated. Up until 1980 there were three extensive sources on the behaviour of prestressing steel under practical conditions.[9-11] To obtain more recent information numerous institutions, administrations, and companies were asked for their assistance. In total, reports on 23 structures containing prestressing steel of the quenched and tempered 'old type' strength class St 1420/1570 were received. Table 1 gives a summary of the steel types and the causes of investigation.

Fracture prior to stressing in the non-grouted ducts was caused by corrosion attack initiated or accelerated by improper storage or the presence of corrosion stimulating substances such as chlorides. The lifetime of the other 18 structures was between 11 and 32 years at the time of the investigation.

Table 1   *Evaluation of external reports*

Cause of Investigation	Number			
		Steel Type		
	Total	Neptun	Sigma	No Data
Damage prior to grouting	5	1	1	3
Chloride induced damage	6	-	6	-
Planned demolition	8	1	7	-
Inspection due to actual problems	4	4	-	-
Sum	**23**	6	14	3

Only in one case were signs of hydrogen-induced stress corrosion cracking found. The structure was a bridge which was demolished due to extension of road width. The prestressing steel was Neptun N40. Cracks and intercrystalline corrosion were detected by metallographic investigations, and fracture analysis by scanning electron microscopy, after tensile testing showed low values of reduction in area. The tensile strength was rather high with values between 1654 and 1779 N/mm^2.

Our own investigations of structures were carried out as part of planned demolitions, rehabilitation measures, damage analyses or precautionary inspections. The main aim was to assess the prestressing members, i.e. prestressing steel, quality of injection, and properties of the injection grout. Generally, the following parameters were investigated:

Prestressing member: Corrosion condition of the duct, injection condition
Prestressing steel: Macroscopic assessment, chemical analysis, crack detection, metallographic examination, tensile testing, residual stress measurements
Injection mortar: Chloride-, sulphate-, and water-content, degree of water saturation, open porosity, sulphide- and nitride-content.

In total, investigations were carried out on 21 structures. Table 2 gives information on the different steel types and the causes of the investigation.

The lifetime of the structures was between 30 and 41 years at the time of the investigation. Three buildings showed chloride-induced damage of single prestressing steel wires. In these cases fractures were only caused by reduction of cross section and subsequently exceeding of the residual strength. Even in areas with deeper corrosion pits no cracks could be found.

Table 2   *Overview of our own investigations*

Cause of Investigation	Number			
		Steel Type		
	Total	Neptun	Sigma	Henningsdorf
(Partial) collapse	2	2	-	-
Chloride induced damage	3	-	3	-
Planned demolition	8	3	5	
Inspection due to actual problems	8	2	5	1
Sum	**21**	7	13	1

In four structures damage in the form of wire fractures due to hydrogen-induced stress corrosion cracking were observed. In one case this resulted in the total collapse of a roof beam; in a further case a partial collapse occurred. For all these structures Neptun steel was used. Figure 3 shows an opened duct in such a building. Several wire fractures can be observed without any of the distinctive features of bad injection conditions or corrosion attack. On the macroscopic scale the fracture surfaces show dark-coloured incipient crack areas (Figure 4). Under scanning electron microscopy these areas show intercrystalline corrosion which is characteristic of hydrogen-induced stress corrosion cracking of high-strength steel of the quenched and tempered type (Figure 5). The magnetic particle test indicated several further cracks (Figure 6), some of them having considerable crack depths (Figure 7). The investigations have shown that cracks and fractures can be expected along the whole length of the prestressing member. Indications of a particular susceptibility are given by the high values of the tensile strength which in some cases were up to 1850 N/mm². This seems to be a necessary but not sufficient criterion as specimens with high tensile strength were found without any signs of cracks. The chemical analyses have shown no significant deviation from the nominal values.

Figure 3   *Fractures of prestressing steel wires (indicated by arrows) in a totally grouted duct*

Figure 4   *Fracture surface with incipient crack*

Figure 5   *Intergranular corrosion in the region of the incipient crack (SEM micrograph)*

Figure 6   *Surface cracks detected by magnetic particle testing*

Apart from isolated pores, which are not unusual, no distinctive features were observed with regard to the injection mortar and the state of injection. Local observations of corrosion at single prestressing steel wires were detected in nearly all the structures investigated. In some cases this was due to uniform corrosion caused by reactions prior to grouting; in other cases local pits were found to result from pores or crevices between adjacent wires. Such attack is not a necessary condition for crack initiation and propagation as often pitting is found without any signs of stress corrosion cracking. From these results it can be concluded that the specific susceptibility of the material is the determining factor with respect to hydrogen-induced stress corrosion cracking.

Figure 7   *Transverse micro-section through prestressing steel wire with crack*

4 LABORATORY INVESTIGATIONS AND THEIR RESULTS

In order to investigate the crack propagation of damaged prestressing steel wires in totally grouted ducts, electrochemically controlled tests and lifetime tests with specimens surrounded by an alkaline injection mortar, or immersed in calcium hydroxide solution, were carried out. The specimens were taken from failed material. Prior to testing the extent of cracking was determined by a non-destructive method. As a reference sample, crack-free wires were also included in the test program. The specimens were stressed to 40% of the ultimate tensile strength. While crack-free specimens did not fail or show crack initiation in the saturated $Ca(OH)_2$ solution within the testing time of 4000 h - even under severe crevice corrosion conditions - specimens with cracks failed after less than 500 h. The greater severity of the test conditions due to crevice effects or superimposed vibration stress led to a further reduction of lifetime. Specimens surrounded by injection mortar showed no indications of crack propagation after about 30000 h. This result is probably due to lower passive current densities in the mortar compared to the aqueous solution and hence a lower rate of hydrogen production.

In further laboratory investigations tests were carried out using different prestressing steels of the quenched and tempered type in order to assess the corrosion risk, (i.e. crack initiation and propagation), under depassivating conditions. Table 3 gives information on the specimens used. The first three materials were taken from structures and represent the 'old type' of prestressing steel which lost its approval in the mid-sixties. Specimens with the indication SIGNEU are steel wires used today. The steel from Henningsdorf is material produced and used in the former DDR. The chemical composition of this steel corresponds to the quenched and tempered steel of the 'old type'. Specimens with the indication NEPWIT are taken from one of the failure cases with several cracks and fractures mentioned in the previous section.

Table 3 *Steel specimens tested*

Steel Type	Internal Indication	Shape/Dimension	Mean Tensile Strength [N/mm²]	Remark
Neptun St 1420/1570	NEPWIT	N40, oval, ribbed	1790	taken from structure (case of damage)
Neptun St 1420/1570	NEPHIL	N40, oval, ribbed	1583	taken from structure
Sigma St 1420/1570	SIGEGI	5.2 mm, round, smooth	1602	taken from structure
Sigma St 1420/1570	SIGNEU	8mm, round, ribbed	1625	new, as received
Henningsdorf St 1370/1570	HENNEU	OV50, oval, ribbed	1696	new, as received

Table 4 *Test conditions*

Corrosion Conditions	Prestressing	Temperature
Condensed water formation by periodical temperature changes	55% of nominal tensile strength	100% relative humidity, 12h at 35°C, and 12 h at 25°C
70% relative humidity	55% of nominal tensile strength	25°C
0.5 g/l chloride, 5 g/l sulphate, 1 g/l rhodanide, pH=7.0 (DIBt-test)	80% of nominal tensile strength	50°C

Three specimens of each of the five steel types were mounted in test rigs, stressed and exposed to the corrosion conditions listed in Table 4. Prior to exposure the wire specimens were inspected to check for cracks. To ensure that the tests include the crack initiation stage only crack-free specimens were used. The DIBt-test solution is normally used in the approval of prestressing steels to assess the behaviour of stressed steel not yet in grouted ducts with respect to hydrogen-induced stress corrosion cracking susceptibility. If the specimens withstand 2000 h without fracture they have passed the test. During the experiment the severity of temperature, medium and mechanical loading are increased in comparison to practical conditions. But, experience has shown that this test allows a practical differentiation to be made with respect to stress corrosion cracking susceptibility.

Figure 8 shows the time to failure of the five steel types in the DIBt-test solution. Unlike in the approval test, the specimens were kept in the test rigs and the test was continued even after 2000 h without fracture. The steel wires from the failure case (NEPWIT) and the material HENNEU did not last for the required 2000 h. The material produced and used today (SIGNEU) has a considerably longer lifetime. Specimens exposed to 70% relative humidity have not shown any signs of cracking during the present testing time of about 20000 h. The time to failure under condensed water conditions is shown in Figure 9. Under these conditions, the material from the failure case failed first while other samples from the same producer, and having comparable composition, did not fail even after long testing times (>20000 h). The only distinguishing features are considerable differences in the strength values (see Table 3).

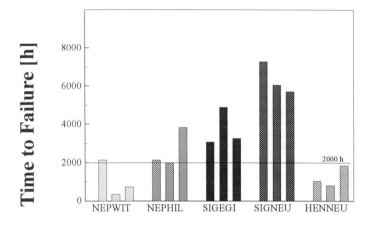

Figure 8   *Time to failure of prestressing steels in DIBt-test solution*

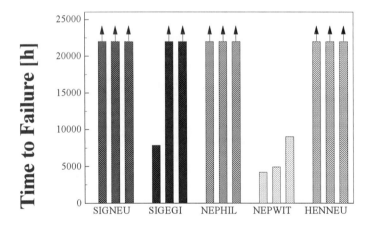

Figure 9   *Time to failure of prestressing steels under condensed water conditions*
(↑ these experiments are still going on)

5 CONCLUSIONS

In the past, damage to prestressed concrete structures with post-tensioned steel members due to hydrogen-induced stress corrosion cracking was often traced back to defects related to non-injected ducts or the presence of corrosion inducing substances within the grout. Recent failures have shown that the failed prestressing steel members were fully grouted and no unusual composition of the injection mortar could be determined.

The investigations on real structures as well as the laboratory tests have shown that in some cases heats of the 'old type' quenched and tempered steel, strength class St 1420/1570 exist, having a tendency to form cracks prior to grouting under certain environmental conditions. High values of the tensile strength are indicative of a certain susceptibility although this seems to be a necessary but not sufficient criterion. According to the present state of knowledge crack propagation in fully grouted ducts cannot be excluded. The laboratory tests suggested that crack initiation within an alkaline milieu is unlikely even after a long time. The specific susceptibility of the prestressing steel is the determining factor with respect to crack initiation and propagation ahead of parameters of the environment. If depassivating conditions exist cracks can be initiated with condensed water as an electrolyte.

From the results obtained so far it can be concluded that a general risk to older, prestressed concrete structures is not proven. Structures with post-tensioned prestressing steel members of the 'old type' of quenched and tempered steel (Neptun St 1420/1570) are particularly predisposed especially if the strength considerably exceeds the nominal strength, and signs of cracking existed before grouting.

Prestressed concrete structures subjected to reconstruction or change of utilization should be inspected prior to these measures.[2] For structures with pre-tensioning it should be ensured that depassivating conditions at the steel surface can be excluded.

References

1. U. Nürnberger, *Otto-Graf-Journal*, 1992, **3**, 148.
2. E. Wölfel, *Beton- und Stahlbetonbau*, 1992, **87**, 155.
3. B. Isecke, K. Menzel, J. Mietz and U. Nürnberger, *Beton- und Stahlbetonbau*, 1995, **90**, 120.
4. U. Nürnberger, K. Menzel, J. Mietz and B. Isecke, 'Spannungsrißkorrosion vorge-schädigter Spannstähle im verpreßten Zustand', Final Report of IfBt Research Project IV 1-5-654/91, FMPA Stuttgart and BAM Berlin, 1993.
5. J. Mietz, U. Nürnberger and W. Beul, 'Untersuchungen an Verkehrsbauten aus Spann-beton zur Abschätzung des Gefährdungspotentials infolge Spannungsrißkorrosion der Spannstähle', Final Report of BMV Research Project FE 15.209 R91D, BAM Berlin and FMPA Stuttgart, 1994.
6. R. Troiano, *Trans. ASM*, 1960, **52**, 54.
7. R. A. Oriani, *Ber. der Bunsengesellschaft*, 1972, **76**, 848.
8. E. Wendler-Kalsch, in: 'Wasserstoff und Korrosion', (Ed. D. Kuron), Verlag Irene Kuron, Bonn, 1986, 8.
9. J. Erdmann, J. Neisecke and F. S. Rostasy, 'Baustoffuntersuchungen an Spannbeton-bauwerken zur Ermittlung des Langzeitverhaltens von Spannstählen', Final Report of IfBt Research Project IV/1-5-134/77, TU Braunschweig, 1982.
10. J. Erdmann, K. Kordina and J. Neisecke, 'Auswertung von Berichten über Abbrucharbeiten von Spannbetonbauwerken im Hinblick auf das Langzeitverhalten von Spannstählen', Final Report of IfBt Research Project IV/1-5-133/77, TU Braunschweig, 1982.
11. U. Nürnberger, 'Analyse und Auswertung von Schadensfällen an Spannstählen', Forschung Straßenbau und Straßenverkehrstechnik, Heft 308, Bundesminister für Verkehr, Bonn, 1980.

THE EFFECT OF BRICK TILING ON THE CARBONATION OF CONCRETE

M.J. Pentti and J.S. Mattila

Tampere University of Technology
Structural Engineering
P.O. Box 600
FIN - 33101 TAMPERE
FINLAND

1 INTRODUCTION

Brick tiles are commonly used in Finland to finish the outer face of prefabricated concrete facade units. In manufacturing the facade panels the tiles are first laid on the bottom of the formwork, and after assembling the reinforcement the concrete is cast on. The tiles are thus fixed by the adhesion of the hardening concrete.

The thickness of the brick tiles normally ranges from 20 to 35 mm. A cross-section of a sandwich facade panel structure with a brick tile finish is shown in Figure 1.

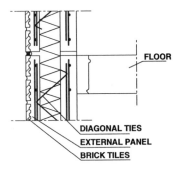

Figure 1 *Sandwich facade panel with clay brick finish*

In several investigations it has been found that a clay brick finish, when fixed by the hardening concrete, can significantly reduce the carbonation rate of the concrete in natural weather conditions. These observations are somewhat surprising because of the relatively high porosity and thus the high permeability of bricks to gases.

It can be assumed that the beneficial effect of brick tiles results mainly from three different mechanisms, effective during different phases of manufacture and service:

1. The capillary suction of tiles immediately after casting, taking up water from the concrete and thus reducing its porosity and permeability.

2. The improvement of curing efficiency due to the ability of tiles to keep the concrete surface moist during hardening.
3. The diffusion resistance of tiles to CO_2 after manufacturing.

2 AIM OF THE STUDY

The aim of this investigation was to examine the ability of different tiles to reduce the rate of carbonation, and to establish the relative significance of the three mechanisms mentioned above.

3 EXPERIMENTAL

3.1 Bricks

Six different types of clay bricks were chosen for testing. The basic properties of the bricks are shown in Table 1. The pore size distributions of the bricks can be seen in Figure 2.

3.2 Concrete Qualities

Two qualities of concrete were used for the test slabs. The concrete qualities were similar to those used for exterior wall structures in Finland. The cement type in both mixes was rapid-hardening Portland cement. Superplastisizing and air-entraining agents were used as admixtures. The nominal compressive strengths and water-cement ratios of the concretes were 30 MPa (w/c = 0.62) and 40 MPa (w/c = 0.52), respectively.

3.3 Other Variables

In addition to brick types and concrete qualities the degree of saturation of the bricks as well as the curing procedure and storage conditions of the test slabs were varied. The description of the variations used in the tests is shown in Table 2.

Table 1   *Properties of bricks*

	BRICKS OF SEPPÄLÄN TIILI CORP			BRICKS OF LOHJA CORP		
	SP (red)	SK (yellow)	SV (white)	LP (red)	LK (yellow)	LV (white)
Density kg/m³	1910	1310	1170	1790	1340	1240
Water suction rate kg/m²min	0.4	0.7	0.9	0.7	3.3	2.2
Water absorption %	6.2	8.4	14.3	7.4	20.5	20.7

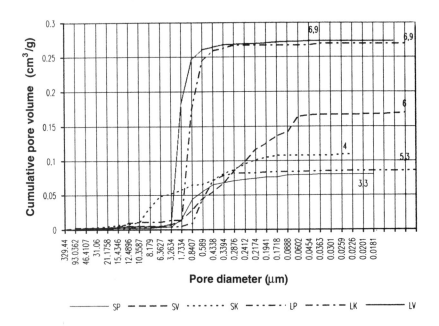

Figure 2   *Cumulative pore size distributions of bricks. The figures on each curve show the mean carbonation depths after the 90 day test*

3.4 Test Procedures

After preparation and curing the test slabs were exposed to accelerated carbonation tests. The conditions in the testing chambers were as follows: CO_2-content 4 %, temperature 20 °C and relative humidity 75 %. In these conditions the rate of carbonation in the chamber can be assumed to be roughly 50 times higher than in natural climate (one week in test ≈ one year of natural exposure).

Two test series were made: the first test series included all materials and two curing variations, as well as moist and dry bricks, and the duration of the exposure was 90 days. In the second test there were three different bricks and one concrete quality (30 MPa). The test slabs were prepared according to four different procedures to discover the influence of the three different carbonation reducing mechanisms - capillary suction, curing and diffusion resistance - both separately and together. The duration of exposure in the second test was 40 days.

The specimens for the second test series were prepared in four different ways as follows:

1.   In order to establish the effect of the **capillary suction** of bricks on the carbonation of concrete the brick tiles were taken off two days after casting.

2.   The pure **curing effect** of the bricks was established by fixing wet tiles onto the concrete surfaces with cement mortar at an age of two days. The tiles were taken off at an age of three weeks, just prior to moving the specimens into the CO_2-chambers.

3. The effect of the **diffusion resistance** of bricks was established by fixing the tiles onto the concrete specimens just before moving them into the CO_2-chambers. The tiles were fixed with silicone around their edges and the joints were sealed with tape and bitumen.
4. The total effect of the bricks was examined by using normal specimens with tiles throughout the whole test. Both wet and dry tiles were used, with two curing conditions (Table 2).

Table 2   *Description of test specimens*

Test series	Concrete strength	Brick types	Wetting of bricks	Storage of slabs	Remarks
I duration 90 days	30 MPa or 40 MPa (symbol B)	LK, LP, LV SK, SP, SV whole tiles or sawn to 10 mm thickness (symbol S)	1 day in water + 6 h in air or in air 30 % RH, + 20 °C (symbol kui)	type I: 95 % RH, 23 °C, 7days + 70 % RH, 21 °C, 21 days type II: 70 % RH, 21 °C, 14 days, + removal of bricks, + 70 % RH, 21 °C, 14 days	
II duration 40 days	30 MPa	LV, LK, SP	in air 30 % RH, + 21 °C or oven dried (symbol kui) or 10 days in water + 6 h in air (symbol mär)	type I: 30 % RH, 21 °C, 21 days type II: 90 % RH, 21 °C, 21 days	four different preparing procedures

3.5 Measurements

After the accelerated carbonation test the carbonation depths were measured from the sawn cross-sections of the specimens with phenolphthalein indicator. The average carbonation depths were determined with a planimeter because of the unevenness of the carbonation fronts both under the bricks and at the uncoated concrete surfaces. The unreliable values in the edge zones were not included in measurements.

The pore-size distributions of bricks and mortars were measured by mercury intrusion porosimetry.

4 RESULTS

The rates of carbonation are expressed in mm/√week so that the results from both test series can be compared.
The results of **test series I** are shown in Figure 3.

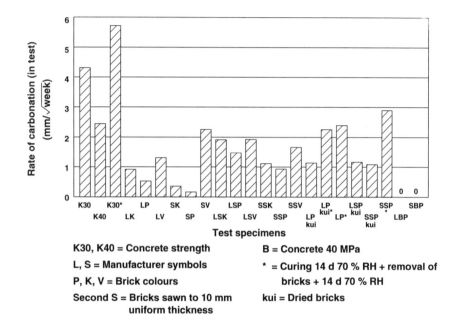

Figure 3   *Results of 90-day accelerated carbonation test (test series I)*

The results show clearly that the brick tiles are able to slow down the rate of carbonation by about 50 - 90 %, compared with the average carbonation of the exposed concrete surfaces.

In addition to this it can also be seen, that the porosity and water absorption capacity of the bricks has an effect so that the bricks with lower porosity are more efficient in slowing the rate of carbonation than those with higher porosity (Figures 4 and 2). Dry bricks were usually somewhat more effective than wet ones.

There were only two slabs of higher strength concrete (40 MPa), with two different brick types. Neither of them showed any carbonation under the bricks, while the average carbonation depth measured from the exposed concrete surfaces was 8.8 mm in the 90 days test.

The results of **test series II** are shown in Figures 5, 6 and 7.

The carbonation coefficients of the specimens where the tiles were taken off at an age of two days after casting (the effect of capillary suction alone), are shown in Figure 5. The rates of carbonation are, depending on the brick type, 54.9 - 94.4 % lower than at the exposed concrete surfaces.

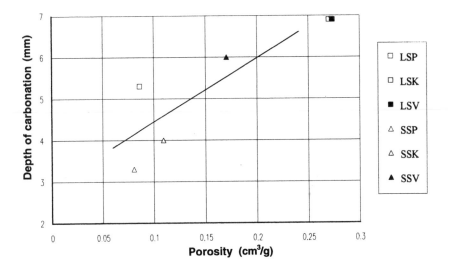

Figure 4   *Measured carbonation depths under sawn 10 mm tiles as a function of brick porosity*

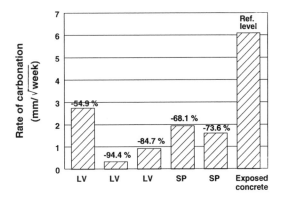

Figure 5   *Effect of capillary suction on carbonation. Bricks were taken off at an age of 2 days. Concrete strength 30 MPa, storage 30 % RH, 21 °C, 21 d; 40-day test*

The curing effect of tiles, obtained by fixing wet bricks on the hardened specimens at an age of two days and removing them prior to the accelerated carbonation test, showed a reduction of 41.0 - 46.5 % in the carbonation rate (Figure 6).

According to the results of this test, fixing brick tiles just prior to the accelerated carbonation test (the effect of diffusion resistance alone) had no significant effect on the carbonation rate of concrete.

The total effect of the brick tiles on the carbonation of concrete in test series II is shown in Figure 7.

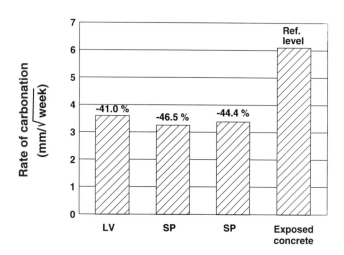

Figure 6   *Curing effect of bricks on rate of carbonation. Bricks fixed on hardened concrete at an age of 2 days and removed prior to testing. Concrete strength 30 MPa, storage 30 % RH, 21 °C, 21 d; 40-day test*

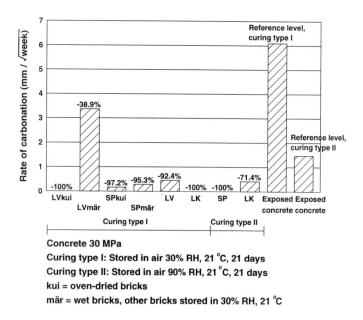

Figure 7   *Results of the 40-day test, specimens with tiles throughout the whole test. Effect of wetting/drying of bricks and curing of specimens on carbonation*

The results show clearly that the most effective mechanism in controlling the carbonation of concrete with brick tile finish is the ability of bricks to absorb water from concrete, thus making it denser. The effect of the capillary suction of tiles can also be seen in Figure 8, which shows how different bricks affect the pore size distribution and porosity of a cement mortar (max. aggr. 3 mm).

When studying the results, the considerable unevenness of the carbonation front should be noticed (Figure 9). Therefore, values measured from a single specimen may be somewhat inaccurate.

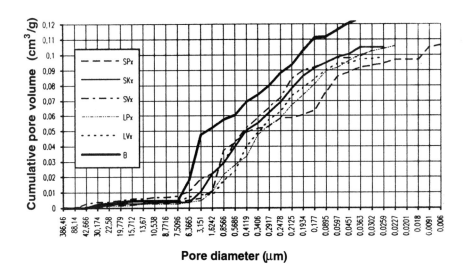

Figure 8 *Cumulative pore size distributions of hardened cement mortar with and without (symbol B) the suction effect of different brick tiles. Bricks oven dried*

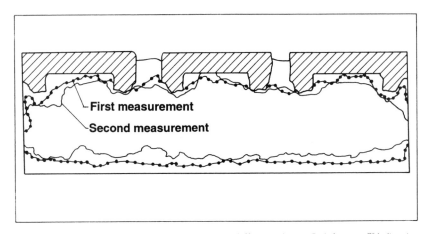

Figure 9 *Measured carbonation fronts at two different times. Brick type SV, 90-day test*

Figure 9 also shows the effect of the brick shape. At the thicker edges of the bricks the carbonation penetrates deeper into the concrete.

5 DISCUSSION OF RESULTS

The purpose of this investigation was to examine the ability of brick tiles to slow the carbonation rate of concrete in structures such as facade units, in which the concrete is cast directly onto the tiles.

The experiments proved that brick tiles are able to significantly slow down the carbonation rate of concrete. The best specimens showed a reduction of over 90 % in comparison to the average carbonation rate of the uncovered concrete surfaces. The beneficial effect of bricks was better than the effect of raising the strength of concrete from 30 to 40 MPa.

Different types of bricks had different abilities to slow the rate of carbonation. The efficiency of bricks to slow carbonation seems to increase with decreasing porosity, but the correlation is not clear. Probably a better correlation would be found with the capillary properties of bricks.

The most important mechanism in slowing carbonation appeared to be the ability of bricks to absorb water from fresh concrete. The beneficial effect of tiles on curing conditions was also notable, but the effect of diffusion resistance of this kind of tiles was weak. For this reason thickness and permeance are not essential properties of bricks in this respect.

It seems evident that brick tiling is beneficial in prolonging the service life of reinforced concrete panels with regard to reinforcement corrosion, but the question of adequate or minimum cover depths between the tiles and reinforcing bars has not been resolved. If smaller cover depths are generally allowed between the bricks and reinforcement than in normal structures, it is likely that the expected design life is not going to be reached throughout the structure. Local or wider damage may arise from the following causes:

-       the unevenness of the carbonation front under the bricks and in the joints between them, being a result of variations in the adhesion of bricks, the wetting of bricks, the quality of mortar joints between the bricks and the compaction of concrete
-       the lack of any protective effect on uncovered parts of the elements, such as edges
-       the variations on the cover thickness.

6 CONCLUSIONS

The ability of brick tiles to reduce the rate of carbonation in reinforced concrete units was examined by accelerated carbonation tests. The test results showed that the brick tiles are able to slow the rate of carbonation significantly, compared with the uncovered concrete surfaces. It was also found that the most important mechanism in retarding carbonation is the ability of tiles to absorb water from the fresh concrete, thus reducing the porosity of concrete under the tiles. The beneficial effect of bricks on curing was also notable, but the effect of the diffusion resistance of brick tiles on carbonation was weak.

Acknowledgements

Financial support from the Finnish Brick Industry Association, Partek Concrete Corp. and Seppälän Tiili Corp. is gratefully acknowledged. The experimental work was carried out by Mr. Henri Sulankivi, to whom we owe many thanks.

LOAD BEARING CAPACITY OF CONCRETE COLUMNS WITH CORRODED REINFORCEMENT

J.Rodríguez, L.M.Ortega and J.Casal

Geocisa (Dragados Group)
Los Llanos de Jerez, 10-12
28820 Coslada (Madrid), Spain

1 INTRODUCTION

It is well known that reinforced concrete structures decay with time, due to different deterioration mechanisms. Urgent measures are required to decide when and how to repair them, owing to the large number of existing damaged structures. However, there is currently a lack of models for assessing these structures.

Some tools are being developed for evaluating the effects of steel corrosion, freeze-thaw and alkali-silica reaction on concrete structures, within a Brite/Euram research project started in 1992.

Corrosion of reinforcing bars induces an early deterioration of concrete structures[1] and reduces their residual service life. When the aggressive agent reaches the reinforcement, corrosion may start affecting: firstly the steel, due to the reduction of both the bar section and the mechanical properties; secondly the concrete, due to the cracking of the concrete cover produced by the expansion of corrosion products; and finally the composite action of both steel and concrete, due to the bond deterioration. Consequently, the safety and serviceability of the concrete structure are affected.

Figure 1 summarizes the main effects of reinforcement corrosion on the behaviour of concrete structures.

Although papers on material deterioration aspects are frequently published,[2] few papers have been written on the structural implications of the steel corrosion. Some studies have been carried out on the effect of corrosion on concrete cracking,[3,4] bond deterioration[5-8] and load carrying capacity of concrete beams.[9,10]

As for the behaviour of deteriorated concrete columns, McLeish[1] pointed out the need to consider aspects such as the reduction of concrete section, due to the cracking and spalling of the concrete cover, and the buckling of the main reinforcement, due to the corrosion of the links. Uomoto[9] carried out experimental work with columns of 400 x 100 x 100 mm. The accelerated corrosion was produced by both adding sodium chloride to the mixing water and applying a constant density current to the reinforcement of 90 and 360 $\mu A/cm^2$ for 2 and 10 days. The failure loads of the corroded columns ranged between 70 and 80% of that of the non-corroded columns. This reduction in load bearing capacity was significantly higher than the value expected from the decrease of the steel section.

This paper summarizes the research carried out under the above mentioned project, to study the behaviour of concrete columns affected by the steel corrosion.

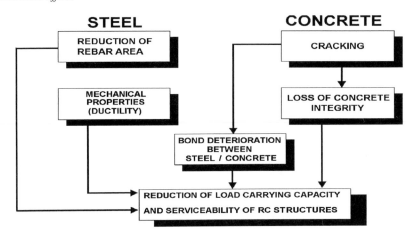

Figure 1 *Effects of steel corrosion on concrete structures*

2 EXPERIMENTAL WORK

An experimental study was carried out using 24 concrete elements to establish a relationship between the level of corrosion and the structural performance of deteriorated columns. Reinforced concrete columns without corrosion and with different levels of corrosion were tested and different reinforcing details were considered.

2.1 Description of the Columns

Three different types of columns were fabricated. The characteristics are summarized in Table 1 and Figure 2. As shown in Figure 2, the column ends were designed to avoid failure during the loading test.

Concrete was produced with siliceous sand and limestone crushed coarse aggregates with particles of 12 mm maximum size. 3% of calcium chloride by weight of cement was added to the mixing water to accelerate the corrosion of reinforcement.

Ribbed bars of Spanish type AEH 500S with yield stress and strength values ranging between 550-590 MPa and 600-670 Mpa were used.

Table 1 *Characteristics of the Concrete Columns*

COLUMNS	CONCRETE		REINFORCEMENT	
TYPE	CHLORIDE(+)	COMPRESSIVE STRENGTH (*)	LONGITUDINAL BARS	LINKS
1	no	30.0	2x2ϕ8	ϕ6/100
	yes	35.8	2x2ϕ8	ϕ6/100
2	no	34.0	2x2ϕ16	ϕ6/150
	yes	35.6	2x2ϕ16	ϕ6/150
3	no	34.3	2x4ϕ12	ϕ6/150
	yes	39.4	2x4ϕ12	ϕ6/150

+ With or without 3% of $CaCl_2$.
* Compressive strength at the date of loading test, in MPa.

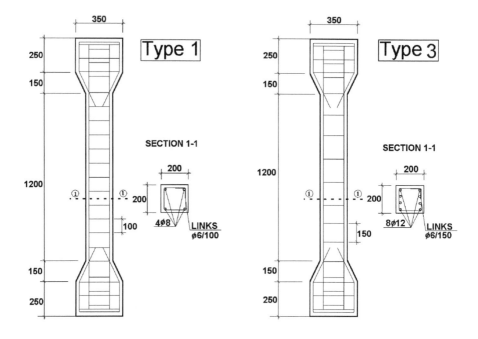

Figure 2 *Scheme of columns types 1 and 3 (dimensions in mm)*

2.2 Corrosion of Concrete Columns

Columns were cast and cured for 28 days, in indoor conditions with a watering system to keep the columns in a wet condition. Columns made with calcium chloride were submitted to an accelerated corrosion procedure, to obtain the required level of corrosion damage in approximately 100-200 days.

Figure 3 shows the arrangement for the corrosion procedure. The reinforcement was submitted to a constant current density of about 100 $\mu A/cm^2$, applied to the concrete surface through the stainless steel counter-electrodes. This current density corresponds to ten times the corrosion intensity I_{corr} measured in highly corroding concrete structures.[11,12] It is worth pointing out that other researchers applied higher values of current density, usually ranging between 300 and 2,000 $\mu A/cm^2$, as was commented elsewhere.[10] Only the reinforcement in the central part of the column with constant concrete section was corroded; the secondary reinforcement at the ends was electrically isolated to avoid corrosion.

Once the corrosion process had finished, a detailed map of the cracks at the concrete surface was obtained for each column. Table 2 shows the time of the accelerated corrosion, the maximum crack width at the concrete surface and the attack penetration, both for main bars and links. The mean value of the attack penetration was obtained by the gravimetric method, weighing the bar before and after corrosion, when the rust products had been removed.

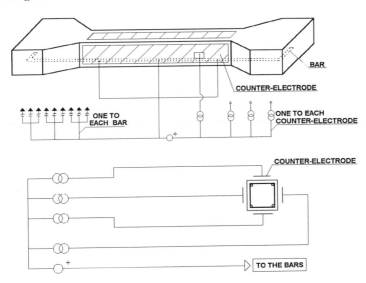

Figure 3 *Scheme of test arrangement for accelerated corrosion of columns*

Table 2 *Accelerated Corrosion Results*

COLUMN TYPE	COLUMN No.	No. OF DAYS	MAXIMUM CRACK WIDTH(mm)	ATTACK PENETRATION (mm) (*)	
				MAIN BARS	LINKS
1	11	——	——	——	——
	12	——	——	——	
	13	106	0.8	0.32	0.30 (1.4)
	14	113		0.31	0.35 (1.4)
	15	143	1.0	0.39	0.40 (2.4)
	16	150	1.0	0.44	0.42 (3.2)
	17	190	1.5	0.52	0.58 (3.9)
	18	204	2.0	0.60	0.56 (3.7)
2	21	——	——	——	——
	22	——	——	——	
	23	106	1.3	0.38	0.28 (2.8)
	24	113	0.9	0.37	0.30 (3.0)
	25	143	1.5	0.45	0.36 (3.9)
	26	150	1.6	0.46	0.41 (3.8)
	27	190	2.0	0.62	0.47 (4.7)
	28	204	2.5	0.63	0.50 (4.7)
3	31	——	——	——	——
	32	——	——	——	
	33	106	1.2	0.30	0.28 (3.0)
	34	113	1.5	0.34	0.30 (3.5)
	35	143	2.2	0.40	0.29 (3.4)
	36	150	1.5	0.39	0.33 (3.4)
	37	190	4.0	0.51	0.41 (4.4)
	38	204	1.8	0.56	0.40 (3.8)

* Values in brackets correspond to maximum values of pitting and were obtained by geometrical measurement on the pits of each bar.

2.3 Loading Test

Columns were axially loaded, according to the test arrangement shown in Figure 4. Loads were applied by means of a hydraulic actuator with a nominal maximum capacity of 2000 kN, centred on a steel plate placed over the top end of the column. The bottom end was simply supported on a bearing made up of two steel plates with a neoprene spacer.

Tests were controlled by: a) a load cell to monitor total axial load; b) an actuator displacement transducer to control the constant displacement rate established for each test; c) four displacement transducers (LVDT) to measure the average compressive strain on each side of the column, (Figure 4); and d) displacement transducers (LVDT) to monitor lateral displacements at the mid-height and bottom sections.

Loading tests were carried out in two phases. In the first phase, the load was increased up to the calculated service load, at a constant displacement rate of 0.5 mm/min. In the second phase, the load was increased from zero up to failure of the column; the load was applied at a displacement rate of 0.5 mm/min up to the service load, and at a rate of 0.25 mm/min until failure.

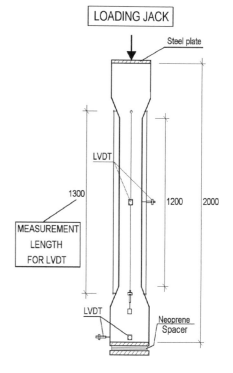

Figure 4 *Loading test arrangement (dimensions in mm)*

3 RESULTS

Although the columns were axially loaded, some eccentricities were detected during the loading test in the measurements of the strains on the four sides of the columns. These eccentricities were significant in most of the corroded columns.

Some delamination of the concrete cover was detected on one or more sides of the corroded columns, at a certain level of the load, and some links failed at load values close to the maximum load.

Table 3 summarizes the test results obtained in both non corroded and corroded columns, at both delamination and ultimate loads. Whereas the delamination load corresponds to the load at which spalling of one side of the concrete cover was initiated, the ultimate load corresponds to the maximum load measured in the tests.

The mean values of the strain on the four sides of the columns are given in this table, for both load levels. The number of broken links observed after the failure of the columns is also included as are the eccentricities values in two directions for the maximum load. These were obtained from the strain on each pair of column sides.

As is shown in Table 3, the load carrying capacity of corroded columns was significantly lower than that of the non corroded columns, in spite of the lower concrete compressive strength of these columns (Table 1). The reduction of the longitudinal bar section had a negligible influence on these results. Conversely the failure of the concrete cover and the corrosion of the links played an important role in the column strength.

The column failure was generally initiated by the cracking and spalling of the concrete cover and the failure of one or more links, which were heavily affected by localized corrosion (pitting), and provoked the buckling of main bars under the applied load.

Figure 5 represents the load-strain curves (P-ϵ) of column types 1 and 2, during the second loading phase. The ϵ values correspond to the mean value of the strain on the four sides of the column, for each loading level. The corrosion of the reinforcement:

· Not only reduced the maximum load but also the strain at this load.
· Reduced the compressive stiffness of the column, as the slope of the first part of the P-ϵ curve in corroded columns was lower than that of non corroded ones.
· Produced a change of slope in P-ϵ curve, for a load value corresponding to the initiation of the delamination of the concrete cover. After this change, the load did not significantly increase although the concrete strain increased.

Table 3 *Summary of Experimental Results in the Loading Tests*

COLUMN No.	DELAMINATION LOAD (*)		ULTIMATE LOAD				
	AXIAL FORCE (kN)	MEAN STRAIN (·)	AXIAL FORCE (kN)	MEAN STRAIN (·)	No. OF BROKEN LINKS	ECCENTRICITY (+) e_x	e_y
11	——	——	1300	2.4	——	4.9	8.8
12	——	——	1320	2.4	——	9.4	1.1
13	934	1.3	990	2.5	1	2.8	10.7
14	——	——	993	2.3	1	13.6	8.1
15	——	——	947	2.0	1-2	13.4	8.1
16	——	——	828	1.7	2-3	9.4	20.5
17	——	——	822	1.6	2-3	24.2	5.4
18	735	1.3	862	2.3	2-3	7.8	10.6
21	——	——	1680	2.7	——	8.2	2.3
22	——	——	1702	2.6	——	2.1	5.4
23	993	1.1	1080	2.2	1-2	22.2	15.6
24	999	1.2	1040	2.0	1	20.5	14.1
25	934	1.0	1091	1.8	3	13.8	16.4
26	890	1.1	1135	2.1	3-4	1.4	7.4
27	847	1.0	973	1.8	4	3.2	4.6
28	975	1.4	997	1.8	4	1.9	5.5
31	——	——	1728	2.5	——	1.2	5.2
32	——	——	1673	2.3	——	2.1	5.4
33	1258	1.3	1274	2.1	1	10.0	7.9
34	1130	1.2	1178	1.7	1	7.8	11.3
35	1164	1.1	1203	2.1	2-3	9.2	20.9
36	1145	1.3	1174	1.9	2-3	4.0	2.0
37	991	1.2	1038	1.7	4	4.8	4.6
38	1089	1.2	1170	1.5	2-3	21.1	4.0

* Load value when initiation of delamination of concrete cover was observed.
· Mean strain at the midspan zone, in ‰, on the four sides of the column, (Figure 4).
+ Eccentricity of the load, in mm, obtained from the strain values on each side of the column.

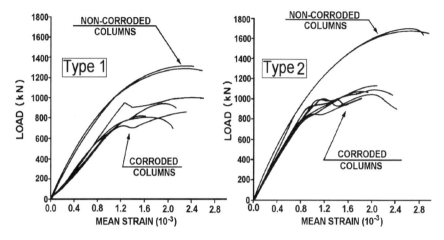

Figure 5 *Load versus mean strain in column types 1 and 2*

4 DISCUSSION OF THE RESULTS

Three main factors have affected the strength of corroded columns and are commented on in this section: a) the increase in eccentricity due to the asymmetric deterioration of the concrete section; b) the likely reduction of reinforcement strength due to premature buckling; and c) the deterioration of the concrete section.

4.1 Increase of Eccentricity

An increase in eccentricity was observed in most of the corroded columns when they were compared with the non corroded ones. This was due to the asymmetric deterioration of the concrete cover on the sides of the column. As these eccentricities were obtained from the mean strain values, measured along 1300 mm on each side of the columns (Figure 4), they only gave a rough value of the eccentricity at the failure section.

The eccentricities increased with the level of corrosion in most of the type 1 columns reinforced with a low ratio of main bar section area to concrete section area (0.5%). On the contrary, in column types 2 and 3, reinforced with a ratio ranging from 2 to 2.2%, the eccentricity decreased in most of the highly corroded columns. Perhaps these ratios of reinforcement resulted in a symmetric deterioration of the column, reducing the concrete section to a core with dimensions close to 140 x 140 mm.

Table 4 shows the mean and standard deviation of these eccentricities and the maximum and minimum characteristic values in each direction, assuming a normal distribution for the experimental values.

An additional mechanical eccentricity e_{corr} should be added to the eccentricities considered in Eurocode 2[13] for non corroded columns when assessing deteriorating concrete columns.

The experimental values in non corroded columns resulted mainly from geometrical imperfections during the loading test. Thus, the difference in eccentricities between corroded and non corroded columns gives a rough estimation of the additional e_{corr} value

Table 4 *Eccentricities in the Loading Tests, in mm*

ECCENTRICITY	NON CORRODED COLUMNS		CORRODED COLUMNS	
	e_x	e_y	e_x	e_y
Mean value	7.1	2.3	13.4	7.1
Standard deviation	1.9	1.4	6.8	4.3
Characteristic values	10.2	4.5	**24.6**	**14.1**
	3.9	**0**	2.3	0

due to corrosion. The conservative values of e_{corr} in both directions would be 20.7 mm (24.6 - 3.9) and 14.1 mm (14.1 - 0) and were obtained from the characteristic values in both corroded and non corroded columns (bold figures in Table 4). The e_{corr} values correspond to values between 0.50 and 1.0 times the mechanical cover (distance between the concrete surface and the main bar axis).

4.2 Buckling of Reinforcement

Although complete evidence was not obtained during the loading tests, several aspects suggest that a premature buckling of the main bars could affect the strength of some corroded columns. These factors are: a) the highly deteriorated concrete cover; b) the failure of some links which were detected during the application of the load, for load values ranging between the delamination and failure loads; and c) the buckling of most of the main bars, observed after the column failure.

An estimation of the reinforcement strength reduction can be obtained from the theoretical stress values corresponding to the critical load on the main bars, calculated from the following expression:

$$\sigma_{cri} = \pi^2 E [0.25 \phi]^2 / [0.75 s]^2 \tag{1}$$

where:

σ_{cri}	is the stress corresponding to the Euler critical load
E	is the modulus of elasticity
ϕ	is the main bar diameter
s	is the link spacing

as 0.75 s was considered to be the effective height of the main bars, for the estimation of the slenderness ratio. To assume 0.5 s as the effective length would have been unsafe because the deteriorated links enabled the bars to exhibit some rotation at the links.

The maximum spacing specified in the Codes for Transverse Reinforcement ranges from 10 to 15 times the diameter of the main bars, and the design yield stress, f_{yd}, is always lower than σ_{cri}. However, this spacing can increase in deteriorated columns when some links fail due to localized corrosion. Then, the maximum allowable stress at the main compression bars affected by the premature buckling can be estimated by $\sigma_s = \sigma_{cri} = k f_{yd}$, with $k \leq 1$. Conservative values for k might be 0.5 for corroded columns when one link may fail, 0.2 when two consecutive links fail, and 0 when more than two consecutive links fail.

4.3 Deterioration of Concrete Section

Figure 6 represents the maximum experimental loads in corroded columns of types 1 and 2 with the corresponding moments (or eccentricities). As biaxial eccentricities were produced in the tests, the values of the bending moments in these figures were obtained by applying a simplified method which reduces the biaxial eccentricities to an equivalent uniaxial one.[14]

Figure 6 also shows the curves for the different pairs of calculated ultimate values (N_u, M_u). Whereas in curves 1, 2 and 3, the non deteriorated concrete section was considered, curve 4 corresponds to a deteriorated section without the concrete cover on the four sides. An intermediate value of deteriorated bar section was also considered in the four curves. Finally, curves 2 and 3 were obtained assuming that two or three consecutive links failed due to corrosion and, consequently, the stress of compression bars was reduced, according to equation (1). The experimental values are placed between those obtained from the non deteriorated and the deteriorated concrete sections.

The spalling was mainly initiated on one side of some tested columns with corroded reinforcement. Therefore, Figure 7 represents the experimental values of the maximum loads and the corresponding moments, obtained from the equivalent uniaxial eccentricities, for type 2 columns.

Figure 7 also represents the curves for the different pairs of calculated ultimate values (N_u, M_u), for a concrete section without the cover on one side. Whereas curve 1 does not take account of likely buckling of compression bars, curves 2 and 3 consider the failure of two or three consecutive links and consequently, the stress of compression bars was reduced, as seen in Figure 6.

Figure 7 shows that some of the experimental values can be predicted when considering this type of deteriorated concrete section together with the premature buckling of the main bars.

In summary, three main effects of reinforcement corrosion on column behaviour have been shown as relevant in experimental work: the reduction of concrete section, the

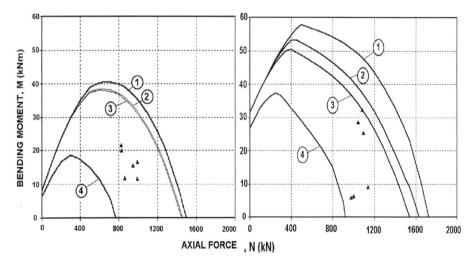

Figure 6 *Experimental and calculated values (N_u, M_u) in corroded columns types 1 and 2*

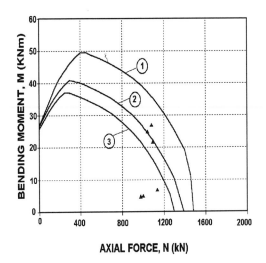

Figure 7 *Experimental and calculated values (N_u, M_u) in Type 2 corroded columns*

additional eccentricity due to the asymmetric deterioration of the concrete section and the reduction of the reinforcement strength due to premature buckling.

5 CONCLUSION

Experimental research work with axially loaded corroded columns, has been summarized. The reinforcement was corroded both by adding calcium chloride to the mixing water and applying a constant current density of about 100 $\mu A/cm^2$. This value corresponds to ten times the corrosion intensity I_{corr} measured in highly corroding concrete structures.

This research work has lead to the following conclusions:

· Corrosion of reinforcement affects the performance of concrete columns, reducing the ultimate load and the strain at this load. The compressive stiffness of the columns at service load is also reduced.

· Damage to concrete cover, due to cracking, delamination and spalling produced by corrosion, has been shown to be the most relevant form of deterioration with significant consequences for the column strength. Pitting at links has also been shown to be relevant, because it facilitates the premature buckling of the main bars when some links have failed.

· An estimate of the value of the ultimate axial force can be made when using reinforced concrete models, by considering the reduced section of the steel and a concrete section with one or more delaminated cover sides.

· Corrosion induces a mechanical eccentricity e_{corr} due to the asymmetric deterioration of the concrete cover on the different sides of the column. A conservative value of e_{corr} in each direction (biaxial eccentricities) is the mechanical cover size of the column in that direction. These eccentricities have to be added to the usual values considered in non corroded columns.

· As soon as some links exhibit a high reduction in section (localized corrosion), premature buckling of the main bars can occur and the maximum strength of the compression bars has to be reduced to values lower than 0.5 times the yield stress. As published elsewhere,[12,13] the reduction of corroded steel section can be estimated by measurement of the corrosion intensity I_{corr} in concrete structures.

It is hoped that some studies will be carried out on the performance of deteriorated sections under eccentric loads (combination of axial force and bending moment), taking account of the experimental work done with beams,[10] where $M \neq 0$ and $N \approx 0$, and with axially loaded columns, where $N \neq 0$ and $M \approx 0$.

More research is needed to study the influence of permanent loads, taking into account the creep effects which produce a load transfer from the concrete to the reinforcement.

ACKNOWLEDGEMENTS

The authors are grateful to both C. Alonso and C. Andrade of Institute E.Torroja (Spain) and to A. M. García, J. A. Bolaño and A. Navarro of Geocisa, for their contribution to the accelerated corrosion procedure. The research project BRITE/EURAM BE-4062 "The service life of reinforced concrete structures", funded by the UE, is being carried out by Geocisa in collaboration with BCA (UK), Institute E.Torroja (Spain), CBI, Lund Institute and Cementa (Sweden).

References

1. A. McLeish, 'Manual for life cycle aspects of concrete in buildings and structures', Taywood Engineering, UK, 1987, B4.1.
2. C. Andrade, C. Alonso, D. García and J. Rodríguez, Int. Conf. Life Prediction of Corrodible Structures. NACE. Cambridge, UK, 1991, 12/1.
3. M. L. Allan and B. W. Cherry, The NACE Annual Conference and Corrosion Show. NACE. Cincinnati, Ohio, USA, 1991, 125/1.
4. C. Andrade, C. Alonso and F. J. Molina, *Material and Structures*, 1993, **26**, 453.
5. J. Al-Sulaimani, M. Kaleemullah, I. A. Basunbul and Rasheeduzzafar, *ACI Structural Journal*, 1990, **87**, 220.
6. Y. Tachibana, K. I. Maeda, Y. Kajikawa and M. Kawamura, Third Int. Symposium on Corrosion of Reinforcement in Concrete Construction. Society of Chemical Industry, Elsevier Applied Science. London, UK, 1990, 178.
7. J. G. Cabrera and Ghoddoussi, Int. Conf. Bond in Concrete. CEB and Riga Technical University. Riga, Latvia, 1992, 10/11.
8. J. Rodríguez, L. M. Ortega and J. Casal, 'Int. Conf. Concrete across borders', Odense, Denmark, 1994, **2**, 315.
9. T. Uomoto and S. Misra, *ACI SP-109*, 1988, 127.
10. J. Rodríguez, L. M. Ortega and J. Casal, 'Int. Conf. Structural Faults and Repairs', London, 1995, **2**, 189.
11. J. Rodríguez, L. M. Ortega, A. M. García, L. Johansson and K. Petterson, *Construction Repair*, 1995, **9**, 27.
12. J. P. Broomfield, J. Rodríguez, L. M. Ortega and A. M. García, *ACI SP-151*, 1994, 163.
13. ENV 1992-1-1: 1991, CEN, 1991.
14. EH-91, MOPTMA, Spain, 1991.

MONITORING AND MEASUREMENT

STATISTICAL STUDY ON SIMULTANEOUS MONITORING OF REBAR CORROSION RATE AND INTERNAL RELATIVE HUMIDITY IN CONCRETE STRUCTURES EXPOSED TO THE ATMOSPHERE

C. Andrade, J. Sarría and C. Alonso

Institute of Construction Sciences 'Eduardo Torroja'.
C/ Serrano Galvache, s/n. Apdo. 19002.
28033 Madrid. Spain.

1 INTRODUCTION

The influence of the concrete moisture content on the rate of corrosion is well known[1,2,3]. In carbonated or chloride contaminated concrete, no significant rate of corrosion is noticed until a certain moisture is present in the concrete pores. Beyond this minimum the rate of corrosion increases until it reaches a maximum at around 95%, where the oxygen control may start to limit the corrosion rate. Therefore, the rebar corrosion rate is a consequence of the balance between electrical resistivity and oxygen availability[4].

In the data reported in the literature, studies were based on the ambient RH. This was inconvenient because of the long time taken to reach equilibrium at the rebar level. Frequently this equilibrium was not achieved in tests in moisture cycles. This is particularly true in the case of concrete structures placed outdoors, where daily and seasonal cycles induce permanent variations in the external RH. Thus, unless very thin covers are used in testing, the internal RH could differ from the external one.

This inconvenience may now be overcome by measuring the "internal relative humidity"[5]. This is the RH established in an artificial cavity at the rebar level, where a humidity probe is inserted,[6] or a humidity sensor permanently embedded.

In this paper, results of simultaneous monitoring of corrosion rate and RH variations, over two years of exposure of a concrete element contaminated with chlorides, are presented. The measurements were made on plain and cracked zones, in order to establish whether cracks induced by the corrosion itself may influence either the internal RH or the corrosion rate. The work has established some concepts on mean and characteristic (95%) corrosion intensities, and could be used in models for assessing the residual life of structures[7].

2 EXPERIMENTAL

A chloride contaminated structure has been selected for the monitoring. In this paper results for a beam with 3% of $CaCl_2$ per weight of cement added in the mix will be presented. The beam is exposed to atmospheric action by placing it outdoors exposed to rain and sun. At the time of starting monitoring the beam was 3 years old and already cracked along the reinforcements as Figure 1 shows. The monitoring was carried out over two years.

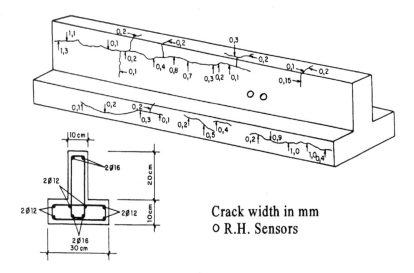

Figure 1 *Crack pattern and shape of the concrete element contaminated with chlorides used for present research*

For the RH measurements special cavities were made in the beam. Initially, the method used was that[5,6] of inserting a humidity probe. However, this operation involves opening the cavity for each measurement, which might change the RH within the cavity. Therefore, permanent humidity sensors and a special measuring device were developed and implemented in a corrosion-rate-meter[8]. These permanent humidity sensors are commercial sensors provided with the necessary electronics to allow convenient measurements. For the measurements of corrosion rate a portable corrosion-rate-meter, Gecor 06, with a controlled guard-ring was used[8,9].

3 RESULTS

Comparing the two methods of recording internal RH, it was found that the permanent sensors enable a quicker reading as their values are usually stable. They give the same information as the humidity probes. That is, both types of measurement give the same internal RH, but the humidity probes take a long time to reach a stable value for each individual reading. With regard to temperature. This was found to be identical within the cavity and in the ambient conditions.

The variations in external and internal RH of the chloride contaminated element are shown in Figure 2, while Figure 3 depicts the variations in temperature. Madrid has a continental climate which is usually dry. It is very dry in summer (the minimum RH may reach 20%) and hot (temperatures around 40°C). In winter it is not too cold (temperatures usually higher than 0°C) and still dry (the annual mean RH is 50%).

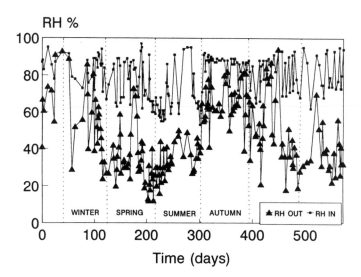

Figure 2   *Variation of record of external (ambient) and internal RH over two years. The seasons are indicated*

From the variation of internal RH shown in Figure 2 it can be deduced that it differs significantly from the external RH. In fact, the internal RH is relatively constant never below about 50% and even falls after the very dry period of the summer.

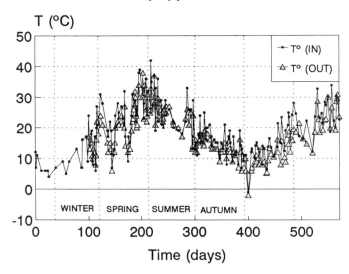

Figure 3   *Variation of temperature over time during the different seasons*

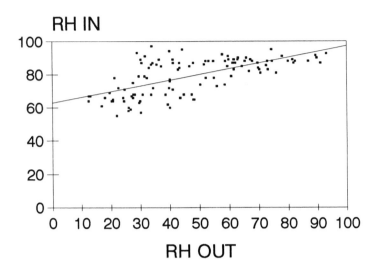

Figure 4   *Relationship between external RH (ambient) and internal RH (inside the cavity)*

No significant difference was found between the cracked and the uncracked zones with regard to the internal RH. This means that cracks up to 0.2 mm in width do not induce a significant change to the ambient RH around the rebar as Table 1 shows.

Table 1 *Internal RH(%) in cracked and uncracked concrete*

RH	Cracked (0.2mm)	Uncracked area
Winter	90	86.6
Spring	86.2	83
Summer	83.6	78.4
Autumn	86.5	86.7

The relationship between the internal and the external RH is shown in Figure 4. A certain relationship exists, although the scatter is high. A similar relationship was noticed between internal RH and temperature. This is shown in Figure 5.

Consequently, neither the external RH nor the temperature may be used alone to predict the internal RH, which seems to be function of both. Nevertheless, a careful study of the results shows that the temperature has a high influence on the internal RH as lower temperature always implies an increase (condensation) in the internal RH. However, rain does not immediately increase the internal RH, at least in this concrete type. Measurements taken during rain did not indicate such a dramatic increase the internal RH as occured during the temperature lowering at nights. The only exception was during long storms in summer.

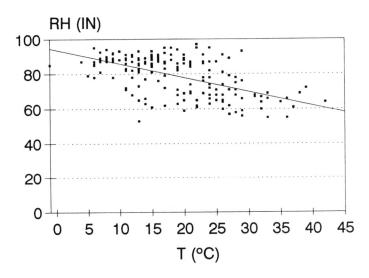

Figure 5 *Relationship between temperature and internal RH*

The parallel evolution of corrosion intensity, Icorr and Ecorr, over time are given in Figures 6 and 7 respectively. The variations recorded in Icorr are within about two orders of magnitude the boundary (Figure 6) established between negligible and significant corrosion[7,9]. This means that the rebar is sometimes actively corroding and sometimes corroding at a negligible rate. The Ecorr values, however, lie in the region indicating active corrosion.

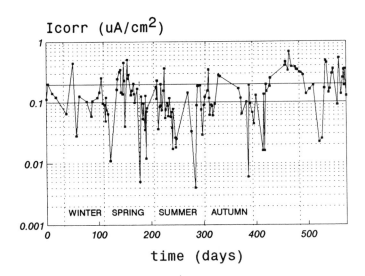

Figure 6 *Variation in Icorr values over time*

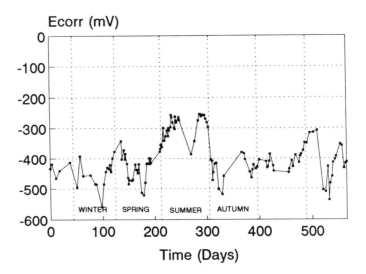

Figure 7   *Variation in Ecorr values over time*

The dependence of Icorr on internal RH and temperature is shown in Figures 8 and 9. No trend could be noticed from these representations. The relationship between Ecorr and Icorr is shown in Figure 10.

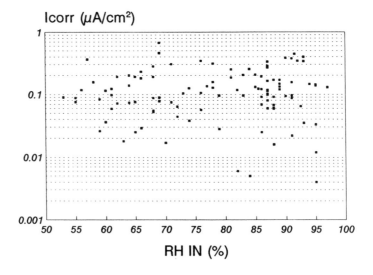

Figure 8   *Relationship between Icorr and internal RH*

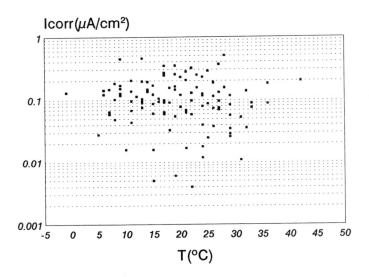

Figure 9 *Relationship between Icorr and temperature*

This apparently random distribution of the corrosion rate, leads to uncertainty in any prediction made based on single or isolated measurements. Therefore, several statistical treatments of the results were undertaken.

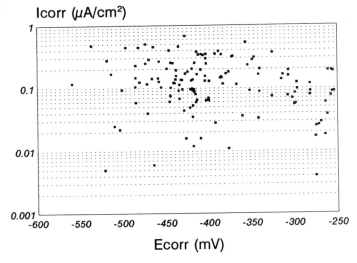

Figure 10 *Relationship between Ecorr and Icorr*

Figure 11 depicts the frequency distribution of the Icorr readings. It can be seen that the majority of the values are grouped around 0.15 μA/cm^2. The 'characteristic Icorr' which is exceeded by only 5% of values higher is 0.35μA/cm^2.

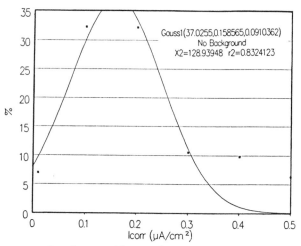

Figure 11 *Frequency distribution of Icorr values*

Another approach undertaken in order to analyze the results was: a) to calculate the 'Mean Icorr' and b) to establish the ranges in which the means are determined. Thus, by integration of the Icorr - time record, a mean Icorr of 0.15 μA/cm^2 was calculated. The standard deviation was 0.12

In order to establish average values, the internal RH was arbitrarily divided into the following ranges < 70%, 70 to 90% and > 90%, and the temperature into the ranges : 0-15°C, 15C° to 25°C and > 25°C. The mean corrosion rate within each range was calculated and is given in Figure 12. By using this kind of approach it can be seen that low temperatures reduce the Icorr values by half, while higher RH values multiply the Icorr values by two, except for temperatures above 25°C where it was likely that the evaporation induced the lowest values of Icorr.

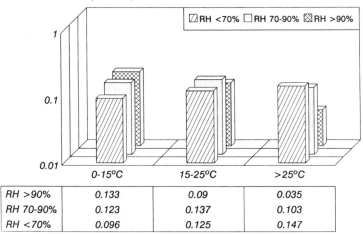

	0-15°C	15-25°C	>25°C
RH >90%	0.133	0.09	0.035
RH 70-90%	0.123	0.137	0.103
RH <70%	0.096	0.125	0.147

Figure 12 *Averaged Icorr values obtained from the ranges of internal RH and temperature indicated in the figure*

4 DISCUSSION

Present results indicate that the corrosion rate distribution when the concrete element is submitted to outdoor climatic action is essentially random. No clear relationship could be found between corrosion intensity and either RH or temperature.

However, this does not contradict previous laboratory experiments showing the influence of external RH upon Icorr. It simply identifies the lack of studies where: 1) internal, and not only external, RH is recorded, and 2) RH variations are studied in conjunction with temperature variations.

The measurement of the internal RH is an important step forward since it indicates better than the external RH, the real humidity at the rebar level.

It should also be noted that no differences were found either in internal RH or in Icorr values, between cracked and uncracked zones. This means that cracks up to 2 mm in width do not necessarily result in intensification of the attack.

Regarding the results from the point of view of their use to predict residual life or future corrosion evolution of the rebar, several deductions can be made. The most evident is the risk of taking an isolated single reading. Either several readings taken during seasonal variations seem to be necessary to characterize a corroding concrete element, or single measurements have to be statistically treated as is done for concrete mechanical strength.

It appears that seasons with the most frequent rain or largest temperature variations, are those which induce the highest Icorr values. Low temperatures (below 10°C) or very high temperatures in dry climates, tend to lead to the lower Icorr values. Therefore, careful selection of the best season to take the Icorr readings seems to be necessary when only a few records are possible. In this case perhaps the use of the 'characteristic Icorr' may be considered together with the mean Icorr in order to make predictions or to be implemented in models of structural residual life.

In the case of continuous monitoring the use of a mean Icorr and its standard deviation may be accurate enough. Dividing the climate (RH and T°C) into ranges and calculating mean values by range, is also an interesting way of climatic characterization. Further studies are needed to optimize these preliminary deductions.

5 CONCLUSIONS

Taking into account that the present results are limited to a single climate and to a concrete element contaminated by 3% of chlorides, the following conclusions may be drawn:
1) The internal RH of concrete behaves differently to the external RH. A high scatter was found in the relationship between internal RH and either external RH or temperature.
2) No differences in internal RH, temperature or Icorr were found between cracked and uncracked zones.
3) No direct relationship was found between corrosion rate and internal RH or temperature. Several avenues were investigated to overcome this apparently random behaviour of Icorr. Calculation of mean Icorr values in RH-Temperature ranges enabled a much better understanding of the process and in consequence, a better Icorr prediction.
4) Research is needed to collect values of on-site Icorr monitored simultaneously with internal RH and temperature.

Acknowledgements

This research is part of the BRITE Project 'The Residual Life of Reinforced Concrete Structures'. The authors are grateful to the DG-XII-Division C for supporting the research. They are also grateful to the partners involved in the project: BCA (U.K.), CBI (Sweden), Lund University (Sweden), Cementa (Sweden) and Geocisa (Spain). They are particularly grateful to Dr Fagerlund for the explanations of the concept of internal RH and how to measure it.

Finally, the authors would like to express their gratitude to Dr L. Parrott (BCA) for providing the humidity probes for measuring the internal RH, to Geocisa for preparing the permanent humidity sensors and to the firm APIA XXI of Spain for supporting the work of J. Sarría with a grant.

References

1. F. Campus, 'Importance des déformation thermo-hygrometriques du béton pour la protection des armatures', CERES (Génil Civil) Memoires, 1966, **16**, 35.
2. J.A. González, A. Algaba and C. Andrade, 'Corrosion of reinforcing bars in carbonated concrete', *British Corrosion J.*, 1980, **15**, 135.
3. K. Tuutti, 'Corrosion on Steel in Concrete', CBI Forskning Research, Swedish Cement and Concrete Research, Stockholm, Sweden, 1982, 486.
4. C. Andrade, C. Alonso, J.A. González and S. Feliú, 'Similarity between atmospheric/underground corrosion and reinforced', Corrosion of reinforcement in concrete, C.L. Page, K. Treadaway and P. Bamforth Editors, SCI, London 1990, 39.
5. G. Fagerlund and G. Hedenblad, 'Calculation of the moisture-time fields in concrete', Div. of Building Materials, Lund, Institute of Technology, Report TVBM-3052, Lund 1993.
6. L. J. Parrott, 'Moisture profiles in drying concrete', *Advances in cement research*, 1988, **1**, 164.
7. C. Andrade, C. Alonso, K. Petterson, G. Somerville and K. Tuutti, 'The Practical assessment of damage due to corrosion', Concrete Accross Borders, Int. Conference, Odense (Denmark), 1994, **I**, 301.
8. J. Rodriguez, L. M. Ortega, A. M. García, L. Johansson and K. Petterson, 'On-site corrosion rate measurements in concrete structures using a device developed under Eureka Project 'EU-401'. Concrete Across Borders, Int. Conference, Odense (Denmark), 1994, **I**, 215.
9. S. Feliú, J.A. González, S. Feliú Jr. and C. Andrade, 'Confinement of the electrical signal for in-situ measurement of Polarization Resistance in reinforced concrete', *Materials Journal ACI*, 1990 September-October, 457.

THE INFLUENCE OF PITTING CORROSION ON THE EVALUATION OF POLARISATION RESISTANCE OF BARS IN CONCRETE STRUCTURES

E.Proverbio and R.Cigna

Department ICMMPM
University of Rome 'La Sapienza'
via Eudossiana, 18 -I-00184 Rome, Italy

1 INTRODUCTION

Penetration of chlorides from the surface is one of the main causes of corrosion of rebars in reinforced concrete structures exposed to either marine environments or the application of deicing salts.[1-3]

Time to initiation of corrosion and corrosion rate (CR) can be monitored by means of corrosion probes embedded in the concrete at the time of casting[4-6] and based on the linear polarisation resistance technique (LPR), as well as by means of equipment based on using an external device on already existing structures.[7,8]

It has been observed, in previous experiments,[9,10] that CR measurements carried out using the LPR technique can sometimes underestimate the residual service life of reinforced concrete structures. The initial stage of corrosion of rebars in chloride permeated concrete has been found to be characterized by very high apparent corrosion rates; during the subsequent propagation stage the corrosion rate decreases down to typical values, in the range of 5 to 15 μm/y (active iron). It has been speculated that the anomalous values, higher than expected, could be correlated with the presence of isolated pits on the rebar surface. In the presence of localised corrosion, when the rebar surface is generally in a passive state (characterized by a low corrosion rate), a high current flows through the bar during polarisation. So, even if this current is mediated over the whole polarised area, the resulting calculated corrosion rate could be of the same order of magnitude and possibly higher than that of uniformally corroding steel.

In the present work a more satisfying explanation of the anomalous phenomenon is put forward; experimental findings supporting the hypothesis are described in detail.

2 EXPERIMENTAL

Fifteen concrete beams, 150 x 150 x 600 mm, were prepared following the mix design reported in Table 1. Seven corrugated steel bars were embedded in each specimen as illustrated in Figure 1, five of which (A,B,C,D,L) acted as working electrodes, the remaining bars acted as counter electrodes.

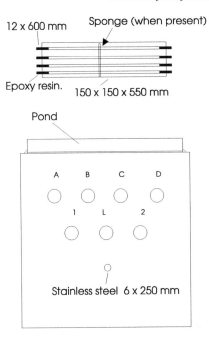

Figure 1 *Scheme of concrete specimen (from A to 2 ⌀12 steel bars)*

A stainless steel bar was also included to act as a reference electrode during LPR measurements. The concrete cover was 30 mm. To promote the initiation of localised corrosion attacks, artificial crevices (obtained by means of 10 mm wide Teflon ribbon wrapped around the central zone of the bar) and artificial cracks (obtained by means of a sponge inserted transversely into the specimen) were provided. In addition, a certain amount of calcium chloride was added to the concrete mixtures of some samples, as reported in Table 2, to accelerate corrosion initiation.

Table 1 *Cement type and concrete mix design*

Cement type	Designation	Notation	S.B.B.(N/mm^2)
III	Slag cement	III/A	400

Concrete Mix Design						
Cement	Cement Factor	Aggregates	Coarse Aggr. (max size)	Mixture Proportion		
Strength class 32.5	$350 \ kg/m^3$	$1700 \ kg/m^3$	12.5 mm crushed	45 % gravel	45% crushed sand	10 % natural sand
Test Results						
W/C s.s.d.	Fresh Density	Slupm	Air	Compressive Strengths		
				7d	28d	
0.64	$2,380 \ kg/m^3$	110 mm	2 %	$190 \ kg/cm^2$	$393 \ kg/cm^2$	

Table 2 *Types of concrete samples*

Sample type	Artificial crevice	Artificial crack	Chloride content (wt % by cement)	Calcium chloride solution ponding	Fresh water ponding
1			2		X
2		X	0	X	
3	X		0.6		X
4		X	0.6	X	
5	X		0.6		

Before being embedded in the concrete, the steel bars were immersed in diluted sulphuric acid, washed and dried, and their ends were covered with epoxy resin for a length of 50 mm. After a curing period of 45 days a 10 % calcium chloride solution was poured periodically onto the upper surface of samples of types 2 and 4; fresh water was used instead for types 1 and 3. Samples of type 5 were only exposed to the air.

After about 70 days of experimentation one sample of each type was crushed in order to observe the surface rebar condition.

3 RESULTS AND DISCUSSION

Corrosion rate measurements carried out by means of a commercial intensiodynamical corrosion meter, acting directly on the whole bars, and, for the first set of measurements, by using equipment based on an external device developed by Andrade and co-workers,[7,8] were performed, as well as corrosion potential measurements, starting after 20-30 days of curing. Results are summarised in Figures 2 and 3.

Except for samples of type 5, corrosion potential and corrosion rate were significantly influenced by the ponding which, due to the high concrete porosity, readily activated corrosion phenomena.

As already suggested in a previous paper[10], initiation of pitting should be responsible for the strong fluctuations of CR, with peaks as high as 50-60 µm/y. Such values are, however, of no practical significance, especially in predicting the structure lifetime, although they are good indicators of the pitting initiation. Moreover, the sudden decrease in these values could be related to the temporary repassivation of the pit surface (more evident in the presence of an inhibiting agent such as calcium nitrite).[11] CR intensity seems, however, to be related to the pit depth rather than to the extension of the total corroded area as evidenced by a visual inspection of the bars. Type 1 samples showed severe pits, with a low density homogeneous distribution like a 'leopard skin', whilst on type 2 samples evidence of a strong localised corrosion attack, in the region of the artificial crack was seen. Type 3 and 4 samples (and, to a limited extent, also type 5) showed a broad corroded area with very small and diffuse pits and a lower average CR.

It should be however stressed that the results reported here are also influenced by the fact that it was very difficult to achieve a truely localized corrosion attack of known extension both by the artificial crevice and the artificial cracks, since quite wide and diffuse

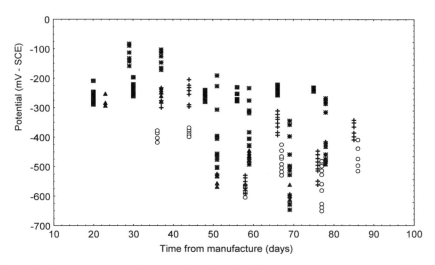

Figure 2 *Corrosion potential vs time for rebars in different concrete specimens.* ○ *Type 1 samples;* ✳ *Type 2 samples;* ✚ *Type 3 samples;* ▲ *Type 4 samples;* ■ *Type 5 samples*

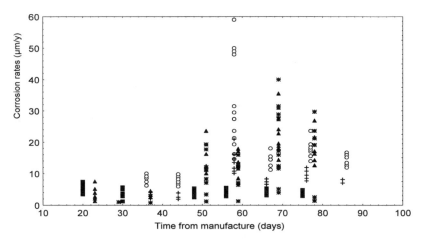

Figure 3 *Corrosion rate vs time for rebars in different concrete specimens.* ○ *Type 1 samples;* ✳ *Type 2 samples;* ✚ *Type 3 samples;* ▲ *Type 4 samples;* ■ *Type 5 samples*

corrosion phenomena occurred on the lower surface of the bars, almost independently of the type of samples. Such phenomena, which were also observed in previous experimentation as well as on real concrete structures, chould be related to the compaction of fresh concrete as a result of mechanical vibration, which could lead to bubble collection or bleeding under the horizontal bars, thus providing a zone richer in uncombined chloride (when chloride is originally present in the concrete mixture) or a preferential path for water percolation and collection.

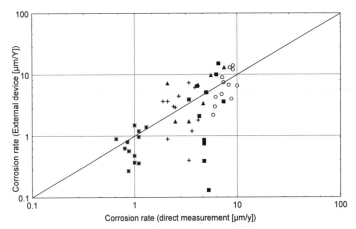

Figure 4 *Relationship between CR measured by means of an external device base equipment and a corrosion meter which considered the whole bar area. ○ Type 1 samples; ✱ Type 2 samples; ✚ Type 3 samples; ▲ Type 4 samples; ■ Type 5 samples*

The presence of spots of localized corrosion or pits is also a typical cause of error when the external corrosion meter is used.[12,13] If a very active zone is present on the bar, current can easily escape from the geometrical area covered by the external counter electrode, even when an electrical confinement is considered. This can result in a broader spread of CR values compared with the values obtained by considering the whole area of the bars (Figure 4).

Pitting corrosion can be responsible for not only altered CR measurements but also erroneous evaluation of the structure's residual service life, as stressed by Gonzales et al. in a recent paper.[14] It was found that the maximum penetration of the localised attack on steel was about four to eight times the average penetration of the attack on the overall reinforcement surface. Since mechanical resistance of steel reinforcement in the structure itself could be severely affected by this type of corrosion attack, these results should be taken into account when predicting the structure's residual life.

In conclusion it should be stressed that the presence of pitting corrosion could alter the CR measurements. So it is necessary to better understand this influence and also to provide a theoretical justification of the phenomenon. A statistical approach is to be considered as the best way to handle CR data rather than to regard them as absolute and real values.

4 CONCLUSIONS

Pit initiation can significantly affect CR measurements obtained by LPR, leading to a strong variation of the CR itself over time. It may therefore be very difficult to predict the residual life of concrete structures in such circumstances. This problem can be considered of minor importance for structures where the CR is monitored from the initial construction. In this case the CR fluctuations with the time will be well known, and the development of dangerous conditions could easily be detected following the stabilisation of moderate to high values.

For existing structures the evaluation of either the optimal maintenance time or the residual life may be strongly influenced by the lack of historical knowledge.

In the case of the half-cell measurements, it is suggested that only several sets of measurements, repeated over time, will give a reliable idea of the corrosion state in concrete reinforced structures.

However, in spite of the above mentioned limits, the CR determination, especially for existing structures, is one of the few tools that can give a good preliminary evaluation of the risk of corrosion: further investigation, to be carried out in parallel (half-cell potential, resistivity, chloride profiles or carbonation depths) can clarify the situation.

5 ACKNOWLEDGEMENT

The authors would like to thank Dr. Giorgio Peroni, Dr. Marcello Luminari and Dr. Guido Bastianelli of the Autostrade S.p.A. for the sample preparation and for the helpful collaboration.

6 REFERENCES

1.   N. J .M. Wilkins and J. V. Sharp in 'Corrosion of Reinforcement in Concrete' C. L. Page, K. W. J. Treadaway, P. B. Bamforth, Elsevier Applied Science, London, 1990, 3.
2.   T. R. Menzies in 'Corrosion Forms & Control for Infrastructure', V. Chaker, ASTM STP 1137, Philadelphia, 1992, 30.
3.   N. S. Berke and M. C. Hicks in 'Corrosion Forms & Control for Infrastructure', V. Chaker, ASTM STP 1137, Philadelphia, 1992, 207.
4.   R. Cigna, G. Familiari and G. Peroni, *Autostrade*, 1991, **33**, 1.
5.   N. R. Short, C. L. Page and G. K Glass, *Magazine Concr. Res.*, 1991, **43**, 156.
6.   P. Schiessl and M. Raupach, in 'Rehabilitation of Concrete Structures', (Ed. D. W. S. Ho and F. Collins), RILEM, 1992, 97.
7.   S. Feliu, J. A. Gonzales and M. C. Andrade, V. Feliu, *Corr. Sci*, 1989, **29**, 1.
8.   C. Andrade, A. Macias, S. Feliu, M. L. Escudero and J. A. Gonzales, in 'Corrosion Rates of Steel in Concrete' N. S. Berke, V. Chaker, D. Whiting, ASTM STP 1065, Philadelphia, 1990, 134.
9.   R. Cigna, G. Familiari, F. Gianetti and E. Proverbio in 'Corrosion and Corrosion Protection of Steel in Concrete', (Ed. R. N. Swamy), Sheffield Academic Press, Sheffield, 1994, Vol. II, 878.
10.   E. Proverbio and R. Cigna, EMCR 94, *Mat. Science Forum*, 1995, **192-194**.
11.   R. Cigna, G. Familiari, F. Gianetti and E. Proverbio, *Ind. Ital. Cemento*, 1995, 10.
12.   J. A.Gonzales, M. Benito, S. Feliu, P. Rodriguez and C. Andrade, *Corrosion*,1995, **51**, 2.
13   S. Feliu, J. A. Gonzales and C. Andrade, Proc. Int. Symposium on the Condition Assessment, Protection, Repair and Rehabilitation of Concrete Bridges Exposed to Aggressive Environments, ACI Fall Convention, Minneapolis, Minnesota, November 1993.
14.   J. A. Gonzalez, C. Andrade, C. Alonso, S. Feliu, *Cem. Concr. Res.*, 1995, **25**, 2.

GALVANOSTATIC PULSE TRANSIENT ANALYSIS FOR DETERMINING CONCRETE REINFORCEMENT CORROSION RATES

K. R. Gowers, J. H. Bungey and S. G. Millard

Department of Civil Engineering
University of Liverpool
Brownlow Street, P.O. Box 147
Liverpool L69 3BX

1 INTRODUCTION

The traditional technique for obtaining information about the corrosion state of steel reinforcement in concrete structures is by measurement of potential. From measurements conducted on highway bridge decks in the United States which had been subjected to deicing salts,[1] a standard was derived relating potential values to the probability of active corrosion.[2] More recently, it has been found that similar standards may be applied to reinforced concrete structures in different environments, but that the threshold potential values defining the different probabilities of corrosion will not be the same.[3]

A method of mapping potential was therefore devised, in which potential measurements are carried out at points on a regular grid over the surface of a structure, and this is then presented as an iso-potential contour map. The positions of most likely active corrosion are the positions of the most negative potential. Potential differences between different locations on a surface which exceed 200mV, particularly where iso-potential contour lines are closely spaced, are also indicative of corrosion activity.[4]

A method has recently been devised of enhancing the potential contours of a potential map by applying a small constant current pulse at each location where the potential is measured and measuring the shift in potential after a pre-determined time period.[5,6] It is claimed that this enables regions of active corrosion to be identified more easily.

A more sophisticated technique, which analyses the potential transient response to a galvanostatic pulse, has been used to provide a method of directly measuring the corrosion rate of steel in concrete. That technique is the subject of this paper.

2 EXPERIMENTAL TECHNIQUE

The potential transient response is recorded during the application of a constant current perturbation. The transient behaviour may then be analysed according to electronic theory.[7] For a simple Randles circuit (Figure 1) the potential response is given by:

$$V_t = IR_\Omega + IR\left[1 - \exp\left(\frac{-t}{RC}\right)\right] \qquad (1)$$

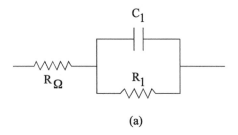

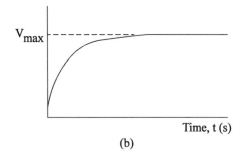

(b)

Figure 1 *Typical potential transient response to a galvanostatic pulse*

(a) Simplified (Randles) equivalent circuit.

(b) Transient response

where V_t is the shift in potential after time t, I is the applied current, R_Ω is the ohmic component of the resistance (normally taken as the concrete resistance), and R and C are the resistance and capacitance at the steel/concrete interface.

Plotting this equation as $\log_e (V_{max} - V_t)$, against t, where V_{max} is the maximum value of potential shift, gives a straight line with an intercept of $\log_e$ IR and slope 1/RC (Figure 2). From this R and C can be determined.

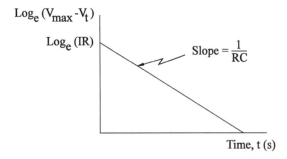

Figure 2 *Logarithmic plot of the transient response to a galvanostatic pulse*

Unfortunately, steel in concrete does not readily lend itself to being modelled as a simple Randles circuit, and a model containing several different RC time constants is more appropriate. However, similar analysis may be used to derive an equivalent circuit. Successive fitting of slopes to gradually shorter time periods of the plot of $\log_e (V_{max} - V_t)$, against t, followed by removal of data corresponding to the fit, yields a succession of RC time constants and incrementally yields a complete equivalent circuit.[7]

Recently published work by the Authors[8] has shown how the results obtained from galvanostatic transient analysis accurately assess the resistor and capacitor values of known circuits of electronic components. Measurements were also carried out on steel in a calcium hydroxide/sodium chloride solution and on steel reinforcement in concrete. It was demonstrated that the galvanostatic pulse technique could be successfully used to separate the different RC time constants of a corroding system.

This work has now been extended to investigate the influence of varying the measurement parameters on the results obtained. The effect of different sampling rates, sampling periods and current magnitudes have all been investigated.

Collection and storage of large amounts of transient data is difficult when operating on site, and the objective of this study was to identify the slowest sampling rate and the minimum period of data acquisition which would give sufficiently accurate results.

Efforts have also been made to assign the interfacial corrosion processes to specific RC time constants. It is intended that the magnitudes of the capacitive components could be used to define which time constants are related to the corrosion processes, and thus obtain a more accurate value of the corrosion rate.

A constant current source has been developed by technical personnel within the University of Liverpool. This gives a constant current of 0.01, 0.1, or 1.0 mA, with a rise time of ≤ 0.2ms. A p.c. based data aquisition system has been used that enables transient voltage changes to be measured at a maximum rate of 5 kHz for a 60 s period with 12-bit resolution.

The three electrode measurement system shown in Figure 3 was employed, whereby current could be applied to the steel reinforcement from the brass plate auxiliary electrode, and the potential change of the steel measured against a central reference electrode.

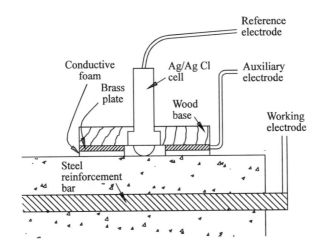

Figure 3 Three electrode measurement for steel reinforcement in concrete

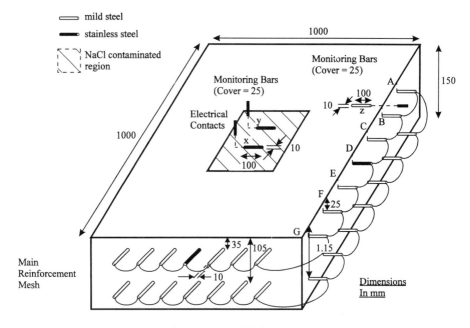

Figure 4 Concrete Test Slab

The applied current should be chosen so that the polarisation remains within the linear ± 30 mV region, and the electronic theory remains valid. However as some of the potential shift is due to an ohmic voltage drop across the concrete cover, an overall potential shift to within 50-100 mV of the rest potential for steel in concrete was considered acceptable.

Following some preliminary tests performed on circuits of known electronic components, measurements were carried out on the monitoring bars located (Figure 4) in a 1000 mm x 1000 mm x 150 mm slab of 50 N/mm² strength concrete positioned outdoors in a coastal urban environment in Liverpool. The measurements were performed approximately three years after casting.

2.1 Variation of Sampling Period, Sampling Frequency, and Current Magnitude

An anodic galvanostatic pulse of 0.01 mA was applied to each of the circuits shown in Figure 5 for a period of 60 s, and the resulting potential transient was recorded using the maximum available sampling frequency of 5 kHz. A similar measurement was also performed on the monitoring bars in the concrete slab, with a current of 0.1 mA employed for the actively corroding mild steel bars and a current of 0.01 mA for the passive stainless steel bar   The potential transient data was subsequently manipulated in order to produce data files of the same data but using sampling times of 30 s and 10 s (by removing the subsequent data) or at sampling rates of 500 Hz and 50 Hz (by only using every tenth data point or every hundredth data point from the original data set).

The effect of using currents of 0.01 mA or 1.0 mA for the mild steel bars and 0.1 mA for the stainless steel bar was also investigated in a similar manner.

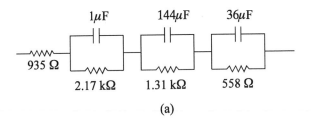

(a)

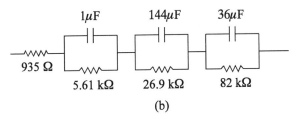

(b)

Figure 5 Electrical circuits to simulate corrosion
(a) Circuit A: Active corrosion simulation
(b) Circuit B: Passive corrosion simulation

2.2 Identification of Time Constants Associated with the Corrosion Processes

Galvanostatic pulse transient analysis was carried out on each electrically isolated reinforcement bar with the auxiliary electrode/reference electrode assembly positioned at different lateral distances along the surface of the slab away from the bar. This was because the response due to the steel/concrete interface (i.e. due to the corrosion processes) should remain constant, whereas the response due to the concrete resistance should change as the auxiliary electrode/reference electrode assembly is moved away from the bar. Hence it should be feasible to identify which components of the equivalent electronic circuit are associated with corrosion activity.

3   EXPERIMENTAL RESULTS

3.1 Electronic Circuits

The results obtained for two circuits of electronic components are shown in Table 1. The two circuits were devised to simulate active corrosion and passivity.

For the passive simulation circuit, with larger values of resistance and hence larger RC time constants, all the values were quite accurately determined for all sampling rates from 5 kHz to 50 Hz. However, the active simulation circuit, containing smaller values of resistance and therefore smaller time constants, became more difficult to measure as the sampling rate was reduced. Accurate results were obtained at 5 kHz, and a reasonable degree of accuracy was also obtained at 500 Hz, but 50 Hz was found to be too slow a sampling rate to extract the fastest time constant, and hence all the subsequent values derived were inaccurate.

Table 1   Equivalent circuit values measured using the galvanostatic pulse technique on the electronic circuits in Figure 5.

	R_0 (kΩ)	R_1 (kΩ)	C_1 (μF)	R_2 (kΩ)	C_2 (μF)	R_3 (kΩ)	C_3 (μF)	R_{TOTAL}
Actual Values for 'Active' Circuit	**0.935**	**2.17**	**1**	**0.558**	**36**	**1.31**	**144**	**4.97**

Sampling Rate (kHz)	Time (s)	R_0 (kΩ)	R_1 (kΩ)	C_1 (μF)	R_2 (kΩ)	C_2 (μF)	R_3 (kΩ)	C_3 (μF)	R_{TOTAL}
5.00	10	1.16	1.90	1.05	0.648	23.2	1.35	148	5.06
0.500	10	2.32	1.02	2.69	0.533	61.7	1.17	192	5.04
0.050	10	3.66	-	-	1.17	119	0.192	22000	5.02

	R_0 (kΩ)	R_1 (kΩ)	C_1 (μF)	R_2 (kΩ)	C_2 (μF)	R_3 (kΩ)	C_3 (μF)	R_{TOTAL}
Actual Values for 'Passive' Circuit	**0.935**	**5.61**	**1**	**82.0**	**36**	**26.9**	**144**	**115**

Sampling Rate (kHz)	Time (s)	R_0 (kΩ)	R_1 (kΩ)	C_1 (μF)	R_2 (kΩ)	C_2 (μF)	R_3 (kΩ)	C_3 (μF)	R_{TOTAL}
5.00	40	1.04	5.18	0.908	93.6	31.8	16.0	369	116
0.500	40	2.62	3.14	1.03	93.3	31.4	16.9	347	116
0.050	40	7.14	2.76	0.479	87.9	34.9	18.6	284	116

A sampling period of 10 s was found to allow complete stabilisation of the response for the active simulation circuit, but a longer period of 40 s was found to be necessary for the passive simulation circuit.

As the sampling rate was reduced, a larger value was obtained for both circuits for the ohmic component of the resistance, and this became significantly overestimated when the sampling rate was reduced to 50 Hz.

3.2 Mild Steel in Concrete

The effect of varying the sampling rate and the time period are shown in Table 2 for a mild steel bar in concrete. Four time constants were readily separated when using faster sampling rates of 5 kHz or 500 Hz and longer time periods of 30 s or 60 s. However, the faster time constants became more difficult to isolate and quantify as the sampling rate was reduced, as would be expected. A minimum sampling rate of 500 Hz was needed for a minimum acquisition period of 30 s to give adequate resolution of the data.

AC impedance results are given for comparison in Table 2. These were obtained from the Nyquist plot in Figure 6. The results are not identical, but are encouragingly similar. However, it was much easier to separate the different time constants using the galvanostatic pulse technique than it was using AC impedance data.

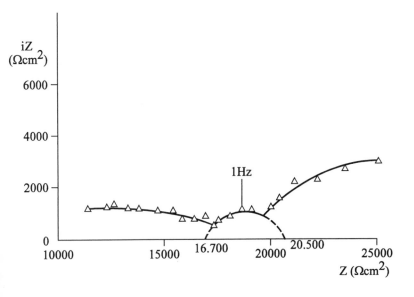

***Figure 6 Nyquist plot of AC impedance measurement
of corroding mild steel bar in concrete***

Table 2   *Equivalent circuit values obtained from a corroding mild steel bar in concrete (0.1 mA pulse)*

Sampling Rate (kHz)	Time (s)	R_0 (kΩcm²)	R_1 (kΩcm²)	C_1 (μF/cm²)	R_2 (kΩcm²)	C_2 (μF/cm²)	R_3 (kΩcm²)	C_3 (μF/cm²)	R_4 (kΩcm²)	C_4 (μF/cm²)	R_{TOTAL}
5.00	60	10.6	2.30	1.62	2.64	79.5	9.06	2180	9.06	2180	27.6
0.500	60	13.0	1.26	13.1	2.31	80.5	8.80	2360	8.79	2360	28.9
0.050	60	14.2	-	-	1.95	165	8.79	2460	8.80	2460	28.2
5.00	30	10.6	2.31	1.63	2.23	76.8	7.23	1610	7.23	1610	25.4
0.500	30	13.0	1.93	11.5	2.39	173	7.18	1640	7.18	1640	26.7
0.050	30	14.2	-	-	1.51	320	6.96	1900	6.96	1900	25.8
5.00	10	10.6	2.43	1.69	2.22	48.5	5.66	690	5.66	690	22.6
0.500	10	13.0	-	-	2.33	9.39	5.48	740	5.48	740	23.3
0.050	10	14.2	-	-	-	-	6.34	496	6.34	496	23.0
AC Impedance Results		**9**	**8**	**0.0074**	**3.8**	**42.4**	**15**	**1060**	**15**	**1060**	**35.8**

3.3 Stainless Steel in Concrete

Table 3 shows the results for a passive stainless steel bar in concrete. Separation of time constants was again only effectively achieved with relatively fast sampling rates and long time periods. A minimum sampling rate of 500 Hz was required for reliable results.

The sampling period was very important for this specimen. Particularly for the second, larger, resistance value, R_2, the value obtained was considerably reduced by a decrease in the measurement time from 60 s to 30 s or to 10 s. In fact, the value of >1000 kΩcm^2 obtained from AC impedance suggests that even after a 60 s galvanostatic pulse duration, the potential response has not fully stabilised and hence the resistance may be underestimated. It was concluded that a minimum sampling period of 60 s is needed for measuring passive steel in concrete.

3.4 Effect of the Magnitude of the Applied Current

The magnitude of the applied current pulse has a significant effect on the results obtained from galvanostatic pulse transient analysis (Table 4). The use of higher currents permits the resolution of the response into more time constants, because the signal is larger compared with the background noise. However, if the current is large enough to give polarisation outside the linear region, an erroneously low value of total polarisation resistance, R_{TOTAL}, is obtained. This is particularly significant for an applied current of 0.1 mA to the stainless steel bar, but also occurs when using a 1.0 mA applied to the mild steel bar.

Conversely if a relatively low current pulse is applied, then it is difficult to separate the response from the background noise, and separation into different time constants can not be obtained. This has been found to be the case if the total polarisation including the ohmic part of the response is less than 10 mV, for example the result obtained for the mild steel bar using 0.01 mA current.

3.5 Identification of Time Constants with the Corrosion Processes

Galvanostatic pulse results obtained for the auxiliary electrode/reference electrode assembly positioned at different distances away from the mild steel bar are presented in Table 5. The resistances are tabulated adjacent to the range of capacitances with which they were linked in parallel.

The resistances which remain constant as the distance from the measurement location to the corroding steel bar is increased are those which are assumed to be related to the steel/concrete corrosion interface. R_1 and R_2, corresponding to $C_1 < 1$ μF/cm^2 and $C_2 = 1$-10 μF/cm^2, comply with this. R_3, R_4, and R_5, corresponding to capacitance values > 20 μF/cm^2, increase with distance, and are therefore assumed to be related to electrical properties of the concrete rather than to the steel/concrete interface. They could be due to impedances arising from transport/diffusional properties of the concrete. Capacitance values of the order of several mF/cm^2 have previously been related to diffusional impedances.[9] R_0, arising from the ohmic resistance of the concrete, also increases with distance as was expected.

From previous galvanostatic pulse work in solution, double layer capacitance values (i.e. corresponding to the steel/concrete corrosion interface) of between 10 and 500 μF/cm^2 have been suggested.[8] However, the results in Table 5 suggest that 1-20 μF/cm^2 is a more appropriate range here. This is in agreement with the experimental value of 17 μF/cm^2

Table 3   *Equivalent circuit values obtained from a passive stainless steel bar in concrete (0.01 mA pulse)*

Sampling Rate (kHz)	Time (s)	R_0 (kΩcm^2)	R_1 (kΩcm^2)	C_1 (μF/cm^2)	R_2 (kΩcm^2)	C_2 (μF/cm^2)	R_{TOTAL} (kΩcm^2)
5.00	60	25.5	54.9	31.3	623	36.1	678
0.500	60	27.5	54.1	47.7	623	36.3	677
0.050	60	29.5	-	-	630	34.6	630
5.00	30	25.5	88.7	41.1	454	34.7	543
0.500	30	27.5	90.1	58.7	472	35.1	572
0.050	30	29.5	-	-	479	33.4	479
5.00	10	25.5	56.5	63.1	220	29.7	277
0.500	10	27.5	-	-	221	29.1	221
0.050	10	29.5	-	-	221	28.8 -	221
AC Impedance Results		17	-	-	>1000	?	>1000

Table 4.   *Equivalent circuits derived from galvanostatic pulse transient analysis using different applied currents*

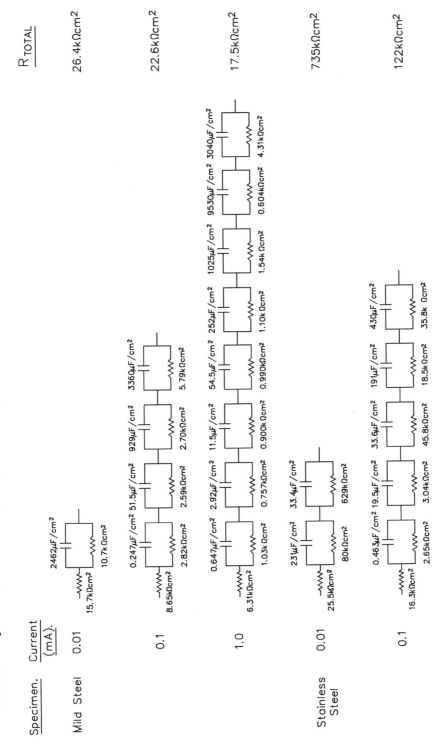

obtained for steel in a corrosive 0.001M potassium chloride solution.[10]

If R_2 is the interfacial charge transfer resistance at the steel/concrete interface, in parallel with the double layer capacitance, then the rate of corrosion activity can be directly evaluated.

The resistance R_1, in parallel with a capacitance $C_1 < 1$ $\mu F/cm^2$, also stays constant with distance, and may therefore also be assumed to be related to the steel/concrete interface. This resistance and capacitance may be due to film formation at the interface. It seems reasonable to assume that this resistance also affects the corrosion processes.

Incidentally, this division of time constants for the system into three distinct regions is also seen on the AC impedance plot shown in Figure 6. The values do not correspond exactly, as the measurements were not taken at exactly the same time, but the electrical behaviour is analogous.

A similar result is obtained for stainless steel when the auxiliary electrode/reference electrode assembly is again moved laterally to different distances (Table 6) from the embedded bar. Both the measured resistance values R_1 and R_2 clearly stay constant with distance, showing that capacitance values of 10-50 $\mu F/cm^2$ are related to the steel/concrete interface for this bar.

4   DISCUSSION

For the use of galvanostatic pulse transient analysis, parameters need to be chosen to allow the correct interpretation of measurements on active and passive steel in concrete.

The applied current should be chosen to give sufficient polarisation to permit analysis of the data over any background noise, and yet remain approximately within the linear polarisation region. A current sufficient to give rise to a non-ohmic component of the polarisation between 5 and 50 mV is suggested.

A minimum data acquisition rate of 500 Hz was found to be necessary if the fastest time constants are to be evaluated, and a minimum sampling period of 60 s is recommended, in order to allow reasonable stabilisation of the slower time constants. Faster acquisition rates and longer sampling periods can be used, but for site use there may then be significant problems with data handling and storage with portable data acquisition equipment. The use of a graduated sampling rate is one possible solution to this difficulty.

Subsequent analysis of the transient response to a galvanostatic pulse may be used to derive an equivalent electrical circuit for the corroding system. As a general rule, it would seem that the corrosion processes are related to the resistances in parallel with capacitances <100 $\mu F/cm^2$. (This value is a compromise in order to take into account the results obtained for both actively corroding and passive steel in concrete. The use of this value should not lead to significant error in either situation.) Summing these resistances would yield a value equivalent to the polarisation resistance. A corrosion penetration rate may be readily calculated from this.[8]

The identification of the time constants associated with the actual corrosion processes of steel in concrete is a major step forward. Only the galvanostatic pulse and AC impedance techniques allow separation of the response to an applied signal into different time constants. AC impedance is a very time consuming system to use if a complete frequency sweep is employed, and therefore the advantage of using the galvanostatic pulse technique in practice is significant, because of its ability to acquire enough data in a relatively short time.

The major problem in practice is one which is common to all direct in-situ measurement techniques for the corrosion of steel in concrete, and this is in quantifying the area of steel

Table 5 **Equivalent circuit resistance ($k\Omega\ cm^2$) measured using different lateral distances from measurement location to corroding mild steel bar**

	Lateral Distance From Bar (mm)			
	0	130	260	390
R_o	28.1	36.1	52.2	38.9
R_1 ($C_1 < 1\mu F/cm^2$)	10.7	10.5	8.92	8.26
R_2 ($C_2 = 1 - 10\mu F/cm^2$)	6.75	7.25*	8.92	7.03
R_3 ($C_3 = 10 - 100\mu F/cm^2$)	-	-	7.35	6.15
R_4 ($C_4 = 100 -1000\mu F/cm^2$)	-	3.30	5.16	5.14
R_5 ($C_5 > 1000\mu F/cm^2$)	2.64	5.10	7.86	8.2

(* $12\mu F/cm^2$)

Table 6 **Equivalent circuit resistances measured using different lateral distances**

	Lateral Distance From Bar (mm)			
	0	130	390	520
R_o ($k\Omega.cm^2$)	22.7	20.1	99.8	145
R_1 ($k\Omega.cm^2$)	529	528	532	554
C_1 ($\mu F/cm^2$)	48.0	47.8	44.8	43.9
R_2 ($k\Omega.cm^2$)	83.8	81.0	74.4	77.6
C_2 ($\mu F/cm^2$)	12.2	22.5	38.3	35.4

reinforcement which is actually being measured in a site structure.

It has sometimes been assumed that the area polarised is the area of steel directly beneath the auxiliary electrode. For actively corroding steel, this assumption would probably not produce significant errors.[11] Larger errors would be obtained for passive steel, but this would be less important as the absolute accuracy of the corrosion rate being evaluated has little practical significance.

Another possible way of defining the polarised area is to incorporate monitoring bars of known area into the structure. This has been done previously during the construction of certain structures,[12] but could also be achieved at a later stage by cutting through and electrically isolating parts of the main reinforcement. This would not be expected to significantly weaken the structure if it were only done in a few places.

Probably the best way to quantify the area of polarisation is to use a guard ring surrounding the auxiliary electrode. A second constant current source would be needed to apply the guard ring current, and the two current sources would have to be synchronised and have the same current rise time. Galvanostatic linear polarisation resistance studies have already been performed using this technique,[13] and the method could possibly be adapted for use in galvanostatic pulse measurements.

5   CONCLUSIONS

1.   For galvanostatic pulse transient analysis, an applied current should be chosen so as to give a non-ohmic component of polarisation between 5 and 50 mV.
2.   A minimum data acquisition rate of 500 Hz is recommended.
3.   A minimum sampling period of 60 s is recommended.
4.   When an equivalent electrical circuit has been derived from the transient analysis, the resistances in parallel with capacitances $<100\ \mu F/cm^2$ may be summed to give a value of polarisation resistance corresponding to the steel/concrete interface.
5.   Techniques for use of galvanostatic pulse transient analysis to measure the instantaneous rate of corrosion of steel in concrete have been demonstrated. The way is now open for field equipment for in-situ measurements to be commercially developed.

6   ACKNOWLEDGEMENTS

The authors would like to thank the Science and Engineering Research Council for funding this work (Grant Ref. GR/G53552). They are also indebted to the technical support of Mr D. Cottingham in developing a constant current source and Mr J. Wilkinson in writing the computer software for data processing.

7 REFERENCES

1. R. F. Stratfull, Highway Research Record, 433, 1973, 12.
2. 'Half-Cell Potentials of Uncoated Reinforced Steel in Concrete', American Society for Testing and Materials, ASTM C876-87.
3. B. Elsener and H. Bohni, 'Potential Mapping and Corrosion of Steel in Concrete', 'Corrosion Rates of Steel in Concrete', (Ed: N. S. Berke, V. Chaker, and D. Whiting), ASTM, Philadelphia, 1990, 143.
4. P. R. Vassie, 'The Half-Cell Potential Method of Locating Corroding Reinforcement in Concrete Structures', Transport and Road Research Laboratory Application Guide 9, 1991.
5. B. Elsener, H. Wojtas and H. Bohni, 'Inspection and Monitoring of Reinforced Concrete Structures - Electrochemical Methods to Detect Corrosion', ' Non-Destructive Testing in Civil Engineering, Vol. 2', Northampton, UK, 579. 'The British Institute of Non-Destructive Testing', 1993.
6. J. Mietz and B. Isecke, 'Electrochemical Potential Monitoring on Reinforced Concrete Structures - Electrochemical Methods to Detect Corrosion', 'Non-Destructive Testing in Civil Engineering, Vol. 2', Northampton, UK, 567. 'The British Institute of Non-Destructive Testing', 1993.
7. C. J. Newton and J. M. Sykes, *Corrosion Science*, 1988, **28**(11), 1051.
8. S. G. Millard, K. R. Gowers and J. H. Bungey, 'Galvanostatic Pulse Techniques: A Rapid Method of Assessing Corrosion Rates of Steel in Concrete Structures', Corrosion/95, Houston, TX, USA, 1995.
9. J. G. M. Bockris and D. Drazic, 'Electro-chemical Science', Taylor and Francis, London 1972.
10. A. Sehgal, Y. T. Kho, K. Osseo-Asare and H. W. Pickering, *Corrosion*, 1992, **48**(10), 871.
11. S. G. Millard, K. R. Gowers and R. P. Gill, *British Journal of NDT*, 1992, **34**(9), 444.
12. K. R. Gowers, S. G. Millard and J. H. Bungey, *Corrosion Science*, To be published.

13. J. P. Broomfield, J. Rodriguez, J. Ortega and L. M. Garcia, *ASTM STP 1276*, 1995.

EVALUATION OF LOCALIZED CORROSION RATE ON STEEL IN CONCRETE BY GALVANOSTATIC PULSE TECHNIQUE

B. Elsener, A. Hug, D. Bürchler and H. Böhni

Institute of Materials Chemistry and Corrosion, Swiss Federal Institute of Technology, ETH Hönggerberg, CH-8093 Zurich (Switzerland)

1. INTRODUCTION

Corrosion of rebars due to chloride attack and/or carbonation is a major cause of damage and early failure of reinforced concrete structures incurring enormous costs for maintenance, restoration and replacement worldwide. Maintenance, planning of restoration of these structures and quality control require a rapid, non-destructive inspection technique that detects corrosion of the rebars at an early stage, defines adequately which areas of the structure require repair and provides a measure of the corrosion rate.

The use of electrochemical potentials to determine areas of corrosion risk for reinforcing steel in concrete was pioneered in the United States[1,2] and resulted in the development of an ASTM standard (ASTM C876-91). Today potential mapping is state of the art to locate corroding zones precisely.[3-5] The extent of any corrosion problem in the structure being investigated can be mapped prior to a more detailed and costly examination and repair. Field experience on a large number of bridge decks has shown a clear correlation between the measured potential value and the chloride content in the concrete.[6,7] Potential readings can be misinterpreted (e.g. a lack of oxygen in very wet, dense or polymer-modified concrete leads to negative potentials) and the corrosion rate can only be estimated from the potential gradient and the concrete resistivity.[4,8] The galvanostatic pulse technique introduced for field application in 1988[9] is a rapid, non-destructive technique designed to overcome these difficulties, and other authors are also using it as on site monitoring technique.[10,11]

1.1 Corrosion Rate

The DC polarization resistance technique with IR compensation with calculation of the instantaneous corrosion current, i_{corr}, with the Stern Geary equation[12] has been applied extensively since 1970 by Andrade, Feliù and coworkers to study the corrosion of steel in concrete.[13,14] A strong correlation between the electrochemical weight loss calculated by integration of Rp data and the gravimetric weight loss was reported. AC impedance measurements[15-17] and transient techniques operating in the time domain have been used to determine the corrosion rate of steel in concrete.[18,19] The study of the response signal requires a model for the transfer function which describes the steel / concrete interface.[15,18,19]

In real structures the area of the counter electrode is much smaller than that of the working electrode (rebars) and the electrical signal tends to vanish with increasing distance. As a result, the measured effective polarization resistance can not be converted to a corrosion rate. Two ways to overcome these problems have been proposed: Feliù and coworkers[20,21] introduced the concept of a critical length, L_{crit}, based on the transmission line approach. L_{crit} is a function of the true polarization resistance, the concrete resistivity and geometrical parameters.[21] The experimentally measured Rp' can then be converted to a specific polarization resistance and thus the corrosion rate can be determined. The second approach is to

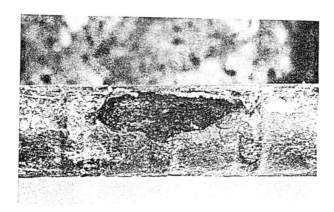

Figure 1 *Localized corrosion attack on a 20 mm rebar*

use a second concentric CE (guard ring) to confine the current to the area of the central CE.[22] Different commercially available instruments for on site corrosion rate determination have been tested[23] and found to show large differences in the results depending on whether a guard ring was used or not. Experimental data from on-site measurements have shown[24] that in the frequent case of chloride induced, very localized corrosion attacks (Figure 1) occuring in heavily chloride contaminated concrete, the average corrosion rates determined from R_p measurements underestimate the real, local penetration rates by a factor of five to ten. From an engineering point of view such local reduction in the cross section of the rebars is dangerous for the safety of structures when the rebars are located in a zone of high tensile or shear forces. For lifetime prediction it is thus essential to know the true local penetration rates.

In this work the dependance of the macrocell current on conductivity, cover depth and time is studied using model-macrocells in water and in mortar. A new approach to measure the localized corrosion rates experimentally is presented: the counter electrode on the slab surface is segmented and connected via 10 Ω shunts to the galvanostat, thus a direct assessement of the current distribution in the counter electrode and in the macrocell is possible. This will enable a distribution function to be used to determine the size of the localized corroding spot. The new approach can greatly improve the accuracy of corrosion rate measurements and life time predictions for locally corroding rebars.

2 EXPERIMENTAL

2.1 Model Bar for Macrocell Corrosion

The experiments simulating localized corrosion in an active/passive macrocell were performed on a macrocell bar (total length 0.4 m) with a central anode of mild steel (3 cm²) and 12 lateral cathodes (16 cm²) of stainless steel 1.4301 on both sides (Figure 2) resulting in a surface/length ratio of 6.3 cm²/cm for a bar diameter of 2 cm. The anode and the 12 cathodes were mounted on an electrically isolated plastic bar with a slot for the connecting wires. Anode and cathodes could be short-circuited externally. The total macrocell current and the individual cathode segment currents were measured with a zero resistance ammeter, and the switching was performed on a programmable multimeter (Keithley). A special switchboard guaranteed a complete short circuit during the measurements. Data were recorded on a personal computer.

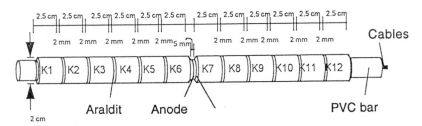

Figure 2 *Macrocell bar with anode (steel) and segmented cathodes (SS 1.4301)*

2.2 Experiments in Water and Mortar

One of the bars was mounted in a small basin (0.4 x 0.2 x 0.2 m) in order to perform experiments in water of different conductivities where the cover could be varied easily by filling the basin to different heights, usually 20 and 30 mm above the bar. The macrocell bar was fixed at a height of 70 mm. In water, in addition to the macrocell current, potential profiles were measured at different heights above the bar. Two bars were embedded in PC mortar using 500 kg of cement/m^3 and a water cement ratio of 0.5. 3% chloride as NaCl were added to the mixing water. The motar blocks were cured for one week in a humidity cabinet and then exposed to the laboratory atmosphere. The dimensions of the mortar blocks were 0.4 x 0.1 x 0.1 m and the cover over the bar was 20 mm (MPK1) and 30mm (MPK2) respectively.

2.3 Galvanostatic Pulse Measurements with Segmented CE (Patent Pending)

The galvanostatic pulse measurements were performed as described elsewhere.[25] In addition to the change in potential (measured with a reference electrode on the surface, positioned over the anode) the currents in the counter electrode and in the macrocell were recorded. To measure the distribution of the imposed galvanostatic current, the counter electrode, CE, used was segmented, consisting of 11 copper segments (2.5 x 2.5 cm) mounted on a plastic bar (Fig. 2). The individual CE segments were connected to a computer controlled galvanostat over a 10 Ω shunt to measure the segment currents during the time a current was imposed. The anodic and the cathodic currents in the macrocell bar, WE, were measured simultaneously.

3. RESULTS

In section 3.1 the results from the model-macrocells under open circuit conditions for different conductivities, geometries and times are presented, in section 3.2 the behavior of a macrocell under an anodic current imposed from an external counter electrode is described.

3.1 Macrocell Under Open Circuit Condition

The resistance between the central anode and the lateral cathodes, multiplied by the conductivity of the electrolyte, is shown in Figure 3 for different specific conductivities of the electrolyte including the mortar blocks. All the curves measured in water coincide and have the same characteristic, symmetrical shape, the resistances increase with distance (to cathode 1 or 12) and with decreasing conductivity of the electrolyte. The two mortar blocks with a slightly different geometry show higher normalized resistance values, MPK1, with the lower cover of 20 mm shows the highest values. The resulting potential profiles measured along the macrocell bar are shown in Figure 4 for the different electrolytes (cover 20 mm). The lower the conductivity of the electrolyte the more pronounced is the potential

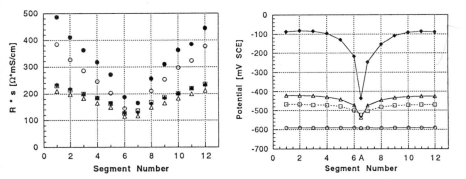

Figure 3 *Normalized resistance RΩ*σ between anode and cathode segments in distilled water , ♦, tap water, Δ, dilute sulfate solution, ⊡, and in the two mortar blocks MPK1 (cover 20 mm) , o, and MPK2 (cover 30 mm) ●*

Figure 4 *Potential profiles measured along the macrocell bar (cover 20 mm) for the same solutions as in Figure 3.*

difference between anode and cathode. Cathodes close to the anode are polarized to more negative potentials, this is especially clear for the potential profile close to the bar surface (Figure 5). The current distribution in the macrocells is shown in Figure 6. The anode dissolution current increases with increasing conductivity and with increasing cover depth, current densities between 60 and 450 $\mu A/cm^2$ are measured in the water, in contrast to the mortar samples where only 5 - 6 $\mu A/cm^2$ were measured. The current distribution on the cathodes is symmetrical, the cathodic currents are highest close to the anode and decrease with increasing distance from the anode. For electrolytes with medium to high conductivity, the macrocell bar used in this work is too short for a significant decrease of the cathodic current to be observed.

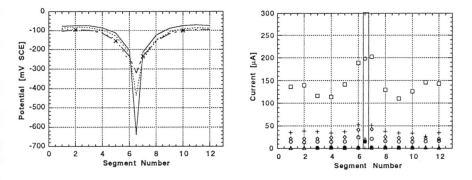

Figure 5 *Potential profiles measured anlong the macorcell bar (cover 20 mm) in distilled water at different heights above the bar (0.5 ——, 5 ·····, 10--- and 20 mm —·—)*

Figure 6 *Distribution of the cathodic currents in the macrocell for different conductivities of solution and cover (tap water, o20mm, tap water + 30mm, dilute sulfate solution, ⊡,20 mm and in the two mortar blocks MPK1 (cover 20 mm) ●, and MPK2 (cover 30 mm) Δ The anode current for tap water as 200 μA, tap water with 30 mm cover 390 μA and for dilute sulfate solution 1400 μA*

Table 1: *Resistance between CE and Macrocell and total resistance Rtot between anode and cathodes for different electrolytes and covers*

Medium	Cover d [mm]	σ [mS/cm]	$R_{ME/CE}$ [Ω]	R*σ	R_{tot} [Ω]	R_{tot}*σ
DW	20	0.02	567	11.3	4700	94
	30	0.02	860	17.0	-	-
	50	0.02	1146	22.5	-	-
LW	20	0.37	30.4	11.25	250	92.5
	30	0.37	41.1	15.2	-	-
	50	0.37	56.5	21.7	-	-
MPK1	20	0.432	563	-	235	101.5
MPK2	30	0.408	752	-	225	97.2

3.2 Macrocell Under Anodic Current Pulse

In this work the current distributions on the segmented counter electrode macrocell and on the anode/cathodes of the macrocell were measured simultaneously. The resistance between the CE and the macrocell in a parallel plane arrangement (Table 1) is proportional to the conductivity of the electrolyte and increased with increasing cover. Identical behaviour was observed for the experiments in water with different conductivities; for the mortar blocks the resistance $R_{ME/CE}$ is much higher than would correspond to the conductivity due to the coupling of the CE to the concrete surface. The resistance of the path from the anode to the CE segments changes in a similar way to that shown in Figure 3.

3.2.1 Current Distribution at the Macrocell. When an anodic current is imposed on the macrocell, the currents in the different segments change as shown in Figure 7. The current measured at the anode increases, the current at the cathodes decreases with time of the pulse. In mortar (Figure 7a) all the currents are anodic due to the low cathodic currents in the macrocell, in aqueous electrolytes (Figure 7b) the currents measured at the cathodes remain cathodic. The most important result to be seen from Figure 7 is the fact that most of the imposed current reaches the (small) anode, indicating that the pulse current distribution is governed to a large extent by the secondary current distribution. For covers of 50, 30 and 20 mm between 70 and 44 % of the imposed current flows to the anode, it is only for low and very low covers, 10mm, 1 mm, that appreciable current enters the cathodes (Table 2). This is further demonstrated by Figure 8 where the current distribution in the macrocell is shown for different cover depths.

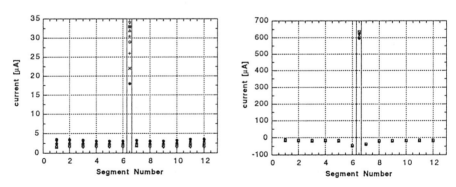

Figure 7 *Dependance of the currents through anode and cathodes of the macrocell on the time of an imposed anodic current pulse. a) macrocell in mortar ($I_{puls} = 0.05$ mA), b) macrocell in dilute sulfate solution ($I_{puls} = 0.4$ mA)*

Table 2: *Distribution of pulse current I_{puls} between anode and cathodes of the macrocell in distilled water (7 µS/cm) for different covers*

Cover [mm]	I puls [µA]	I Anode [µA]	I Cathodes [µA]	% Anode
50	50	35	15	70
30	80	55	25	68
20	80	35	45	44
10	200	60	140	30
1	300	60	240	20

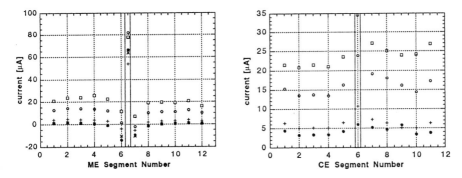

Figure 8 *Dependance of the ME currents through anode and cathodes of the macrocell on the cover depth for macrocell in distilled water. Cover depth 50 mm, ●, 30 mm, x, 20 mm, +, 10 mm, o, and 1 mm, □.*

Figure 9 *Distribution of the counter electrode current on the different CE segments (segment #6 over the anode of the macrocell) measurements in distilled water, pulse currents see Table 2. Cover depths as in Figure 8.*

3.2.2 Current Distribution at the Counter Electrode. The current distribution is measured at the segmented counter electrode. Figure 9 shows the results of experiments in distilled water with different cover depths (as Figure 8 and Table 2): the currents in the CE segments show practically no variation with the time of the pulse, the central segement (#6) shows the highest currents, with the lateral segments showing decreasingly lower currents. The higher currents in segments #1 and #12 are due to the larger volume of electrolyte at the end of the bar. The pulse measurements with segmented CE on mortar samples showed the highest currents in the centre (CE #4, #5, #6, #7 and #8), while the currents for the end segments (CE #1, 2 and #10, 11) were very low.

4. DISCUSSION

Macrocell corrosion between actively corroding areas on rebars and large passive areas (either beside the active spot or behind in a second layer of reinforcement) is of great concern because it results in very high local anodic current densities with corrosion rates up to 0.5 to 1 mm/year.[4,6,24] The resulting local loss in cross section has dangerous implications for the structural safety if the corroded rebars are located in a zone of high tensile or shear stresses. Furthermore, these dangerous attacks very often do not manifest themselves at the concrete surface by cracking or spalling because soluble chloride complexes are formed. It is thus of great importance to know more about the mechanism of these corrosion attacks as well as to develop a method which might allow these high local corrosion rates to be assessed.

4.1 Mechanism of Macrocell Corrosion

The driving force for macrocell corrosion is the potential difference between the anode (local corrosion spot, Figure 1) and the surrounding cathode area. Several laboratory studies[26-28] and results from potential mapping on reinforced concrete structures[4,6,7] have shown that the cell voltage may reach $\Delta U = 0.5$ V. In this work values of $\Delta U = 0.5$ V have been measured in aqueous electrolytes; in chloride containing wet mortar a value of about 0.3 V was found. The resulting current distribution (Figure 6) and the potential profiles (Figures 4, 5) are governed by the resistance of the current path between anode and cathode(s), the polarization resistance of the anode and cathode and the throwing power of the cathodic reduction reaction.

The experiments in aqueous electrolytes (no limitation of oxygen access to the cathodes) show evidence of mainly ohmic control. The primary resistance distribution between anode and cathode segments (Figure 3) is in agreement with the total resistance between the anode and cathode (Table 1) and is proportional to the electrolyte resistivity, which governs the current and the potential distribution in the macrocell. The more restricted electrolyte volume in the mortar blocks compared with the water basin results in higher resistance values for the mortar blocks. The primary current distribution also means that the macrocell bars in electrolytes with medium to high conductivity are too short to reach unaffected cathodic areas. Only in distilled water is the resistance high enough to focus the macrocell action in the region of ± 15 cm, which represents an anode to cathode ratio of 1:30 on each side.

Despite the high conductivity of the mortar blocks (σ between 0.76 and 0.43 mS/cm) and their nearly identical geometry (primary resistance) the macrocell currents measured are about 10 times lower than in aqueous solutions (Figure 6). This could be explained by the restricted oxygen access to the cathodes in the very humid, chloride containing mortars. Evidence to support this explanation is given by the high currents measured a short time after coupling the macrocell. Thus in mortar or concrete the primary resistance distribution is controlling the macrocell current only when the conductivity decreases and oxygen is readily available, under other circumstances the cathodic reaction is dominant.

4.2 Measuring Localized Corrosion Rates

The evaluation of the corrosion rate in real structures is of great importance for the assessement of corrosion risk to existing structures or for life time predictions. Different electrochemical techniques, all based on the determination of Rp (as shown in the introduction), give comparable results on a laboratory scale (small samples, homogeneous current distribution) and in field tests with homogeneous current distribution.[24] With new instruments using a sensorized guard ring the problem of non uniform current distribution in real structures has been partly overcome, and from the measured polarization resistance $R_{p,eff}$ a corrosion rate related to the rebar area under the CE can be calculated. The significance of these i_{corr} values in the case of chloride induced localized corrosion attacks with local loss in cross section of the rebars (Figure 1) is questionable: from site investigations corrosion rates five to ten times higher than those based on R_p measurements have been found,[4,24] from examination of reinforced concrete slabs a factor of five was reported.[29] From a theoretical point of view the Stern Geary relationship is not applicable to localized corrosion unless the anode area is known.

The results of this work, imposing an anodic current onto an active/passive macrocell, further confirm the doubts regarding R_p measurements in heterogeneous conditions: nearly all the current imposed from the counter electrode flows to the small anode, regardless of the quite large cathode used and CE currents also being high beside the local anode (Figure 7). This indicates - as has been shown already with calculations based on electrical network[25] - that the current distribution is governed by the secondary current distribution and the very low value of $R_{p,Anode}$ determines the current path. The potential / current relationship in LPR measurements[13-16] and the potential time curves measured in galvanostatic pulse measurements are related to the polarization of the local anode. This is confirmed by Figure 10 where

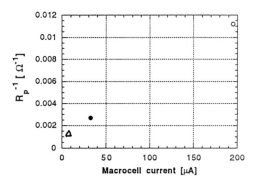

Figure 10 *Reciprocal polarization resistance, R_p^{-1}, as determined from galvanostatic pulse experiments versus macrocell current in tap water, o, distilled water, ●, and mortar, Δ.*

the macrocell currents measured under open circuit conditions are plotted against the reciprocal polarization resistance determined by the GPM technique.[25] A clear linear relationship is seen, which does not pass through the origin due to the self corrosion of the anode.

Another important consequence of these results (Figure 7) is that for localized corrosion attacks in an otherwise passive rebar no guard ring is necessary due to the dominating effect of the secondary current distribution. It might even be possible that the use of a guard ring modifies the electric field in such a way as to induce higher Rp values in the case of localized corrosion attacks.[30]

5. CONCLUSIONS

In real structures with very localized chloride induced corrosion the average corrosion rates determined from Rp measurements underestimate the local penetration rates by a factor of five to ten, and so life time predictions based on average corrosion rates become questionable. From this work on model macrocells in water and mortar using a new approach with a segmented counter electrode can be concluded

1. The resistance, potential and current distribution of the macrocell in water are governed by the primary current distribution (ohmic control) whereas in very humid, chloride containing mortars the oxygen diffusion limits the current.
2. Most of the current applied by an external counter electrode to the active / passive macrocell flows to the local anode despite the large cathode area (anode / cathode area ratio 1:60). Signal confinement for local corrosion attacks is thus not necessary.
3. The experiments have shown that the polarization resistance, R_p, determined by the galvanostatic pulse technique (GPM) agrees well with the macrocell current measured.
4. The segmented counter electrode (patent pending) permits the determination of the current distribution of the CE. The size of the corroding area (anode) can be approximated from this distribution.

6. ACKNOWLEDGEMENTS

This work was financially supported by the Swiss Federal Highway administration within the research program "Brückenunterhaltsforschung".

7. REFERENCES

1. J. R. Stratfull, *Corrosion NACE*, 1957, **13**, 173t.
2. J. R. Van Daveer, *Journal American Concrete Institute* , 1975, **12**, 697.
3. B. Elsener and H. Böhni, *Schweiz. Ingenieur und Architekt*, 1987, **105**, 528.
4. B. Elsener and H. Böhni, 'Potential Mapping and Corrosion of Steel in Concrete', in "Corrosion Rates of Steel in Concrete", ASTM STP 1065, eds. N. S. Berke, V. Chaker and D. Whiting, American Society for Testing and Materials, Philadelphia, 1990, 143.
5. J. P. Broomfield, P. E. Langford and A. J. Ewins, in "Corrosion Rates of Steel in Concrete', ASTM STP 1065, eds. N. S. Berke, V. Chaker and D. Whiting, American Society for Testing and Materials, Philadelphia, 1990, 157.
6. F. Hunkeler, *Schweiz. Ingenieur und Architekt,* 1991, **109**, 272.
7. B. Elsener and H. Böhni, *Materials Science Forum* , 1992, **111/112**, 635.
8. C. C. Naish, A. Harker and R. F. Carney, in 'Corrosion of Reinforcement in Concrete', eds. C. L. Page, K. W. Treadaway and P. B. Bamforth, Elsevier Applied Science, London, 1990, 314.
9. B. Elsener, 'Elektrochemische Methoden zur Bauwerksüberwachung', Zerstörungsfreie Prüfung an Stahlbetonbauwerken, SIA Dokumentation D020, Schweizer Ingenieur- und Architektenverein, Zürich, 1988, 27.
10. J. Mietz and B. Isecke, 'Electrochemical Potential Monitoring on Reinforced Concrete Structures using Anodic Pulse Technique', in 'Non destructive Testing in Civil Engineering', ed. H. Bungey, The British Institute of NDT, 1993, **2**, 567.
11. K. R. Gowers and S. G. Millard, in 'Corrosion and Corrosion Protection of Steel in Concrete' (Ed. R. Swamy) Sheffield Academic Press, 1994, 186.
12. M. Stern and A. L. Geary, *J. Electrochem. Soc.,* 1957, **104**, 56.
13. C. Andrade, V. Castelo, C. Alonso and J. A. Gonzales, in 'Corrosion Effect of Stray Currents and the Techniques for Evaluating Corrosion of Rebars in Concrete', ASTM STP 906, ed. V. Chaker , Philadelphia, 1986, 43.
14. J. A. Gonzales, S. Algaba and C. Andrade, *Br. Corros. J.,* 1980, **15**, 135.
15. L. Lemoine, F. Wenger and J. Galland, in 'Corrosion Rates of Steel in Concrete', ASTM STP 1065, eds. N. S. Berke, V. Chaker and D. Whiting, American Society for Testing and Materials, Philadelphia, 1990, 118.
16. D. C. John, P. Searson and J. L. Dawson, *British Corrosion J.,* 1981, **16**,102.
17. B. Elsener and H. Böhni, *Materials Science Forum* , 1986, **8**, 363.
18. S. Feliu, J. A. Gonzales, C. Andrade and V. Feliu, *Corros. Sci.,* 1986, **26**, 961.
19. C. J. Newton and J. M. Sykes, *Corrosion Science,* 1988, **28**, 1051.
20. S. Feliu, J. A. Gonzales, C. Andrade and V. Feliu, Corrosion/87 Paper Nr. 145, San Francisco CA, March 1987.
21. S. Feliu, J. A. Gonzales, C. Andrade and V. Feliu, *Corrosion NACE*, 1988, **44**, 761.
22. S. Feliu, J. A. Gonzales, C. Andrade and I. Rz-Maribona, in 'Corrosion of Reinforcement in Concrete', eds. C. L. Page, K. W. Treadaway and P. B. Bamforth, Elsevier Applied Science, London, 1990, 293.
23. J. Flis, A. Sehgal, D. Li, Y. T. Kho, S. Sabol, H. Pickering, K. Osseo and P. D. Cady, Condition Evaluation of Concrete Bridges Relative to Reinforcement Corrosion, Vol. 2: Method for Measuring the Corrosion Rate of Reinforcing Steel, SHRP-S-324, National Research Council, Washington DC, 1993.
24. B. Elsener, *Materials Science Forum* , 1995, **192-194**, 857.
25. B. Elsener, H. Wojtas and H. Böhni, in 'Corrosion and Corrosion Protection of Steel in Concrete' (Ed. R. Swamy) Sheffield Academic Press, Sheffield, 1994, Vol. 1, 236.
26. J. Nöggerath, PhD Thesis ETH Zürich Nr. 9310, 1990.
27. M. Raupach, Deutscher Ausschuss für Stahlbeton, Heft 433, 1992.
28. M. C. Alonso, C. Andrade, J. Farina, F. Lopez, P. Merino and X. R. Novoa, *Materials Science Forum* , 1995, **192-194**, 899
29. J. A. Gonzalez, C. Andrade, P. Rodriguez, C. Alonso and S. Feliù, in 'Progress in the Understanding of Corrosion', (Eds. D. Costa and A. D. Mercer), The Institute of Materials, London, 1993, Vol.1, 629.
30. H. Wojtas, B. Elsener and H. Böhni, submitted to CORROSION/NACE

EVALUATION OF THE CONCRETE CORROSIVITY BY MEANS OF RESISTIVITY MEASUREMENTS

Stefania Fiore[*], Rob B. Polder[**] and Ranieri Cigna[*]

[*]Department ICMMPM
University "La Sapienza"
Via Eudossiana 18, 00184 Roma
Italia

[**]TNO Building and Construction Research Delft
Lange Kleiweg 5 Rijswijk
The Netherlands

1 INTRODUCTION

The electrical resistivity of concrete ranges from 10 to 10^5 Ωm and depends on the concrete composition and moisture content: the resistivity is higher for dry materials, for a low water/cement ratio and for composite cements (blast furnace slag, fly ash, silica fume).[1]

From the electrochemical nature of rebar corrosion, a relationship may be expected between concrete resistivity and the reinforcement corrosion rate. Such a relationship has been investigated by some authors and a linear relationship between inverse resistivity and corrosion rate was found for BFS concrete,[2] for carbonated mortars with and without chloride,[3] for non-carbonated mortars with and without chloride[4] and for carbonated mortars made with various (blended) cements.[5]

This relationship can be explained theoretically using the concept of anodic resistive control[3] which also predicts a linear relationship between the steel potential and the logarithm of the corrosion rate. The aim of this paper is to contribute to this study by analysing the influence of the relative humidity in specimens with mixed-in chloride and after accelerated carbonation. The work was carried out as a part of the European Union Programme COST 509 'Corrosion of metals in contact with concrete' as joint project NL-1 I-1 'Concrete resistivity and corrosion rate'.

2 EXPERIMENTAL

The first part of the investigation[6] started in 1991. Concrete specimens were cast varying the following factors: type of cement (OPC and BFSC with 70% slag), chloride content (0% and 2%), water/cement ratio (0.45 and 0.65) and concrete cover to the steel bars (10 and 30 mm). The original set of specimens involved 32 prisms (300 x 100 x 100 mm^3), with embedded steel bars and brass electrodes, and 16 cubes (150 mm diameter) with the same characteristics as the prisms: one half of the specimens were exposed in a 20°C - 80% R.H. climate room and the other half were exposed in a fog room.

In November 1992 eight prisms from the 80% R.H. room were exposed to accelerated carbonation in a carbonation cabinet with 5% (by volume) CO_2 and 50% R.H. for about 100 days.[7] The cubes of the same composition as the prisms were cut into two halves and the halves were exposed together with the prisms in the 80% R.H. room or in the carbonation cabinet. After a period of 105 days of accelerated carbonation, the specimens were stored again in the 80% R.H. room.

In September 1994, in the last part of the investigation, all prisms and half cubes in the 80% R.H. room were transferred to a climate chamber and subjected to stepwise humidity increases. The relative humidity content was increased from 75% to 80%, 90% and 95%, each level being maintained for a period of at least three weeks.

Concrete resistivity was calculated from resistance measurements using two different methods. The two-point method (Volt-Amperometric method) involved taking resistance measurements between pairs of embedded brass bars using a GEOHM resistance meter (AC, 108 Hz). The same instrument was used in the four-point method (Wenner method) which was applied using two surface probes with an electrode spacing of 30 and 50 mm.

The corrosion rate was measured using the polarization resistance technique, with three embedded steel bars as working, reference and counter electrodes: a shift of 10 mV was applied by an ELAB POTENTIOSTAT P134 to the steel bars supposed to be in an active state of corrosion and a shift of 50 mV was applied to the steel supposed to be in a passive state. The resulting current was recordered after 20 seconds. No correction was made for Ohmic drop.

The measurements of free corrosion potential were carried out using a Ag/AgCl reference electrode.

3 RESULTS AND DISCUSSION

As a result of the addiction of chloride and the various exposures, in some specimens the steel bars may be assumed to have lost passivation; this was confirmed by R_p and potential measurements. In the 80% R.H. climate, the corrosion rates were relatively low, which could be attributed to the relatively dry state of the concrete (high resistivity).

In order to test the influence of increasing humidity, selected specimens were subjected to controlled stepwise increase of humidity.

For reasons of simplicity, this paper concentrates on specimens where corrosion was assumed to be taking place, either due to the presence of chloride or to carbonation. The specimen features are summarised in Table 1.

Table 1 *Characteristics of specimens*

Specimen code	Cement type	Chloride added	w/c ratio	Exposure
B1	BFSC	2 %	0.45	80% R.H.
B2	BFSC	2 %	0.45	80% R.H.-C.C.
C2	BFSC	0 %	0.65	80% R.H.
D2	BFSC	2 %	0.65	80% R.H.-C.C.
F1	OPC	2 %	0.45	80% R.H.
F2	OPC	2 %	0.45	80% R.H.-C.C.
H1	OPC	2 %	0.65	80% R.H.
H2	OPC	2 %	0.65	80% R.H.-C.C.

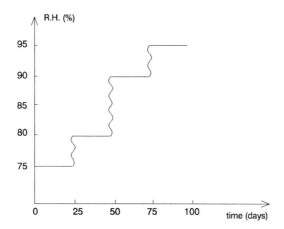

Figure 1 *Change in the cabinet humidity over time*

In Figure 1 the change of humidity with time is shown for the period from September 1994 to December 1994.

In Figure 2 the changes in corrosion rate are plotted as a function in the increase of relative humidity; in Figure 3 the same plot is shown for the concrete resistivity and in Figure 4 for the corrosion potential.

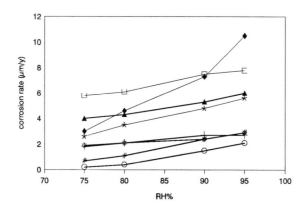

Figure 2 *Corrosion rate versus relative humidity for specimens B1 (+), B2 (*), D2 (□), H2 (o), H1 (♦), F1 (▲), F2 (¤), C2 (#)*

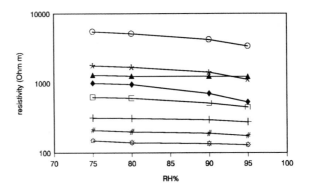

Figure 3 *Logarithm of resistivity versus relative humidity for specimens H1 (+), B2 (*), D2 (□), H2 (o), B1 (♦), C2 (▲), F1 (✿), F2 (#)*

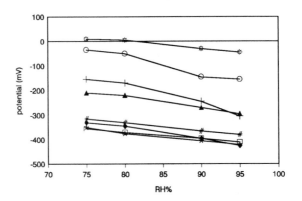

Figure 4 *Steel potential versus relative humidity for specimens H1 (+), B2 (*), D2 (□), H2 (o), B1 (♦), C2 (▲), F1 (✿), F2 (#)*

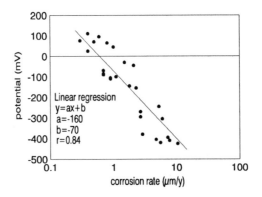

Figure 5 *Potential versus logarithm of corrosion rate*

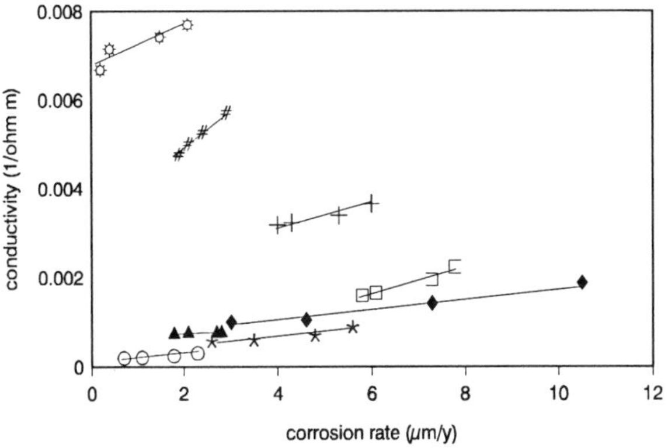

Figure 6 *Conductivity versus corrosion rate for specimens H1 (+), B2 (*), D2 (□), H2 (o), B1 (♦), C2 (▲), F1 (¤), F2 (#)*

A consequence of the increase in relative humidity is that the corrosion rate increases, the resistivity decreases and the free corrosion potential shifts to more negative values for all the specimens.

The prisms not exposed to accelerated carbonation and without chlorides show a CR equal to or lower than 1 μm/y, which is reasonable because the steel in these specimens remains in a passive state. The same low CR was found in the carbonated specimens without chlorides. This suggests that in these specimens the steel is also in a passive state and that such quality of concrete protects the steel even after carbonation.

All specimens with 2% Cl^- show a significant increase in CR with the increase of R.H. In OPC specimens the CR rises more strongly for w/c=0.65 than for w/c=0.45; in BFSC specimens both w/cs react similarly.

The corrosion rate-corrosion potential relationship (Figure 5) shows that, as the corrosion potential shifts to more negative values, the corrosion rate rises in an approximately exponential fashion.

Plotting conductivity as a function of the corrosion rate for each (single) specimen showing active corrosion, and using a linear regression to fit the data points, a linear relationship was found between conductivity and CR for each specimen (Figure 6).

A linear relationship seems to exist only within particular groups of data, not for all the data taken together.

In fact, using the same method to correlate the data points only for BFSC specimens, the following linear relationship was found:

$$\text{Conductivity} = 0.000255\ \text{CR} - 0.00005 \qquad r = 0.83 \qquad (1)$$

For OPC specimens, the results clearly show that CR is increasing as given by the following ranking:

- w/c=0.45, 2% Cl
- w/c=0.45, 2% Cl, accelerated carbonation
- w/c=0.65, 2% Cl
- w/c=0.65, 2% Cl, accelerated carbonation

but there is no common relationship with conductivity.

Similar results have been found by other researchers,[2, 3, 4, 5] and the linear relation between steel potential and Log (CR) found in the experimental results confirms the mechanism of anodic resistive control.

A statistical analysis of the experimental results at the different values of R.H. was performed to see the effects of the factors varied in the casting.

The stepwise increase in relative humidity gives the following results (Figures 7, 8, 9, 10, 11, 12, 13, 14, 15).

For the corrosion rate, chloride content is the only significant factor: at 80% R.H. CR is 10 times higher in specimens with 2% Cl than in those without Cl, at 90% R.H. 13 times and at 95% 9 times.

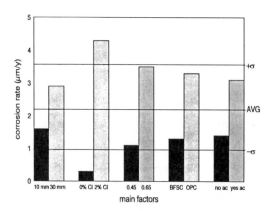

Figure 7 *Corrosion rate at 80% relative humidity; 0.45 indicates specimens with water/cement ratio=0.45; 0.65 indicates specimens with water/cement ratio=0.65; no ac indicates specimens not exposed to accelerated carbonation; yes ac indicates specimens exposed to accelerated carbonation*

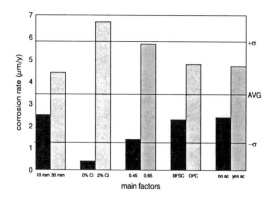

Figure 8 *Corrosion rate at 90% relative humidity; 0.45 indicates specimens with water/cement ratio=0.45; 0.65 indicates specimens with water/cement ratio=0.65; no ac indicates specimens not exposed to accelerated carbonation; yes ac indicates specimens exposed to accelerated carbonation*

For the resistivity, the increase in relative humidity gives the following results:
- all resistivities decrease;
- the effect of cover depth (higher resistivity at 10 mm) is stable from 80% R.H. to 90% R.H. and is slightly reduced from 90% to 95% R.H.;
- the effect of chloride content and of accelerated carbonation decreases with the increase in humidity;
- the cement type is significant only at 90% and 95% R.H.

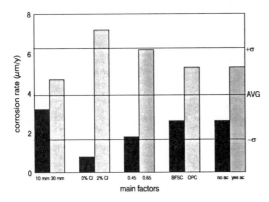

Figure 9 *Corrosion rate at 95% relative humidity; 0.45 indicates specimens with water/cement ratio=0.45; 0.65 indicates specimens with water/cement ratio=0.65; no ac indicates specimens not exposed to accelerated carbonation; yes ac indicates specimens exposed to accelerated carbonation*

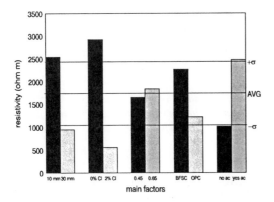

Figure 10 *Resistivity at 80% relative humidity; 0.45 indicates specimens with water/cement ratio=0.45; 0.65 indicates specimens with water/cement ratio=0.65; no ac indicates specimens not exposed to accelerated carbonation; yes ac indicates specimens exposed to accelerated carbonation*

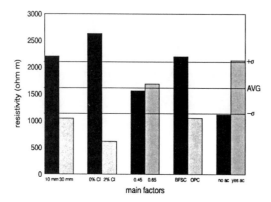

Figure 11 *Resistivity at 90% relative humidity; 0.45 indicates specimens with water/cement ratio=0.45; 0.65 indicates specimens with water/cement ratio=0.65; no ac indicates specimens not exposed to accelerated carbonation; yes ac indicates specimens exposed to accelerated carbonation*

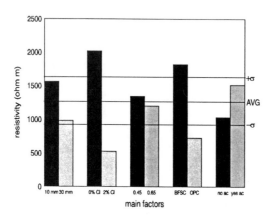

Figure 12 *Resistivity at 95% relative humidity; 0.45 indicates specimens with water/cement ratio=0.45; 0.65 indicates specimens with water/cement ratio=0.65; no ac indicates specimens not exposed to accelerated carbonation; yes ac indicates specimens exposed to accelerated carbonation*

For the corrosion potential, the increase in relative humidity leads to a strong increase in the effect of chloride content, with a potential 280 mV more negative at 90% R.H. and 300 mV more negative at 95% R.H. with 2% Cl as compared to 0% Cl.

The influence of w/c ratio changes from 160 mV at 80% R.H. to 105 mV at 90% R.H. and to 95 mV at 95% R.H.; a similar effect is due to the cement type. The same trend, but smaller, was found for accelerated carbonation and cover depth.

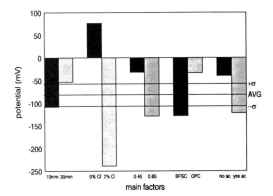

Figure 13 *Potential at 80% relative humidity; 0.45 indicates specimens with water/cement ratio=0.45; 0.65 indicates specimens with water/cement ratio=0.65; no ac indicates specimens not exposed to accelerated carbonation; yes ac indicates specimens exposed to accelerated carbonation*

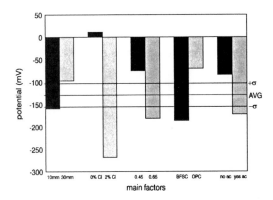

Figure 14 *Potential at 90% relative humidity; 0.45 indicates specimens with water/cement ratio=0.45; 0.65 indicates specimens with water/cement ratio=0.65; no ac indicates specimens not exposed to accelerated carbonation; yes ac indicates specimens exposed to accelerated carbonation*

4 CONCLUSIONS

Keeping in mind that the specimens in this investigation were exposed to constant humidity (for at least three weeks), the following conclusions can be drawn.

1. The corrosion rate of steel in semi-dry to wet concrete is a complicated function of Cl content, the environment of exposure and the w/c ratio; but does not depend on the cement type or cover depth.

The main controlling factors of the corrosion rate at an age of about three years are the relative humidity, the chloride content and the w/c ratio. The effect of chloride content increases with the increase on relative humidity.

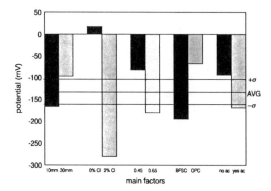

Figure 15 *Potential at 95% relative humidity; 0.45 indicates specimens with water/cement ratio=0.45; 0.65 indicates specimens with water/cement ratio=0.65; no ac indicates specimens not exposed to accelerated carbonation; yes ac indicates specimens exposed to accelerated carbonation*

2. The corrosion rate in semi-dry concrete (constant climate of 80% R.H.) is a function of the Cl content but the carbonation depth does not influence it.

3. The resistivity of concrete is strongly influenced by the cement type, the environment of exposure and the chloride content. The increase in relative humidity results in a decrease in the effects of cover depth, chloride content and accelerated carbonation, but an increase in the effect of cement type.

4. The free corrosion potential of steel in the specimens is mainly influenced by the environment of exposure, the chloride content, the w/c ratio and the cement type. The increase in relative humidity leads to a significant increase in the effect of chloride content.

5. A linear relationship was found between conductivity and corrosion rate for single specimens which were actively corroding. However, different relationships appear to exist for different cement types.

6.These results suggest that in semi-dry concrete, the corrosion rate is under anodic resistive control, and this type of control is confirmed by the exponential rise in CR when the potential shifts to more negative values.

5 REFERENCES

1. O. Gjorv and O. Vennesland, 'Electrical resistivity of concrete in the oceans', OTC paper 2803, Houston, 1977.
2. R. B. Polder, P. B. Bamforth, M. Basheer, J. Chapman-Andrews, R. Cigna, M. I. Jafar, A. Mazzoni, E. Nolan and H. Wojtas, COST 509 NL-1& Resistivity group 1993 report, TNO report 94-BT-R0458.
3. G. K. Glass, C. L. Page and N. R. Short, *Corrosion Science*, 1991, **32** (12), 1283.
4. W. Lopez and J. A. Gonzales, *Cement and Concrete Research*, 1993, **23** (2), 368.
5. C. Alonso, A. Andrade and J. A. Gonzales, *Cement and Concrete Research*, 1988, **8**, 687.
6. M. Valente, R. B. Polder, R. Cigna and T. Valente, TNO report BI-92-173, and extracted paper TNO report BI-92-139, 1991.
7. A. Tondi, R. B. Polder and R. Cigna, TNO report 93-BT-R0170, 1993.

ELECTRICAL RESISTIVITY AND DIELECTRIC PROPERTIES OF HARDENED CEMENT PASTE AND MORTAR

D. Bürchler, B. Elsener and H. Böhni

Institute of Materials Chemistry and Corrosion,
Swiss Federal Institute of Technology, ETH Hönggerberg,
CH-8093 Zurich, Switzerland

1 ABSTRACT

The resistivity and dielectric properties of hardened cement pastes and mortars were investigated. The techniques applied were impedance spectroscopy, mercury intrusion porosimetry and plasma emission spectrometry. Changes in water content, porosity and pore solution composition were measured over a period of 256 days and related to resistivity and dielectric properties. The results show the influence of these parameters on the resistivity related to the ionic current flow in the bulk of the pores and the dielectric properties related to conduction at the interface between pore solution and solid phase.

2 INTRODUCTION

Corrosion of reinforcement in concrete is associated with ionic current flow, thus the electrical resistivity of the concrete controls to a great extent the corrosion rate. Nondestructive methods such as the galvanostatic pulse technique or polarization resistance measurements have shown a good correlation between corrosion rate and resistivity.[1-5] On site measurements of the resistivity might give an indication of the corrosion risk.[6,7] The permeability of concrete has also been related to the resistivity.[8,9]

Several works have studied the electrical resistivity of hardened cement paste,[10,11] mortar [12-15] and concrete.[16,17] In some of these works only resistance values were determined and the poorly defined geometry does not allow the calculation of resistivity [10,16]. Variations of resistivity with temperature,[13,15] with the resistivity of the pore solution[12] and with humidity or water content have been reported.

A few authors[10-12] have investigated the dielectric properties of hardened cement paste and mortar using impedance spectroscopy with a maximum frequency between 1 MHz and 110 MHz. They all found that in the complex plane the impedance formed a semicircular arc with the centre below the real axis. Niklasson[12] in the discussion of this result used an equivalent circuit consisting of an ideal resistor and a constant phase element (CPE) in a parallel combination. This circuit was interpreted by assigning the resistor to conduction of the pore water and the CPE to the capacity of the pore walls. All these works focused on the early stages of hydration[10,11] or changes in humidity that were forced by oven drying,[12] but none of the results were obtained from samples in equilibrium with environmental humidity.

The resistivity of cement based materials is related to the movement of dissolved ions in the pore solution. This current flow is therefore expected to be influenced by parameters such as water content (environmental humidity), pore solution composition (ion content and mobility), pore structure (porosity, pore size distribution) and temperature. In this work the influence of the first three parameters on dielectric properties and the resistivity of cement paste and mortar is studied systematically by impedance spectroscopy combined with mercury intrusion porosimetry and chemical analysis of the pore solution.

Table 1   *Chemical composition of Portland cement used (wt.%)*

Al_2O_3	Fe_2O_3	SiO_2	CaO	MgO	SO_3	K_2O	Na_2O
5.24	2.57	19.30	63.1	2.51	2.95	1.22	0.07

Table 2   *Mixtures (wt.%) of hardened cement pastes (#1 and #2) and mortars (#3 and #4)*

Mixture	#1	#2	#3	#4
Cement	71.5	62.5	20.4	16.8
Water	28.5	37.5	10.2	12.6
Sand 0-1 mm	-	-	34.7	35.3
Sand 1-4 mm	-	-	34.7	35.3
w/c ratio	0.4	0.6	0.5	0.75

3 EXPERIMENTAL

3.1 Cement Paste and Mortar Samples

A series of 60 cement paste and mortar samples with a w/c ratio of 0.4 / 0.6 (cement paste) and 0.5 / 0.75 (mortar) were cast. The composition of the Portland cement used is shown in Table 1. The proportions of the ingredients were varied as shown in Table 2. Preliminary HF measurements were performed to optimise the sample size for measurements of the impedance in a frequency range up to 200 MHz. The final sample geometry was cylindrical with a height and diameter of 40 mm. Two parallel steel electrodes (DIN 1.4529) of the same diameter were placed at the top and bottom of the model before vibration of the mixture. These electrodes were in contact with the surface of the samples during the hydration. After curing for 30 days in a humidity cabinet (rh > 95%) three samples of each mixture were exposed to constant relative humidity of 98.5%, 90%, 80%, 70% and 60% (± 2%) at 295 K (± 1 K). At logarithmic time intervals the weight change and the impedance of all the samples were measured. At the beginning of the experiment (end of the curing period) and at selected times, additional samples were used to determine the pore size distribution (by mercury intrusion porosimetry) and the pore water composition (by pore water expression and analysis).

3.2 Impedance Spectroscopy

A sine voltage of small amplitude (10 mV) was applied between the electrodes. The frequency, f, was logarithmically varied in the range 100 kHz to 200 MHz (120 points/decade). The amplitude, the phase angle and the complex impedance $Z_{(f)}=Z'_{(f)}+iZ''_{(f)}$ were calculated. Impedance spectra were recorded using a network analyser (Hewlett Packard Analyser HP 4195 A with HP 41951 A Monoport). No cables were used.

3.3 Weight Measurements

The weight of the samples was measured every time impedance spectroscopy was performed. The water content, u, was calculated as the quotient of the weight loss during drying and the weight of the dried sample. Drying conditions were 378 K over 48 hours. Thus the volume of the liquid phase in the sample was determined during the experiment.

3.4 Mercury Intrusion Porosimetry (MIP)

Measured volumes of mercury were forced into samples of 20 mm diameter and 20 mm height under measured increments of pressure using a Micrometrics autopore II 9220 (appl. max. pressure: 360 MPa, 150 steps).

The relative volumes of the different pore sizes were estimated using the measured volumes of mercury. Results are presented as total porosity and pore size distribution (logarithm of differential intrusion versus logarithm of local average pore diameter). The pore size distributions have been fitted to two Gaussian distributions. These fitted distributions have been assigned to micro- and nanopores.

There are several assumptions involved in MIP,[18] so the measured pore size distribution cannot be interpreted as a true pore size distribution. The most important assumption is that a small pore is always filled through a larger one and not vice versa. This is not true, because there are 'ink bottle' pores connected to the surface only through smaller pores. The fraction of nanopores is therefore overestimated, while the fraction of micropores is underestimated. MIP was performed to investigate changes in the solid phase of the sample during the experiment.

3.5 Pore Solution Expression and Analysis

Samples were placed in a steel cylinder and pressed up to 350 MPa by a hydraulic press (Schenck UPN 1600). The steel cylinder and the parameters were similar to those used in other works.[19-22] The liquid obtained was analysed by plasma-emission spectrometry (ARL 3580 B ICP-Analyser). The pH values were determined by the dynamic equivalence point method (DET) using a Metrohm DMS-Titrino 716.

After preliminary tests this procedure was applied to hardened cement paste w/c 0.4 after 4, 128 and 192 days and on the other three mixtures after 192 days. The chemical analysis of the pore solution was performed in order to observe changes in the composition of the liquid phase.

4 RESULTS

The results are presented in the following order: 1. Water content. 2. Pore solution composition. 3. Porosity. 4. Impedance spectroscopy.

4.1 Water content

Changes in water content (determined from weight measurements) for cement paste samples with w/c 0.4 are shown in Figure 1.

In samples from the humidity cabinet with rh > 95% the water content increased slightly for 98.5% rh and decreased for all other samples. Constant weight was achieved after 256 days for 98.5 and 90% rh only. Each point shown is the average of 3 samples, the standard deviation is less than 1.5%. In Figure 2 the water content after 256 days is presented for all mixtures. The water content is lower for mortar samples and lower for smaller w/c ratios.

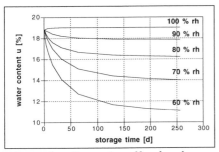

Figure 1 *Desorption of hardened cement paste (w/c 0.4)*

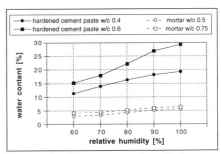

Figure 2 *Water content for all mixtures after 256 days*

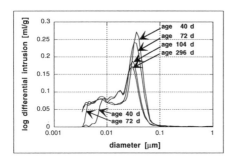

Figure 3 *Pore size distribution of hardened cement paste (98.5 % rh) after different storage times. Hydration of the cement diminishes the pore sizes and reduces the total porosity. The effect of hydration decreases with increasing storge time.*

4.2 Porosity

The pore size distribution (determined by MIP) changed with time as depicted in Figure 3 in a characteristic way: the peak at larger diameters diminished and shifted to smaller values and the area < 0.01 µm increased.

The pore size distribution for hardened cement paste and mortar after 192 days (98.5% rh) is shown in Figure 4. The additional water of the hardened cement paste with w/c 0.6 leads to larger pore sizes and to a higher total porosity. Only one distribution of pore sizes was detected for w/c 0.4. This distribution is assigned to nanopores. The higher w/c ratio of 0.6 shows another small distribution around 0.3 µm. The mortar samples clearly show two separated peaks. Higher w/c ratios lead to an augmentation of porosity in the distribution of larger pore sizes. This distribution is assigned to micropores. The smaller distribution is similar to the distribution in the hardened cement paste.

Table 3 gives the porosity of the mixtures. The total porosity (vol.%) of the cement paste samples is higher than that of the mortar samples. A higher w/c ratio leads to a higher total porosity. Fractions of nano- and micropores are determined as mentioned in section 3.4.

4.3 Pore Solution Composition

The compositions of the pore solutions from hardened cement paste and mortar are summarised in Table 4. The main components are K^+, Na^+ and OH^-. The concentrations are higher for cement paste samples. The values for hardened cement paste samples (mixture #1) after 4 and 128 days were the same as those given in Table 4. The resistivity of the pore solution and the ionic strength are given in Table 5. In agreement with the composition, the resistivity is higher for mortars and increases with higher w/c ratios.

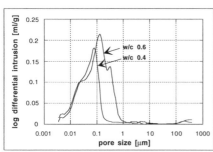

Figure 4a *Pore size distribution of cement paste after 192 days (98.5% rh)*

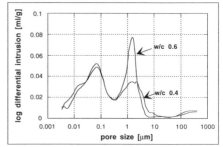

Figure 4b *Pore size distribution of mortar after 192 days (98.5% rh)*

Table 3 *Total porosity of the mixtures after 192 days (4 samples per mixture, max. standard deviation 4.2%) determined with MIP and fitted fractions of nano- and micropores*

Mixture		#1	#2	#3	#4
Total Porosity	[vol.%]	25.8	35.2	21.3	23.1
Nanopores	[vol.%]	23.7	32.2	9.2	8.9
Micropores	[vol.%]	0	2.1	10.9	13.6

Table 4 *Chemical composition of the pore solution after 192 days of storage (mmol/l) for 98.5% rh. The values for hardened cement paste w/c 0.4 exposed to 98.5% rh after 4 and 128 days were the same as after 192 days. Standard deviation for concentration of K^+, Na^+ and OH^- was 5% (6 samples).*

Mixture	Al^{3+}	Ca^{2+}	Fe^{2+}	K^+	Na^+	SO_4^{2-}	pH
#1	0.3	0.28	0.01	594.6	60.09	5.59	13.7
#2	1.03	0.14	0.03	402.5	44.35	7.81	13.6
#3	1.0	0.14	0.05	241.2	37.91	7.67	13.4
#4	1.7	0.13	0.05	217.9	34.78	9.69	13.3

Table 5 *Ionic strength and measured resistivity of pore solutions (98.5% rh)*

Mixture		#1	#2	#3	#4
Ionic Strength	[mmol/l]	587	455	284	233
Resistivity	[Ωm]	0.08	0.11	0.17	0.20

4.4 Impedance Spectroscopy

A typical impedance spectrum is shown in Figure 5 in the bode plot and in the complex plane. At frequencies below 100 kHz the resistance is constant (phase angle = 0). This plateau remains until electrochemical reactions on the steel electrodes interfere at frequencies below 1 kHz. At high frequencies (>10 MHz) capacitive behaviour is observed. As shown in Figure 5b the corresponding time constant is distributed (depressed semicircle).

The impedance, Z, of a sample was estimated by the impedance of an equivalent circuit made up of an ideal resistor (Z_R) and a constant phase element (Z_{CPE}) in a parallel combination:

$$Z = \left(1/Z_R + 1/Z_{CPE}\right)^{-1} \qquad (1)$$

$$Z_{CPE} = A \cdot (i \cdot f)^{-\alpha} \qquad (2)$$

where A is the amplitude [F^{-1}], i is $\sqrt{-1}$, f is the frequency [Hz] and α is the power law exponent taking values between unity and zero.

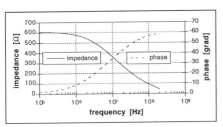

Figure 5a *Typical impedance spectrum presented as a bode plot (hardened cement paste w/c 0.4 exposed to 80% rh after 128 days)*

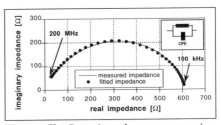

Figure 5b *Same impedance spectrum in the complex plane. Calculated values of the equivalent circuit (inlay) are in excellent agreement with measured impedance*

A very good fit to the experimental impedance was always found. The resistivity, ρ [Ωm], of the hardened cement paste or mortar samples was calculated from the fitted resistance of the resistor, Z_R, taking into account the sample geometry of height, h [m], and cross sectional area, a_{CS} [m²]:

$$\rho = \frac{Z_R \cdot a_{CS}}{h} \tag{3}$$

Figure 6 shows the change in resistivity of hardened cement paste with time at different humidities. Values for 98.5% and 90% rh coincide and increase slightly. Samples at lower rh have an asymptotic increase in resistivity. Samples exposed to 60% rh show a different behaviour: after a certain time of exposure a sudden drastic increase occurred. For mortar samples this drastic increase took place after 64 days and at 70% rh.

Figure 7 shows the resistivity of all samples after 256 days. Each point is the mean of three values, the maximum standard deviation is 6% for hardened cement paste and 8% for mortar (both maximum standard deviations are for 60% rh). The resistivity is higher for mortars, for low relative humidities and for higher w/c ratios. Below 70% rh the resistivity increases drastically. The shape of the curve is more continuous for the mortar samples.

The evolution of the amplitude A and the exponent α with time is shown in Figure 8. In both examples shown - and for all other samples - the exponent α remained constant during the whole exposure period. The mean values of exponents are presented in Table 6.

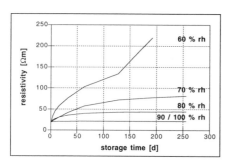

Figure 6   *Resistivity of hardened cement paste (w/c 0.4) for different storage times and relative humidities. The development is inverse to figure 1*

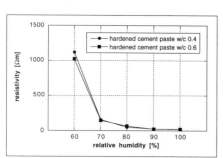

Figure 7a *Resistivity of hardened cement paste versus relative humidity*

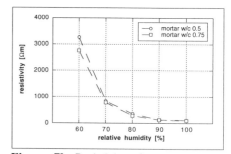

Figure 7b *Resistivity of mortar versus relative humidity after 256 days*

Table 6   *Mean and standard deviation of the CPE exponents for each mixture (a mixture is represented by 15 samples) and all mixtures after 256 days.*

Mixture	#1	#2	#3	#4	all
Mean	0.72	0.66	0.74	0.76	0.72
Standard Deviation	0.04	0.05	0.06	0.04	0.08

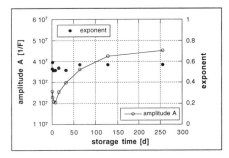

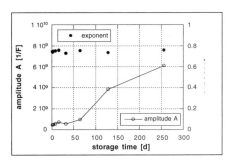

Figure 8a *Development of the amplitude, A, and the exponent, α, of the CPE for hardened cement paste (w/c 0.4) exposed to 80% rh*

Figure 8b *Development of the amplitude, A, and the exponent, α, of the CPE for mortar (w/c 0.75) exposed to 60% rh*

Samples exposed to relative humidities below 70% showed an increase in amplitude of over one order of magnitude. The standard deviation for the presented amplitudes was up to 13% and therefore much higher than for values of resistivity (up to 6%).

5 DISCUSSION

The changes in water content, porosity and pore solution composition and the interactions of these parameters are discussed in the first part of this section. In the second part these parameters are related to the dielectric behaviour of the samples.

5.1 Change of Water Content

Adsorption and desorption and thus the water content are dominated by pore size effects and the relative humidity of the environment. The mechanisms that lead to an increasing water content are capillary condensation at over 60% rh and molecular adsorption between 0% and 80% rh. The smaller the pore sizes the lower is the vapour pressure over the liquid in those pores. This leads to capillary condensation, where small pores are filled with condensed humidity. For lower relative humidities BET[23] found that after the building of a monolayer several other layers are formed according to the relative humidity of the atmosphere. These layers are removed during desorption (there is a hysteresis observed in experiments of adsorption and desorption). Thus the water content is reduced when the relative vapour pressure of a distribution of pore sizes as depicted in Figure 4 is higher than the pressure assigned to the relative humidity of the environment. The rate of reduction of water content is dependent on the diffusion length to the surface of the sample. After 256 days the samples (diameter 40 mm) below 90% rh are still not in equilibrium.

5.2 Change of Porosity

The advancing hydration of the cement leads to the pore size distributions depicted in Figure 3. The influence of hydration is strongest immediately after casting and diminishes with increasing age. The total porosity for hardened cement paste exposed to 98.5% rh changes from 29% (age 40 days including 30 days cured in the humidity cabinet) to 25.8% (age 296 days). This change in total porosity and pore size distribution results in a change in water content regardless of the environmental relative humidity.

The compositions of the mixtures were adapted in preliminary tests to achieve four different porosities, one pair with different nanoporosity and one pair with different microporosity as depicted in Figure 4. The volume fractions of nano- and micropores given in Table 3 based on MIP and have been determined as mentioned in section 3.4 ; other methods of separating the total pore volume, e.g. the Powers-Brownyard model,[24] will lead to other fractions. One advantage of the use of MIP in this work is that the 'ink bottle' pores have the same effect on MIP and on desorption: the volume fraction of nanopores appears greater than it actually is.

5.3 Change of Pore Solution Composition

The chemical composition is important for the conductivity of the liquid phase (see Table 4 and Table 5). After curing for 30 days the hydration does not have any influence on the chemical composition of the pore solution of hardened cement paste. This is in agreement with the observations of Longuet,[19] Diamond,[20] Gunkel[21] and Byfors.[22] The pore solutions they analysed were of cements with different compositions, which led to different pore solution compositions, but all authors reported a constant composition over a certain time.

Due to the hydroxide content of the pore solution, the resistivity given in Table 5 must increase under the influence of carbonation, because the mobility of hydroxides is about five times greater than the mobility of carbonates or bicarbonates.

5.4 Interactions

The interaction between porosity, pore size distribution and relative humidity leading to capillary condensation can be expressed by the Kelvin-Thompson equation:

$$\ln\left(\frac{p}{p_0}\right) = -\frac{\sigma \cdot \eta}{R \cdot T \cdot r} \tag{4}$$

were p/p_0 is the relative humidity, σ is the surface energy [N/m], η is the viscosity [Pa·s], R is the gas constant [J/(K·Mol)], T is the temperature [K] and r is the radius of the pore [m].

Despite this relation between rh and the largest pore radius filled with humidity, the results in Figure 4 can not be used to calculate the water content presented in Figure 2. The reasons for this are:
- the pores do not have a cylindrical shape
- the influence of 'ink bottle' pores
- the limitations of MIP[18]
- the unknown changes in surface energy and viscosity of the pore solution under the influence of concentration during desorption.

There has been an attempt to consider the first three points with the help of a multiscale model that describes the connection of the pores.[25]

5.5 Change in resistivity and dielectric properties

Resistivity and dielectric properties are discussed using the equivalent circuit given in section 4.4 and depicted as an inlay in Figure 5b. This circuit consists of a resistance and a CPE in a parallel combination. In agreement with Niklasson[12] we propose that the resistance is related to the ionic current flow in the bulk of the pores and the distributed element is due to conduction at the interface between the pore solution and the solid phase.   Although both elements are related to the liquid phase, the contribution of the solid phase cannot be neglected, because the structure (porosity) of the solid phase determines the water content and the distribution of this water in the hardened cement paste. The solid phase itself has a resistivity over 10^4 Ωm and acts as an insulator.

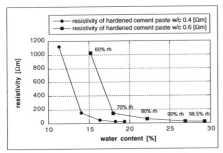
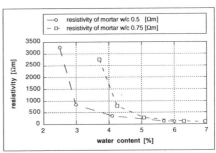

Figure 9a *Resistivity of hardened cement paste versus water content after 256 days*

Figure 9b *Resistivity of mortar versus water content after 256 days*

5.5.1 Resistivity. The relationship between water content and resistivity is shown in Figure 9. The shape of curves is similar to those presented in Figure 7. The resistivity increases with decreasing water content. The trend of the values for mortar samples is more continuous than for the samples of hardened cement paste. This is due to the dual pore size distribution of the mortar samples depicted in Figure 4b. Still there is a salient change of slope visible for samples exposed to relative humidities below 70%. This change is not limited to samples exposed to different relative humidities. Samples stored below 70% rh show the same development of resistivity as that depicted in Figure 6. The volume fraction filled with pore solution at 70% rh is 10.7% / 11.5% / 4% / 3.5% for mixtures #1 to #4 (determined from weight measurements). If one compares these values with the volume of nanopores presented in Table 3, it can be seen that the electrical network of conducting pores must be interrupted when the pore solution evaporates in the nanopores. That means that resistivity is not only dependent in the water content, but also on the way the pore solution is distributed in the pore space.

The influence of desorption on the pore solution composition and the resistivity has not yet been studied, because there are no additional samples at humidities lower than 98.5% on which to perform pore solution expression. At the end of the experiment this point will be reconsidered. It is probable that the resistivity of the pore solution decreases with decreasing water content. The values of resistivity versus relative humidity in Figure 7 therefore are a superposition of the desorption shown in Figure 1 and the decreasing resistivity of the liquid phase. The values increase with decreasing water content so the effect of the amount of the liquid phase is much stronger than that of the changing resistivity.

5.5.2 Dielectric properties. McCarter[10] suggested the presence of a charge on the hardened cement paste that leads to the formation of a double layer in the pore solution. Even if the surface of the cement paste does not carry any ordered charges the diffusion of the ions towards the surface under the influence of an electric field can lead to a charge separation in the pore solution. When an electric field is applied the ions move in the direction of the field and become blocked by the solid surface or the interface between pore solution and vapour phase. This is similar to charging up a capacitor. While the resistivity is related to the ionic current flow in the bulk of the pore, the dielectric properties are related to effects on the walls of the pores.

The amplitude of the CPE is inversely related to the area working as a capacitance.[26] This area is perpendicular to the electric field, E. In Figure 10 the amplitude, A, is plotted against the relative humidity. It is clear that with the exception of mortars stored below 80% rh and less clear for hardened cement pastes stored below 70% rh, there is not the same dependence of relative humidity or water content as in Figure 7. This is explained as follows; during desorption it is mainly the pore solution in the bulk of the pores that evaporates. Water at the surface is more strongly bound.[23] Therefore the ionic current must flow in thinner and thinner layers. The conduction continues until the last layers are removed. If this happens the amplitude, A, increases because there is no further polarisation possible.

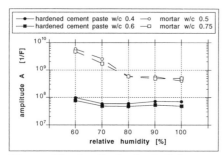

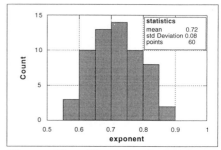

Figure 10 *Amplitude, A, of all the mixtures versus relative humidity after 256 days*

Figure 11 *Histogram of α for all mixtures after 256 days*

The exponent, α, showed a behaviour that was not dependent on water content or porosity. The 60 values of α obtained after 256 days from all the samples are shown as a histogram in Figure 11. The statistical data is given as an inlay. The values seem to follow a Gaussian distribution and must be related to an attribute that does not change with the mentioned parameters.

Many authors have observed that the surface of cement gel has a fractal structure.[27-29] This structure is determined by the fractal dimension, d_f, and the correlation length. Allen[28] found a correlation length of 40 nm using dynamic small angle neutron scattering. Flückiger[29] suggested 50 nm for the correlation length by analysing transport phenomena. Liu[30] proposed that for diffusion to a fractal surface the exponent of a CPE is related to the fractal dimension by $d_f = 3 - \alpha$. He studied the rough interface between two materials of very different conductivities and used a fractal model called Cantor bar to calculate the fractal dimension from a CPE. Using the equation of Liu[30] the values presented in Figure 11 lead to a fractal dimension, d_f, of 2.28 ± 0.08. The value Flückiger[29] suggested was 2.3. This indicates that the exponent of the CPE is related to the nanostructure of the C-S-H-gel and explains the independence of the exponent and the composition of the mixture or other parameters. It would be interesting to investigate the development of the structure in the early stages of hydration.

In this paper the influence of the temperature on dielectric properties has not been presented although this parameter was investigated. Further details will be published. The reported experiment has been continued until the samples stored below 80% rh were in equilibrium.

Acknowlegments

The authors would like to thank the following persons and organisations:
Swiss Federal Highway Administration, ASB, for financial support
A. Neiger, Institute of Electronic, ETH Zurich, for providing impedance spectroscopy
M. Guecheva, Departement of Analytical Chemistry, EMPA, for analysing the pore solutions
Dr M. Faller, Departement of Corrosion, EMPA, for determination of the pH values.

References

1. B. Elsener, S. Müller, M. Suter and H. Böhni, 'Corrosion of reinforcement in concrete', Edited by C. L. Page, K. W. J. Treadaway and P. B. Bamforth, Elsevier Applied Science, London, 1990, 358.

2. C. C. Naish, A. Harker and R. F. A. Carney, 'Corrosion of Reinforcement in Concrete', Edited by C. L. Page, K. W. J. Treadaway and P. B. Bamforth, Elsevier Applied Science, London, 1990, 333.
3. S. G. Millard, M. H. Ghassemi and J. H. Bungey, 'Corrosion of Reinforcement in Concrete', Edited by C. L. Page, K. W. J. Treadaway and P. B. Bamforth, Elsevier Applied Science, London, 1990, 314.
4. C. Alonso, C. Andrade and J. A. Gonzales, *Cem. Conc. Res.*, 1988, **18**, 687.
5. L. J. Parrott. *Mat. Struct.*, 1990, **23**, 230.
6. M. Raupach, *DAfStb*, 1992, **433**, 88.
7. P. G. Cavalier and P. R. Vassie, *Durab. Build. Mat.*, 1982, **1**, 113.
8. C. Andrade, C. Alonso and S. Goni, 'Concrete 2000', E &FN Spon, London, 1993.
9. E. J. Garboczi, *Cem. and Conc. Res.*, 1990, **20**, 591.
10. W. J. McCarter, S. Garvin and N. Bouzid, *J. Mater. Sci. Lett.*, 1988, **7**, 1056.
11. P. Gu, P. Xie, J. Beaudoin and R. Brousseau, *Cem. and Conc. Res.*, 1992, **22**, 833.
12. G. A. Niklasson, A. Berg and K. Brantervik, *J. Appl. Phys.*, 1991, **79**, 93.
13. F. Hunkeler, *Schweiz. Ingenieur und Architekt,* 1993, **43**, 767.
14. C. L. Page, N. R. Buenfeld and J. B. Newman, *Cem. and Conc. Res.*, 1986, **16**, 511.
15. B. B. Hope, A. K. Ip and D. G. Manning, *Cem. and Conc. Res.*, 1985, **15**, 525.
16. J. Tritthart and H. Geymeyer, *Zement und Beton*, 1985, **30**, 23.
17. S. G. Millard, K. R. Gowers and J. H. Bungey, 'NDT in civil engineering', Brit. Inst. of NDT, 1993.
18. R. A. Cook, PhD Thesis, Cornell University, 1991.
19. P. Longuet, L. Burglen and A. Zelwer, *Rev. mat. constr.*,1973, **676**, 35.
20. S. Diamond, *Cem. Conc. Res.*, 1975, **5**, 607.
21. P. Gunkel, *Beton-Information*, 1983, **23**, 3.
22. K. Byfors, C. M. Hansson and J. Tritthart, *Cem. Conc. Res.*,1986, **16**, 760.
23. S. Brunauer, P. H. Emmet and E. Teller, *J. Amer. Chem. Soc.*, 1938, **60**, 309.
24   T. C. Powers and T. L. Brownyard, 'Studies of the Physical Properties of Hardened Portland Cement Paste', Portland Cement Association, Chicago, 1947.
25. J. F. Daïan, K. Xu, D. Quenard, 'IUPAC Symposium proceeding', Laboratiores des Ponts et Chaussées, Prêsqu'île de Giens, 1994.
26. J. R. Macdonald, 'Impedance Spectroscopy: Emphasizing Solid Materials and Systems',Wiley-Interscience, New York, 1987.
27. D. Winslow, *Cem. Conc. Res.*, 1995, **25**, 147.
28. A. J. Allen, R. C. Oberthur and D. Pearson, *Phil. Mag.*, 1987, **56**, 263.
29. D. Flückiger, PhD Thesis, Swiss Federal Institute of Technology Zurich, 1993.
30. S. H. Liu and T. Kaplan, *Solid State Ionics,* 1986, **18&19**, 65.

REBAR CORROSION MONITORING USING EMBEDDED MINISENSORS

Hiroshi Tamura, Masaru Nagayama and Kazuyuki Shimozawa

General Building Research Corporation of Japan
5-8-1, Fujishirodai,
Suita, Osaka,
Japan

1 INTRODUCTION

In order to accurately monitor rebar corrosion in concrete, embedded sensors are desirable for not only preventing concrete cover effect on monitoring measurement,[1,2] but for continuous long-term monitoring. Minisensors have been developed which can be installed in both newly constructed structures and existing ones.The sensors developed are only 10mm in diameter and 12mm in depth so they never affect the durability of structures of the load carrying capacity. They comprise both a counter electrode and a pseudo-reference electrode which are made of gold. Using them three kinds of electrochemical characteristics can be measured ; natural potential, polarization resistance and electrolyte resistance . This paper reports one part of experimental results on rebar corrosion monitoring tests using embedded minisensors.

2 MINISENSORS

Newly developed minisensors have the following features.
① They have a round pseudo-reference electrode 4 mm in diameter in the center and a circular counter electrode 6 mm in internal diameter and 10 mm in external diameter. These electrodes are made of gold.
② The minisensors are embedded in the vicinity of rebars so that three kinds of electrochemical characteristics ; natural potential, polarization resistance and electrolyte resistance , can be measured with minimum effect of the concrete cover.
③ The minisensors are only 10 mm in diameter and 12mm in depth so they never affect the durability of structures or the load carrying capacity.
④ The minisensors can be installed in both newly constructed structures and existing ones.

3 EXPERIMENTAL PROCEDURES

The experiments presented here comprise two parts. The shape and size of reinforced concrete prism specimens are shown in Figure.1. A concrete mix of water/cement ratio 65% was used for both the experiments. Concrete was cast in two separate parts for two days so that rebars could be set without using any spacers.

3.1 Experiment I

Two hundred and twenty day corrosion tests were conducted varying both the chloride ion content in concrete between 0 kg/m³ and 10 kg/m³ and environmental conditions from dry (20°C, 60±5% R.H.), to wet (40°C, 100% R.H.) and alternately dry and wet(irregularly). Minisensors were installed 1 mm from the surface of rebars at two points in each specimen as shown in Figure.1(a) before casting the concrete. Electrochemical measurements were obtained every two hours just after demolding. After the corrosion test, the rebars were taken out from chloride-bearing control specimens, and both weight loss and corroded area were measured.

3.2 Experiment II

The 10 kg/m³ chloride-bearing concrete specimens were subjected to forty-eight day corrosion tests under dry (20°C, 60±5% R.H.) or wet (40°C, 100% R.H.) conditions. Twelve minisensors were installed in each specimen in varying orientations against a rebar as shown in Figure.1(b). Electrochemical measurements were performed during the corrosion test and weight loss and corroded area measurements after the test.

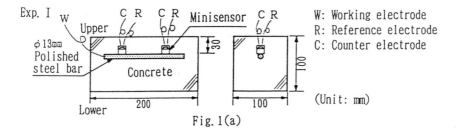

Fig. 1(a)

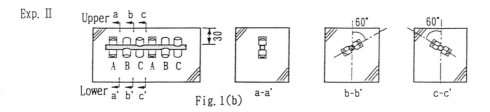

Fig. 1(b)

Figure 1   *Reinforced concrete specimen*

4 EXPERIMENTAL RESULTS AND DISCUSSIONS

4.1 Experiment I

By means of visual inspection, as shown in Table 1 and Figure.2, in the case of chloride-bearing concrete, broad general corrosion and local pitting corrosion were identified under dry and wet conditions, respectively. Under alternately dry and wet condition, both general and local corrosion were observed, and weight loss was the greatest of the three conditions. In every condition, corrosion predominantly occurred in the bottom parts of horizontal rebars.

As minisensors were embedded adjacent to the upper surface of rebars, even in the case of chloride contaminated concrete, no signs of corrosion were obtained from electrochemical measurements by minisensors under dry conditions alternately dry and wet conditions. However, under wet condition, after about 100 days, one minisensor recorded the following particular measurements. Natural potential was gradually changing toward less noble values, and polarization resistance was steadily decreasing. From the measurements, local corrosion in an adjacent area to the minisensor was estimated. Representative electrochemical measurements are presented in Figures.4.1~4.3, and 5.1~5.3.

The effect of concrete cover on the values of the measured polarization resistance has already been identified as being a few percent of the electrolyte resistance.[3] Electrolyte resistance measurements obtained in the experiment were much smaller than polarization resistance measurements even under dry conditions. Therefore, polarization resistance measurements using minisensors adjacent to rebars are not affected by concrete cover and therefore they can be satisfactorily used for the estimation of the corrosion rate of rebars.

Table 1   *Weight loss and corroded area*

Exp.	Condition	Age (days)	Corroded Area(mm^2)	Weight loss(mg)
I	Dry	220	3598	348
	Wet		245	247
	Alt. dry & wet		1203	602
II	Dry	48	151	48
	Wet		77	39

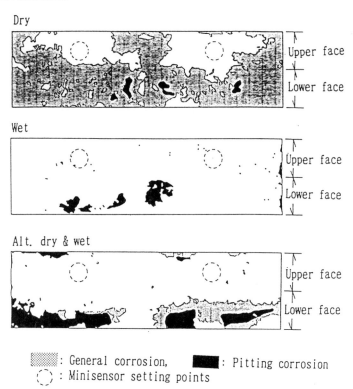

Figure 2 *Corrosion map*

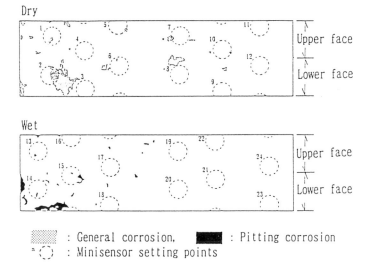

Figure 3 *Corrosion map*

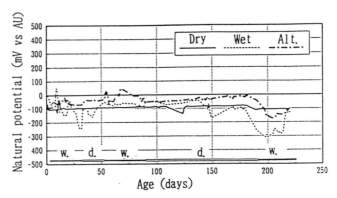

Figure 4.1   *Natural potential with age (non chloride-bearing concrete)*

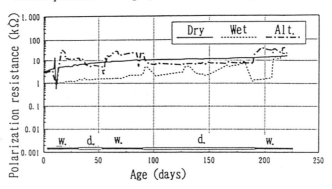

Figure 4.2   *Polarization resistance with age (non chloride-bearing concrete)*

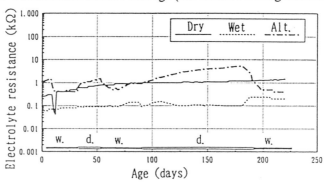

Figure 4.3   *Electrolyte resistance with age (non chloride-bearing concrete)*

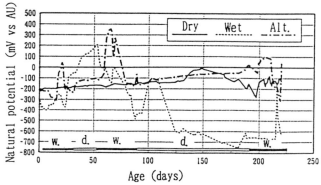

Figure 5.1   *Natural potential with age (chloride-bearing concrete)*

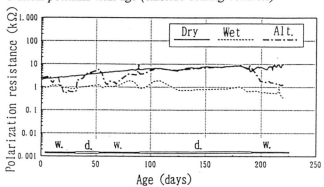

Figure 5.2   *Polarization resistance with age (chloride-bearing concrete)*

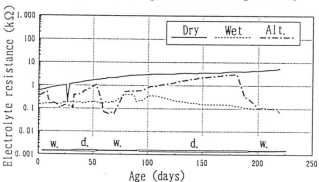

Figure 5.3   *Electrolyte resistance with age (chloride-bearing concrete)*

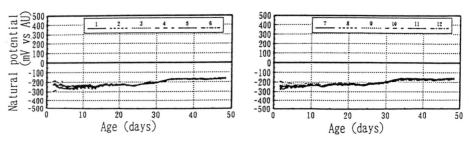

Figure 6.1   *Natural potential with age (dry conditions)*

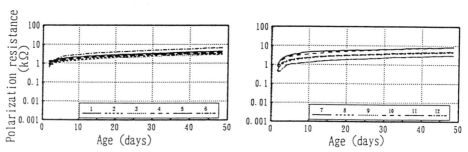

Figure 6.2   *Polarization resistance with age (dry conditions)*

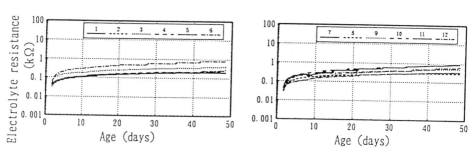

Figure 6.3   *Electrolyte resistance with age (dry conditions)*

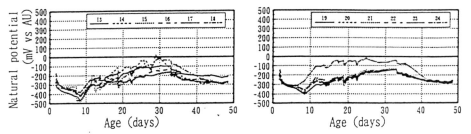

Figure 7.1   *Natural potential with age (wet conditions)*

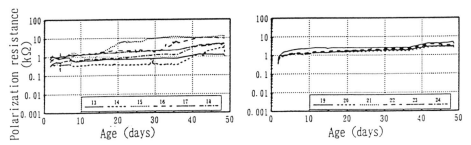

Figure 7.2   *Polarization resistance with age (wet conditions)*

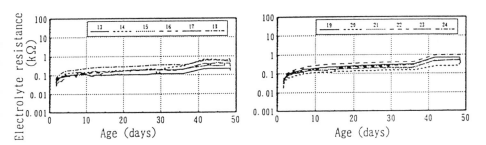

Figure 7.3   *Electrolyte resistance with age (wet conditions)*

4.2 Experiment II

From the electrochemical measurements presented in Figures.6.1~6.3, and 7.1~7.3, general corrosion and pitting corrosion were estimated under dry conditions and wet conditions, respectively. By means of visual inspection, as shown in Figure.3, corrosion was slightly observed under both conditions and local corrosion was identified only under wet condition, corresponding to the electrochemical measurements.

Since local corrosion was observed to correspond to the minimum polarization resistance measurements recorded by minisensors under wet conditions, it was concluded that corrosion monitoring can be satisfactorily conducted by minisensors (See Figures.3 and 7.2).

As in Experiment I, electrolyte resistance measurements obtained in the experiment were much smaller than polarization resistance measurements even under dry conditions (See Figures.6.2 ~6.3, and 7.2~7.3). Therefore, polarization resistance measurements using minisensors adjacent to rebars are not significantly affected by concrete cover and therefore can be satisfactorily used for the estimation of corrosion rate of rebars.

5 CONCLUDING REMARKS

According to the experimental results presented in this paper, electrochemical measurements obtained using minisensors embedded adjacent to rebars can be used for a more accurate estimation of the corrosion state and rate of rebars in concrete without having to take into account concrete cover effects. Further research on the threshold values for evaluation of corrosion rate and state will be needed before establishment of corrosion monitoring using embedded minisensors is possible not only for newly constructed concrete structures, but also for existing ones.

References

1. H. Tamura, M. Nagayama and K. Shimozawa, 'Corrosion of Reinforcement in Concrete', Edited by C. L. Page , K. W. J. Treadaway and P. B. Bamforth , Elsevier Applied Science, London, 1990, 372.
2. H. Tamura, M. Nagayama and K. Shimozawa, 'Durability of Building Materials and Components', Edited by S. Nagataki , T. Nireki and F.Tomosawa , E & FN Spon, London, 1993, 983.
3. H. Tamura, M. Nagayama and K. Shimozawa, Proc. of JCI Ann. Meeting, 11, 1989, 575.

REFLECTOMETRIC MEASUREMENTS ON PRESTRESSING CABLES

M. Donferri* , A. Gennari Santori , G. Nava and M. Tommasini

C.N.D. S.r.l
*Autostrade S.p.a.
Roma, Italy

1 INTRODUCTION

Controlling and guaranteeing the quality of installation of construction materials in structural engineering works represents an increasingly important aspect in any assessment of the reliability of the latter. Of particular concern, for example, is the installation of prestressing cables in the beams of bridge decks or tie rods in rock. In fact, incomplete filling of the sheath opens the road in an overall sense to a series of potential channels along which aggressive agents can reach the cable and produce undesirable effects such as localized oxidation of the reinforcing under tension and consequent loss of cross-section. On the other hand, besides defective installation, it is possible to cite defects caused by the fretting or fatigue of the steel strands due to contact or rubbing against the metal sheath.

At present the diagnostic methods most commonly used to evaluate the state of the cables can be summarized as follows:

- **X-ray:** this allows to spot voids and, in exceptional cases, severe corrosion and/or breaking of the cables; the limits of this technique include the need to have access to the two lateral sides of the element under examination, the thickness of the latter, the cost and bulkiness of the equipment and the radiation exposure risks.
- **Radar:** this allows to pinpoint voids and marked loss of section in the cables; the main problems with this method are its cost, the bulkiness of the equipment and frequent difficulties in interpreting the signals.
- **Ultrasonic method:** in certain cases and under various limitations, this method allows to identify the presence of voids.
- **Electro-chemical potential:** this provides a qualitative estimate oxidation's degree of prestressing steel.
- **Endoscopic inspections:** these provide clear visual data on the state of the cables, but is a purely punctual and partially destructive technique.

The purpose of this report is to describe a relatively new diagnostic method, called **RIMT**, a **time domain reflectometric method**. This method consists of strictly non-destructive electrical measurements on the cables, which permit identifying, quantifying and pinpointing possible corrosion phenomena, loss of cross-section and injection voids.

Between 1991 and 1995, the AUTOSTRADE Company, in collaboration with the CND Company, conducted a series of on-site and laboratory experiments using the reflectometer survey method to control the quality of prestressing cables.

The present technical report, which provides a summary and technical critique of all the experimentation conducted up to 1995, is divided into the following parts:
1) first, the theoretical basis of the method is set out, with a description of the various stages of the measuring operations and processing of RIMT signals;
2) the second part describes the experiments carried out on several sample beams especially constructed in collaboration with the AUTOSTRADE Company laboratory at Romagnano Sesia;
3) finally, the third section explains the survey conducted on 1500 prestressing cables located on 4 viaducts of the Udine-Carnia-Tarvisio motorway.

2. THEORETICAL PRINCIPLES OF THE METHOD

2.1 Impedance Measurements

The method consists of the application of electrical impulses of very brief duration (2-10 ns) at one end of a cable, and recording at the same time, at the same point, the electrical signals reflected by the opposite end, as well as by any possible defects present along the cable. From an examination of the time history of the reflected signal, it is possible to deduce the length of the cable and the location, type and entity of the possible defects.

The signal that travels along the cable is partially reflected back to its origin, and its frequency is modified and it overlaps the emitted signal as a result of local variations in the impedance, due essentially to dishomogeneities encountered during its course.

Laboratory experimentation permitted correlation of particular types of defects to the three components of the impedance: the ohmic resistance, the inductive reactance and the capacity reactance. By analyzing the reflected signal for each defect encountered, it was thus possible to assess:
· the **type** of defect, by noting variations in the impedance and determining its resistance, capacity reactance and inductive reactance;
· the **entity** of defect, as a function of the amplitude of the signal;
· the **distance** (s) of defect from the point of measurement, by:

$$s = \frac{V \cdot t}{2} \tag{1}$$

where: t = time between the instant of transmission and arrival of echo signal; V = speed of propagation of the electrical impulse along the cable.

2.2 Transmission Line consisting of the Prestressing Cable - Concrete System

In the context of reflectometer measurements, the prestressing cable is analogous to a transmission line of concentrated constants, consisting of a set of RLC elementary circuits similar to one another, with resistance and inductance parameters (R and L) depending on the electrical, mechanical and geometrical characteristics of the strands and with capacity components (C) depending on the similar characteristics of the injection material.

In RIMT measurements, the transmitter sends a series of current impulses with duration from 2 to 10 ns, so as to excite a spectrum of frequencies sufficiently large as to contain the resonance frequencies of the various elementary circuits and thus to put the entire line in resonance. Moreover, the high frequency components of the signal transmitted permitt the pinpointing of even very slight surface defects, thanks to the **Kelvin** or "**skin**" **effect**, whereby the higher the frequency of the current, the more it tends to be localized along the external surface of the conductor.

In the absence of defects along the cable, the impedance of each elementary circuit of which it is made up, will be equal to the characteristic impedance Γ_0 of the line. In such conditions there is no reflection back toward the generator.

In the presence of a defect, the impedance of the corresponding elementary circuit differs from Γ_0 and a reflection of the signal bounces back to the generator. The amplitude and the phase displacement of the current depend on the characteristics of the defect (variations in capacity impedance or the inductive impedance, magnitude of the variation, etc.). This reflection is recorded by the measuring device in d.d.p. terms, as a function of the current reflected and the impedance of the circuit.

Expressed in very simple terms, one could compare the behaviour of the elementary circuit of the cable affected by the defect to that of RLC-series circuits subjected to transitory phenomena. Capacity type defects can thus be identified on the reflectometer signal (reflected d.d.p./distance) by an upward slope, located at a distance greater than that at which the defect is actually located. In the case of inductive type defects, one notes downward sloping variations in the reflectometer signal, located at distances shorter than those of their actual positions.

In practice, for the normal L, R or C values, while the temporal phase displacements owing to inductive type variations are irrevelant, those caused by capacity type variations are situated at +10% of the distance between the generator and the defect.

2.3 Relationship of Electrical Anomalies to Physical Defects of the Cable-Injection System

The relationships between the electrical characteristics of the equivalent circuit and the physical properties of the prestressing cable-injection mortar were determined on the basis of a considerable number of experiments on prestressing structural elements conducted both in the laboratory and in situ.

Following parameter criteria and varying at a time whilst one parameter leaving the others unchanged (with respect to a properly integral cable/mortar system) the following relationships emerged:

• **Loss of strand useful cross-section** results in increase in R and L;
• **Corrosion of strand** results in increase in R, sharp rise in L, slight increase in C;
• **Defective injection (lack of injection material)** results in increase in C;
• **Physical-Chemical Changes in Injection Material** results in variations in L, in C and in the impulse propagation speed.

Under real conditions, the method showed limited sensitivity in identifying one or more defects which entailed simultaneous increases in both L and C; for example, in the case of a loss of useful cross-section of the cable due to corrosion, or in cases of injection defects combined with loss of section. This loss of sensitivity, however, is partly limited by the fact that upward variations in the amplitude of the reflectometer signal (owing to changes in C and characteristic of injection defects) tend to be shifted some 10% toward the final end of the cable with respect to their actual position.

3   EXECUTION,   PROCESSING   AND   INTERPRETATION   OF   THE
MEASUREMENTS

3.1 Instruments Utilized

The measurements were performed with the following apparatus:
· impulse emitter with automatic impedance adaptor;
· reflected signal reception/acquisition unit;
· digital multimeter with entry impedance of > 20 MOhm;
· digital oscilloscope for recording the disturbance signals.

3.2 Processing of Signals

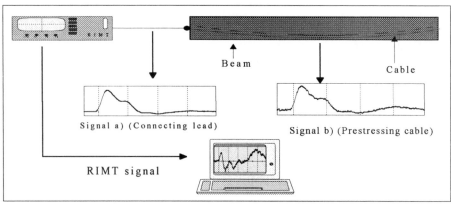

Figure 1 - *Rimt signal processing*

The various steps in processing the RIMT signals are illustrated in Figure 1.
1) Signal a) represents the reflectometer signal relative to just the connecting lead.
2) Signal b) represents the reflectometer signal relative to the sequence of connecting lead
    + prestressing cable on which the measurements are being performed. This signal can
    be considered as composed of the overlapping of several signals, regarding just the
    connecting lead, the prestressing cable, various disturbances, noise, etc.
3) The next step is to subtract signal a) from signal b), normalized in such a way that the
    maximum amplitudes of the two signals are equal (signal c)).
4) To correct the attenuation, signal c) is multiplied by an exponential function.
5) Finally, the derivative with respect to the length of the reflectometer signal is obtained.
    This permits determination of the gradient of the possible defects.

3.3 Determining the Type and Position of the Defect

Generally, for simplicity, several similar and closely positioned defects will be treated as
a single one.
One difficulty frequently encountered in identifying defects is the so-called **masking
effect**, whereby a defect of notable magnitude may conceal an immediately successive
defect of lesser importance. This inconvenience can generally be overcome by effecting the
measurements from the both ends of the beam.

Table 1 *Scale of corrosion defects*

TYPE	Level			
	1	**2**	**3**	**E**
	Oxidation	Corrosion	Corrosion	Crack
Diffuse Phenomenon	**A** (Powder)	**C** (Corrosion)	**C** (Corrosion)	
	B (Crusts)			
Concentrated	**D** (Pitting)	**D** (Pitting)	**D** (Pitting)	

3.4 Defect Rating Scale

In defining an appropriate scale of defect levels which permit assessing the risks in a repetitive and objective manner, account was taken of the different structural reliability requirements established by the various control's departments responsible for maintenance of the structures.

Given the qualitative character of these levels, it is advisable to implement calibration procedures so as to obtain greater reliability in assessing the defects. This calibration can be achieved through two methods:

1) in situ calibration, using endoscope readings to assess visually the magnitude of particular defects spotted using the RIMT method;
2) laboratory calibration, conducting a series of measurements on tensioned reinforcing, similar to those used in the structures.

3.4.1 Scale Adopted in Italy. With respect to injection defects, the only distinction is between partial corrosion (P) and absence of corrosion (A). Regarding corrosion, defects are broken down as shown in Table 1.

3.5 Surveillance Measures

Reflectometer measurements are widely used on prestressed cables as a surveillance technique (in Switzerland, for example), for the purposes of following the evolution of defects over time, and to obtain information regarding the speed of deterioration and the reliability of the cables under examination. In this context, the performance capabilities of the method can be fully exploited (replicability, sensitivity, precision in pinpointing the position and type of defect), without particular concern for its present limits regarding the qualitative aspects of the defect. The operating costs in terms of both time and money are essentially limited to opening and closing the beam heads. The surveillance measurements are repeated at regular intervals, and become more frequent as the level of defects increases.

3.6 Advantages and Limits of the Method

The main advantages of the method are:
1) Possibility of obtaining information on the global state of the cables.
2) Possibility of application also on cables difficult to access with other methods (e.g. tie rods in rock).
3) Use of the method for surveillance of prestressing cables over time.
4) Reduced costs of equipment, measurement operations and analysis of the data acquired.

On the other hand, at the present stage of experimentation, the method presents the following limitations:

1) The measurements may be influenced by the presence of structural defects located near the cable and of significant dimensions with respect to the cable itself.
2) The position of the defects and the measurement of the length of the cable is subject to an error of some 3%, owing to the tollerance of the chain of measurement.
3) Disturbances which cannot be completely eliminated.
4) Masking effect.
5) The level of the defect is not strictly linked, other than qualitatively, with loss of cable cross-section or absence of injection mortar. This is due to the current limits in both the measuring apparatus and the methods of analysis.

4 DESCRIPTION OF THE EXPERIMENTS

4.1 Description of the Beams

In order to develop certain aspects of the RIMT method, three sample PRC beams were constructed, simply supported, 8 m in length, with rectangular cross-section.

Each beam is reinforced lengthwise with four slack 10 cm steel bars arranged in correspondence with the corners of the section, and a post-tensioned prestressing cable of parabolic pattern, consisting of two 15.2 mm standard strands. The cable was tensioned in a single operation and from one end (drawn to 13000 kg/strand; anchorings of TENSACCIAI 2 PTA and 2 PTC 15).

Of the three beams, one was free of defects, whereas the other two had a series of artificial defects inserted. Specifically:

• Beam N°1: absence of defects.
• Beam N°2: injection defects, located in three well determined zones of 1 m extent: two zones of partial injection and one zone lacking injection mortar altogether.
• Beam N°3: corrosion defects, located in three well determined zones of 1m extent: one zone of uniformly reduced section, one zone with concentrated corrosion phenomena (pitting), and a mixed zone (uniform corrosion + pitting).

In each of the three beams was inserted a copper cable, 10 mm in diameter, lodged within a straight PVC sheath, 15 mm in diameter, to permit comparing the behaviour of an "ideal" system (copper cable - concrete) with the "real" system of steel cable - injection - sheath - concrete.

Furthermore, each of the three beams also contained a slack reinforcing bar 24 mm in diameter, suitably insulated from the other reinforcing cage, so as to be able to study the differences in cases of non-prestressed steels.

4.2 Description of the Corrosion Defects

As far as the **generalized corrosion** was concerned, the strand sections involved were soaked in vats containing HCl at 75° C for a period of about 4 hours. The loss of section resulted was non-uniform, ranging between 0.1 mm and 0.9 mm (account was not taken of the loss of section on the internal surfaces of the wires making up the strands).

In the case of the **corrosion by pitting**, a rod of stainless steel was positioned parallel to the strand along the sections involved; this served as a counter-electrode, connected to the negative pole of a continuous current generator (5 Amp), whereas the strand was connected to the positive pole. All of this was placed in a calcium saturated solution doped with 2% NaCl for 15-16 hours.

In constructing the beams, small conduits were inserted to permit the injection of an aggressive solution so as to assess the development of corrosion over time with the RIMT method.

4.3 Direct Transmission Measurements

A series of measurements by direct transmission was conducted: by transmitting the impulse from one end of the cable and receiving the response at the opposite end.

4.3.1 Measuring the Transmission Speed. The speed of propagation of the electrical impulses was measured experimentally along both the prestressing cable and the copper strand.

The signal transmitted propagates along the a path of length D consisting of the cable under examination (8 m) and the connecting cables linking the beam to the measuring instruments; the impulse transmission speed is thus calculated as follows:

$$V = D/T_{cable} \tag{2}$$

$$T_{cable} = T_{tot} - T_m \tag{3}$$

with T_m being the impulse transmission time along the instrument leads.

From these measurements it was found that the impulse travels through the steel elements at an average speed of 143,200,000 m/s, whereas it travels through the copper wire at an average speed of 141,333,000 m/s.

4.3.2 Examination of the Transfer Functions. The transfer function (the ratio between the FFT of the signal received and the FFT of the signal transmitted) was then calculated, with the following results:

• useful information can be deduced from the transfer function once over the threshold of 100 MHz;
• the structure seems to behave as an oscillating Marconian antenna;
• the amplitude of the various oscillation frequencies depend on the position of the defects in the element under examination, the mathematical correlation function of which remains to be defined;
• with the increase in the humidity of the concrete, the impulse propagation speed drops by some 10-12%;
• a transfer of energy between cable and concrete occurs, calculated to be in the order of 20-25% of the energy passing through the cable;
• analyzing the signals and disturbances in the realm of frequency permits better filtering and subsequent reconstruction of the useful signals.

4.4 Reflection Measurements

A series of measurements by reflection was conducted: by transmitting the impulse from one end of the cable and receiving the response at the same end.

4.4.1 Copper Cables and Slack Reinforcing Rods. All the signals showed similar patterns. For the copper cables the transfer function indicated a peak of 9.9 MHz, whereas for the reinforcing rods the peak was found at 7.05 MHz, but this was due to the difference in the impulse speed on the steel rods.

4.4.2 Prestressing Cables. The signals relative to the two stands of the same beam, measured at the same end showed no significant variations, and these latter were greatest in the case of beam N°3. From the standpoint of signal processing, however, some interesting variations were encountered taking the signals from the strands of beam N°1 as a reference, and comparing these with the signals from the strands of beams N°2 and N°3. This confirms that accuracy in signal processing is much greater when there is a reference signal available for comparison (the theoretical signal of the cable, the experimental signal of the sample cable, or the mean values of the signals of cables similar to one another).

Thus the program devised to process the RIMT signals took as a reference the signal obtained from beam N°1 (free of defects). On the basis of these, and taking into account that the starting point of the cable is situated in correspondence with the 1m abscissa, one obtains the defects reported in Table 2.

4.5 Future Objectives

From what has been set out above, the following objectives have top priority:
· Widening the frequency band examined.
· Improvement of the impedance matching.
· Development of easy to use instruments to determine the moisture level of the concrete and the preparation of comparison tables showing signal speeds as a function of the moisture level of the concrete.
· Study of the quantitative effects of a mechanical-physical defect on variations of the capacity impedance and the inductive impedance.
· The development of a new apparatus and data analysis instruments capable of separating the inductive component from the capacity component of the reflectometer signal.
· Studies to determine the possibilities of measuring the state of tension of the prestressing cables.
· Studies to evaluate the possibilities of identifying the presence of hydrogenation phenomena in the prestressing cables.

Table 2 - *RIMT Analysis of the Sample Beam*

Beams	*Defects*			*RIMT Analysis*		
	TYPE	From (m)	To (m)	TYPE	From (m)	To (m)
2	Partial Injection (P)	0.3	0.9	P	0.7	1.3
	Absent Injection (A)	2.5	3.5	A	2.7	4.0
	Partial Injection (P)	5.0	6.0	P	4.5	5.8
3	Corrosion (1B)	0.0	1.0	1A	0.0	1.0
	Corrosion (1B/D)	2.5	3.5	1A	1.4	1.8
	Corrosion (1D)	5.0	6.0	1B	2.2	3.2
				1B	4.3	5.2
				1A	7.7	8.0

5   FIELD   SURVEY   PERFORMED   ON   VIADUCTS   OF   THE   A23
UDINE-CARNIA-TARVISIO MOTORWAY

In the period from 22 November 1994 to 6 February 1995 an RIMT survey was conducted
on the prestressing cables of four viaducts: Malborghetto, Fella VI, Fella VII and Fella IX.
Each of the viaducts consisted of two flanking, continuous beam decks, simply supported
on abutments and piers, with closed, rectangular box cross-section.

The measurements involved a sampling of 1500 cables, equivalent to more than 25% of
the total number of cables present. The purpose of the measurement was to perform a
quality control check on the prestressing cables, providing indications of the locations,
extent and magnitude of possible anomalies due to both corrosion phenomena and injection
defects.

5.1 Measurements and Processing

To assess the defects, the scale described in section 3.4.1 was adopted (Table 1).   An
RIMT signal was processed for each cable (Fig. 2) to produce "deterioration profiles":
diagrams reporting the level of deterioration of the various corrosion or injection defects
encountered (Fig. 3). Tables were then prepared showing the information on the cables and
the related defects encountered (Table 3).

Given the quantity of measurements taken, a statistical analysis was made of the defects
recorded, thus providing a better interpretation of the results and a qualitative overview of
the general state of each viaduct (Fig. 4 and 5).

Acknowledgements

The Romagnano Sesia laboratory, coordinated by Eng. Familiari, superintended and
managed the construction of the beams, participated in the signal acquisition operations at
the various stages of the operation, and collaborated in the examination of the effects.

Professor Cigna of the University of Rome studied and proposed the various methods for
accelerated corrosion of the prestressing steel.

The anchor headpieces, the strands and the metal sheathes were kindly provided by the
TENSACCIAI Company of Milano, which also performed the related stressing operations.

Table 3 - *Example of an RIMT Analysis Table*

Cable	*L Proj.*	*L Meas.*	*Corrosion*			*Injection*		
			From (m)	To(m)	Lev	From (m)	To(m)	Lev
MSS11T17	43.8	43.8	0.2	1.8	1B	2.0	6.2	P
			6.8	8.0	1B	8.8	12.0	P
			14.4	20.2	1A	18.6	22.6	P
			25.0	28.8	1A	27.2	30.6	P
			34.2	38.2	1A	35.0	39.0	P
			43.6	43.8	1A			

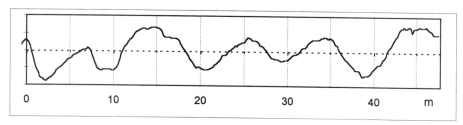

Figure 2 *Example of an RIMT Signal (reflection coefficient/distance)*

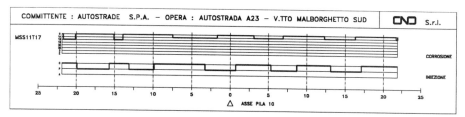

Figure 3 *Example of a Deterioration Profile*

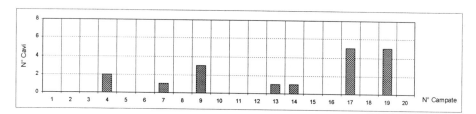

Figure 4 *Distribution of grouting level A (absence of mortar) - Viaduct Fella IX (North).*

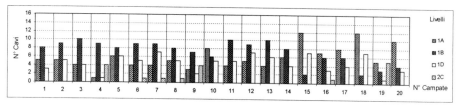

Figure 5 *Distribution of the **worst levels of corrosion** - Viaduct Fella IX (North).*

MEASURING THE HALF-CELL POTENTIAL OF STEEL EMBEDDED IN IMMERSED CONCRETE : PRINCIPLES AND APPLICATION

Gilbert GRIMALDI
Laboratoire Régional de l'Est-Parisien, Melun, France

André RAHARINAIVO
Laboratoire Central des Ponts et Chaussées,
58 Bd Lefebvre, F-75732 Paris CEDEX 15, France

1 SCOPE

1.1 The Rusting of Steel in Concrete

Steel in concrete is usually protected by a film formed by the reaction between steel and cement. When this film is damaged, metallic corrosion occurs at some locations.[1] However, this corrosion begins only the agressive agents have come from the environment, penetrate the covering concrete and contacted the rebar, in high enough quantities.

The rust formed after this initiation period can grow and result in the spalling of the concrete cover. When such deteriorations occur, the costs of repairs are high. Therefore, a non-destructive diagnosis of the rebar condition, at any point of the structure, is very often useful.

1.2 The Steel Half-Cell Potential

For assessing the condition of rebar in concrete, the procedure which is most commonly applied consists of determining eventual rusting by using electrochemical measurements.[2,3] It involves measuring the steel half-cell potential, E_c , which is the potential drop between a rebar and a reference cell which is placed on the concrete surface. This reference cell is often either a copper-copper sulphate electrode, or a silver-silver chloride electrode (Figure 1).

For such a measurement, the concrete cover is removed by boring a hole of small diameter (about 50 mm or less), in such a way that the rebar is visible. This steel is cleaned, then, by using clips and wires, it is connected to an electronic millivoltmeter, the other pole of which is connected to the reference cell. This reference cell is placed at chosen points on the concrete surface.

The distance between two measuring points depends on the accuracy required for locating the corrosion areas, while taking into account the total number and the cost of the measurements. This distance is commonly in the range between 30 mm and 1 meter.

The half-cell potential values are meaningless if an electric insulator is lying between the rebars and the reference cell, or if a stray current exists which makes the potential fluctuate.

In the absence of any stray current, the half-cell potential usually ranges between -500 and +100 mV_{CSE} (reference cell : copper-copper sulphate saturated).

The half-cell potential E_c values are recorded as a function of the locations of the measuring points. A map of the structure surface can then be drawn and the E_c results written on it.

They are processed as follows :

a) the Ec values at the location of the measuring points are plotted on the map,

b) "equipotential" contours are drawn, corresponding to the points having the same E_c values (Figure 1),

c) some critical E_c values are selected for drawing these equipotential curves. They correspond to "corrosion risk" of the rebars, according to some recommendations, such as ASTM C 876. So, they separate areas where the rebars are likely to corroded or not. More precisely, the E_c values are divided into "classes" whose meanings are :

- "class S" : ≥ -200 mV_{CSE} : rebars are unlikely to be corroded (passivation),
- "class M" : $-350 \leq E_c < -200$ mV_{CSE} : rusting is possible
- "class R" : $E_c < -350$ mV_{ESC} : < -350 mV_{CSE} : rebars are very likely to be corroded.

So, it appears hat in the areas where rebars are corroded, the half-cell potential , E_c, values are low.

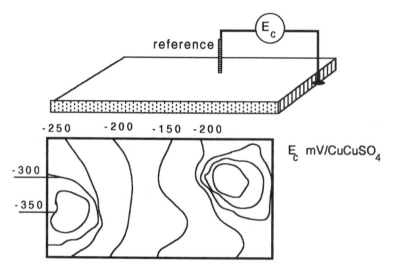

Figure 1 *Principle of the half-cell potential E_c measurement concerning steel in concrete*

When the reinforced concrete is in water, some additional difficulties occur. First, the reference cells must be stable in the surrounding water. For example, their junction liquids must not react with the surrounding water or leak through the porous material at the end of the cell.

Secondly, half-cell potential value does not correspond to a short length of rebar: a kind of short-circuit occurs at the immersed concrete surface. It makes the potentials, E_c, of all the rebars even (uniform). The reason for this result is that the half-cell potential is measured on the concrete surface, while inside the concrete, the equipotential lines are distributed according to :

- the locations of the dissolution areas (way out of the electric field from which the potential derives) and of the cathodic areas (the electric field entering the steel),
- the electrical resistivity and surface of the concrete : if this material is wet, the equipotential lines are parallel to the concrete surface. Therefore, locating the areas having a precise half-cell potential value is less precise.[4]

Thirdly, the half-cell potential E_c describes the rebar condition when it touches the concrete. The oxygen content in the concrete influences the E_c value. In other words, the potential (of a given rebar in concrete) changes, depending on the aeration conditions of concrete. More precisely, the higher the potential E_c, the higher the oxygen content of the surrounding concrete (Figure 2). Conversely, in a highly deaerated water, the E_c value of a rebar is lower than the value corresponding to the same concrete in atmosphere.

However, up to now, no procedure had been validated for measuring the half-cell potential of rebars embedded in concrete immersed in various natural waters.

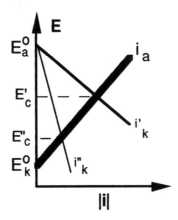

Figure 2 *Influence of the oxygen content around the rebar, on its E_c value for which the cathodic current i''_k (oxygen reduction)is equal to anodic current i_a (iron dissolution). For the cathodic reactions, the oxygen content is higher for i'_k curve than for i''_k curve.*

2 DEVELOPING THE METHOD

2.1 Principle of the Procedure. Design of the Device

The principle of the device for measuring the half-cell potential of a rebar in an immersed concrete, consists of restoring the conditions of atmospheric concrete, i.e. of placing an artificial atmosphere on the concrete surface.

The reference cell is placed on the concrete surface which is surrounded by an artificial atmosphere, confined in a vessel (Figure 3), during a period which is is needed for reaching a steady state, as it will be detailed later.

This device includes :
- a reference cell,
- a vessel (jar) around the reference cell, tightly placed on the concrete surface,
- an oxygen (or compressed air) supply to blow the water out of the vessel and to introduce oxygen into concrete,
- possibly, a system for fixing the vessel on the concrete.

2.2 Experiments in the Laboratory

The object of this part of the investigation was to show how an immersion could change the rebar half-cell potential, the potential distribution and its value at a given point.

Tests were carried out on three concrete slabs, 0.50 x 0.50 x 0.10 m, reinforced with a steel mesh (wire diameter : 10 mm). The concrete mix was designed with 400 kg ordinary Portland cement per cubic meter of concrete. The mesh (0.10 x 0.10 m) was embedded under a concrete cover 25 mm thick.

The embedded rebars were partially corroded by anodic polarisation. For this purpose, the slabs were stood in a vessel full of sodium chloride solution, and immersed to a depth of about 300 mm of them was immersed. A titanium mesh placed 100 mm from the concrete surface was connected to the negative pole of a DC power supply (12 volts). The positive pole of the power supply was connected to the rebars. The polarization lasted for 66 hours.

Figure 4 shows schematically the pretreatment procedure and the locations of the E_c checking points.

To obtain an E_c value corresponding to a small rebar area, the reference cell was isolated from the surrounding liquid, by using a vessel from which the water was displaced with compressed air.

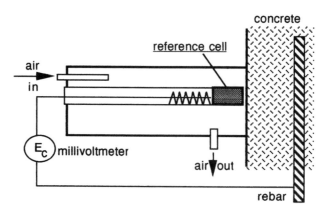

Figure 3 *Principle of the device for measuring the half-cell potential of rebars in immersed concrete*

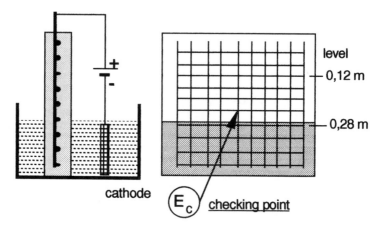

Figure 4 *Reinforced concrete slabs : previous corrosion procedure and E_c checking points*

Figure 5 shows the results obtained. The E_c values measured immediately after the water was blown out of the vessel are different from the values obtained in the absence of the vessel. However, they are not equal to the values corresponding to concrete in air. This means that the presence of an artificial atmosphere during a short period, changed the distribution of the equipotential lines, but was not sufficient to instantaneously modify the electrochemical behaviour of the rebars in their close environment.

These first tests were complemented with other experiments to show the influence of the aeration duration. The artificial atmosphere was blown into the vessel on the concrete surface for a longer period of time.

Figure 6 shows a typical "half-cell potential E_c - time t" curve which includes several stages

a) "initial condition" : the reference cell is immersed, no artificial atmosphere being in the vessel,

b) "water free" stage : the water is blown out of the vessel, the half-cell potential E_c value decreases,

c) "progressive drying" stage : the half-cell potential E_c value increases,

d) "plateau": the half-cell potential E_c value is constant,

e) "re-immersion" stage : the artificial atmosphere is removed, water comes back into the vessel, the half-cell potential E_c value still increases.

f) "hard drying" stage : if the progressive drying stage is too long, the half-cell potential E_c value increases quickly, up to very high levels, because the pore water in concrete can be removed.

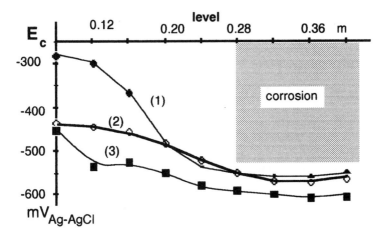

Figure 5 *Half-cell potential of rebars in slabs, after previous corrosion, measured :*
(1) in air, immediately after the corrosion, (2) in air, after the slab was saturated with water,
(3) in water after about 5 minutes under an artificial atmosphere.

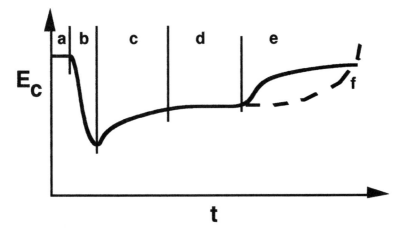

Figure 6 *Half-cell potential change versus time of rebars in the presence of an artificial atmosphere a) stage of the "initial condition" b) "water free" stage c) "progressive drying" stage d) "plateau" e) "re-immersion" stage f) "hard drying" stage*

The following conclusions can be drawn, concerning the half-cell potential E_c value at a given point :
- for a rebar which was not in the sodium chloride solution during the pretreatment, the half-cell potential initial shift was about -150 mV, when the reinforced concrete had been immersed for about 4 hours,
- this potential decay was observed even after 6 days (after the immersion),
- the half-cell potential value at the "plateau" stage is close to the value obtained when the reinforced concrete is in natural atmosphere.

So, when an artificial atmosphere is placed on the concrete surface,
- the half-cell potential E_c of the rebars is no longer uniform,
- the E_c values in an artificial atmosphere are equal to the values obtained in natural air atmosphere.

Therefore, the procedure for interpreting the E_c values developed for rebars in atmospheric concrete is still valid. But it was noted that concretes designed for immersed structures are not identical to concretes for atmospheric structures, such as bridge decks.

3 APPLYING THE METHOD

3.1 The Operational Equipment

About one hour is needed for measuring the half-cell potential at an immersed point. So, the operational facility must make it possible to obtain several results simultaneously and include several reference cells, each of them being in a vessel. The artificial atmosphere is given by compressed air that the operators (who are divers) need to breathe.

The vessels are fixed onto the concrete surface by several means. One of them is to use the Venturi effect. In such a method, auxiliary bells are linked to the vessels which include the reference cells. The water in these bells is removed by compressed air, which results in a decrease of local pressure. So, the water pressure around a bell pushes it to the concrete surface, and fixes the half-cell measurement equipment.

One reading of the steel potential takes many minutes, so several half-cell devices with their artificial atmospheres are applied simultaneously. The equilibrium state of steel potential can be reached after various durations, so this potential is recorded continuously during the oxygenation of the concrete. This potential is read until it becomes constant over time.

3.2 Applications to Structures in Service

The half-cell measurement method was applied to several reinforced concrete boats. These boats were used either for river cruising or for permanent housing. For example, it dealt with a 70 m long and 7.75 m wide boat, which was in service.

The concrete proportioning and the cement used were not known. The concrete cover was about 0.30 m. The immersed parts of the boat, which were assessed were covered with no paint or coating.

Visual examinations showed that some reinforcing steel was corroded. Steel half-cell potential was measured to determine the areas where other rebars were likely to be corroded. The connections between rebar and voltmeter were made far above the water level. It should be noted that no drilling was allowed in the concrete, for verifying the non destructive assessment of the submerged parts. So, no direct verification of the potential measurements could be made, as is usual for atmospherically exposed structures, such as bridge decks.

The results are shown in Figure 7. All the values of steel half-cell potential,.E_c,were lower than -200 mV$_{CSE}$. But only a few E_c values were lower than -350 mV$_{CSE}$, which corresponds to corroding steel. As a rule, steel corosion is due to large corrosion cells, i.e. it was corroded in some areas but it was sound in adjacent areas. So, this boat, as a whole, was assessed to be in no immediate danger.

This was not the case of some other immersed reinforced concrete structures which were assessed by applying half-cell potential measurements.

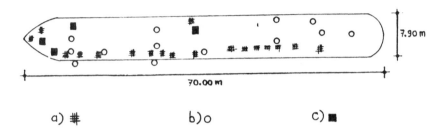

Figure 7 *An example of reinforced concrete boat in which the steel condition in immersed part, was assessed with half-cell measurement a) steel apparently corroded, b) steel potential between -200 and -350 mV$_{CSE}$, c) steel potential lower than -350 mV$_{CSE}$.*

4 CONCLUSION

The usual method for measuring the half-cell potential of steel in concrete is not valid when the concrete element is immersed. The method does not give reliable results when the reference cell is simply placed on the concrete surface.

This study shows that the condition of steel embedded in immersed concrete can be correctly assessed by measuring its half-cell potential. Using the new method which has been developed, a reference cell is in a vessel where an artificial atmosphere is introduced.

The validity of this method was determined with laboratory tests. It appears that the half-cell potential values in immersed concrete specimens are almost identical to the values in natural atmosphere. These tests also indicated how long the artificial atmosphere must be maintained on the concrete surface to achieve a similar result to that in natural atmosphere.

Operational equipment has been developed. This method has been applied to immersed reinforced concrete structures, such as reinforced concrete boats.

References
1. A. Raharinaivo, Mme G. Arliguie, P. Brevet, F.Derrien, J. Galland, L. Lemoine, G. Taché and F. Wenger, Proccedings Journée Durabilité et Protection des Armatures du Béton AFREM - CEFRACOR, Paris, 1988, 30.
2. L. Lemoine, A. Raharinaivo and G.Taché. Proceedings 8° Congrès Européen de Corrosion, Nice 1985, **1**, paper 29.
3. G. Grimaldi, R. Berissi, P. Brevet and A. Raharinaivo, (Ed A.S. NOWAK) "Bridge Evaluation, Repair and Rehabilitation", NATO ASI Series, Series E: Applied sciences , 1990, **187**, 249.
4. C. Andrade, I. R. Maribona, S. Feliu, J. A. Gonzalez and S. Jr Feliu , *Corrosion Science* 1992, **33**, 237.

CORROSION MONITORING OF CONCRETE PILLARS IN MARINE ENVIRONMENT

Øystein Vennesland
Norwegian Institute of Technology, 7034 Trondheim, Norway

Per Austnes
Public Road Administration, Møre and Romsdal Road Office, 6022 Ålesund, Norway

Olav Ødegård
Ødegård og Lund A/S, Etterstadkroken 6, 0659 Oslo, Norway

1 INTRODUCTION

Concrete damage due to chloride induced corrosion is difficult to repair and the repairs generally become very expensive. Most of the coastal bridges in Norway are subjected to a very severe environment and various types of trials for maintenance and repair have been made, including cathodic protection with different systems, electrochemical chloride removal, ordinary mechanical repairs and different types of surface treatments. This report describes a test at Stokksund bridge where a dense epoxy coating was applied to the surface.

Stokksund bridge is a part of the coastal road system in the county of Møre and Romsdal and, accordingly, subject to almost continuous sea water splashing. A survey of the bridge showed that in some of the pillars the reinforcement had started to corrode while in other pillars the chloride content was so high that the onset of corrosion could be expected any time.

It was decided to undertake a trial involving the application of an epoxy membrane to two pillars. Rescon A/S, a supplier of materials for concrete maintenance and repair, came up with a proposal that was approved. Three pillars of the bridge were included in the trial, one reference, one coated from the top structure down to the foundation and one coated from the top structure, beyond the foundation to below the lower water level. In all three pillars probes of various types were embedded.

The main objective of the trial was to study the effect of the dense coating on central corrosion parameters, especially the oxygen content. By reducing oxygen transport to the reinforcement the cathodic activity might be reduced to such a degree that corrosion activity is considerably less than in the untreated references. Theoretically a dense membrane might function both by acting as a diffusion barrier for oxygen and therefore directly affecting the oxygen transport or by acting as a diffusion barrier for water and indirectly affecting the oxygen transport as a result of the increased water content. It was also important to see whether it was practical to apply a dense epoxy coating below low water level.

2 INSTALLATION

2.1 Condition Survey

A very thorough condition survey was undertaken before installation of the measuring probes. The survey included measurement of steel potentials, cover thicknesses and chloride profiles. The potential values indicated that the reinforcement had only locally started to corrode, with typical values around -100 mV vs $Cu/CuSO_4$ reference electrode, with spots down to -500 mV showing the onset of corrosion. The cover thickness was generally between 30 and 35 mm and a typical value for the chloride content at the depth of reinforcement was 0.15 % by weight of concrete.

2.2 Monitoring System

All monitoring probes were installed in zones of about 10 cm in width with a distance of one metre between each zone to monitor the effect of increasing height above sea level. The probes consisted of reference electrodes for potential measurements, stainless steel rods for resistance and oxygen diffusion measurements and thermocouples for temperature recording.

The installation was made just before application of the coating. It is very important that the connecting leads were not acting as channels for diffusion of oxygen. Special care was therefore taken in making the connections between leads and the installed electrodes.

2.3 Coating

The epoxy coating was delivered and applied by Rescon A/S. Before coating, the concrete surface was cleaned by sand blasting. The epoxy was then applied in three layers with a glass fibre reinforcement between the second and third layers. The first layer, an Epoxy L-L primer, was rolled. The second layer of Epoxy MS was sparkled and the third layer of Epoflex, a slightly elastic epoxy with some polyurethane, was rolled. In the underwater zone the procedure was different. The glass fibre reinforcement was soaked with Epoxy UV-C, a special underwater epoxy. The soaked material was pressed against the concrete surface by hand and more Epoxy UV-C was sparkled on top. The coating was applied in October 1993.

3 MEASUREMENTS

3.1 Potential vs Embedded Reference Electrodes

The corrosion state of the reinforcement is measured by embedded electrodes. Two types of electrodes were used, a manganese dioxide electrode and a titanium electrode. The electrodes were installed in the pillars by drilling holes slightly bigger than the electrode diameter and embedding the electrodes with a repair mortar.

3.2 Potential Difference Between Two Types of Reference Electrodes

The potential difference between the two types of reference electrode should be affected by the oxygen content. The manganese dioxide electrode[1] should be independent of the oxygen content while the titanium electrode should be affected by the partial pressure of oxygen. Changes in oxygen content should therefore be reflected in changes in potential between the two types of electrode.

3.3 Oxygen Transport

The measurements are based on the assumption that when a negative potential within certain limits is applied to steel embedded in concrete the only cathodic reaction taking place is reduction of oxygen at the steel surface:

$$O_2 + 2H_2O + 4e = 4OH^-$$ (1)

Therefore, by applying a constant negative potential to an embedded electrode, the oxygen transport (or rather the rate of oxygen reduction at the electrode) through the concrete may be recorded as an electrical current. By applying Faraday's law the mass transport per unit time (Flux) can be calculated:

$$J = I_{steady-state}/nF$$ (2)

where:

J = Flux $(mole/m^2s)$
I = Current density (A/m^2)
n = Valence   (4)
F = Faraday constant (As)

The measurements are normally made by means of a potentiostat, reference electrode, counter electrode and the working electrode. A constant voltage is applied between the reference electrode and the working electrode, and the resulting current is a measure of the oxygen reduction current.

Here a variant of this method was used. Instead of using one potentiostat for each working electrode - which is required by the ordinary set-up - all working electrodes and all counter electrodes were paralleled and a constant voltage of 600 mV applied between the working and counter electrodes. The current for each working electrode was recorded by means of a 60-ohm resistor in each measuring circuit.

Both the working and counter electrodes were stainless steel rods. The electrodes had a diameter of 18 mm and a length of 50 mm. The distance between the electrodes was 50 mm and they were installed by drilling holes slightly bigger than the electrode diameter and embedding the electrodes with a repair mortar. The electrode cover was 20 mm.

This way of measuring introduces one uncertainty but two significant advantages. The uncertainty is the stability of the applied potential to the working electrode. While a reference electrode has a stable potential, the potential of a counter electrode is affected by the current passing through the system. The advantages are firstly that many measuring circuits may be coupled in parallel and controlled by one voltage source, while by using a potentiostat and reference electrode each potentiostat can only control one measuring circuit and secondly that very simple voltage sources (e.g. batteries) may be used.

The main point is the stability of the potential. The range of the limiting current for oxygen reduction is so wide, however (about 500 mV^2), that it should be possible to select a voltage that creates a working electrode potential within this range. This type of measurement should, however, only be used for short durations as the anodic polarisation of the counter electrode induces a risk of pitting corrosion in chloride contaminated concrete

3.4 Resistance

The resistance was measured between two embedded stainless steel electrodes. The electrodes had a diameter of 18 mm and a length of 50 mm. The distance between the pairs of electrodes was 50 mm and they were installed by drilling holes slightly bigger than the electrode diameter and embedding the electrodes with a repair mortar. The electrode cover was 20 mm. Each location was carefully selected to maximise the distance to reinforcing steels. The measurements were made with an AC bridge at 1000 Hz.

3.5 Temperature

The temperature was measured by using an ordinary thermocouple. In each pillar thermocouples were installed at the top and bottom. The thermocouples were installed in drilled holes at a depth of 30 mm.

4 RESULTS

Two sets of measurements were made, one in spring 1994 and one in spring 1995, 10 and 12 of May, respectively. On both occasions the coating was inspected to detect any disbonding or other defects. In 1994 a few spots had to be recoated but the bonding seemed very good, even in the zone below water. In 1995 a very careful inspection showed spots of delamination between the second and third epoxy layer. This was probably caused by insufficient soaking of the rather thick glass fibre reinforcement. A closer inspection showed that all epoxy layers were intact and there was no disbonding between the concrete and the epoxy. During this inspection the concrete surface was exposed in some locations and it was observed that the concrete was surprisingly dry. The colour was light grey and drops of water were immediately absorbed.

4.1 Potential Measurements

The results are shown in Tables 1 and 2. The presented metres are measured from the mean water level. The tidal zone is about two metres. The steel potentials presented in Table 1 show only very small differences between the two measurements.

The potential differences between the two types of reference electrodes presented in Table 2 are almost identical, except for the top levels in the reference and in the pillar coated below water.

Table 1 *Potential measurements at Stokksund bridge between reference electrode ERE10 (manganese dioxide) and the reinforcement (mV)*

Metres above water	Reference		Coated to foundation		Coated below water	
	1994	1995	1994	1995	1994	1995
5					-204	-124
4			-95	-111	-146	-153
3	-163	-164	-112	-128	-111	-118
2	-161	-167	-142	-142	-128	-135
1	-190	-200	-137	-138	- 98	-118

Table 2 *Potential differences (mV) at Stokksund bridge between reference electrodes ERE10 (manganese dioxide) and titanium*

Metres above water	Reference		Coated to foundation		Coated below water	
	1994	1995	1994	1995	1994	1995
5					-119	-208
4			-211	-206	-227	-233
3	-149	-100	-229	-244	-210	-206
2	-174	-157	-225	-219	-241	-277
1	-179	-172	-209	-202	-206	-211

4.2 Oxygen Transport

The results are shown in Table 3. There is a marked difference in the oxygen reduction currents measured in 1994 and in 1995.

Table 3 *Oxygen reduction currents at Stokksund bridge (mA/m^2)*

Metres above water	Reference		Coated to foundation		Coated below water	
	1994	1995	1994	1995	1994	1995
5					5.01	0.100
4			9.08	0.206	9.02	0.017
3	15.10	5.60	7.72	0.294	5.66	0.460
2	13.09	0.90	5.60	0.076	12.27	0.153
1	7.55	0.312	2.95	0.330	4.89	0.235

In the coated pillars the values are between one and two orders of magnitude lower in the later measurements. In the reference pillar the values in the lower part are also much lower while the oxygen reduction current in the upper part is in the same range for both measurements.

4.3 Resistance

The results are shown in Table 4. The resistivity has increased in both the coated and the reference pillars. The increase is generally higher for the coated pillars, between 36 and

91% for the pillar coated below water and between 60 and 73% for the other coated pillar. For the reference pillar the increase is about 25% in the two upper levels and 75% in the lower level. No attempt has been made to calculate the resistivity. Although formulas based on geometry exist,[3] they are not very accurate. Here changes in resistivity are of greater interest.

Table 4 *Resistance measurements at Stokksund bridge (ohms)*

Metres above water	Reference		Coated to foundation		Coated below water	
	1994	1995	1994	1995	1994	1995
5					3067	4741
4			2404	3845	2521	4099
3	1905	2384	2681	4569	2484	4749
2	1592	2023	4953	8476	2815	4059
1	4865	8508	5662	9830	2263	3085

4.4 Temperature

The results are shown in Table 5. In 1994 the temperature was between 8 and 9 °C and in 1995 it was about three degrees lower.

Table 5 *Temperature measurements at Stokksund bridge (°C)*

Metres above water	Reference		Coated to foundation		Coated below water	
	1994	1995	1994	1995	1994	1995
5					8.0	5.3
4			8.3	5.1		
3	8.6	4.9				
2						
1	Unstable	Unstable	9.0	6.2	8.7	6.2

5 DISCUSSION

The data from the measurements taken in 1994 were unsurprising. The oxygen reduction currents are within the range that might be expected[2,4-10] as were the resistance values.[3,11]

The steel potentials showed that the reinforcement close to the embedded reference electrodes did not corrode and the potential differences between the reference electrodes were unsurprising. The changes in results between 1994 and 1995 are rather unexpected, especially the data on oxygen reduction. The oxygen reduction currents became very low, suggesting an impoverishment of oxygen in the coated pillars. On the other hand, the potentials between the two different types of reference electrodes do not support this indication.

The increase in resistance suggests a drying of the concrete rather than an increased humidity. The lower temperature (about 3 °C) accounts for 10% of the increase in resistivity[12] but the major part of this increase must be due to a drier concrete. Increased dryness is also indicated by the visual inspection.

We are, however, reluctant to draw conclusions. At the moment there are few measurements and some of the data seem inconsistent.

References

1.   H. Arup and B. Sørensen, Paper no 208, Corrosion/92, NACE.
2.   Ø. Vennesland, *Nordic Concrete Research,* 1991, **10**, 139.
3.   R. B. Polder, NAMES., Proceedings of the International Conference on Corrosion and Corrosion Protection of Steel in Concrete, Sheffield, 1994, 571.
4.   O. E. Gjørv, Ø. Vennesland and A. H. S. El-Busaidy, *Material Performance,* 1986, **25**, 39.
5.   K. Tuutti, Corrosion of Steel in Concrete, Research Report Fo 4:82, CBI, 1982.
6.   C. L. Page and P. J. Lambert, *J. Mat. Sci.,* 1987, **22**, 942.
7.   H. Arup, «Galvanic Action of Steel in Concrete», Spec. Publ., Korrosioncentralen, Denmark, 1977.
8.   H. J. Fromm, *Materials Performance,* 1977, **16**, 57.
9.   B. Heuzè, *Materials Performance,* 1965, **4**, 57.
10.   D. A. Hausmann, *Materials Protection,* 1969, **8**, 23.
11.   O. E. Gjørv, Ø. Vennesland and A. H. S. El-Busaidy, OTC paper 2803, 1977, Houston.
12.   G. A. Woelfl and K. Lauer, *Cement, Concrete and Aggregates,* 1979, **1**, 64.

DEVELOPMENT OF A GALVANIC CORROSION PROBE TO ASSESS THE CORROSION RATE OF STEEL REINFORCEMENT

J. Gulikers

Delft University of Technology
Department of Civil Engineering
Stevinweg 1
2628 CN Delft, The Netherlands

1 INTRODUCTION

One of the major factors affecting the long term performance of reinforced concrete structures is the extent of corrosion of the steel reinforcement, which may be induced by chloride attack or carbonation. The presence of a non-transparent concrete covering of a certain thickness and strength means that the corrosion process becomes noticeable only after a considerable amount of time has passed and serious damage has occurred. There is, therefore, a great need for more accurate and sensitive non-destructive monitoring techniques which can provide reliable information on the condition of the steel rebars. Concrete structures are expected to last for several decades. Therefore, it is important that corrosion-testing should be an on-going activity over several years rather than a one-time project.

Based on experimental research into macrocell corrosion,[1] a galvanic corrosion probe has been devised which can be incorporated in new as well as existing structures. The probe can be used as an early warning system since it can reveal incipient corrosion as early as possible. Moreover, changes in the actual corrosion rate over time can be assessed. Continuous monitoring of the corrosion rate from the time of depassivation, allows the extent of metal loss to be quantified.

In this paper the principle of the galvanic measuring technique will be elucidated and the design of the galvanic corrosion probe to be incorporated in actual reinforced concrete structures will be described in detail.

2 THE PRINCIPLE OF THE GALVANIC MEASURING TECHNIQUE

In laboratory experiments, galvanic testing is frequently used to assess the instantaneous corrosion rate of steel reinforcement. The magnitude of the galvanic current is considered to be a direct measure of the corrosion rate of the anodic member of the galvanic couple. However, difficulties are usually encountered in interpreting the measurement results from galvanic testing. Only under ideal conditions, can the magnitude of the galvanic current be converted directly into the actual corrosion rate.[2,3] Most practical situations are far from ideal and in these cases the galvanic current can severely underestimate the true corrosion rate. This is due to the fact that the anodic member of a galvanic couple can still operate as a mixed electrode whereupon anodic and cathodic reactions occur simultaneously. Only when the cathodic reactions can be eliminated completely, will the galvanic current be a measure of the true corrosion rate. Nevertheless galvanic testing can be employed to calculate the actual corrosion rate even in non-ideal systems.

The instantaneous corrosion rate is normally derived from the resistance to a small shift from the original steady state potential, E_{corr}. Applying a small overvoltage to a rebar and measuring the resulting current indicates the corrosion condition. The associated corrosion current, I_{corr}, can be estimated using a conversion factor, B:

$$I_{corr} = B \left(\frac{dI}{dE} \right)_{E = E_{corr}} = \frac{B}{R_p} \tag{1}$$

For practical purposes, conversion of polarisation resistance (R_p) measurements to corrosion rates is valid only for the case of general corrosion. Based on numerous laboratory experiments on actively corroding steel, an empirical value for the Stern-Geary constant, B, of 26 mV has been estimated.[4,5] However, in the case of strongly localised attack the calculated average corrosion current density, i_{corr} ($\mu A/cm^2$), could severely underestimate the local dissolution rate. Furthermore, the success of the polarisation measurements is strongly dependent on the ability to compensate for the ohmic concrete resistance between the reference electrode and the steel bar.[6] This can be achieved by using a potentiostat equipped with a positive feedback or current interrupt option.

An excursion from the steady state corrosion potential, E_{corr}, can also occur as a consequence of galvanic interaction between actively corroding and passivated steel. In this case a natural open-circuit potential difference acts as a driving voltage and both the corroding steel (anodic member) and the passive steel (cathodic member) will be polarised. The magnitude of the galvanic current can be expressed mathematically by:

$$I_{gal} = \frac{E_{corr}^c - E_{corr}^a}{R_p^a + R_{con} + R_p^c} = \frac{U_{oc}}{R_{cell}} \tag{2}$$

The integral value of the polarisation resistance, R_p^a, of the anodic member (corroding steel) can be estimated to be the ratio between the potential shift, ΔE_{corr}^a, and the galvanic current, I_{gal}. Introducing the value of R_p^a into Equation (1), the actual corrosion rate can be determined. The value of ΔE_{corr}^a can be obtained only when the ohmic concrete resistance can be adequately compensated for.

Due to galvanic interaction, the anodic member of a galvanic couple is always polarised in the anodic direction and this is normally accompanied by an increase in the anodic reaction (iron dissolution) rate. Since active/passive galvanic cells frequently occur in actual concrete structures, galvanic corrosion testing may provide additional useful information, in particular the concrete resistance and the susceptibility to galvanic reinforcement corrosion. However, the galvanic interactions in reinforced concrete structures can be very complex and numerical analysis is normally required to quantify the possible effects.[7]

3 LABORATORY RESEARCH

The validity of estimating corrosion rates based on galvanic interaction for active/passive cells was investigated experimentally. Concrete specimens were designed with a steel reinforcement arrangement as shown in Figure 1. The design was aimed at the simulation of actual reinforcement corrosion in which actively corroding steel embedded in carbonated concrete is connected to passivated steel encased in alkaline repair mortar. This will lead to a strong galvanic interaction, which increases the dissolution rate of the corroding steel rebars.[8]

Two specimens with cover depths of 10 mm and 25 mm, respectively, were used in the experiments in order to estimate the actual corrosion rate using data obtained from galvanic testing. The eleven electrode bars contained within each specimen were manufactured from corrugated reinforcing steel with a nominal diameter of 12 mm. Five bars were positioned on each vertical side of the specimen, and one bar was located in the centre. Six bars (designated B, D, E, G, H and L) were surrounded by carbonated concrete and therefore these rebars will be depassivated. The passive state of four bars (designated A, C, F and K) was accomplished by enclosing them in an alkaline environment of cementitious repair mortar. Given the electrochemical condition of the rebars, the six depassivated steel bars will act as the anodic member in a galvanic couple, and the passivated steel bars will act as the cathodic member. The centre bar, denoted M, was encased in non-carbonated concrete. Due to the relatively stable environmental conditions prevailing in this part of the specimen, the centre bar can act as a reference electrode during galvanic testing.

All of the steel bars were epoxy-coated at both ends giving an unshielded exposition length of 100 mm, which corresponds to a gross exposed surface area of 3770 mm². At one end of each steel bar a copper wire was attached to allow electrochemical measurements to be made. The centre-to-centre distance between two neighbouring bars was 62 mm.

The concrete was designed with a blast furnace slag cement content of 300 kg/m³ and a water to cement ratio of 0.60 to encourage rapid accelerated carbonation.

The two specimens, denoted C10R and C25R, were exposed to periods of alternate wetting and drying (14 days at 50% R.H. followed by 14 days spray water exposure).

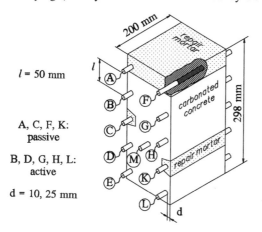

Figure 1 *Concrete corrosion cell used for laboratory experiments*

4 EXPERIMENTAL TEST RESULTS

Prior to galvanic coupling, the electrical potential difference and concrete resistance between the active and passive steel rebars was monitored for over 100 days. Galvanic testing to estimate the instantaneous corrosion rate took place during a wetting period, along with linear polarisation resistance measurements.

4.1 Open-circuit Potential Differences

The potential difference, U_{oc}, between active and passive steel bars acts as the driving force for the galvanic interaction. Over several cycles of wetting and drying, the development of U_{oc} was monitored on a daily basis. Although the magnitude of the potential difference is dependent on the individual steady state potentials, it was expected that the rest potential of the passive steel would remain essentially constant, whereas the developing potential of the corroding steel would be strongly influenced by the changing environmental humidity. Consequently, the fluctuations in the potential difference would be dictated by the electrochemical state of the corroding steel.

Figure 2 shows the development of the voltage, U_{oc}, between active and passive steel for cover depths of 10 mm and 25 mm. For 10 mm cover depth the potential differences range from approximately 200 mV at the end of the drying period, to 450 mV at the end of the wetting period. The influence of the alternating humidity conditions is clearly reflected in the development of the potential difference. For 25 mm cover depth the periodic fluctuations due to changing air humidity are less pronounced. The values recorded vary from 320 mV to 610 mV, and are therefore some 100 mV higher than those recorded for the 10 mm cover depth.

Prior to galvanic testing, the open-circuit potential differences of all the steel bars were measured with respect to the reference bar M located in the centre of the specimen. For specimen C10R the calculated values for U_{oc} between the active and passive steel bars varied from 425 mV to 518 mV, with an average value of 470 mV. In the case of specimen C25R lower values for U_{oc} were determined, ranging from 391 mV to 489 mV, with an average value of 437 mV.

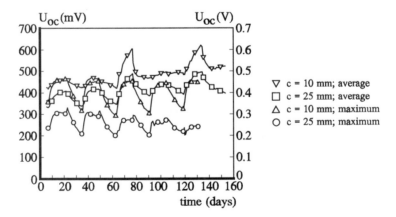

Figure 2 *Development of the open-circuit potential differences, U_{oc}, between actively corroding and passive steel rebars over time*

A sufficiently high potential difference will be necessary to shift the potential of the corroding steel over at least 10 mV, in order to accurately calculate the corrosion rate. Experience has shown that the voltage differences which develop between actively corroding and passive steel are usually high enough to achieve this.

4.2 Electrical Concrete Resistance

The concrete resistance can play an important role in controlling the magnitude of the galvanic interaction. The humidity content of the concrete cover has a significant influence on the concrete resistivity. Resistivity values may range from 100 Ωcm to 100,000 Ωcm, but 10,000 Ωcm is considered typical.

The concrete resistance was determined by measuring the AC impedance between two neighbouring rebars, one actively corroding and the other passive. The resistance values reported refer to concrete manufactured with blast furnace slag cement and an electrode spacing of 50 mm (centre-to-centre distance of 62 mm).

In Figure 3 the results of measurements performed during several wetting and drying cycles are shown. For both cover depths the changing humidity conditions can be seen to strongly affect the concrete resistance, R_{con}. For 10 mm cover depth the concrete resistance varies from 6000 Ω to about 21000 Ω, whereas for the 25 mm cover depth the values range from 2000 Ω to 6000 Ω. The resistance values obtained for a cover depth of 10 mm appear, therefore, to be significantly higher than those recorded for 25 mm cover depth. This is considered to be largely due to the thinner concrete cover limiting the cross section available for ionic current transport.

Although the humidity content has a strong influence on concrete resistance, the environmental factor which has the greatest influence is temperature. The laboratory experiments were performed at a constant temperature of 20°C, and hence the effect of temperature on resistance values was not seen. However, in actual concrete structures temperature variations are likely to occur and hence it is difficult to relate concrete resistance solely to humidity.

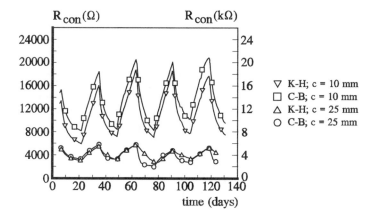

Figure 3 *Development of the concrete resistance, R_{con}, between actively corroding and passive steel over time*

4.3 Corrosion Rate

The similarity between corrosion rates determined from galvanic testing and those calculated from the traditional linear polarisation resistance technique was validated in laboratory experiments. The test procedure involved the galvanic coupling of one of two passivated steel rebars (C or K) with one of the actively corroding steel bars. Thus an active/passive galvanic corrosion cell was created with a surface area ratio $A_{pas}/A_{act} = 1.0$. The corroding steel rebar acted as the anodic member of the galvanic cell, and hence was polarised in the anodic direction, resulting in an increase in the corrosion rate. The coupling was maintained for several hours until relatively stable steady state conditions were achieved with respect to potential and galvanic current. During coupling of the active and passive steel, the galvanic current was measured using a Zero Resistance Ammeter (ZRA). This instrument allowed continuous monitoring of the galvanic current without introducing any additional load to the external part of the circuit. The galvanic currents measured ranged from 8.7 µA to 16.3 µA for 10 mm cover depth, resulting in anodic overvoltages which varied from 5.8 mV to 15.8 mV. Galvanic testing on steel bars with 25 mm cover depth gave substantially higher values for the galvanic currents and anodic overvoltages (galvanic current varied from 16.6 µA to 34.0 µA and anodic overvoltage from 17.5 mV to 31.5 mV). Further results from these laboratory investigations have been described in greater detail in a previous publication.[1]

The calculated values for the overvoltages in the corroding steel are within the region of approximate linear polarisation behaviour. The polarisation resistance, R_p, was approximated from the magnitude of the galvanic current and anodic overvoltage by using Equation 1.

In addition the polarisation resistance was determined using a computer-controlled EG&G Model 273A Potentiostat with automatic IR compensation (positive feedback). The potential scan was applied at a sweep rate of 10 mV/min within the range ($E_{corr} - 20$ mV) to ($E_{corr} + 20$ mV).

Comparison of the values for the polarisation resistance, R_p, obtained from the traditional polarisation measurement procedure (i.e. using a potentiostat) with those

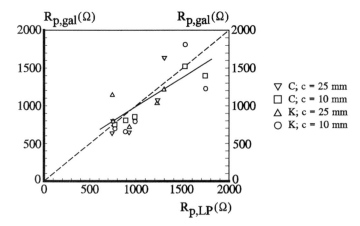

Figure 4 *Comparison between polarisation resistance values, R_p, obtained using the traditional measurement technique and using the galvanic testing method*

calculated using input data derived from galvanic testing, in general show reasonably good agreement, as presented in Figure 4. The average value of R_p obtained from galvanic testing is slightly lower, but the difference is not significant. R_p values generally provide a rough indication of the actual corrosion rate.

Based on the results of these laboratory investigations it can be concluded that galvanic corrosion probes may be used to provide a reliable non-destructive method for determination of the actual corrosion rate of steel reinforcement. However, for use on site, the test arrangement would have to be modified.

5 DESIGN OF THE GALVANIC CORROSION PROBE FOR USE ON EXISTING CONCRETE STRUCTURES

As shown by the laboratory experiments, galvanic testing can be used to determine the instantaneous corrosion rate of the reinforcement provided that the polarised steel area is known.

The experimental laboratory set up was used as a basis for the design of a galvanic corrosion monitoring system that could be used on site. Initially, the objective was to develop a galvanic corrosion monitoring system which could be used in new structures. Here the priority was directed towards predicting the time to depassivation (i.e. the monitoring strategy was designed to follow the progress of chloride ingress and advancing carbonation). To this end the so-called anode-ladder-system has already been developed which permits the determination of the time-to-corrosion by extrapolating electrochemical signals.[9] However, most potential clients are not interested in monitoring the corrosion condition of the steel reinforcement in new structures, since active corrosion is not expected to occur within the intended design life, or at least for several decades after construction. Instead there is more interest in a corrosion monitoring system which could be used on existing concrete structures, where the steel reinforcement is already in the propagation stage of corrosion or where the initiation stage of corrosion is well advanced. In existing structures the emphasis is on determining the corrosion rate and assessing whether deterioration has been arrested. Consequently, the original reinforcement steel has to be used as the anodic member of the galvanic element.

A well defined and representative part of the reinforcement in an actual concrete structure undergoing corrosion is retrieved by drilling cores. Next, the concrete cores are treated in the laboratory and lead wires are attached to both ends of the steel bar contained within the core. In order to allow potential measurements a reference electrode is inserted at a short distance from the steel/concrete interface.

For the cathodic member of the galvanic couple passive steel is required, and this is achieved by embedding a rebar in a precast concrete cylinder. The alkaline environment prevailing in the non-carbonated and chloride-free fresh concrete will ensure that the steel will remain essentially passive for a long period of time. As an alternative passive steel from the original concrete structure may also be used.

The cores containing the anodic steel members are replaced in their original locations in the structure using a cementitious repair mortar. The precast cylinders are positioned near these drilled cores, so that the two steel bars making up the galvanic cell run in parallel. In Figure 5 a preliminary version of the test arrangement is shown.

All the cables are connected to a central measuring unit through precut slots in the concrete surface. The unit contains three devices to monitor the response of the galvanic

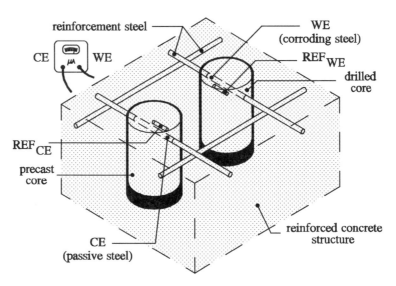

Figure 5 *Isometric view of preliminary test arrangement for galvanic testing on site*

couple prior to, and after, coupling is effected. A voltmeter with a high input impedance (10^{12} Ω) is used to monitor the development of the steady state potential in the individual steel bars. A ZRA with a sensitivity of 0.1 µA is employed to measure the small polarising current produced upon galvanic coupling. The concrete resistance is measured by an ohmmeter which includes an AC current generator.

Once corrosion is initiated, an electrical potential difference will develop and galvanic corrosion testing can be executed. Using a computer program, the corroding steel is coupled to the passive steel to form a galvanic element. The development of the galvanic current and the associated shift of the potential in the anodic direction is monitored continuously. When relatively stable conditions are reached, the corrosion rate can be calculated using the magnitude of the galvanic current and the anodic overvoltage.

This test set up should enable permanent long term monitoring of the corrosion condition of the steel to take place. If necessary, the arrangement can be modified to monitor the effectiveness of repair and protective strategies for a variety of reinforced concrete structures. Using embedded sensors, variations in temperature, corrosion potential, concrete resistivity, galvanic current density and corrosion rate can be followed over time. The data is recorded by on-site loggers that are installed in a secure location close to or within the structure. The data is then either periodically downloaded to a portable computer or transmitted via a modem to the client's office.

The frequency with which galvanic corrosion measurements are performed will depend on the information required by the client. The effect of seasonal variations on the corrosion rate can be assessed by taking measurements once a month. Short term changes can be determined by taking measurements on a daily basis.

A finite element program can be employed to calculate the accelerating effect of the galvanic interaction on the corrosion rate, and to convert the electrical concrete resistance into an effective average value for the concrete resistivity.[7]

6 CONCLUDING REMARKS

Incorporation of galvanic corrosion probes into new as well as existing concrete structures could prove to be an attractive method for long term monitoring of the performance of steel reinforcement. With the probe set up proposed, the moment of depassivation can be pinpointed and the development of the corrosion rate in the subsequent period of time assessed. The availability of these data will allow for a better evaluation of the remaining service life and possible repair strategies. The corrosion characteristics can be monitored using half-cell potential, galvanic rate-of-corrosion, AC concrete resistance and temperature measurements. In addition to the instantaneous corrosion rate, useful information regarding the susceptibility of a structure to galvanic corrosion can be obtained (e.g. active/passive cells set up by patch repairs). This will permit a quantification of the accelerating effect on the corrosion rate.

At present this system has been evaluated in the laboratory only. For practical application of this electrochemical technique to structures in the field, numerous problems must still be solved. In addition to the measurement problem itself, choosing a measurement schedule to provide a representative sampling of long and short term environmental conditions is another problem that remains to be addressed. Work is continuing to improve the galvanic corrosion monitoring system for use on site.

At the beginning of 1996 a prototype of the galvanic corrosion probe described will be incorporated into an existing bridge structure suffering from chloride-induced reinforcement corrosion, but showing no serious deterioration. Critical zones of the structure will be monitored by galvanic corrosion probes. If it is possible to detect early the signs of an increasing corrosion risk, for instance advancing carbonation or chloride ion penetration then preventive action can be taken before it is too late, which should be more cost effective.

7 ACKNOWLEDGEMENTS

The author wishes to express his thanks to Mr Kees van Beek for his help in developing the measurement equipment and many useful discussions of the work. This research project is supported by the Dutch Technology Foundation (STW), which is gratefully acknowledged.

References

1. J. Gulikers, 'Determination of the corrosion rate of steel reinforcement by galvanic testing', in 'Corrosion and corrosion protection of steel in concrete' (Ed: R. Swamy), Sheffield Academic Press, Sheffield, 1994, 247.
2. P. Schießl and M. Schwarzkopf, *Betonwerk + Fertigteil-Technik*, 1986, **52**, 626.
3. M. Raupach, 'Zur chloridinduzierte Makroelementkorrosion von Stahl in Beton', Beuth Verlag, Berlin, 1992, 433.

4. C. Andrade, V. Castelo, C. Alonso and J. González, 'The determination of the corrosion rate of steel embedded in concrete by the polarization resistance and ac impedance methods', in 'Corrosion effect of stray currents and the techniques for evaluating corrosion of rebars in concrete' (Ed: V. Chaker), American Society for Testing and Materials, STP906, Philadelphia, 1986, 43.

5. C. Alonso, C. Andrade and J. González, *Cement and Concrete Research*, 1988, **18**, 687.

6. N. Berke and M. Hicks, 'Electrochemical methods for determining the corrosivity of steel in concrete', in 'Corrosion testing and evaluation' (Eds: R. Baboian and S. Dean), American Society for Testing and Materials, STP1000, Philadelphia, 1990, 425.

7. J. Gulikers and E. Schlangen, 'Numerical analysis of galvanic interaction in reinforcement corrosion', in 'Corrosion of reinforcement in concrete construction' (Eds: C. Page, K. Treadaway and P. Bamforth), Society of Chemical Industry, London, 1996, published in this volume.

8. J. Gulikers and J. van Mier, 'Accelerated corrosion by patch repairs of reinforced concrete structures', in 'Rehabilitation of concrete structures' (Eds: D. Ho and F. Collins), RILEM, Melbourne, 1992, 341.

9. P. Schießl and M. Raupach, *Concrete International*, 1992, **7(14)**, 52.

MONITORING AND DRAINAGE METHOD FOR STRAY CURRENTS GENERATED BY A METRO SYSTEM

F. M. Bajenaru, V. Beldean and G. Rozorea

Engineering and Design Institute for Underground - S.C. "METROUL S.A."
Bucharest
Romania

1 INTRODUCTION

The underground reinforced concrete construction of a metro system can be affected by leakage of the d.c. currents generated by the metro traction system and/or the tram traction system. Stray current corrosion at anodic zones will put these structures at risk.

This phenomenon must be strictly surveyed and if possible controlled in order to extend the working life of these constructions.

This paper proposes a survey method for stray currents generated by a metro traction system and a method of monitoring the voltage loss in the track (the negative bar). It also presents a control and drainage procedure for these currents.

2 PRESENTATION OF BUCHAREST METRO TRACTION SYSTEM

The Metro in Bucharest is totally underground, has a total network of 60 km (double line) length and was constructed between 1975 and 1986. In some areas of the tunnels which have a single lining of reinforced concrete, because of abundant groundwater has resulted a high levels of concrete humidity and water leakage.

The traction supply system used in Bucharest is different from the four rail London underground system (two rails for the track, one rail for the positive bar and one rail for the negative bar). The metro system consists of three rails: two rails for the track and for the negative bar and one rail for the positive bar. With this system, it is impossible to use porcelain insulators for the negative bar and it is expensive to produce good insulation between the track and the reinforced concrete.

Furthermore, mechanical wear of the track has a detrimental effect on the electrical resistance along the rails. The increase in resistance generates an increase in voltage loss in the track, according to the following relation:

$$U_r = r \cdot I \tag{1}$$

where: U_r = voltage loss in the track;

r = equivalent electrical resistance along the track;

I = traction current (maximum 6000 Amps for a train with six cars)

These two factors - the level of the track insulation and the voltage loss in the track (or the resistance along the track) - in addition to two other factors - the level of the concrete humidity and the water leakage - generate stray currents through the reinforced concrete of the tunnel.

2.1 First System for Stray Current Corrosion Protection

For the first metro line, constructed in the late 1970s, the main method for stray current corrosion protection was by ensuring reinforcement continuity and the mounting of a drainage copper band connected to the reinforcement bars every 30 m along the tunnel. In the electrical substations, this drainage band was linked to the rectifier's negative terminal with a special separator which turns on when the voltage loss in the rail exceeds 45 Volts (Figure 1).

This electrical drainage system has the disadvantage that the stray currents are not drained when the voltage loss is lower than 45 Volts.

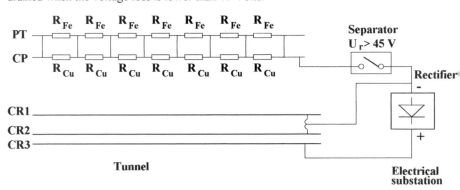

R_{Fe} = the electrical resistance of a tunnel's reinforcement section

R_{Cu} = the electrical resistance of a drainage copper band segment

PT = the tunnel's electrical potential (the reinforcement)
CP = the drainage copper band
CR1, CR2 = the two rails of the track (negative bar)
CR3 = the third rail (positive bar)

Figure 1 *The first system for stray current corrosion protection (First Metro Line)*

2.2 Second System for Stray Current Corrosion Protection

For the second and third metro lines, constructed in the early 1980s, the stray current corrosion protection system was improved. The separators have been replaced by groups of power diodes as shown in Figure 2.

This electrical drainage system has the disadvantage that for a low value of the resistance between track and reinforcement, only the stray currents near to the electrical substation are efficiently drained.

3 NEW SYSTEM FOR STRAY CURRENTS CORROSION PROTECTION

In 1993, a corrosion study for the underground reinforced concrete structure was carried out and was presented at the Bournemouth Corrosion Conference in 1994.

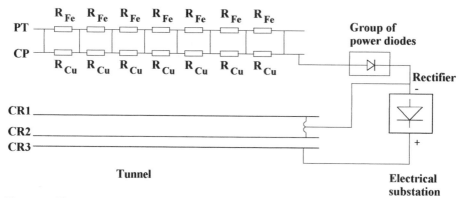

Figure 2 *The second system for stray current corrosion protection*

A conclusion of this study was that, under similar conditions of concrete humidity and track insulation, the corrosion rate (lost material/year) is smaller for the second and third metro lines because the stray current drainage system is more efficient than the one used for the first metro line.

Furthermore, measurements of the electrochemical potentials of the reinforcement bars at the first metro line, and the calculations performed, emphasized that the values of the stray currents are important (50 - 100 A on average with a maximum value of 250 A, for the parallel circuit comprising the tunnel's reinforcement and the grounding copper band). Therefore, the simple drainage system with a grounding copper band is not safe.

Based on these conclusions it is proposed that the diodes drainage system be improved and to achieve a survey method for the stray currents. The experimental system is presented in Figure 3.

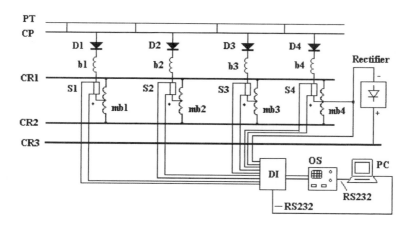

Figure 3 *Monitoring and drainage system for stray current corrosion protection*

The real problems in achieving this system are: the diodes selection, the shunt values and the drainage devices mounting positions.

The drainage devices are composed of a power diode and a shunt resistor for the diode protection. The connections between the drainage devices and the track are made with joint coils (noted with mb1, mb2, mb3 and mb4 in Figure 3) in order to avoid disruption of the automatic traffic control signals.

In order to monitor the drained currents using the diodes (noted with D1, D2, D3 and D4 in Figure 3), measuring shunts S1, S2, S3 and S4 were connected between the drainage devices and the grounding copper band.

The shunt signals were measured using a memory oscilloscope (OS) serially linked by a computer (PC) on RS232 cable. The oscilloscope has an internal RAM memory which permits storage of four wave forms.

It is also possible to use a newer version of this oscilloscope with internal EPROM, RAM and floppy-disk. In this case the computer (PC) is not required.

In order to record more measurements using an oscilloscope, an interface device (DI) was used. This device comprised ten analog inputs, one analog output, an internal RAM and an RS232 connector.

Using the DI it is possible to store a measurements sequence programme in RAM and to display the wave forms on the oscilloscope (or the PC monitor). It is also possible to control the measurements using the computer.

Furthermore, the system permits the measurement of voltage loss in the track at various locations. The voltage loss values at these points are determined by measuring the voltage between the marked points of the shunts S1, S2, S3 and S4 (see Figure 3) and the rectifier's negative terminal.

Better precision is obtained by first measuring the diode drained current and then calculating the voltage between this point and the track (the half joint coil voltage). The real voltage loss in the track at the measurement point is the difference between the measured voltage and the half joint coil voltage.

The track voltage loss values are stored by the computer and compared with the normal values in order to identify interruptions or poor electrical connections where the track resistance and voltage loss are increasing.

Figures 4 and 5 show the drained currents for a low and a high voltage loss in the track. The diodes are charged one by one.

To analyse the method and to evaluate the effects of the stray current drainage on the decrease in corrosion, the electrochemical potential curves along the tunnel, before and after assembling the drainage devices, have been plotted (see Figure 6).

The electrochemical potential diagram shows that in areas with a poor insulation of track (or a high level of water leakage) the potentials are positive (anodic zones) and in areas with a better insulation the potentials are negative (cathodic zones).

After three months of the new drainage system experiments the potentials decreased, because stray currents are better drained and returned to the negative bar.

4 CONCLUSIONS

The presented monitoring method offers the advantage of an efficient survey of values of stray currents and voltage loss in the track. Also, it is useful for the optimal selection of the diodes and shunts (for the drainage system achievement).

The effective drainage of stray currents generated by the metro traction system and the decrease of the electrochemical potentials after three months of experiments (-370 mV Cu/CuSO$_4$ in the critical areas - the anodic zones and -300 mV in general) confirms the applicability of the proposed method.

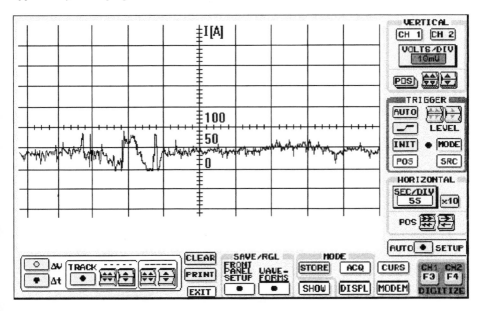

Figure 4 *Drained current diagram for a low voltage loss in track*

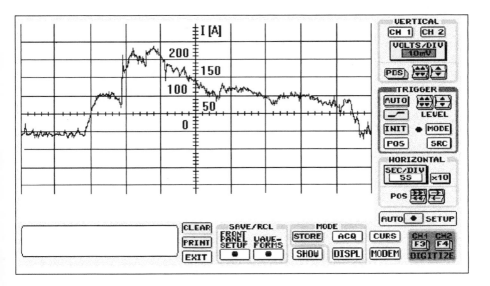

Figure 5 *Drained current diagram for a high voltage loss in track*

Figure 6   *The reinforcement electrochemical potentials diagram along to Dristor-Titan interstation before and after the drainage system experiment*

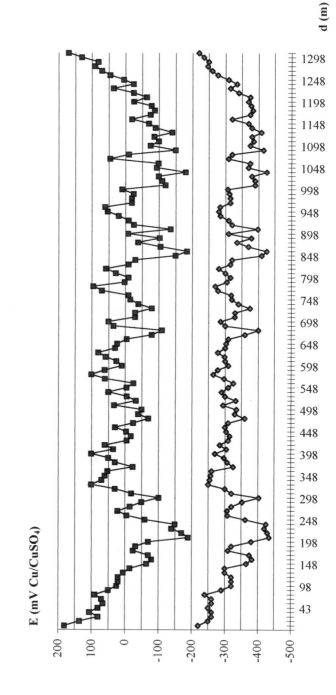

CATHODIC PROTECTION

THE TAY ROAD BRIDGE CATHODIC PROTECTION MONITORING SYSTEM: AN ANALYTICAL TOOL TO ASSIST THE ENGINEER

J. Crerar and F. Wilson

Tay Road Bridge
Marine Parade
Dundee
DD2 3JB

R. L. Jefferson and R. J. R Gibson

Integrated Design Techniques Ltd
Mountjoy Research Centre
Stockton Road
Durham DH1 3SW

1 INTRODUCTION

The Tay Road Bridge is the longest road bridge in Great Britain and the fifth longest in Europe, spanning the River Tay between Dundee and Newport-on-Tay, some 2300 metres. The bridge was designed by William A. Fairhurst and took three years to build, opening in 1966. It is a toll bridge, providing a link to the centre of Dundee from Fife, and is used mainly by commuters.

The bridge is the responsibility of the Tay Road Bridge Joint Board which comprises seven Regional Councillors from Tayside and five from Fife. The Board is responsible for the management and maintenance of the bridge which is financed primarily from tolls.

1.1 Bridge Construction

The bridge comprises 42 spans supported by twin reinforced concrete columns which are parabolic in cross-section. The columns vary in height from 5.5 m at the Dundee end to 30.5 m at the Fife end. The columns are supported by reinforced concrete piers built into cofferdams in the river bed. The columns support twin hollow steel box girders which in turn support a 300 mm thick composite concrete slab carrying the roadway.

1.2 Inspection

During the 1980s a detailed survey of the bridge columns[1] revealed large areas of delaminated concrete and corrosion of the reinforcement in the bottom 4 metres of the columns. Measurement of the chloride levels in the bottom 4 metres were 2 - 3 % by weight of cement on the column surface and 1 - 2 % at the reinforcement. These levels fell significantly above 6 metres. It was evident from these results that corrosion of the reinforcement was chloride induced and that the source of chlorides was sea water splashed on to the column surface by wave action and airborne sea water spray.

The survey also assessed the strength of the columns and found that there were still adequate reserves of strength despite the presence of corrosion in the reinforcement. A review of repair options was carried out in 1986 and cathodic protection (CP) of the column reinforcement was chosen as offering the most suitable long term protection.

2 CATHODIC PROTECTION

In reinforced concrete, corrosion of the reinforcement is naturally prevented by the surrounding highly alkaline environment which leads to the formation of a passive oxide layer on the surface of the reinforcement steel. However, when chloride ions penetrate through the concrete to the reinforcement they cause local depassivation of the steel leading to the formation of anodic and cathodic sites. An electrochemical cell is formed with rust occurring at the anodic site and current flowing through the cell to the cathodic site.

Impressed current cathodic protection provides protection to the reinforcement by applying a small DC current through a surface mounted anode. This forces the reinforcement to act entirely as a cathode and therefore suppresses the formation of rust. Impressed current systems are preferred for the protection of reinforced concrete structures due to the high resistivity of concrete. The impressed current cathodic protection system installed on the Tay Road Bridge was developed over a number of years and is the result of testing both on and off site. Due to the harsh operating environment a titanium expanded mesh anode coated with mixed metal oxides was fixed to the column surface and covered with a 50 mm thick overlay of shrinkage compensated flowable concrete.

The anode mesh covers the lower 6 metres (Splash zone) of each column and is divided into three 2 metre sections. The centre mesh section is linked to either the upper or lower section to give two independent panels. This allows flexibility in setting the applied current to different parts of the column to maintain the best protection.

In general the current is set to provide a current density of 5 mA/m^2 in the lower section of each column, where corrosion activity is greater, and 3 mA/m^2 in the upper section. To coincide with the installation of the mesh anodes a number of reference cells (half cells) were placed in close proximity to the reinforcement. The reference cells monitor the polarisation potential of the reinforcement and are used to determine the required impressed current. Reference cells were placed throughout the entire splash zone of each column to ensure that sufficient data was available to enable the "optimum" current density to be set.

The mesh anodes, reinforcement steel and reference cells are cabled back to a cabinet mounted on the central pedestrian walkway. Large cross sectional area cable was used to connect each circuit to minimise ohmic losses and hence improve the accuracy of any measurements. The cabinet contains DC power supplies (Transformer/Rectifier) which are used to supply the anode current and terminals to measure applied voltage, applied current and reference cell potential.

To provide CP for the entire bridge there are 14 cabinets (T/R cabinets) placed along the bridge and spaced approximately 165 metres apart. Each cabinet contains 12 power supply units and terminals for up to 48 reference cells and can drive up to 6 columns.

When the CP system was installed the performance of the system was monitored by performing manual measurements at each of the cabinets. Test measurements were carried out with a multimeter at three-monthly intervals with test periods of up to 24 hours. For each power supply unit the current density was calculated and a graph was plotted of reference cell potential versus time for all reference cells.

In total there are 160 supply units and approximately 600 reference cells. The time taken to make the measurements, the exposed nature of the bridge, the volume of data produced and the time taken to collate and analyse the results dictated that a method of automating the above process was required.

3 CATHODIC PROTECTION MONITORING SYSTEM (CPMS)

In 1993 the Tay Road Bridge Manager, in consultation with consulting engineers Oscar Faber TPA, drew up a functional specification for a cathodic protection monitoring system. Given the large number of monitoring points on the bridge the following features were considered essential:

- to be flexible and configurable to allow for possible changes in cathodic protection practice;
- to be expandable to allow for extra measurement channels;
- to be capable of performing scheduled measurements;
- to be able to run unattended for at least a year; and
- to provide data management, analysis and archiving functions.

A review of available systems was carried out and none was found to meet all of the criteria and therefore the Tay Road Bridge Manager went out to tender for a design and build contract. The contract was awarded to Integrated Design Techniques Ltd in October 1993 and work commenced with a detailed design study. The findings of the study were reviewed and a system was built to meet the criteria described above.

3.1 Transformer/Rectifier Interface Unit (TRIU)

The TRIU was designed to interface to the transformer/rectifier units and reference cells and provide timed measurements. The TRIU is a modular system utilising plug-in boards and modules. The main sub systems communicate through STEbus, an industry standard computer bus offering rugged construction and high noise immunity. The TRIU is housed in a light weight high load bearing aluminium enclosure and comprises:

- microprocessor running custom software;
- real time clock;
- battery backed memory;
- 12 bit analogue to digital converter;
- signal conditioning and multiplexing; and,
- diagnostics and status display.

The microprocessor controls all operations of the TRIU and allows for timed measurements to be programmed. The microprocessor unit also provides serial communications for downloading of stored data and command signals to and from a portable PC. A real time clock is included on the microprocessor unit and keeps track of the current time and date. This clock is used by the TRIU software to schedule measurements and time and date stamp all measurements. Sufficient battery backed memory is provided to store in excess of one year's worth of measurement, error and system data.

All measurements are performed by a 12 bit signed integrating analogue to digital converter. The converter makes each measurement in 30 ms with a best resolution of 1 mV. A signal conditioning circuit was developed for interfacing the T/R and reference cell signals to the analogue to digital converter. The transformer/rectifier units use phase control of a 50 Hz rectified sine wave and produce waveforms as shown in Figure 1. The TRIU has to measure the average level (or DC equivalent) of this signal and must average the signal over a suitably long time period to obtain an accurate measurement.

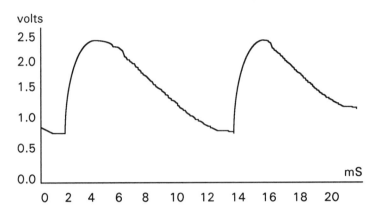

Figure 1 *T/R driving voltage from Column 30 Upstream*

The design criteria required that measurements are to be made within 0.1 to 0.5 seconds (instantaneous off) and therefore this dictated the use of a high performance filter.

Each input incorporates a switched capacitor filter as part of an instrumentation amplifier offering a very high input impedance (> 100 MΩ). Twelve instrumentation amplifier circuits were placed on a printed circuit board with two circuits dedicated to T/R driving voltage measurement, two circuits dedicated to T/R driving current measurement and eight circuits dedicated to reference cell measurement. This printed circuit board is sufficient to monitor the CP system in one column. The outputs of the twelve circuits were multiplexed into the analogue to digital converter with channel selection made under the control of the microprocessor unit. Two relays are included on the board to interrupt the CP current for instantaneous off measurements.

There are seven signal conditioning cards mounted in each TRIU. This allows each TRIU to measure six columns with one spare board for future expansion. The signal conditioning cards are daisy-chained together and their outputs are multiplexed into the analogue to digital converter.

The TRIU is capable of detecting and recording a number of system faults. Front panel indication is also given when a fault is detected and can be used in diagnosing the nature of the fault. The unit also drives an external indicator lamp which indicates system operation. This lamp is extinguished on detection of any logged fault.

3.2 Measurement Types

In order to determine the effectiveness of the cathodic protection the TRIU provides two types of measurement; routine measurements and depolarisation tests.

3.2.1 Routine Measurements. Routine measurements measure anode driving voltage and current, the reference cell "on" potential and the reference cell "instantaneous off" potential. The driving voltage and current are used to determine the drive level and current density applied to the anodes. The "instantaneous off" potential is a measurement taken with the CP driving current interrupted. Within a very short time (<10 μs) the voltage gradient induced by the CP driving current is dissipated to leave the polarisation potential.[2] The reference cell "on" potential provides a measure of the combined effect of the polarisation potential and the ohmic losses resulting from the CP driving current.

Table 1 *Minimum and maximum times for depolarisation tests*

Level	Length	Measurement step size
1	Minimum 10 minutes	10 minutes
	Maximum 24 hours	
2	Minimum 30 minutes	30 minutes
	Maximum 1 week	
3	Minimum 12 hours	12 hours
	Maximum 1 month	

From the interruption of the driving current the polarisation potential will begin to decay, however, it has been observed that for a structure such as the Tay Road Bridge the polarisation potential remains constant for up to 1 second.

In a typical routine measurement there are 12 T/R driving voltage measurements, 12 T/R driving current measurements, 48 reference cell on measurements and 48 reference cell instantaneous off measurements making a total of 140 measurement points. These measurements are known collectively as one record which is time and date stamped. The measurement type (routine measurement) is added and a location description (TRIU ID) is added to the record for identification.

The TRIU is capable of scheduling routine measurements to occur at step sizes of 12 hours between 12 hours minimum and 1 week maximum. Typically, routine measurements are performed once a day.

3.2.2 Depolarisation Tests. A depolarisation test requires the removal of the T/R driving current for a period of time and measuring the reference cell potential as the bridge steel depolarises. The TRIU is capable of performing a depolarisation test in three stages or levels. The time, date and frequency of depolarisation tests can be programmed into the TRIU and the length of the test is determined by programming the three levels. The first level allows frequent measurements to be made to accurately define the shape of the depolarisation curve when it is changing most frequently. Levels 2 and 3 allow for more infrequent measurements where the depolarisation value changes slowly and allow for tests to be run for up to 5 weeks. The levels have minimum and maximum time lengths as shown in Table 1.

At present depolarisation tests are carried out once a month over a 48 hour period. Measurements are taken at 10 minute intervals for the first hour, hourly intervals for the next 11 hours then at 12 hourly intervals thereafter.

4 DATA RETRIEVAL AND TRIU CONFIGURATION

On the front of each TRIU is a serial port to which a portable PC can be connected. The portable PC is used to run a dedicated Windows software package. This package allows an operator to perform the following functions:
- retrieval and reconfiguration of the time and date of the next routine measurement;
- retrieval and reconfiguration of the time and date of the next depolarisation test;

-    retrieval and storage of measurement and error data;
-    carrying out system tests with the display of diagnostic information; and,
-    calibration of the TRIU.

All of these functions are accessed through a menu system with dialog boxes to prompt the user and an on-line help should the user require assistance. Measurement and error data received from a TRIU are stored on the portable PC's hard disk for transfer to a computer running the analysis package.

5 DATA STORAGE AND ANALYSIS (WinCPMS)

Analysis of data collected from the TRIUs is carried out by a dedicated software package running under Windows called WinCPMS. WinCPMS provides the following functions:

-    data retrieval from the portable PC and updating of a CP database;
-    viewing and printing of the database;
-    querying of data and reports generation; and,
-    archiving for long term storage.

WinCPMS reads the data collected by the TRIUs and automatically sorts this data by time and date, by TRIU and by measurement type into the CP database. An identifier is added to each channel and specifies pier number, column, panel, zone and a user definable comment. The database also holds values for the area of each mesh anode and "native potentials" for the reference cells. Once the data has been transferred to the database it can be viewed, printed or exported to another analysis package. The primary function of WinCPMS is to query and produce reports from the data held in the database and to this point two queries have been implemented. Queries allow the user to customise report production and concentrate on a particular range of channels.

5.1   Routine Measurement Query and Report

A routine measurement query can be applied to a single T/R circuit or a complete T/R cabinet with selection made by a number of dialog screens. The user may individually select or omit channels as required. Once the query has been defined the database is scanned to extract all data meeting the query and this data is then presented in the form of a report. The report also displays the channel identifier and calculates the applied current density for each measurement. Upper and lower limits can be programmed into this calculation and any values falling outside the range are highlighted. Once prepared the report can be printed out for inclusion in a test results report file. An example of the type of report produced is shown in Table 2.

5.2   Depolarisation Test Query and Report

A depolarisation test report (as shown in Table 3) can be generated for a single reference cell or up to a complete T/R cabinet with selection via a series of dialog screens. Once the query has been defined the database is scanned for depolarisation test data meeting the query and is presented in the form of a report. The report includes channel identifiers, the "native" reference cell potential, the reference cell "on" and "instantaneous off" potentials, the reference cell potential during depolarisation, the polarisation potential and the change in reference cell potential after 4, 24 and 48 hours.

Table 2 *Transformer/Rectifier report for CP system at Piers 21, 22 and 23*

Tay Road Bridge Transformer/Rectifier Report							
Created 05/12/95				Date	05/05/95		
				Time	09:30:00		
CPMS	Description	Pier	Col	Zone	Voltage (mV)	Current (mA)	Density (mA/m2)
P21U1	Pier 21 Upstream	21	U/S	1	954	110	3
P21U2	Pier 21 Upstream	21	U/S	2	925	55	3
P21D1	Pier 21 Downstream	21	D/S	1	909	88	2
P21D2	Pier 21 Downstream	21	D/S	2	929	64	3
P22U1	Pier 22 Upstream	22	U/S	1	983	169	4
P22U2	Pier 22 Upstream	22	U/S	2	943	41	2
P22D1	Pier 22 Downstream	22	D/S	1	981	159	4
P22D2	Pier 22 Downstream	22	D/S	2	852	38	2
P23U1	Pier 23 Upstream	23	U/S	1	1058	178	4
P23U2	Pier 23 Upstream	23	U/S	2	945	51	3
P23D1	Pier 23 Downstream	23	D/S	1	1071	172	4
P23D2	Pier 23 Downstream	23	D/S	2	988	56	3

The user may set upper and lower limits on the 4, 24 and 48 hour shift calculations to highlight data out of range. These shift values are important in determining the effectiveness of the CP system and can influence the setting of the driving current. If, for example, the shift values are lower than expected then this would indicate that the current should be increased. The actual shift values vary between reference cells; historical data and experience dictate the optimum values.

Once the report has been produced reference cell depolarisation data can be displayed in the form of a graph. The graph allows the user to assess the rate of depolarisation and verify that the shape of the depolarisation curve is as expected. A typical report would be performed on a complete T/R cabinet and twelve sets of graphs would be produced to show the operation of each mesh anode (zone).

Figure 2 shows the results for a depolarisation test carried out at pier 33 on 4 February 1995. Some of the reference cells are located within the pier head and although not submerged at high tide are subject to wave action. The depolarisation curve for these cells appears characteristic until high tide where an unexpected peak appears. This peak is caused by the ingress of sea water into the pier which modifies the resistance of the concrete and hence the electrochemical cell. Reference cells located further up the columns do not show this tidal effect. Although tidal effects on certain reference cells have been noted in the past, these were generally confined to noting unstable reference cell readings. This effect had not been recorded prior to the implementation of the CP monitoring system.

6   FURTHER WORK AND CONCLUSIONS

The data analysis software is modular and therefore allows other queries and reports to be added. Currently under development is a query to display T/R driving voltage, current and reference cell potentials for a specified date range and across the entire bridge.

Table 3 *Depolarisation Test Report for CP System at Piers 33, 34 and 35*

Tay Road Bridge Depolarisation Test Report
Created 06/12/95

CPMS	Description	Pier	Col	Zone	Panel	Hours: 0 Date 04/02/95 Time 09:00:00 Native (mV)	On Val (mV)	Pol'n Pot (mV)	Instant. Off (mV)	0.5 04/02/95 09:30:00 (mV)	8 04/02/95 17:00:00 (mV)	24 05/02/95 09:00:00 (mV)	48 06/02/95 09:00:00 (mV)	Change in reference cell potential over 4, 24 and 48 hours 4hr (mV)	24hr (mV)	48hr (mV)
P33U11S1	CELL1 SOUTH	33	U/S	0		-372	-531	-148	-520	-418	-336	-309	-303	162	211	217
P33U11N1	CELL2 NORTH	33	U/S	0		-391	-462	56	-335	-222	-157	-114	-100	158	221	235
P33D11N1	CELL1 NORTH	33	U/S	0		-341	-601	-236	-577	-414	-340	-301	-278	204	276	299
P33D11S1	CELL2 SOUTH	33	U/S	0		-331	-549	-210	-541	-447	-343	-306	-288	170	235	253
P34U12N1	CELL1 SOUTH	34	U/S	0		-301	-267	33	-268	-266	-264	-259	-261	-2	9	7
P34U12S1	CELL2 NORTH	34	U/S	0		-286	-252	33	-253	-256	-255	-250	-250	-4	3	3
P34D11N1	CELL1 NORTH	34	U/S	0		-335	-452	-119	-454	-420	-341	-296	-279	87	158	175
P34D11S1	CELL2 SOUTH	34	U/S	0		-410	-552	-125	-535	-425	-351	-325	-317	166	210	218
P35U22N1	CELL1 NORTH	35	U/S	2	BOTTOM	-433	-564	-125	-558	-442	-344	-316	-308	193	242	250
P35U23S1	CELL2 SOUTH	35	U/S	2	TOP	-391	-785	-303	-694	-516	-418	-361	-329	244	333	365
P35U11S1	CELL1 SOUTH	35	U/S	1		-416	-510	-81	-497	-439	-340	-285	-258	121	212	239
P35U11N1	CELL1 NORTH	35	U/S	1		-421	-532	-113	-534	-532	-473	-397	-301	29	137	233
P35D22N1	CELL1 NORTH	35	D/S	2	BOTTOM	-475	-547	-64	-539	-426	-342	-302	-285	182	237	254
P35D22S1	CELL2 SOUTH	35	D/S	2	BOTTOM	0	-525	-518	-518	-444	-345	-301	-275	144	217	243
P35D11S1	CELL1 SOUTH	35	D/S	1		-442	-411	39	-403	-367	-334	-319	-306	54	84	97
P35D11N1	CELL2 NORTH	35	D/S	1		0	-373	-368	-368	-294	-335	-337	-335	41	31	33
P33PN1	PIER CELL1	33		0	NORTH	-408	-760	-285	-693	-568	-532	-466	-453	179	227	240
P33PN2	PIER CELL2	33		0	NORTH	-252	-836	-580	-832	-670	-568	-491	-481	248	341	351
P33PS1	PIER CELL1	33		0	SOUTH	-430	-860	-326	-756	-554	-525	-472	-475	270	284	281
P33PS2	PIER CELL2	33		0	SOUTH	-393	-653	-223	-616	-504	-497	-427	-430	163	189	186
P34PN1	PIER CELL1	34		0	NORTH	-405	-771	-299	-704	-558	-495	-435	-435	218	269	269
P34PN2	PIER CELL2	34		0	NORTH	-433	-596	-131	-564	-486	-422	-387	-385	135	177	179
P34PS1	PIER CELL1	34		0	SOUTH	-362	-606	-223	-585	-474	-406	-378	-376	173	207	209
P34PS2	PIER CELL2	34		0	SOUTH	-357	-636	-277	-634	-477	-410	-377	-375	226	257	259

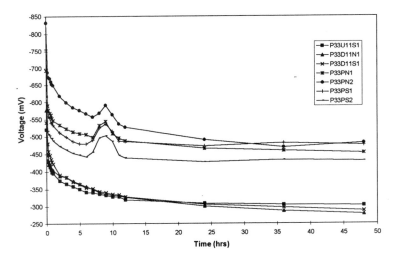

Figure 2 *Depolarisation Test results for Pier 33 on 4 February 1995*

This facility will be used to monitor long term changes in reference cell potential under steady state conditions. A second query under development will allow the automation of depolarisation test results and the automated production of twelve graphs per TRIU (one per zone). This will greatly reduce the time spent analysing results and allow the user to examine far more data than was previously possible.

A major addition to the system currently under consideration is to network the TRIUs back to the Bridge office. This will allow the user to obtain measurement data in real time and allow system tests and diagnostic functions to be performed from the Bridge office. This will greatly reduce the time spent in both collection of data and monitoring the CP system and allow a faster response to changing conditions.

This paper has demonstrated that all of the original criteria set out by the Tay Road Bridge Manager have been met, resulting in a system that is easy to use and which provides accurate, repeatable measurements. The system outlined is capable of storing, analysing and reporting on far more data than was previously possible and will save time in monitoring the CP system. The science of cathodic protection of reinforced concrete structures is still developing and it is hoped that the data collected and the analysis tools described in this paper will help with its future development.

References

1. J. Crerar, F. Wilson and A.T. Coulson, 'A large scale cathodic protection monitoring installation on the Tay Road Bridge', PTRC Conference European Transport Forum, Warwick, UK, September 1994.

2   Von Baeckmann et al, 'New developments in measuring the effectiveness of cathodic protection', Australian Corrosion Association conference 22, Hobart, Australia, November 1982.

QUALIFICATION OF CATHODIC PROTECTION FOR CORROSION CONTROL OF PRETENSIONED TENDONS IN CONCRETE

Stanislas Klisowski and William H. Hartt

Center for Marine Materials
Department of Ocean Engineering
Florida Atlantic University
Boca Raton, Florida 33433
USA

1 INTRODUCTION

A number of factors, including the extent of existing corrosion induced concrete cracking and spalling and the anticipated remaining service life, influence applicability of cathodic protection (cp) for a reinforced or prestressed concrete structure or component. For prestressed concrete, however, two additional concerns; first, hydrogen embrittlement and, second, possible loss of bond in the case of pretensioned members, have been identified, neither of which is generally important in the case of cathodic protection of reinforcing steel.[1,2] Hydrogen embrittlement data for notched, polarized specimens suggest that, if tendon wire is locally corroded such that notch-like pits exist, then susceptibility to hydrogen embrittlement and brittle fracture is greater than if corrosion damage is modest and uniform.[3-6] This is illustrated by comparison of Figures 1 and 2, which plot fracture results from constant extension rate tests (CERT) as a function of potential for smooth[3,4] and notched[3,5-7] specimens, respectively. In the former (smooth specimen) case earlier data[3] indicated only about a ten percent strength

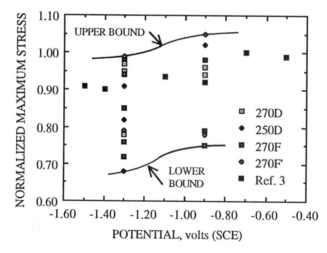

Figure 1 *Normalized fracture load as a function of potential from previous CERT experiments involving smooth specimens and material from four lots.[3,4]*

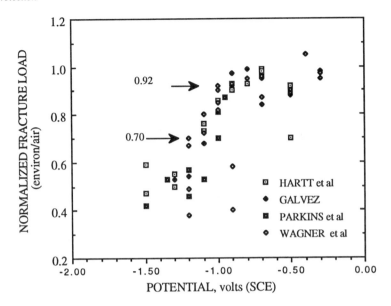

Figure 2 *Normalized fracture load as a function of potential for notched Grade 270 prestressing steel specimens as reported from previous research.*[3,5-7]

decrease upon transitioning from a relatively positive to negative potential. However, more recent results show greater reduction in maximum stress at -0.90 V, where hydrogen evolution and, hence, embrittlement should not occur. However, this reduced strength at -0.90 V was noted only for specimens which contained 0.24 percent chromium; and, as such, this more susceptible material was microalloyed, whereas chromium concentration for the other specimens was 0.02 percent. These data indicate for non-chromium microalloyed material that design and operation of cp systems should be such that potentials below -1.00 V (SCE) do not occur; and a lower potential limit of -0.90 v has accordingly been proposed.[3,7] An appropriate lower potential limit for chromium microalloyed material remains to be defined, however.

The data in Figure 2 (notched specimens) indicate that the average fracture load upon polarization within the ideal range (potential equal to or greater than -0.90 V) was reduced to about 92 percent of the value without cp. While this seems like a modest reduction, it must be assumed that the prestress is 70 percent of the fracture load; and on this basis the lowered strength reduces the safety margin by 27 percent. The situation is further compounded by the fact that the 70 percent prestress is referenced to the original wire and tendon cross section, but with the occurrence of pitting this stress increases. Also, factors such as the level of prestress and the remaining wire fracture strength are statistically distributed parameters such that occurrence of some failures is not unexpected upon application of cp even though consideration of the mean strength suggests integrity should be maintained.

An important question that arises from the above considerations pertains to how a prestressed structure or structural member can most appropriately be qualified for cathodic protection. In this regard a technology termed Electrochemical Proof Testing or ECPT has been proposed for establishing the likelihood that tendon failures will not occur upon application of cathodic protection.[1] This involves 1) identification of a worst case area(s) of the structure or component, 2) mounting of a temporary cp system and polarizing one or more tendons to a current-off potential of at least -1.20 V (SCE) for a period of about one day, and 3) monitoring for strand failure during the polarization by a procedure such as (a) electrical continuity, (b) acoustic emission or (c) component shape change. Qualification then is based upon no tendons failing, either during the polarization or within one or two days thereafter. Shortcomings of the procedure are the requirement for intentional excessive cathodic polarization and the possibility of inducing damage to the structure as a consequence.

The present paper reports results from CERT experiments that were performed on smooth, notched and pitted prestressing wire for the purpose of further assessing the influence of potential upon fracture properties in the presence of a geometrical discontinuity and, from this, developing an improved procedure for cp qualification.

2 EXPERIMENTAL PROCEDURE

2.1 Materials and Specimens

Seven wire, nominally 12.70 mm (0.500 inch) diameter Grade 270 tendon, which conformed to ASTM Standard A416-90a,[8] was employed for the experiments. The composition and mechanical properties are presented in Tables 1 and 2, where for the former the two sets of elemental composition values indicate results from samples taken at different locations.

Three different types of specimens, smooth, notched and pitted, were fabricated from the straight central tendon wire according to the geometries in Figures 3 and 4. These show that notched specimens included six different configurations; and pitted specimens reflected four depth/width combinations, as summarized in Table 3, each of which was intended to represent a unique geometrical irregularity and stress concentration. Smooth specimens were prepared by centerless grinding followed by

Table 1 *Chemical composition of the prestressing wire.*

ELEMENT	C	Mn	Si	Cr	Ni	Mo	S	P	Cu	Al	Fe
WEIGHT(%)	0.85	0.78	0.2	0.02	0.03	0.01	0.02	0.02	0.07	0.004	Bal
	0.7	0.7	0.18	0.02	0.02	0.01	0.02	0.02	0.06	0.004	Bal

Table 2 *Mechanical properties of the prestressing wire.*

HARDNESS, Rockwell C	YOUNG'S MODULUS, GPa	ULTIMATE TENSILE STRESS, MPa	ELONGATION, percent
51	199.3	1985	5.2

Table 3 *Dimension parameters of pits in CERT specimens.*

PIT TYPE	PIT DEPTH, L mm	PIT DIAMETER d, mm	DEPTH TO DIAMETER. RATIO (L/d)	AREA FRACTION*
P1	1.3	1.0	1.3	0.74
P2	0.8	0.5	1.6	0.92
P3	0.45	0.5	0.9	0.93
P4	0.2	0.35	1.75	0.98

    * Ratio of cross section area (reduced) in the plane of
      the pit to the nominal (unreduced) area.

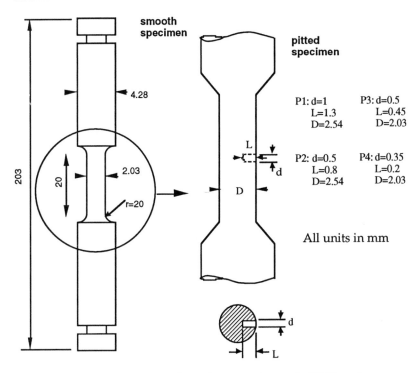

Figure 3 *Geometry of smooth and pitted prestressing steel CERT specimens.*

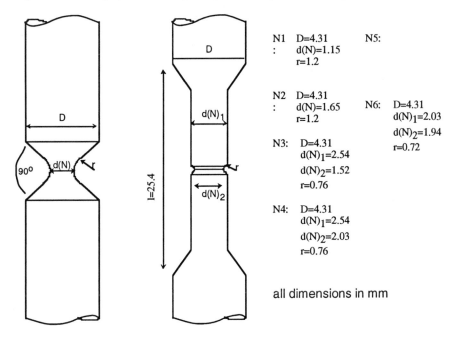

Figure 4 *Geometry of notched prestressing steel CERT specimens.*

longitudinal polishing through 600 grit SiC paper. Notched specimen types N1 through N4 were prepared using circular surface grinding and N5 and N6 by electrochemical machining. The procedure for the latter involved masking all but the circumferential notch surface area with polytetrafluoroethylene (PTFE) tape and anodically polarizing in a saturated $Ca(OH)_2$ solution at 0.1 A for 30 (N5) and 60 (N6) seconds. Preparation of pitted specimens began using a smooth specimen with pit types P1 and P2 being produced by electrochemical machining using the same procedure as for notched specimens, but here a pinhole of diameter less than 0.2 mm in the PTFE tape was used instead of a circumferential band to expose bare metal. Mechanical machining with a drill bit was employed for pit types P3 and P4.

2.2 Test System

The experimental set-up was essentially the same as described previously[3] and consisted of 1) a four station constant extension rate unit, 2) electrochemical cells and 3) data acquisition system. The CERT machine, which was patterned after that of Nutter et al.,[9] consisted of four stations and was operated at an extension rate of $4.7 \cdot 10^{-6}$ cm/sec, as was judged appropriate from earlier experiments.[3] The electrochemical cells were fabricated from PTFE and utilized platinum coated niobium counter and saturated calomel reference electrodes. Potentiostats were locally fabricated according to the circuit diagram proposed by Baboian et al.[10] The electrolyte was saturated $Ca(OH)_2$-distilled water which was deaerated by nitrogen purging. The data acquisition system consisted of an eight channel signal amplification unit and a personal computer with software that permitted load and time to be recorded during testing and for data to be subsequently analyzed.

2.3 Test Procedure

Prior to testing specimens were degreased with acetone, and the portions of the surface for which the cross section was not reduced were covered with PTFE tape. The pH of the test electrolyte (saturated $Ca(OH)_2$) was measured both before and after testing and was found to be nominally 12.5 with little or no change occurring with time. Nitrogen purging commenced about 90 minutes prior to testing and continued during the experiments. Specimens tested in the solution were polarized to -1.30 V (SCE), which corresponded to the lower plateau in Figure 2 and was considered to provide a severe test of embrittlement susceptibility.

3 RESULTS AND DISCUSSION

3.1 Notched Constant Extension Rate Testing Specimens

Table 4 presents air test results for notched specimens of the six different geometries (Figure 4); and Figure 5 plots average normalized maximum stress in air against notched area cross section, where the latter is expressed as a percentage of the smooth specimen area. This indicates that maximum stress was greatest for the N1 geometry, that it generally decreased as the specimen cross section increased and that it merged with the smooth specimen data as the notch size approached zero (stress calculated based upon the reduced cross sectional area). Table 5 and Figure 6 present companion data for specimens polarized to -1.30 V with the previous air data (Figure 5) also shown for comparison. In this case the normalized maximum stress for the notched specimens was invariably less than for the smooth specimens. The two curves (air and -1.30 V) exhibit a similar trend for the relatively severe N1-N4 geometries but differ for N5 and N6, the latter being necessitated since the two data sets must merge at the smooth specimen extreme.

Specimen ductility was represented in terms of two parameters, Plastic Elongation Before Necking (PEBN) and Plastic Elongation After Necking (PEAN), as illustrated

Table 4 *CERT results for notched specimens tested in air. Smooth specimen data are included for comparison.*

NOTCH TYPE	SPECIMEN NUMBER	A/Ao* percent	MAXIMUM STRESS, MPa (ksi)	PEBN, mm	PEAN, mm
N1	N1-1	7.5	2624 (380)	0.03	0
	N1-2	7.5	2638 (382)	0.02	0.02
	N1-3	7.5	2621 (380)	0	0.01
N2	N2-1	15	2390 (346)	0	0
	N2-2	15	2328 (337)	0	0
	N2-3	15	2722 (394)	0	0
N3	N3-1	36	2405 (349)	0	0
	N3-2	36	2398 (348)	0	0
N4	N4-1	64	2310 (335)	0.12	0
	N4-2	64	2395 (347)	0.04	0
N5	N5-1	86	2155 (312)	0.25	0
	N5-2	86	2130 (309)	0.32	0
N6	N6-1	92	2-31 (294)	0.3	0.04
	N6-2	92	2140 (310)	0.37	0.12
Smooth		100	2040 (296)	0.8	0.2

* A/Ao is the area ratio of the notched versus smooth specimen
cross section based upon the parameters d and D (Figure 4).

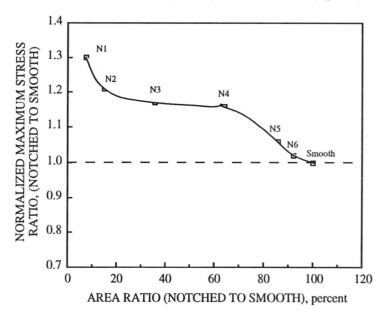

Figure 5 *Normalized maximum stress CERT results as a function of normalized specimen cross section area for each of the six notch geometry specimens and for smooth specimens tested in air.*

schematically in Figure 7. On this basis the data in Tables 4 and 5 also indicate that ductility of the notched specimens was less than for the smooth and greater in air than

Table 5 *CERT results for notched specimens tested in deaerated, saturated Ca(OH)₂-distilled water and polarized to -1.30 V (SCE). Smooth specimen data are included for comparison.*

NOTCH TYPE	SPECIMEN NUMBER	A/Ao, percent	MAXIMUM STRESS, MPa (ksi)	PEBN, mm	PEAN, mm
N1	N1-4	7.5	1680 (243)	0	0
	N1-5	7.5	1782 (258)	0	0
	N1-6	7.5	1524 (221)	0	0
N2	N2-4	15	1628 (236)	0	0
	N2-5	15	1580 (229)	0	0
	N2-6	15	1160 (168)	0	0
N3	N3-3	36	1340 (194)	0	0
	N3-4	36	1414 (205)	0	0
N4	N4-3	64	1400 (203)	0	0
	N4-4	64	1330 (193)	0	0
N5	N5-3	86	1740 (252)	0.11	0
	N5-4	86	1647 (239)	0.09	0
N6	N6-3	92	1804 (261)	0.15	0
	N6-4	92	1848 (268)	0.08	0
Smooth		100	1850 (268)	0.3	0

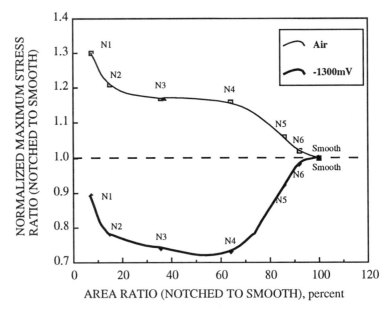

Figure 6 *Normalized maximum stress CERT results as a function of normalized specimen cross section area for each of the six notch geometry specimens and for smooth specimens tested in deaerated, saturated calcium hydroxide-distilled water while polarized to -1.30 v (SCE). Air data from Figure 5 are included for comparison.*

when polarized to -1.30 V. This point is also illustrated from comparison of the schematically represented stress-strain curves of each of the six geometries as shown in Figure 8 for both air (a) and -1.30 V (b) tests. For notch types N2 and N3 in air and N1-

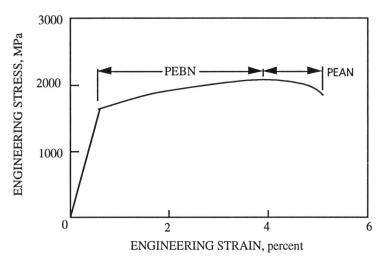

Figure 7 *Schematic illustration of CERT stress-strain curve and the relative regimes of Plastic Elongation Before Necking (PEBN) and Plastic Elongation After Necking (PEAN).*

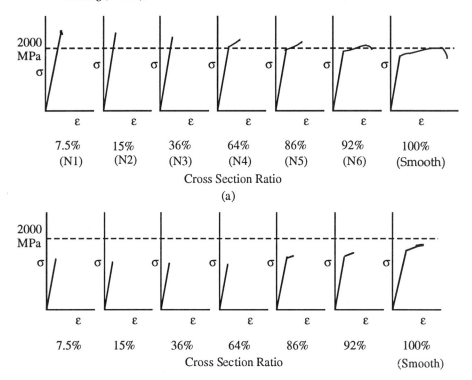

Figure 8 *Schematic representation of the stress-strain curve for each of the six notch geometry specimens and for smooth specimens tested in (a) air and (b) deaerated* saturated $Ca(OH)_2$-distilled water while polarized to -1.30 v (SCE).

Table 6 *Calculated values for the stress concentration factor for each of the six notch geometries along with normalized maximum stress data.*

NOTCH TYPE	A/Ao, percent	STRESS CONC. FACTOR	r/d (Figure 4)	NORMALIZED MAX. STRESS (AIR)	NORMALIZED MAX. STRESS (-1.30 V)
N1	7.5	<1.2	1.04	1.29	0.90
N2	15	1.3	0.73	1.22	0.79
N3	36	1.8	0.5	1.28	0.75
N4	64	2.3	0.37	1.15	0.74
N5	86	2.1	0.16	1.05	0.92
N6	92	1.6	0.37	1.02	0.98
Smooth	100	1.00	-	1.00	1.00

N4 at -1.30 V both PEBN and PEAN were nil, and fracture occurred in the elastic range.

Two factors are considered to have influenced specimen strength in the presence of a notch. The first is the magnitude of the stress concentration factor which, in itself, had a negative influence upon strength. Table 6 gives calculated stress concentration factors[11] for the present notch geometries and indicates that this parameter increased from N1 to N4 and then decreased for N5 and N6. This, in the light of the data trend in Figure 5, indicates that there is no simple relationship between the stress concentration factor and the mechanical response of the notched specimens. The second influential factor is plastic constraint, whereby the geometrical irregularity associated with the notch limits the lateral ductility and, as a consequence, renders the material stronger than if the geometrical irregularity were not present. This is referred to as notch strengthening.[12] For the representation in Figure 5 (air tests) the latter factor (notch strengthening) dominated over the former (stress concentration factor), particularly for geometries where the notch was relatively severe. Values of R/d (ratio of notch root radius to reduced specimen diameter, see Figure 4) are also listed in Table 6, since the magnitude of triaxiality and, hence, notch strengthening has been predicted to increase in proportion

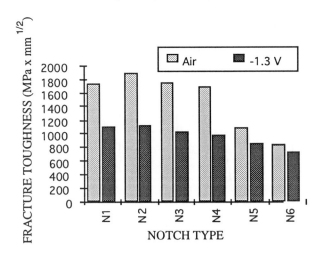

Figure 9 *Fracture toughness of the six notched specimen types as calculated from CERT results in air and in deaerated, saturated Ca(OH)2-distilled water while polarized to -1.30 v (SCE). The analysis assumed that the notch was equivalent to a sharp crack of the same depth.*

to this parameter.[13]   A correlation between maximum stress and R/d is generally apparent for specimens tested in air. For the -1.30 V results the stress concentration in conjunction with an embrittling influence of the excessive polarization apparently outweighed any notch strengthening contribution.

An attempt has been made to determine if the notched specimen behavior was conducive to a fracture mechanics representation, where the notch was assumed to be the same as a sharp crack. Figure 9 shows fracture toughness, as computed from the stress intensity factor (K) value at the point of fracture using a classical expression for a circumferential crack in a cylindrical member,[14] for each of the six notch types and for the two test conditions. Of particular significance is the fact that the calculated fracture toughnesses for both air and -1.30 V, respectively, are approximately constant for geometries N1-N4, for which ductility was minimal or nil (see Tables 4 and 5). The observation that fracture toughness for specimen types N5 and N6 was lower than for

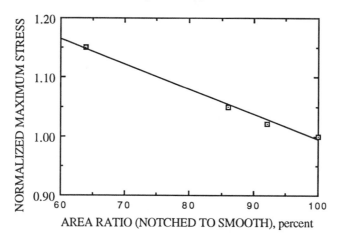

Figure 10 *Replot from Figure 5 of the normalized maximum stress versus normalized cross section area for smooth and N4-N6 notch type specimens tested in air.*

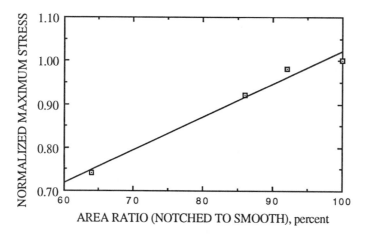

Figure 11 *Replot from Figure 6 of the normalized maximum stress versus normalized cross section for smooth and N4-N6 notched type specimens polarized to -1.30 v (SCE).*

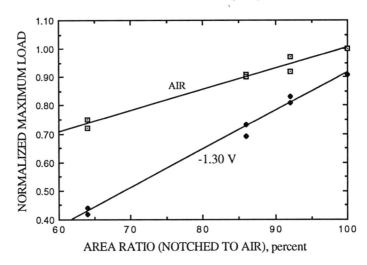

Figure 12 *Normalized maximum load versus normalized cross section area for N4-N6 notched and smooth specimens tested in air and polarized to -1.30 V (SCE).*

N1-N4 is attributed to the small 'crack' size for these geometries (~0.10-0.15 mm) and to more extensive cross section yielding, each of which compromise the applicability of linear elastic fracture mechanics.

If the normalized maximum stress versus normalized cross section area data from Figure 5 for the notch specimen types N4-N6 and smooth specimen geometries are reploted as shown in Figure 10, then it is apparent that these conform to a linear relationship with a slope of -0.0043. Application of this same rationale to the -1.30 V data based also upon a linear representation of the N4-smooth specimen results, as replotted in Figure 11 from Figure 6, indicates a slope of -0.0075. Figure 12 provides an alternative representation of the data in Figures 10 and 11 in terms of normalized maximum load, which is a more revealing parameter with respect to the integrity of tendon wire with locally reduced cross section than is strength based upon remaining cross sectional area. Thus, this latter representation (Figure 12) shows that for exposure of notched tendon in air or in an environment that does not result in embrittlement it is projected that a localized area reduction of 40 percent is required to reduce the load bearing capacity to the 0.7 value (magnitude of the assumed pretensioning). At -1.30 V the corresponding area reduction is by about 15 percent.

3.2 Pitted Constant Extension Rate Testing Specimens

Experimental results for pitted specimens tested in air are listed in Table 7. Plasticity was restricted in this case with fracture occurring in the elastic region in the most severe geometry case (P1). Some PEBN was apparent for specimen types P2 and P3, and necking and PEAN also occurred in P4. Specimen types P2 and P3 had essentially identical area ratios; however, the pit for the former was relatively deep and narrow (depth-diameter ratio 1.6, see Figure 3 and Table 3) while for the latter it was shallower and wider (ratio 0.9). In contrast to the notched specimen case, presence of a pit lowered the maximum stress (reduced cross sectional area basis) in air for all four of the geometries investigated. This indicates that, irrespective of the extent of plasticity, little or no notch strengthening resulted from the presence of the pits. Correspondingly, Figure 8 presents results at -1.30 V. These exhibited the same trend as the air results, but the maximum stress for a particular pit geometry was lower and the ductility less.

Figure 13 illustrates the above results graphically in perspective to the -1.30 V

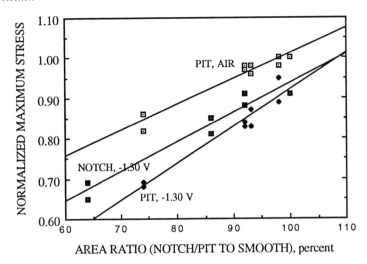

Figure 13 *Normalized maximum stress versus normalized cross section area for pitted and smooth specimens in air and at -1.30 V (SCE). Notched specimen data at -1.30 V are included for comparison.*

notched specimen data and in the same format as Figures 5 and 6 but with all data normalized with respect to the smooth specimen maximum stress in air. This makes it apparent that the average maximum stress of pitted specimens in both air and solution with polarization to -1.30 V decreased with decreasing cross sectional area according to the same general trend as for similarly polarized notched specimens. A close correspondence is seen between the notched and pitted specimen data at -1.30 V with data for the latter in air being positioned at a slightly higher stress. Of particular significance is that, first, cross sectional area is further confirmed as a parameter with which maximum stress is simply dependent and, second, the relative pit dimensions (depth and breath) exhibited little influence upon maximum stress compared to the magnitude of area reduction, as least to the extent that these factors were represented by

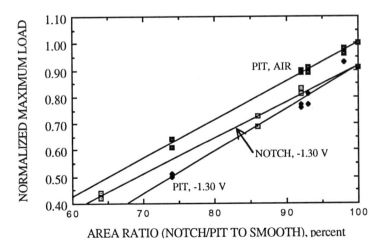

Figure 14 *Normalized maximum load versus normalized cross section area for pitted and smooth specimens in air and at -1.30 V (SCE). Notched specimen data at -1.30 V are included for comparison.*

the present geometries.

Figure 14 reproduces the maximum load representation in Figure 12 for the notched specimens at -1.30 V and adds to this the data for pitted specimens both in air and at -1.30 V. These data indicate that a pit encompassing about 20 percent of the cross sectional area would be required to reduce the load bearing capacity to 0.7 of the original value upon stressing in air or under environmental exposure conditions that do not change the fracture properties. Alternately, for the -1.30 V case the magnitude of this area reduction was 13 percent. The relatively modest wire cross section area reduction to reduce the load bearing capacity of notched and pitted specimens to the prestress amount (13-15 percent) is within the range where fracture mechanics cannot be relied upon to provide a geometry independent assessment of fracture strength (see Figure 9). Hence, the fracture mechanics approach to evaluation of residual strength of tendon with a locally reduced cross section is not considered to be of any practical significance.

Note also with regard to the pitted specimen that the trends for both the air and -1.30 V data are relatively close to one another. This suggests that interpretation of the results would not have been significantly different if experiments had been performed at a more positive, less embrittling potential.

3.3 Structure Qualification for Cathodic Protection

The fact that the maximum load of the notched and pitted specimens could be correlated with the remaining specimen cross sectional area lends itself to a new approach whereby a particular pretensioned concrete component or structure can be qualified for cathodic protection. This proposed protocol involves the following steps:

1. If no corrosion induced concrete cracking or spalling is evident, then the structure is automatically qualified. The rationale for this is that corrosion of tendons in sound (uncracked, non-spalled) concrete is expected to be modest in comparison to that in areas which have undergone corrosion induced damage (see below), and the residual strength for tendons here should be essentially unchanged from the original value.

2. In cases where cracking and spalling are evident, unsound concrete should be removed from these corrosion damaged areas thereby allowing visual inspection of the already exposed tendon. Such concrete removal should be performed with caution and in consultation with a structural engineer. It is assumed, however, that the concrete being removed is in such a poor condition that load bearing capacity is not further compromised by this removal process.

3. Locations on the exposed steel tendon where corrosion is uniform and where it is localized are identified. Attack at such sites is expected to represent the worst cases. For locations where uniform corrosion has occurred, residual strength should be proportional to the remaining cross section. Thus, a 30 percent loss in cross sectional area for a wire whose tensile strength corresponds to the specification minimum (1863 MPa or 270 ksi for Grade 270 material,[8]) should result in a residual strength equal to the assumed prestress (70 percent of ultimate). Where corrosion is localized, the diameter, d, of the largest pit is measured. From this and assuming a semi-circular pit cross section it is possible to determine the normalized remaining cross section area, A, of the wire from the expression:

$$A = \left[1 - \left(\frac{d}{D} \right)^2 \right] \cdot 100, \tag{1}$$

where D is the original wire diameter as illustrated in Figure 15 (recall from Figure 14 that a 13 percent reduction in cross section should reduce the residual strength to that of the same assumed prestress).

4. The structural member is qualified for cathodic protection if it conforms to the following criteria:

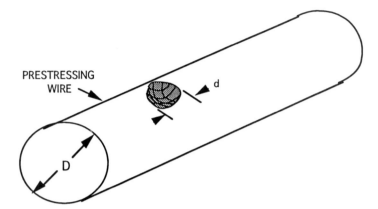

PRESTRESSING
WIRE

d

D

Figure 15 *Schematic illustration of the procedure for field estimation of remaining wire cross section determination.*

A. The remaining wire cross section at locations of uniform corrosion is not less than 85 percent of the original (uncorroded) area.
B. The diameter of the largest pit is such that the remaining localized area is ≥ 90 percent of the original value. For locations where there is both uniform and localized corrosion the same 90 percent net remaining wire cross section should apply.

These proposed qualification criteria incorporate a safety factor from several sources. Firstly, the maximum permissible localized area reduction (10 percent) is less than the 13 percent indicated from the experimental data (Figure 14). Secondly, the data upon which the local corrosion penetration criterion is based were acquired at a potential of -1.30 V, whereas a cathodic protection system for prestressing steel in concrete should be designed for a lower potential limit of -0.90 v or more positive (this factor may not be of minimal significance as discussed above). Third, the assumed wire strength is the specified minimum and in practice is likely to be higher. The above procedure for qualification is simpler and more straightforward than Electrochemical Proof Testing.[1]

4 CONCLUSIONS

1. Fracture properties of prestressing steel for which the cross section has been locally reduced by either a notch or pit were governed by the relative influence of a) the magnitude of stress concentration, b) notch strengthening and c) hydrogen embrittlement. While notch strengthening appeared to be the dominant factor for notched specimens tested in air, at -1.30 V and for pitted specimens under both environmental conditions (air and in solution with polarization to -1.30 V), the combined influence of stress concentration and hydrogen embrittlement was most influential.
2. For a prestressing steel wire containing a relatively modest notch or pit (remaining cross section approximately 75 percent or more of the original) the ultimate stress in air and with polarization in saturated $Ca(OH)_2$ to -1.30 V (SCE) was linearly dependent upon the remaining cross sectional area, independent of notch or pit shape. Based upon these results a procedure for qualification of a particular prestressed concrete member for cathodic protection is proposed as follows:
    A. If no corrosion induced concrete cracking and spalling are evident, then the structure is automatically qualified.
    B. In cases where corrosion related cracking and spalling are evident concrete should be removed from the damaged areas thereby exposing the tendon directly.

  C. Locations should be identified on the exposed steel tendon where corrosion is
   uniform and where it is localized. The structural member is qualified for
   cathodic protection if the following criteria are met:
   1. The remaining wire cross section at locations of uniform corrosion is not less
     than 85 percent of the original (uncorroded) area.
   2. The remaining wire cross section at the locations of any localized attack is at
     least 90 percent of the original area. For locations where there is both
     uniform and localized corrosion this same 90 percent remaining cross
     sectional area limit applies.
This new approach constitutes a straightforward procedure for qualifying a particular
structure for cathodic protection and is simpler than the previously proposed
Electrochemical Proof Testing methodology.

5 ACKNOWLEDGMENTS

The authors are indebted to the Federal Highway Administration for their financial
support of this research through Contract No. DTFH61-92-00058 and to Mr. Peter Clark
for machining assistance.

6 REFERENCES

1. W. H. Hartt, 'A Critical Evaluation of Cathodic Protection for Prestressing Steel in
  Concrete', (Eds. C. L. Page, K. W. J. Treadaway and P. B. Bamforth), in 'Corrosion
  of Reinforcement in Concrete', SCI, London, 1990.
2. R. Pangrazzi, L. Ducroq and W. H. Hartt, 'An Analysis of Strain Changes in
  Cathodically Polarized Pretensioned Concrete Specimens', paper no. 554 presented
  at CORROSION/91, Cincinnati, March 11-15, 1991.
3. W. H. Hartt, C. C. Kumria and R. J. Kessler, *Corrosion*, 1993, **49**, 377.
4. S. Kliszowski and W. H. Hartt, 'Mechanical Properties and fracture Strength of
  Cathodically Polarized Prestressing Wire', paper no. 96314 presented at
  CORROSION/96, Denver, March 24-29, 1996.
5. R. N. Parkins, M. Elices, V. S. Galvez and L. Caballero, *Corrosion Sci.*, 1982, **22**,
  379.
6. W. S. Galvez, L. Caballero and M. Elices, ASTM STP 866, American Society for
  Testing and Materials, Philadelphia, 1985, 428.
7. J. Wagner, W. Young, S. Scheirer and P. Fairer, 'Cathodic Protection Developments
  for Prestressed Concrete Components', Report No. FHWA-RD-92-056, Federal
  Highway Administration, Washington, D. C., March, 1993.
8. Standard Specification for Steel Strand, Uncoated Seven Wire for Prestressed
  Concrete, A 416-90a, Annual Book of ASTM Standards, American Society for
  Testing and Materials, Philadelphia, 1994.
9. W. T. Nutter, A. K. Agrawal and R. W. Staehle, ASTM STP 665, American Society
  for Testing and Materials, Philadelphia, 1979, 375.
10. R. Baboian, L. McBride, R. Langlais, G. Haynes, *Materials Performance*, 1979,
  18(12), 40.
11. R. D. Cook and W. C. Young, 'Advanced Mechanics of Materials', Macmillan
  Publishing Company, New York, 1985, 28.
12. R. W. Hertzberg, 'Deformation and Fracture Mechanics of Engineering Materials',
  third edition, J. Wiley and Sons, New York, 1989, 246.
13. R. Hill, The Mathematical Theory of Plasticity, Oxford University Press, Oxford,
  1956, 245.
14. H, Nisitani and N. Noda, Stress Intensity Factors Handbook, Society of Materials
  Science, Japan, Vol. 1, 1990, 643.

INFLUENCE OF TEMPERATURE ON THE POTENTIAL OF REINFORCING STEEL IN CONCRETE, UNDER CATHODIC PROTECTION

André RAHARINAIVO
Laboratoire Central des Ponts et Chaussées, Paris (France)

Ioan PEPENAR
INCERC, Bucharest (Romania)

Gilbert GRIMALDI
Laboratoire Régional de l'Equipement Est-Parisien, Melun (France)

1 INTRODUCTION

Cathodic protection is a method for mitigating corrosion of steel in concrete. Several criteria have been proposed for assessing the effectiveness of a cathodic protection system. According to the French Standard, such protection is effective when the steel potential is between -850 and -1050 mV_{CSE}. This criterion has been established after several laboratory experiments.

The object of this investigation is to determine how climatic factors, mainly temperature, can influence steel potential under cathodic protection.

2 EXPERIMENTAL

2.1 Test Pieces

2.1.1 Reinforced Concrete Slabs. Six reinforced concrete specimens were studied: they were made of four Portland cements. Table I gives the concrete mixture proportions and properties. These slabs were referenced as D1, D2, D3, D4, D5 and D6 [1].

Figure 1 shows a reinforced concrete test piece. It consists of a slab 900 mm x 900 mm and with a variable thickness. Reinforcement was plain carbon steel deformed bars, 6 mm in diameter, parallel to a side of the slab. A frame made of steel rebar, 10 mm diameter, was also embedded in the concrete. The concrete cover on rebar varied from 20 to 80 mm. The total area of the rebar to be protected was 0.20 m^2.

In every slab, a silver-silver chloride reference electrode was embedded in the concrete. Temperature probes (thermocouples) were also embedded in every specimen under 2, 4, 60 or 80 mm thick cover.

2.1.2 Prestressed Concrete Pipes. Two prestressed concrete pipes (T1 and T2), 7 m long, 1 m internal diameter, were tested [1], the anodes on them were not continuous. They were placed as shown in Figure 2, in simulating real cases where the structure geometry is complex.

2.2 Applying Cathodic Protection

The rebars in the slabs were connected in series to the power supply. Hence the same current flows in each of them. The half-cell potential of rebars under cathodic protection was not constant in every slab. Some of them were well protected, others were less protected. The rebars in prestressed concrete pipes were polarised either at -800 $mV_{Ag-AgCl}$.(for T1) or at -1000 $mV_{Ag-AgCl}$.(for T2).

The current I was applied, increasing in steps. The steel potential, E, was checked at each step. The final polarization step corresponded to about $E = -800$ mV$_{Ag-AgCl}$.in each slab. During long-term polarisation, the voltage between anode and rebar was kept constant.

Table 1 : *Concretes proportions and properties*

Slab		D1	D2	D3	D4	D5	D6
Cement		C1	C1	C2	C3	C4	C4
Cement content	kg/m^3	350	350	350	400	400	400
Sand (0/4mm)	kg/m^3	753	720	720	700	700	700
Gravel (4/12 mm)	kg/m^3	1150	1130	1130	1100	1100	1100
Silica fume	kg/m^3	30	0	35	0	45	0
Admixture Agilplast	kg/m^3	11	0	12	0	14	0
Air entrainer MBVR	kg/m^3	0.33	0	0.385	0	0.44	0
Water	l/m^3	97	170	115	190	131	190
Compressive strength at 28 days	MPa	51.0	40.5	56.3	52.0	62.2	62.0

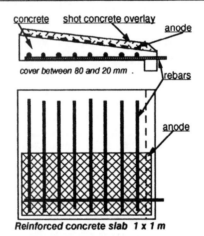

Figure 1 *Reinforced concrete slabs tested*

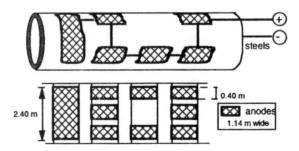

Figure 2 *Prestressed concrete pipes tested (8 m long and 1 m in diameter)*

2.3 Accelerating Concrete Damage

The specimens were stored outdoors, lying on two parallel walls, about 1 meter high. They were submitted to ageing cycles (Figure 3), which lasted one day and included :
- freezing by pouring liquid nitrogen (at -196°C) on the upper concrete surface for two hours, during this freezing, the maximum temperature gradient in the concrete was 0.4 °C/mm.
- thawing, which started two hours after the end of freezing. Then a solution saturated with sodium chloride (at about 20°C) was poured onto the upper surface of the specimen.

Hence an ageing cycle corresponds both to freezing and thawing and salt contamination of concrete.

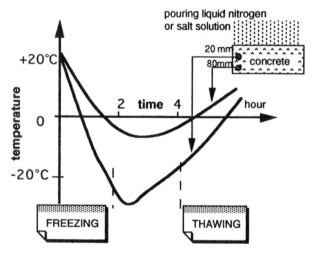

Figure 3 *Freezing and thawing of the test pieces*

2.4 Measurements

Every test piece was subjected to a freezing-and-thawing cycle every week for three years. The main parameters which were measured were steel potential, applied current and temperature.

2.4.1 Long-term Measurement of Steel Potential. Approximately every two months, a rebar potential map of every specimen was drawn. The half-cell potential of a given steel bar was checked at various ages.

2.4.2 Short-term Measurements. Measurements were made during one day (every hour) or five days (every four hours), to determine the effect of ambient temperature on the value of steel potential under cathodic protection. For these measurements, the test pieces were under natural conditions, the sodium chloride solution being poured onto them. These measurements were carried out on only three reinforced concrete slabs and one prestressed concrete pipe.

During the 5 day experiment, the air moisture (relative humidity) was also measured.

3 RESULTS

3.1 Results of Long-term Measurements

Figure 4 gives typical potential-time curves of rebars under cathodic protection with constant voltage between anode and rebar. No apparent effect of the temperature is visible on the curve.

This result could be due to the fact that steel potential was read under various climatic (temperature, humidity) conditions. Further results confirm this assumption.

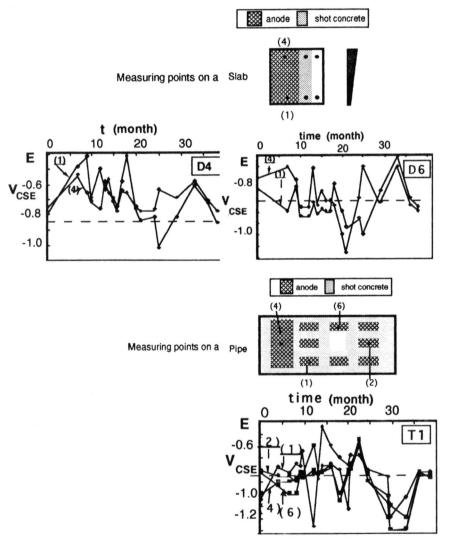

Figure 4 *Potential of rebars at "measuring points" (X) under cathodic protection, measured against copper-copper sulphate electrode*

3.2 Results of Short-term Measurements

3.2.1 Results of Measurements Read During 24 Hours A first series of measurements was made over 24 hours with slab D4 and pipe T1 in which the thermocouple was near the reference electrode. Figure 5 shows the temperature-time curves which were obtained. Readings started early in the afternoon (at 14.00), the maximum temperature was observed at the end of the afternoon, the minimum was read in the morning (at 9.00).

Figure 6 shows the steel potential-time curves for slab D4 and pipe T1. It appears that steel potentials fluctuate in the same way, for D4 and T1, they were almost constant at the end of the afternoon, they decrease during the night and increase in the morning.

Figure 7 was obtained by combining the curves of Figures 5 and 6. It shows the relationship between temperature and steel potential under cathodic protection. It appears that there is no close relationship between concrete temperature and steel potential. For a given temperature (higher than 15°C), steel potential measured during natural warming conditions is higher than potential during a cooling period.

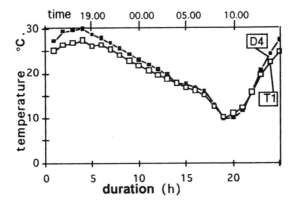

Figure 5 *Temperature in concrete versus time, during the short-term measurements*

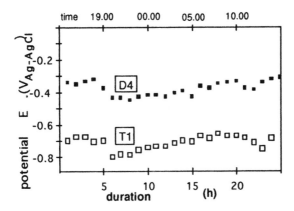

Figure 6 *Steel potential versus time, in slab D4 and pipe T1, under cathodic protection with constant voltage between anode and rebar*

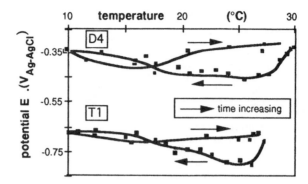

Figure 7 *Steel potential versus temperature, for 24 hours, in slab* D4 *and pipe* T1, *under cathodic protection with constant voltage between anode and rebar*

3.2.2 Results of Measurements Read Over Five Days A second series of measurements was made over five days with reinforced concrete slabs D5 and D6. These slabs were designed almost identically but only D5 contained silica fume. Figure 8 shows the temperature change versus time, for both slabs D5 and D6 (the two curves coincide).

Figure 9 shows the relative humidity (RH) of the air near slab D5, during the same experiment. It appears that RH decreases when temperature increases and vice-versa.

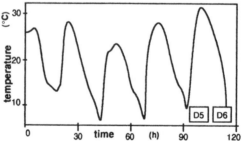

Figure 8 *Concrete temperature in slabs* D5 *and* D6 *during the short-term experimentation*

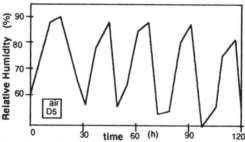

Figure 9 *Relative humidity of the air around slab* D5, *during the second series of short-term measurements*

Steel potential under cathodic protection was almost constant during the five day measurements : it varied between -580 $mV_{Ag\text{-}AgCl}$ and -620 $mV_{Ag\text{-}AgCl}$ for slab D5 and between -800 $mV_{Ag\text{-}AgCl}$ and -860 $mV_{Ag\text{-}AgCl}$ for slab D6. However, some "potential peaks" were observed for the two slabs, i.e. for a few minutes, this potential was less negative than the constant value. The height of these peaks was up to 200 mV (against the constant value of the potential).

By combining temperature-time curves and steel potential-time curves, it appears that temperature has a little effect on the change of steel potential (Figure 10). For slab D5 the steel potential increases slightly with temperature, but it decreases slightly for slab D6.

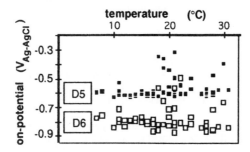

Figure 10 *Relationships between temperature and steel potential and temperature for slabs D5 and D6 (D5 contains silica fume and D6 does not)*

4 INTERPRETATION, APPLICATION AND CONCLUSION

4.1 Effect of Temperature on Electrochemical Reactions

Steel potential was measured against a reference electrode. It is known that this reference potential changes with temperature. In this case, the rate of change is equal to 1 mV/°C. The steel potential change was usually higher than 30 mV. This means that the steel potential change was not due only to the reference electrode.

Steel potential can also change because the rates of electrochemical reactions under cathodic protection are influenced by temperature. As no simple relationship exists between reaction rates and steel potential, the protection current was measured during the short-term measurements. In this case, cathodic protection was applied at constant voltage (between anode and rebar).

The results of the first series of measurements (over 24 hours) show that the protection current, at a given temperature, is slightly higher when concrete is spontaneously warming than when it is cooling. Figure 11 gives these results for slab D4 and pipe T1.

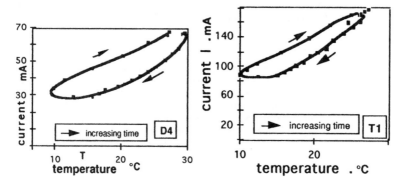

Figure 11 *Protection current versus temperature, under cathodic protection at constant voltage for slab D4 and pipe T1*

4.2 Effect of the Concrete Humidity on Steel Potential

It is usually agreed that rebar potential under cathodic protection is influenced by the oxygen content in the cement pore solution. The measured rebar potential is due both to steel immunity and oxygen reduction. The oxygen content in concrete depends on temperature: its maximum value (solubility) decreases with temperature. However, this oxygen content could not be determined during the experiments, but its influence was studied in an indirect way.

If oxygen content is constant, steel potential is also constant. So, the periodic change of steel potential is likely to be due to a semi-cyclic change of oxygen content. The oxygen diffusivity in concrete varies between 10^{-9} and 10^{-6} $m^2.s^{-1}$, depending on the humidity and other properties of the concrete cover. [2 - 4] So, the oxygen content near embedded steel under a cover about 30 mm thick, can change periodically. This semi-periodical change in oxygen content agrees with the semi-periodical change in rebar potential under cathodic protection.

Oxygen can come in and out of concrete, depending on the relative humidity (or temperature) change. But these two processes are not quite equivalent: during "wetting", water is adsorbed on the pore walls and during "drying" it is desorbed. Adsorption and desorption have different rates, depending on the pore system in the concrete.[5] This means that for a given value of air humidity, the water content is higher during desorption than during adsorption. This result is to be compared with the hysteresis effect of temperature on steel potential.

Moreover, if, for example, the pore solution is saturated with (dissolved) oxygen at 10°C, when the concrete is warming up, to 30°C, gaseous oxygen can be evolved from this solution. Such an evolution can explain the "potential peaks" which were observed, and which correspond to bubbles reaching the steel surface.

4.3 Application to a Cathodic Protection Installation

Cathodic protection has been applied on rebars in reinforced concrete columns of a bridge over a sea pass.[6] The protected areas were in the tidal zone. Steel potential was recorded over two days, at two locations (referenced S and W). The results obtained are shown in Figure 12.

It appears that steel potental fluctuates periodically. The potential amplitude is about 60mV, which is similar to the values read during the experiments. The length (time) of the period corresponds to that of the tide. Between the times of total immersion (high tide) and emersion (low tide), concrete near the embedded steel is always wet. This means that the local concrete humidity does not change. But the oxygen content in the pore solution changes over time and influences steel potential.

4.4 Conclusion

Reinforcing steel potential under cathodic protection is not constant if it is not controlled. The main factor influencing this potential change is temperature. This does not deal with the effect of the reference electrode, whose potential value changes slowly (1 mV/°C).

Temperature has two indirect effects on rebar potential. First, it influences almost periodically the relative humidity around the concrete, inducing oxygen movement into and out of the concrete. Second, it influences the saturation value of oxygen in concrete. These two effects result in a (semi-periodical) change in the oxygen reaction on steel under cathodic protection.

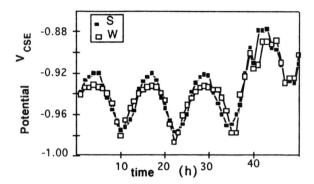

Figure 12 *Potential of two rebars (S and W) in a reinforced concrete column in a tidal zone and under cathodic protection (Reference electrode : copper-copper sulphate)*

References

1. I. Pepenar, G. Grimaldi and A. Raharinaivo, 'Fluctuation du potentiel des aciers dans le béton et sous protection cathodique', Laboratoire Central des Ponts et Chaussées, Paris, France, 1994, série Etudes et Recherches Ouvrages d'art OA 14, chap 4, 37.
2. O. Gjørv, Ø. Vennesland and H. S. Busaidy, *Materials Performance*, 1986, **25** (21), 39.
3. K. Kobayashi and K. Shutto, *Cement and Concrete Res.* 1991, **21**, 273.
4. Y. F. Houst, 'Influence of microstructure and water on diffusion of CO_2 and O_2 through cement paste', Second CANMET-ACI Intern. Confer. 'Durability of Concrete', Montreal, Canada, ACI SP 126, Supplementary papers, 141.
5. V. Baroghel-Bouny, 'Caractérisation des pâtes de ciment et des bétons' Laboratoire Central des Ponts et Chaussées, Paris, 1994, 293.
6. G. Grimaldi and A. Raharinaivo, 'Monitoring the effectiveness of a cathodic protection system : case of the Noirmoutier bridge', Internat. Conference Strategic Highway (SHRP) and Traffic Safety on Two Continents, The Hague, NL, Sept. 1993, vol. SHRP 23/9.

LABORATORY EVALUATION AND FIELD INSTALLATION OF AN EXPERIMENTAL SACRIFICIAL ANODE CATHODIC PROTECTION SYSTEM FOR REINFORCED CONCRETE BRIDGE DECKS

J. P. Broomfield[1], D. A. Whiting[2] and M. A. Nagi[2]

[1]Corrosion Consultant, 78 Durham Road, London SW20 0TL, U.K.
[2]Construction Technology Laboratories Inc., 5420 Old Orchard Road, Skokie, Illinois 60077-4321, U.S.A.

1 INTRODUCTION

The corrosion of reinforced concrete bridges due to exposure to deicing salt is a multi billion dollar problem in the USA and other Western Countries where salt is used to deice roads in winter. Most OECD countries use waterproofing membranes to protect their bridge decks. This deflects the salt onto the substructure. The USA, Canada and the Netherlands are exceptions in that they routinely built bare concrete decks until the 1980s, even when deicing salts were present. In North America this has lead to a massive program of bridge deck rehabilitation and the development of impressed current cathodic protection systems to stop the corrosion caused when chlorides diffuse to the rebar/concrete interface and then attack the passive layer normally sustained by the high alkalinity (pH 12 to14) resulting from the excess alkaline hydroxides in the concrete pores.

In a survey completed in 1990 it was reported that there were 287 cathodic protection systems in place on bridge decks of the Interstate system in North America[1]. The total area covered was 840,000 m^2, approximately 99% of this was on bridge decks. These cathodic protection systems were of the impressed current type consisting of an inert distributed anode (catalytically coated titanium mesh, conductive asphalt with flat silicon iron anodes, titanium or conductive polymer in slots). A recent report by the Transportation Research Board[2] said that for the next 10 years a quarter to half a million square metres of bridge deck surface area will have to be rehabilitated each year. This will cost about $200 to $400 per square metre, giving an annual cost of $50 to $250 million per year for bridge decks alone.

Conventional methods such as repairing potholes, removing the top 25 mm of cover and placing of a low permeability concrete overlay do not remove all the chlorides. Cathodic protection is a technically superior solution as it prevents corrosion across the whole area, not just in damaged areas that have been patched. However, cathodic protection is initially expensive although it can be justified by life cycle costing. It is also technically alien to many bridge and highway engineers. Given the impact of budget shortfalls in recent years in highway agencies, it is questionable whether the skilled and trained staffing necessary for proper specification, oversight and maintenance of conventional impressed current cathodic protection systems can be developed and sustained in state highway agencies. Evaluations of 287 CP systems in 1990[1] showed significant numbers of trivial problems with installed systems due to inadequate maintenance.

The other form of cathodic protection is 'galvanic' or 'sacrificial anode' cathodic protection (SACP). Instead of using a power supply and an inert or slowly consumed anode, the anode is of a less noble metal than the cathode (rebars), such as aluminium, magnesium or zinc. As the metal corrodes it generates the electric current required to protect the steel.

One of the main advantages of SACP is that there is no power supply. This makes it cheaper to design, install and operate. The anode is installed with direct connections to the rebar and the system operates without further intervention until the anode is consumed and requires replacing. There is minimal risk of damage as there are no exposed wires, no instrumentation and no power supplies.

The main limitation of this system is that there is no way to control the output. Sacrificial anodes have limited driving voltages and these may be inadequate if the corrosion rate is too high or the concrete resistance is too high as the system cannot be 'turned up' to generate more power. It also takes some time to fully polarise the system, which makes early readings difficult to interpret. Another limitation is that the anode is consumed during the protection process. This means that the anode has a finite lifetime. SACP systems have been used for many years for buried and submerged steel structures and on reinforced concrete bridge substructures in marine environments[3].

2 EXPERIMENTAL LABORATORY WORK

The details of the experimental work are given in the full interim report to the Federal Highway Administration[4] so the results will only be summarised here. Comparative laboratory testing was carried out on six materials:

1.	Scrap zinc alloy used for making U.S. one cent coins	(Zn-1)
2.	High grade Zn alloy ASTM B 6	(Zn-2)
3.	Mg/Al/Zn alloy ASTM B90 Alloy AZ31B	(Mg)
4.	CP grade Al/Zn/Hg alloy	(Galvalum I)
5.	CP grade Al/Zn/In	(Galvalum III)
6.	5005 Al alloy meeting ASTM B209	(Al-3)

Specimens measuring 150 mm by 50 mm were cut from the samples obtained. These were mounted as shown in Figure 1, in an electrolyte of Ottawa sand and simulated pore water solution as shown in Table 1. This is equivalent to a chloride level of $6kg.m^{-3}$ of sand (or concrete).

This experimental setup has been used previously to simulate impressed current cathodic protection of reinforced concrete[4]. However, the resistivity of the electrolyte is critical for SACP so the sand was saturated with the solution, and then it was heated in an oven to dry the mixture to predefined levels of resistivity as shown in Table 2.

There were 36 specimens in all (6 anodes x 3 resistivities x 2 replications). Measurements were made of current flow, A.C. resistance, "on" and "instant off" half cell potential, 4 hour potential decay and the driving voltage between anode and cathode over 18 weeks. The temperature was maintained at $23\pm1.5°C$ and no appreciable weight loss of the buckets was observed, indicating that there had been no change in moisture content.

At the end of 18 weeks, all specimens were removed, and cleaned with a camel hair brush. Specimens were weighed (including corrosion products). One of the two duplicate specimens was then cleaned in an ultrasonic bath of water and surfactant. Corrosion products were removed by scrubbing with a tooth brush while immersed. This was not according to any specific ASTM or other standard method as the difference in materials would have lead to different extents of removal of the Al, Zn and Mg alloys using standard solutions to remove the corrosion products. Specimens were examined with a low power microscope to check that all (or most of) the corrosion product had been removed before drying and weighing.

Table 1 - *Composition of Simulated Pore Water Solution*

Constituent	Weight Percent of Solution
Calcium Hydroxide	0.20
Potassium Hydroxide	1.00
Sodium Hydroxide	2.45
Potassium chloride	3.20
Distilled Water	93.15

Table 2 - *Characteristics of Sand Media*

Batch	Resistivity (ohm.m)	Moisture Content (% by weight)
A	23	0.79
B	100	0.51
C	260	0.37

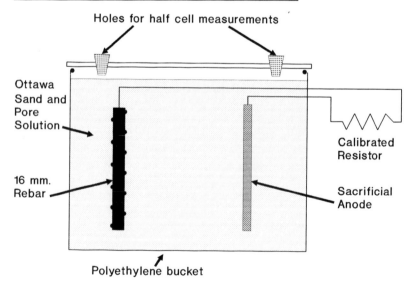

Figure 1 *Test Setup for anode materials tests*

3 RESULTS AND DISCUSSION - LABORATORY WORK

Tables 3 and 4 summarise the data on current flow and four hour depolarisation. The full results and discussion are given in reference 4. Anodes Zn-2, Al-2 and Al-3 were chosen for further testing based on their ability to pass current and polarise the steel over the range of resistivities used.

Table 3 *4 hour depolarisation in mV for different sand resistivities*

	23 ohm.m	100 ohm.m	260 ohm.m
Zn-1	320	100	100
Zn-2	380	260	170
Mg	500	80	140
Al-1	420	280	150
Al-2	450	280	160
Al-3	420	300	210

Table 4 *Current density mA.m² at Steel Surface for different sand resistivities*

	23 ohm.m	100 ohm.m	260 ohm.m
Zn-1	1	1	1
Zn-2	1.5	2.5	2
Mg	9.5	0.5	2
Al-1	3	4	2
Al-2	3	5	2
Al-3	3	3	2

3.1 Slab Tests

Anode materials were fabricated in the form of an expanded metal mesh. 1020 x 510 x 180 mm slabs were cast with two layers of reinforcement, 38 mm cover to the top surface, with 16 mm. bars on 200 mm. centres. Chloride was added to the top batch of concrete to give 6 kg.m^{-3}. The slabs were left for 18 weeks, and macrocell corrosion currents of between 282 and 772 μA, with an average of 462 μA. Half cell potentials show average readings between -350 mV and -467 mV vs. copper/copper sulphate. Corrosion rate readings were taken with proprietary linear polarisation devices (GECOR 6 and 3LP). Readings with the Gecor showed corrosion currents between 0.72 and 0.32 μA.cm^2, which are in the moderate corrosion rate range.

Table 5 lists the anode and overlay combinations used. The anode was applied to the slabs, then covered with a chloride laden bedding mortar, and then the overlay. The overlay was based on the mix used in the first SACP trials on a bridge deck in 1981[5]. The slabs were wet/dry cycled on a weekly basis alternating 4% NaCl solution with tap water for 24 weeks.

Table 5 *Anodes and overlays applied to Slabs*

Slab No.	Anode Material	Overlay Type
1	Galvalum III	Normal Weight Concrete
2	Al-5005 commercial alloy	Normal Weight Concrete
3	High grade Zinc	Normal Weight Concrete
4	Galvalum III	Lightweight Concrete
5	Al-5005 commercial alloy	Lightweight Concrete
6	High grade Zinc	Lightweight Concrete
7	Galvalum III	Free draining OPC concrete
8	Al-5005 commercial alloy	Free draining OPC concrete
9	High grade Zinc	Free draining OPC concrete
10	Galvalum III	Free Draining LMC Concrete
11	Al-5005 commercial alloy	Free Draining LMC Concrete
12	High grade Zinc	Free Draining LMC Concrete

Figures 2 to 9 summarise the depolarisation and current flow data. It can be seen that the zinc did not pass much current, or polarise the steel. The galvalum III cp anode alloy performed better than the structural grade aluminium but given the relative prices and difficulty in obtaining galvalum in the form of an expanded mesh, it was decided that the structural grade aluminium should be tested in field trials.

After discussions with the Illinois Department of Transportation (IDOT), it was agreed that the normal weight concrete overlay would be used as there was insufficient information on the performance of the free draining mixes.

4 DESIGN OF TRIAL CATHODIC PROTECTION SYSTEM

After discussions with IDOT, bridge No. 071-0060, on Interstate 39, which carries the north bound lanes over C&NW railway line, was allocated to the trial. The deck is approximately 70 m long by 12 m wide. The hard shoulder and driving lane (6.5 m wide) was designated a fully instrumented trial area to be applied in phase 1. The overtaking lane (5.25 m) was installed as phase 2, without instrumentation.

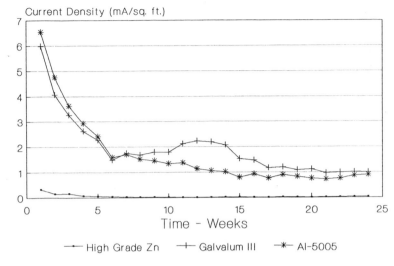

Figure 2 *Current density for normal weight concrete overlay*

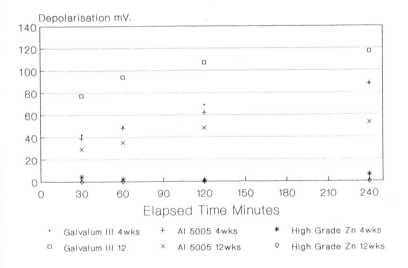

Figure 3 *Potential Decays for normal weight concrete overlay*

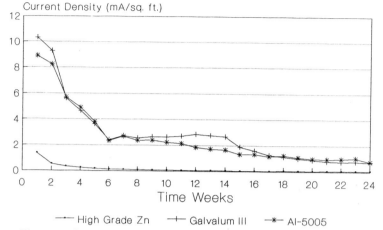

Figure 4 *Current density for lightweight concrete overlay*

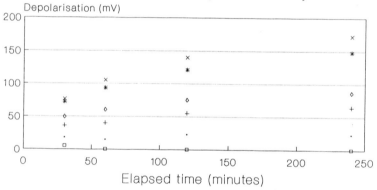

Figure 5 *Potential Decays for lightweight concrete overlay*

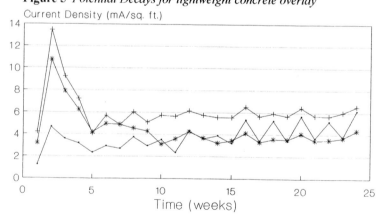

Figure 6 *Current density for free draining concrete overlay*

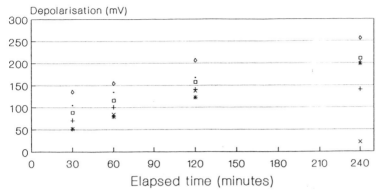

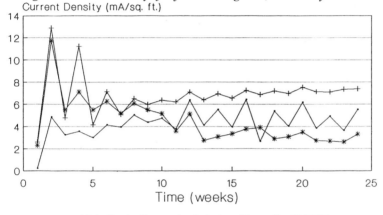

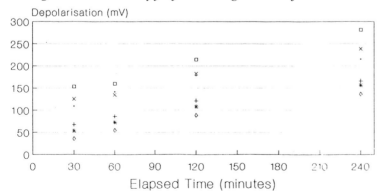

Although the anode was continuous for each phase, four "zones" were defined and instrumented with null probes, half cells and ground connections which ran through an instrumentation box. The instrumentation box contained terminals for taking voltage measurements of half cells to rebar ground connections and across shunt resistors for the null probes[6] and the anode to rebar connections.

The wiring and monitoring installations were done between 12 and 15 June 1995, with the overlay applied the following week. Figures 10 and 11 show schematics of a monitoring location and a null probe. Figure 12 shows the anode being installed after installation of the monitoring system and the rebar connections, and prior to applying the concrete overlay.

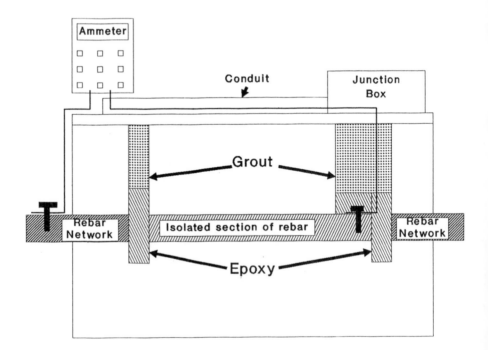

Figure 10   *Null Probe - Isolated section of steel at anodic location, will monitor macrocell currents while corroding and cathodic protection current reversal*

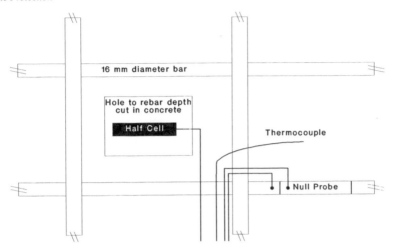

Figure 11 *Schematic of a monitoring location showing an embedded half cell, null probe, thermocouple and local ground connection. Cables go back to a monitoring box containing terminations and shunt resistors to monitor currents and voltage*

Figure 12 *Installation of the expanded aluminium metal sheet anode prior to overlaying*

5 CONCLUSIONS

- Aluminium, Zinc and Magnesium sacrificial anodes were tested in a simulated pore water solution in sand at different resistivities to determine their effectiveness for sacrificial anode cathodic protection of steel in concrete.
- This technique was successful in showing that, while at low resistivities magnesium was an effective anode, it was not effective at higher resistivities.
- When tested in concrete slabs, the aluminium anodes were more effective than zinc, which gave negligible levels of current after a few weeks.
- An 840m^2 field trial has been undertaken using an expanded structural grade aluminium anode in a mortar bedding layer with a Portland cement concrete overlay, conventionally specified by Illinois Department of Transportation.
- Instrumentation includes four half cells, four null probes and four anode rebar connections for current monitoring.

6 ACKNOWLEDGEMENTS

The authors would like to thank Illinois DoT for their support in allowing trial installation on their bridge. This work was carried out as part of a contract with the U.S. Department of Transportation, Federal Highway Administration, DTFH61-92-C-00064. This document is disseminated under the sponsorship of the Department of Transportation in the interest of information exchange. The United States Government assumes no liability for its contents or the use thereof. The contents of this paper reflect the views of the authors who are responsible for the facts and accuracy of the data presented herein. The contents do not necessarily reflect the official views or the policy of the U.S. Department of Transportation.

7 REFERENCES

1.   J. P. Broomfield and J. S. Tinnea, 'Cathodic Protection of Reinforced Concrete Bridge Components', SHRP-C/UWP-92-618, Strategic Highway Research Program, National Research Council, Washington, D.C., 1992.
2.   'Highway Deicing - Comparing Salt and Calcium Magnesium Acetate', Special Report 235, Transportation Research Board, National Research Council, Washington, D.C., 1991.
3.   R. J. Kessler, R.G. Powers and I.R. Lasa I.R. 'Update on Sacrificial Anode Cathodic Protection on Steel Reinforced Concrete Substructures in Seawater', Paper No. 516, Corrosion 95, NACE International, Orlando, Florida, 1995.
4.   D. Whiting, M. Nagi and J. P. Broomfield, 'Evaluations of Sacrificial Anodes for Cathodic Protection of Reinforced Concrete Bridge Decks', FHWA-RD-95-041, Federal Highway Administration, VA, U.S.A., 1995.
5.   D. Whiting and D. Stark, 'Cathodic Protection for Reinforced Concrete Bridge Decks - Field Evaluation', NCHRP Report 234, Transportation Research Board, National Research Council, Washington, D.C., U.S.A., 1981.
6.   J. Bennett and T. Turk, 'Technical Alert - Criteria for the Cathodic Protection of Reinforced Concrete Bridge Elements', SHRP-S-359, Strategic Highway Research Program, National Research Council, Washington, D.C., U.S.A., 1994.

NEW EXPERIENCES ON CATHODIC PREVENTION OF REINFORCED CONCRETE STRUCTURES

L. Bertolini[1], F. Bolzoni[1], T. Pastore[2] and P. Pedeferri[1]

1. Dipartimento di Chimica Fisica Applicata, Politecnico di Milano, piazza L. da Vinci 32, 20133 Milano, Italy
2. Facoltà di Ingegneria, Università degli Studi di Bergamo, via G. Marconi 5, Dalmine (BG), Italy

1 INTRODUCTION

Cathodic prevention is an electrochemical technique aimed at improving the corrosion resistance of new reinforced concrete structures subjected to chloride penetration of the concrete cover. [1-2] This is achieved by applying a very low cathodic current density (lower than 2 mA/m^2) to the steel in the concrete. The passage of a direct current through the concrete towards the reinforcement has several beneficial effects: [3] firstly the potential of the cathodic structure shifts to a lower value, secondly the cathodic reaction occurring at the steel surface leads to the production of alkalinity and thirdly the electrophoretic transport of ions through the pore liquid of the concrete moves the chloride ions away from the steel surface. The electrochemical techniques applied to repair corroding structures, namely cathodic protection, electrochemical realkalisation and electrochemical chloride removal, are based mainly on one of these effects.

Cathodic prevention is a different case, it is applied to new structures before chloride ions enter the concrete cover. In spite of the very low current density applied to the reinforcement, all three mechanisms above are expected to contribute to increasing the resistance of steel to chloride induced corrosion, by delaying initiation of corrosion. Since the critical chloride content for corrosion initiation increases as the potential of the structure shifts to lower values and as the alkalinity of the concrete increases, both cathodic polarisation (i.e. the decrease in steel potential) and cathodic reaction allow the steel to remain in a passive condition in the presence of chloride contents much higher than the values needed to promote corrosion in structures where cathodic prevention is not applied. Moreover, the electrophoretic migration of chloride ions could induce a local reduction of chloride ion concentration in the vicinity of the steel surface, further delaying the initiation of corrosion. Nevertheless, the beneficial effects of this technique and, particularly, their mutual interaction have never been thoroughly investigated.

This work deals with a long term research project aimed at studying the effectiveness of cathodic prevention and defining suitable protection criteria. [4] The effects of cathodic polarisation on the critical chloride content for initiation of pitting corrosion and the effect of the electrophoretic migration are studied using reinforced concrete slabs subjected to galvanostatic polarisation. The results of potential monitoring and chloride penetration analyses carried out during the first 15 months of testing are presented along with the results of laboratory tests carried out to evaluate the influence of the current on the diffusion of chloride ions through the concrete and the effect of chloride content on pitting potential.

2 EXPERIMENTAL

Galvanostatic polarisation tests were carried out on seven reinforced concrete slabs as illustrated in Figure 1. Concrete was mixed with 350 kg/m^3 of ordinary 425 Portland cement, 0.5 w/c ratio, 1950 kg/m^3 of alluvial aggregates, and addition of a superplasticizer.

A reinforcing steel mesh with a spacing of 25 mm was obtained by overlapping two 50×50 mm carbon-steel meshes made of 5 mm diameter wires. Mixed metal activated titanium mesh anodes were embedded in the concrete. The slabs were wet cured for 48 hours and then exposed outdoors under a roof. About six months later, tests began and a ponding cycle was initiated with 3% NaCl solution, consisting of a one-week wetting period followed by a two-week drying period. Current densities of 0.5, 1, 2, 5, 10 and 20 mA/m^2 referred to concrete surface (0.4, 0.8, 1.7, 4.2, 8.4 and 16.8 mA/m^2 at the steel surface) were applied to six slabs. Anode and reinforcement potentials were monitored both versus Mn/MnO$_2$ and activated titanium fixed reference electrodes placed in the vicinity of the steel surface and anode before casting; four-hour depolarisation tests were regularly carried out.

Seven and fourteen months after starting the polarisation, total chloride content profiles were measured by coring a 28 mm diameter sample from each slab and cutting it into slices with a nominal thickness of 7.5 mm. The slices were then ground and dissolved in nitric acid. Chloride content was measured with an ion-sensitive electrode.

The effect of the electric field on the diffusion of chloride ions through the concrete was studied using modified diffusion cells (Figure 2). A concrete disk of diameter 100 mm and thickness 10 mm separated two solutions with different chloride contents and a direct current flux was imposed from the side containing the high concentration to that with the low concentration. The disks were cored from a cylindrical specimen cast from the same concrete used for the reinforced slabs and cured for 1 year at 95% R.H., and were saturated with distilled water before testing. Initially, a 0.5 M NaCl solution was poured into one side of the cell (volume = 650 cm^3) and distilled water was poured into the other side (volume = 400 cm^3). Current densities of 5 mA/m^2 and 20 mA/m^2 were imposed and pure diffusion tests were performed. The exposed surface was 64 cm^2. Temperature was maintained at 30°C. Chloride concentration in each half of the cell was regularly measured using an ion-sensitive electrode. Potential across the specimen was monitored using two SCE reference electrodes and two Lugging probes.

In order to evaluate the pitting potential of the steel as a function of the chloride ion concentration, potentiodynamic polarisation tests were performed in saturated calcium hydroxide solution on 2 cm^2 specimens finished with 6 μm diamond paste.

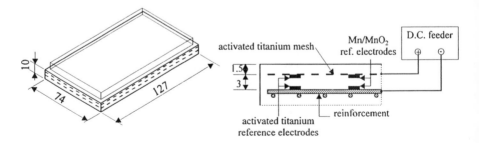

Figure 1 *Schematic illustration of the slabs and arrangement for galvanostatically controlled polarisation tests and potential monitoring (measurements in cm)*

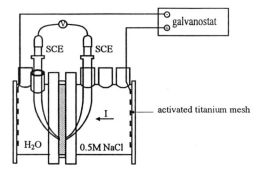

Figure 2 *Modified diffusion cells*

The specimens were passivated for 24 hours in chloride free solution and then were immersed in the test solution for 24 hours. Tests were also performed on specimens covered by a thin layer (1.5 mm) of mortar of portland cement (cement/water/sand = 1/0.5/2) and left for one month in the test solution after 24 hours of wet curing. The polarisation started 200 mV below the free corrosion potential with an anodic scan rate of 20 mV/min.

3 RESULTS AND DISCUSSION

3.1 Cathodic Polarisation on Reinforced Concrete Slabs

Before applying the cathodic current, the reinforcement was in the passive condition and had a free corrosion potential of around -250 mV vs Mn/MnO_2 (-100 mV vs external SCE). The potential of rebars in the control slab shifted to about -350 mV vs Mn/MnO_2 two months after the ponding cycles began and then did not show any further variation (Figure 3).

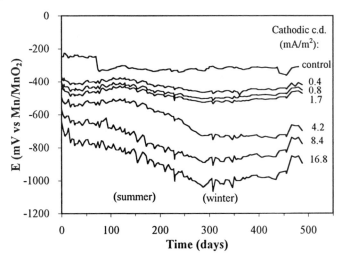

Figure 3 *Rebar potential (on potential) as a function of the cathodic current density*

Due to the application of the cathodic current, the potential of rebars in protected slabs quickly dropped to lower values; and the higher the current density the lower the potential. However, even a very low cathodic current density of 0.4 mA/m² was sufficient to shift the potential to about -400 mV vs Mn/MnO₂ after a few days, resulting in a cathodic polarisation of roughly 150 mV. A potential value of about -700 mV vs Mn/MnO₂ was quickly reached in the slab protected with the highest cathodic current density (16.8 mA/m²).

The cathodic polarisation varied with time, reflecting temperature variations; lower values of polarisation were reached during summer periods and higher ones were reached during the wintertime. This trend, already observed on new bridge decks subjected to cathodic prevention,[3] is more evident with lower current densities. However, with higher current densities (5-20 mA/m²) the average polarisation also increased with time during the first year. In wintertime, minimum potential values of around -1000 mV vs Mn/MnO₂ (-850 vs SCE) were reached with a cathodic current density of 16.8 mA/m².

The cathodic polarisation curve of rebars and its variation with time is shown in Figure 4, where the instant off potential of steel is plotted against the applied cathodic current density. Owing to its high polarisability, low potentials were reached by the passive steel in a very short time after switching on. The slope of the curve increased with time at high current densities (>2 mA/m², i.e. the maximum value for cathodic prevention), and the curve measured at 10000 hours exhibits an abrupt change in the slope.

3.2 Feeding Voltage

Figure 5 shows the effect of the current density on the feeding voltage and also shows the different contributions, namely the anodic and cathodic polarisation, the initial potential difference between the reinforcement and anode (ΔE_o), and the ohmic drop in the concrete cover. The feeding voltage also changed slightly with time, reflecting the changes in the anodic and cathodic polarisation and the ohmic drop in the concrete cover.[4]

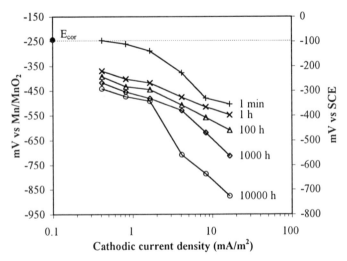

Figure 4 *Cathodic polarisation curves for the reinforcement as a function of time (potential measured with instant off method)*

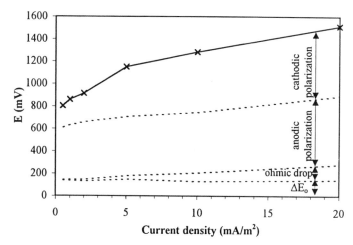

Figure 5 *Feeding voltage (continuous line) in the reinforced slabs, measured 15 months after the start up (summertime)*

Current densities in the range 0.5 to 2 mA/m², i.e. the typical values for cathodic prevention, produced feeding voltages lower than 1 V. Slabs protected with higher current densities exhibited higher feeding voltages and, for instance, a maximum value of 1700 mV was measured in wintertime in the slab protected with 20 mA/m². This increase in the feeding voltage was mainly due to the higher values both of cathodic polarisation (which became comparable to the anodic polarisation) and the ohmic drop. The polarisation of the mixed metal oxide titanium mesh anode was higher than the polarisation of rebars in the slabs protected with the lowest current densities, and lower than the cathodic polarisation in the slabs protected with the highest current densities.

3.3 Cathodic Prevention Monitoring

The 100 mV decay criterion developed for monitoring cathodic protection of corroding structures is usually also used for cathodic prevention. In fact, it guarantees that the rebar polarisation is higher than 100 mV, and this is considered sufficient to increase the critical chloride content for initiation of corrosion by about one order of magnitude.[3]

Figure 6 shows the four hour decay of rebar potential, and indicates that slabs protected with cathodic current densities between 0.4 and 4.2 mA/m² exhibited a minimum value of depolarisation in summertime but, even with the lowest current density, the depolarisation was always higher than 100 mV (only once was it slightly lower). Depolarisation of slabs polarised with current densities of 8.4 and 16.8 mA/m² increased with time and reached values of about 500 mV. The four hour depolarisation reflects the actual polarisation of the rebars, showing a similar dependence on current density, time and temperature. Nevertheless, the potential decay during the four hours is lower than the real polarisation of the reinforcements with respect to the initial free corrosion potential.

3.4 Safety Regarding Hydrogen Embrittlement of High Strength Steel

In presence of high strength steel, i.e. in prestressed structures, the application of cathodic polarisation to the reinforcement can promote hydrogen embrittlement.[5]

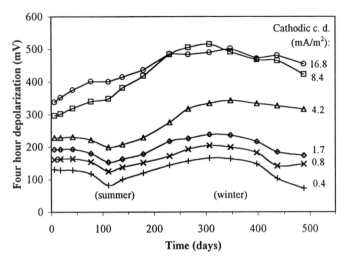

Figure 6 *Four-hour depolarisation of rebars as a function of time*

Results from this work show that cathodic prevention, which applies current densities lower than 2 mA/m², takes the rebar potential of aerial structures to values higher (more positive) than -500 mV vs Mn/MnO₂ even in wet conditions, and thus is much higher than the critical value of -900 mV vs SCE (roughly equal to -1050 mV vs Mn/MnO₂) required for hydrogen evolution. The low and stable feeding voltages confirm the intrinsic safety of cathodic prevention with regard to hydrogen embrittlement. In fact, even assuming that there is no ohmic drop and the anode is not polarised (and thus it operates at -100 mV vs Mn/MnO₂, as observed on the control slab), in order to reach the conditions under which hydrogen evolution becomes possible a minimum potential difference of 950 mV between anode and cathode is required. Since the previous conditions are not verified (in particular anode potential was above 400 mV vs Mn/MnO₂ in all the tested slabs) higher values are necessary. Furthermore, since the ohmic drop is negligible compared to cathodic and anodic polarisation for current densities below 2 mA/m² (Figure 5), a favourable current distribution can be obtained and overprotection can be easily avoided. Consequently, the results confirm the possibility of safely utilising cathodic prevention in prestressed structures, where cathodic protection of already corroding structures cannot be applied.[6]

3.5 Chloride Penetration

Significant quantities of chloride ions migrated into the concrete cover of the slabs (Figure 7); nevertheless, at the reinforcement depth (nominally 45 mm) the chloride contents were lower than 0.05% by weight concrete (about 0.35% by weight cement) and this explains why corrosion was not initiated even in the control slab.

The chloride profiles do not show any appreciable effect of the current flowing between the anode and the rebars on the penetration of chloride ions. As a first approximation an apparent diffusion coefficient, D, has been calculated from the experimental data by using the equation:

$$C_x = C_s(1 - \text{erf} \frac{x}{2\sqrt{D \cdot t}})$$

(1)

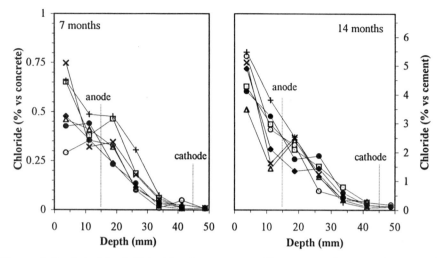

Figure 7 *Total (acid soluble) chloride content profiles measured at 7 and 14 months (Cathodic current density, mA/m²: △ 0.5, ◆ 1, × 2, □ 5, ○ 10, + 20, ● control)*

which represents the solution of Fick's second law under the hypotheses that both the diffusion coefficient and surface chloride level are constant during the period of exposure (Table 1).

The apparent diffusion coefficient, D, has the same order of magnitude in the different tested slabs. The average value decreases with time from 14.9×10^{-8} cm²/s after 7 months to 6.1×10^{-8} cm²/s after 14 months. The average value of surface chloride content, C_s, increased from 0.64% to 0.77% by weight cement.

However, only in the vicinity of the cathode can the electrophoretic migration oppose chloride ion penetration (conversely, in the anode region enrichment in chloride ions is expected). Indeed, experimental data show that the chloride content in the neighbourhood of the steel tends to be lower if the current density is increased, whereas it increases slightly in the upper layer (Figure 8), although a considerable scatter is observed in the results.

The variation in chloride concentration within the diluted chambers of the modified diffusion cells is represented in Figure 9.

Table 1 *Apparent diffusion coefficient (D, cm²/s×10⁸) and surface chloride content (Cₛ, % by weight concrete) calculated from experimental profiles*

C. D.	7 months		14 months	
(mA/m²)	D	C_s	D	C_s
0	13.8	0.56	7.8	0.71
0.5	14.7	0.58	8.4	0.53
1	12.0	0.58	4.2	0.80
2	10.2	0.82	5.0	0.79
5	15.3	0.74	7.8	0.71
10	20.3	0.42	4.3	0.91
20	18.2	0.78	5.2	0.97

Table 2 *Chloride diffusion coefficient (D, cm²/s×10⁸) and average potential difference across the concrete specimen (ΔE) during the diffusion tests*

C.D.	D	D	ΔE
(mA/m²)	0-30 days	30-60 days	(mV)
0	1.5	3.7	30
0	2.5	9.9	38
5	2.1	15.3	34
5	2.5	8.5	36
20	1.3	4.5	62
20	3.7	9.7	42

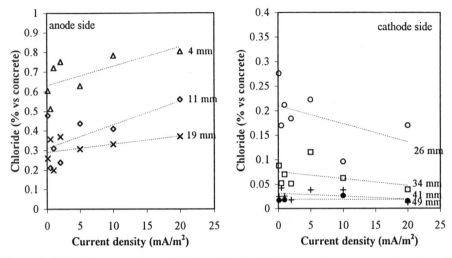

Figure 8 *Chloride penetration at different depths in the concrete cover as a function of the applied current density: (a) upper layer (anode side), (b) inner layer (cathode side)*

After about a week, chloride concentration began to increase linearly but a steeper slope was seen after about one month. Monitoring of the chloride concentration over two months of testing shows a significant scatter of results and no explicit effect of applied current density. Table 2 reports the diffusion coefficients evaluated according to Fick's first law from these data: two different coefficients have been calculated, before and after the slope of the curve changed.

According to the Nernst-Planck equation,[7] the electrophoretic contribution to chloride penetration is proportional to the chloride concentration.

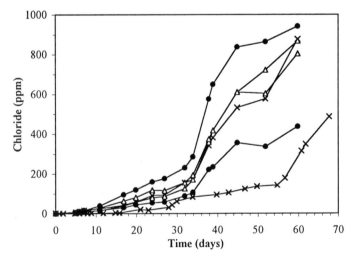

Figure 9 *Change in chloride ion concentration on the diluted side of the modified diffusion cells (current density, mA/m²: × control, Δ 5, ● 20)*

Initially, since chlorides are practically absent in the concrete near the rebars (or in the diluted chamber of the cell), the migration is mainly the result of diffusion processes irrespective of the applied current. Afterwards, as the chloride content increases, the influence of the electric field becomes significant. This effect has already been observed by applying current densities much higher than those utilised for cathodic protection and cathodic prevention.[8] The results of this work suggest that current densities in the range used in cathodic prevention (namely, lower than 2 mA/m^2) do not affect chloride penetration in the concrete cover when the chloride content at the steel surface is below 0.4% by weight cement. However, further investigation is required in order to assess whether this effect might become significant as the chloride content increases. In fact, it should be noted that, owing to the cathodic polarisation, steel reinforcement in structures where cathodic prevention is applied can withstand high chloride concentrations, as discussed below.

3.6 Critical Chloride Content

Tests on the reinforced slabs were also aimed at studying the critical chloride content as a function of the cathodic polarisation; however these results are likely to require several years testing. Hence, a first approximation has been made using potentiodynamic tests on steel samples in saturated Ca(OH)$_2$ solution (Figure 10).

High values of pitting potential were measured for high chloride concentrations in samples of steel exposed to saturated Ca(OH)$_2$ solution. However, tests on samples covered with a thin layer of mortar, which are more representative of steel behaviour in concrete, show lower values above 0.2% chloride. This suggests that the polarisation observed on the protected slabs, by taking the steel potential below -300 mV SCE even with 0.4 mA/m^2 cathodic current density greatly increases the critical chloride content with respect to the free corrosion conditions (-100/-150 mV SCE). These results were obtained in anodic polarisation tests and higher threshold values are likely when a cathodic polarisation is imposed (i.e. when alkalinity is produced at the steel surface).

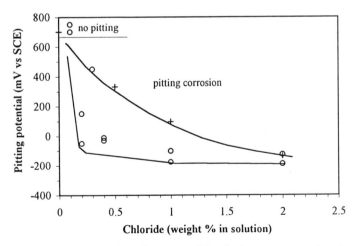

Figure 10 *Pitting potential of steel versus chloride ion concentration in saturated Ca(OH)$_2$ (○ samples with mortar layer, + samples without mortar layer)*

4 CONCLUSIONS

The application of cathodic current densities in the range 0.4-1.7 mA/m^2 to reinforcement in chloride free concrete depressed the steel potential by about 150-200 mV with respect to the free corrosion value. This polarisation was attained quickly and only changed slightly during the first 15 months of testing, although a progressive penetration of chloride into the concrete cover was observed. However, the chloride content at the steel surface was below 0.4% by weight cement and corrosion was not initiated even in the control slab.

Both the potential of the rebars and the feeding voltages showed that cathodic prevention can be regarded as a safe technique as far as hydrogen embrittlement of high strength steel is concerned. In fact, the steel potential was higher than -300 mV vs SCE for current densities up to 2 mA/m^2.

Investigations into the effects of current flowing through the concrete cover on the penetration of chloride ions did not show any clear evidence after 14 months of testing. A slight tendency for chloride content to decrease in the cathodic region was observed. However, especially for the lower current densities, further results are necessary to clarify the role of electric transport of chloride as the concentration of these ions at the steel surface increases. In fact, preliminary results indicate that the application of cathodic prevention allows the reinforcement to remain passive even in concrete with a significant chloride ion concentration, where the electrophoretic transport effect might be significant.

Acknowledgements

The authors are grateful to Società Autostrade, Nuova Polmet SpA and DeNora Permelec SpA for their assistance and support to this work.

References

1. P. Pedeferri, 'Cathodic Protection of New Concrete Construction', Proc. Int. Conf. on 'Structural Improvement through Corrosion Protection of Reinforced Concrete', Institute of Corrosion, London, 1992.
2. L. Bertolini, F. Bolzoni, A. Cigada, T. Pastore, P. Pedeferri, *Corr. Sci.*, 1993, **35**, 1633.
3. P. Pedeferri, 'Cathodic Protection and Cathodic Prevention of Aerial Concrete Structures', in 'Corrosion in Natural and Industrial Environments: Problems and Solutions', NACE International, Houston, 1995, 45.
4. L. Bertolini, F. Bolzoni, T. Pastore and P. Pedeferri, 'Investigation on Reinforced Slabs Exposed to Chloride Solution Ponding under Cathodic Polarization', in 'Corrosion in Natural and Industrial Environments: Problems and Solutions', NACE International, Houston, 1995, 291.
5. W. H. Hartt, 'A Critical Evaluation of Cathodic Protection for Prestressing Steel in Concrete', in 'Corrosion of Reinforcement in Concrete', (Eds. C. L. Page, K. W. J. Treadaway, P. Bamforth), SCI, London, 1990.
6. P. Pedeferri, 'Cathodic Protection of Prestressed and Post-Tensioned Structures', 'Elektrochemische Shutzverfahren fur Stahlbetonbauwerke', SIA, ETH, Zurich, 1990.
7. J. O. M. Bockris and A. K. N. Reddy, 'Modern Electrochemistry', Plenum Press, New York, 1974.
8. G. K. Glass, J. Z. Zhang and N. R. Buenfeld, 'A Cathodic Protection Criterion for Steel in Concrete Based on the Chloride Barrier Properties of an Electric Field', Proc. of 'UK Corrosion & Eurocorr 94', The Chameleon Press Ltd, London, 1994, **3**, 294.

PHYSICO-CHEMICAL BEHAVIOUR OF CONCRETE COVER AND REINFORCING STEEL UNDER CATHODIC PROTECTION

T. Chaussadent[a], G. Arliguie[b] and G. Escadeillas[b]

a) Laboratoire Central des Ponts et Chaussées
58, bd Lefebvre F-75732 Paris cedex 15 France.
b) Laboratoire Matériaux et Durabilité des Constructions
INSA-UPS génie civil Complexe scientifique de Rangueil
F-31077 Toulouse cedex France.

1 INTRODUCTION

Steel in reinforced concrete structures is normally protected against corrosion by the alkalinity and the physical barrier of the surrounding concrete. Nevertheless, carbonation or chloride contamination can cause depassivation and hence corrosion of steel. A valuable technique which will stop further deterioration is cathodic protection.

The objective of this experimental investigation was to study the influence of cathodic protection on the soundness of reinforcing steel and concrete cover. The test pieces were six concrete slabs manufactured with three cements and with and without silica fumes. Cathodic protection with impressed current system had been applied to the reinforcements for three years. To accelerate ageing, the slabs were subjected to freezing and thawing cycles and sprayed with sodium chloride solution. A first part of the study, concerning electrochemical measurements, was presented in 1994.[1]

After cathodic protection testing, the slabs underwent physico-chemical analysis in order to show the final condition of steel and concrete.

2 EXPERIMENTAL PROCEDURES

2.1 Test Pieces

The characteristics of the cements used are given in Table 1. BPC and OPC correspond respectively to Blended Portland Cement (blast furnace slag 11%, pulverized fly ash 8%, limestone 8%) and to Ordinary Portland Cements low C_3A (1) and high performance (2). In this investigation locally available river sand and gravel (cherty limestone from the Seine river) were used. The details of the concrete mixes are presented in Table 2.

Table 1 *Chemical compositions of cements (Wt%)*

Cement	SiO_2	Al_2O_3	Fe_2O_3	CaO	MgO	TiO_2	Na_2O	K_2O	SO_3	Insol.	Ig-loss
BPC	18.58	5.47	2.32	55.25	5.96	0.29	0.20	1.35	3.20	3.30	3.64
OPC 1	21.67	2.89	3.29	67.18	0.21	0.17	0.25	0.38	1.76	0.63	1.24
OPC 2	20.39	5.51	2.20	62.83	0.99	0.31	0.15	0.51	3.36	1.50	1.91

Table 2 *Mixture compositions (kg.m^{-3}) and compressive strengths (MPa) of concretes*

	D1	D2	D3	D4	D5	D6
Cement*	300	350	350	400	400	400
Sand 0-4 mm	753	720	720	700	700	700
Gravel 4-12 mm	1150	1130	1130	1100	1100	1100
Silica fumes	30	-	35	-	40	-
W/C	0.32	0.48	0.32	0.48	0.32	0.48
Superplasticizer	11	-	12	-	14	-
Air entraining agent	0.33	-	0.385	-	0.44	-
Strength at 28 days	51	40.5	56.3	52	62.2	62

* *BPC for slabs D1 and D2 - OPC 1 for slabs D3 and D4 - OPC 2 for slabs D5 and D6*

The reinforced concrete test pieces were six prismatic slabs (Figure 1). The reinforcements were perpendicular to the trapezoidal cross-section, at a distance of 100 mm from each other. In every slab, a transverse steel bar was welded to the others. Slabs were one year old at the beginning of the testing which lasted three years.

Every slab had three distinct areas: a zone (a) covered with the anode and shotcrete, a zone (b) with only shotcrete, and a control zone (c) where the concrete surface remained uncovered. It should be noted that these last two zones can be influenced by cathodic protection. The anodes consist of titanium mesh fixed on the concrete surface and covered with shotcrete (cement OPC 1), approximately 10 mm thick.

Cathodic protection of rebars was obtained with an impressed power supply. The nominal value of the applied potential was equal to -850 mV$_{CSE}$ (Cu/CuSO$_4$ reference electrode). The average current density was 50 mA per m^2 of reinforcement.

Once a month, the test pieces were subjected to freezing and thawing cycles, induced by pouring liquid nitrogen for 2 hours onto the surface of the concrete. The measurements showed that the temperature was approximately -6 °C at a depth of 80 mm and -28 °C at a depth of 20 mm. The test pieces were then warmed up in a natural atmosphere. When the temperature was higher than 0 °C, saturated sodium chloride solution was poured on the slabs, for 5 days per week.

2.2 Study Techniques

The cores were taken from zones (a) and (c) in locations where concrete cover was high. They are labelled respectively as P (protected) and C (control).

Carbonation depth was determined by spraying a solution of phenolphthalein in ethanol, on a fresh concrete surface, previously moistened with pure water. The indicator remains transparent when the pH is lower than about 9. When concrete is not carbonated, the indicator becomes violet pink.

For chemical analysis, concrete samples were sawn (by dry sawing) to obtain slices of concrete cover from different depths. Then they were ground to a grain size lower than 0.315 mm. Chlorides were extracted by using boiling water for 15 minutes. Chloride content was determined by potentiometry using a titrated silver nitrate solution.

The steel-concrete interfaces were observed by scanning electron microscopy. For each sample, reinforcements were taken out of the core. Concrete in contact with each rebar was sampled dry, to avoid any chemical deterioration. Ultimate elementary semi-quantitative analysis was performed by X-ray spectrometry, to determine the presence of chloride and alkalis.

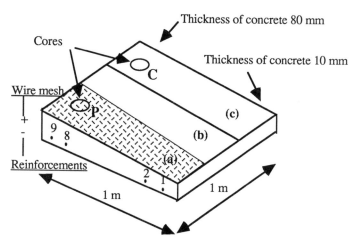

Figure 1 *Schematic representation of slab and position of the cores*

3 RESULTS AND INTERPRETATION

3.1 Macroscopic Appearance of Samples

After three years of exposure, the surfaces of the slabs were visually examined. It appeared that, due to the ageing, these surfaces were eroded and micro-cracked.

In addition, measurements of air permeability were made on the surface of the slabs. The results showed that the pores of the concrete were plugged, probably because of the formation of salt crystals. However, these textural modifications appeared a short depth from the surface as was confirmed by mercury intrusion porosimetry tests. For example, Figure 2 showed no significant difference between ageing shotcrete from slab D5 and shotcrete conserved in sound conditions.

Concrete of the various cores appeared to be sound, except for the two samples taken from slab D1. In this slab, the concrete was very porous, mainly near the reinforcements. In these places, the concrete boundary was difficult to identify because of the presence of aggregate segregation.

3.2 Carbonation Depths

Carbonation depths were measured at various points of every slab. In shotcrete, the depth was 2 mm for every sample.

Uncovered concrete (in contact with the atmosphere) was more carbonated. However, the carbonated concrete cover did not exceed 5 mm, so the rebars, (under a cover at least 10 mm thick), were always far from carbonation.

3.3 Chloride Profiles

Chloride contents in the cores are reported in Table 3. Figure 3 shows chloride profiles in the concrete slabs.

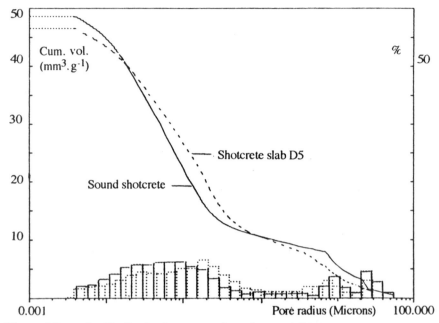

Figure 2 *Pore size distribution by mercury intrusion porosimetry*

It appears that:
- under the anodes, chloride profiles were very similar for the various test pieces. The chloride content was found to be negligible in the concrete just under the shotcrete. This indicates that the anode, positively polarized, attracts chloride ions and fixes them around itself.
- far from the anodes, the chloride contents (at 40 mm depth), were higher respectively in slabs 2, 4, and 6 than in the other slabs. This corresponds to increasing compressive strength of the three concretes for which the water-cement ratio was equal to 0.48. For the other three slabs (W/C = 0.32), the chloride contents were significantly lower.

Table 3 *Chloride contents as a function of concrete depths (% by weight of concrete)*

Position	D1	D2	D3	D4	D5	D6
Protected zone						
Shotcrete	0.61	0.49	0.40	0.50	0.53	0.52
Concrete slab 0-10 mm	0.19	0.14	0.11	0.17	0.04	0.25
35-45 mm	0.01	0.002	<0.001	0.001	0.008	0.02
70-80 mm	0.003	0.002	0.002	0.01	0.006	0.01
Control zone						
Concrete slab 0-10 mm	0.18	0.31	0.35	0.20	0.46	0.23
35-45 mm	<0.001	0.19	<0.001	0.04	<0.001	0.009
70-80 mm	<0.001	0.01	<0.001	0.03	<0.001	<0.001

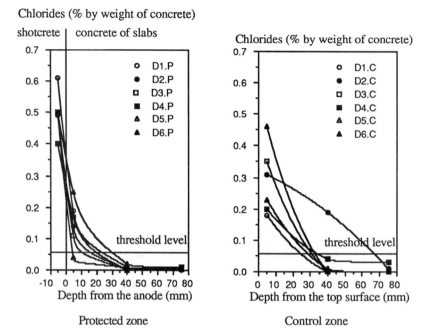

Chlorides (% by weight of concrete)

Figure 3 *Chloride penetration profiles in the slabs*

It has been suggested that the molar [Cl⁻]/[OH⁻] ratio must remain less than 0.6 to prevent steel corrosion.[2-3] The chloride content is determined either as a function of the volume of the pore water or as a function of the cement weight. In this case, this threshold, generally taken to be close to 0.4%,[4] corresponds to a content of 0.06% by weight if referred to the concrete. This critical value shows that, in the test pieces, reinforcements would be expected to remain uncorroded.

3.4 Microscopic Examinations of the Samples

Results of scanning electron microscopy and X-ray spectrometry on the steel-concrete interfaces are detailed below.

3.4.1 Results on Reinforcements. No steel extracted from the cores showed any visible corrosion, except for steels from slab D1.

For slab D1, the corrosion of the reinforcements could be due to the high open porosity of the concrete. In this case, cathodic protection is no longer a decisive means for ensuring the soundness of the reinforcements.

Results of X-ray spectrometry show, for slab D1, a significant amount of chlorine was present in compounds adhering to steel or in concrete at the interface (Figure 4). It should be noted that X-ray spectrometry detects both chemically bonded chlorides as well as "free" chlorides. Free chlorides can play an important role in depassivating the steel (pitting), even in a very alkaline medium like concrete.[5]

Only a few rust spots were observed on steel from slabs D2 and D3. They were located at contacts with the attachments of transverse rebars. This suggests that such rusting can result from local corrosion cells.

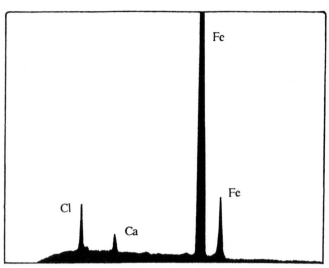

Figure 4 *X-ray spectrometry analysis on the reinforcement (slab D1.C)*

3.4.2 Results on Concrete Covers. The microscopic analysis of the concrete showed that:

- In the samples from slabs D5 and D6, portlandite was predominant in the interface zone.

- In slabs D2, D3, and D4, some alkali-silica reaction gels have been formed (Figure 5). The X-ray spectrometry results show that, as a rule, these gels contain other elements, in particular chlorine which was observed in the interface zone. More detailed observations have shown that such gels are usually located close to the ribs of the rebars.

Table 4 gives the alkali contents released by the cements (per m^3 of concrete). The amount of alkalis contributed by the aggregates or by the admixtures was not determined. However, the totals should not exceed 0.3% and should be similar for all the samples. The alkali contents of the cements were compared with the recommended limits corresponding to the initiation of alkali-silica reaction. It is usually agreed that this threshold is about 3 kg Na_2O equivalent per m^3 of concrete, with a maximum level of 3.3 kg per m^3 of concrete.[6]

Only in slab D2, was the alkali content likely to cause an alkali-silica reaction in the presence of potentially reactive aggregates. On the other hand, the contents in slabs D3 and D4, in which gels were found, were obviously below the allowed limit.

The X-ray spectrometry diagrams show the presence of alkalis, especially potassium, in large amounts at the steel-concrete interface. This suggests that the alkali ions are attracted by the reinforcements which are negatively polarized. This concentration of alkalis near the reinforcements, and the formation of hydroxyl ions, make it possible to create local conditions which favour alkali-silica reactions.

Table 4 *"Calculated" alkali contents released by cements in concretes*

	D1	D2	D3	D4	D5	D6
Proportion of cement (kg.m^{-3})	300	350	350	400	400	400
Na$_2$O eq. content (kg.m^{-3})	3.4	3.9	1.8	2.1	2.0	2.0

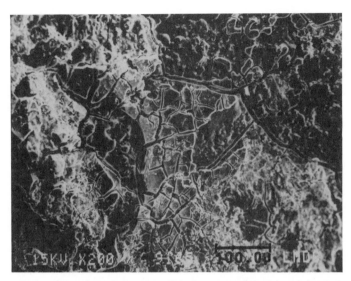

Figure 5 *Alkali-silica gels at the concrete-reinforcement interface (slab D3.C)*

The mineralogical composition of aggregates was studied by petrography, and, in some cases by X-ray diffraction. Some aggregate spectra exhibited peaks corresponding to opal $SiO_2.nH_2O$ (Figure 6 a) which is soluble in high pH media, and is the most reactive form of silica. The presence of opal was confirmed by heating the sample to 1100 °C for 12 hours,[7] which converts opal into cristobalite (Figure 6 b). As opal was a minor component of the aggregates, they must not be considered to be unduly reactive. So, alkali-silica reactions could develop only very locally, close to the ribs and ties of the reinforcements.

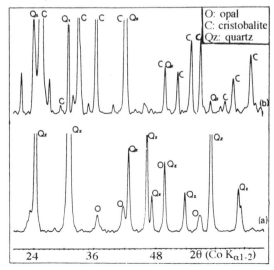

Figure 6 *X-ray diffraction spectra of aggregates   (a) in natural state*
(b) after heating to 1100 °C

Moreover, the occurrence of alkali-silica reaction gels in a relatively short time (less than three years) confirms the influence of cathodic protection in inducing[8] or accelerating[9] the formation of these products under conditions of high local current density.

Samples taken from slabs made with OPC 2 cement exhibit no alkali-silica reaction gels, even though the alkali content was similar for all the cements. Moreover, it should be noticed that concretes of slabs D5 and D6, made with OPC 2 cement, were more compact. So migration of ionic species was more difficult in these slabs. However, it is difficult to draw conclusions about the role of W/C ratio and that of silica fume addition.

4 CONCLUSIONS

In this investigation cathodic protection was applied to rebars in test pieces under very severe conditions. The results obtained have shown that the reinforcements were in a satisfactory condition. Rusting of rebar was observed only in a very porous concrete, where the access of chloride ions was facilitated.

Examinations of the steel-concrete interface zones showed the formation of alkali-silica reaction gels near the ribs of the reinforcements. This reaction occured even when the initial alkali content was less than the recommended critical value. This means that alkali ions migrate under polarization and they accumulate near the polarized rebars.

However, no deleterious effect on concrete cover was observed after a three year cathodic protection of embedded rebars.

Acknowledgments

The authors would like to thank G. Grimaldi and J. C. Languehard (LREP Melun, France) for the setting up of the site test program and maintenance, and A. Raharinaivo (LCPC Paris, France), chief of the project on cathodic protection of steel in concrete, for his helpful discussions and suggestions.

References

1. T. Chaussadent, A. Raharinaivo, G. Grimaldi, G. Arliguie and G. Escadeillas, Actes des Journées JSI94 , Edited by LCPC Paris, **1**, 257.
2. D. A. Hausmann, *Materials Protection*, 1967, **6**, 19.
3. A. Raharinaivo and J.-M. Genin, *Materiales de construcción*, 1986, **36**(204), 5.
4. K. C. Liam, S. K. Roy and D. O. Northwood, *Magazine of Concrete Research*, 1992, **44**(160), 205.
5. T. Chaussadent and R. Dron, *Materiales de construcción*, 1992, **42**(226), 49.
6. Recommendations for the Prevention of Alkali-Reaction Disorders, LCPC Project Report, France, 1994 (in french).
7. G. Bareille, M. Labracherie, N. Maillet and C. Latouche,*Clay Minerals*, 1990, **25**, 363.
8. G. Sergi, C. L. Page and D. M. Thompson, *Materials and Structures,* 1991,**24**, 359.
9. M. G. Ali and Rasheeduzzafar, *ACI Materials Journal,* 1993, **90**(3), 247.

INTERMITTENT CATHODIC PROTECTION OF CONCRETE STRUCTURES

A. M. Hassanein, G. K. Glass and N. R. Buenfeld

Department of Civil Engineering
Imperial College
London SW7 2BU
UK

1 INTRODUCTION

Reinforced concrete structures are particularly vulnerable to chloride-induced corrosion in wetting and drying environments, such as in, or just above, the tidal zone of marine structures. Such locations are also difficult to protect or repair. Cathodic protection (CP) is now well established as a technique for protecting submerged structures (using remote anodes) or above ground structures (using large area surface applied anodes). In the intermediate situation of reinforced concrete being periodically submerged, applying a surface anode is difficult while the electrolyte required for a remote anode system is only present intermittently.

There is strong evidence suggesting that the protection afforded by a CP system will last for a significant period after the current is interrupted.[1] Thus the protective effects of the CP current can be divided into immediate and long term effects. The immediate effect is to shift the steel potential in a negative direction which reduces the rate of dissolution of positive ions. The longer term persistent protective effects of the current are associated with the repassivation of the steel due to the removal of chloride ions resulting from the flow of ionic current,[2,3] the increase in pH resulting from the reduction reactions[4,5] and the decrease in potential resulting from the consumption of oxygen.[6]

The persistent effects are partly exploited in the relatively new techniques of chloride extraction and realkalisation.[7,8] They may also be harnessed to develop an intermittent cathodic protection system. This may be useful when one of the components required by the CP system is present only intermittently. For example sea water which forms the conductive path between a remote anode system and the concrete surface is not always present in a tidal zone or dry dock situation. In addition a system relying on solar power may operate only during daylight hours.

Most CP systems are designed on the basis of achieving a negative potential shift to some absolute value or by a specified amount. For reinforced concrete this will reduce the potential gradient across the passive film which will minimise the risk of film breakdown and promote its repair at damaged areas. Thus protection criteria are based on the measurement of the steel potential and potential change.

The use of the persistent effects as a basis for CP design would represent a significant change to current philosophy. The charge passed through the concrete to the steel will strongly affect the removal of chloride ions, the production of hydroxyl ions and the consumption of oxygen by the cathodic reduction reaction. Thus consideration may be given to a criteria based on repeatedly achieving a minimum current density for a fixed period of time.[1,9]

This preliminary work is concerned with the application of intermittent cathodic protection to reinforced concrete. The main aim is to establish whether the removal of chloride ions could provide a basis for the intermittent protection of steel in concrete. The effect of oxygen removal in also investigated.

2 EXPERIMENTAL METHOD

2.1 Specimens

The specimens consisted of 100 mm diameter, 40 mm thick concrete cylinders cast inside a plastic tube which extended beyond the specimen on each side to form the test cell (Figure 1). The mix consisted of 400 kg/m³ ordinary Portland cement, 620 kg/m³ fine aggregate (grade M sand), 1160 kg/m³ 10 mm aggregate and a free water:cement ratio (w/c) of 0.5. In this preliminary work, 2% chloride by weight of cement was add as sodium chloride to the mixing water. The specimen was cured for 3 days at 100% RH before being exposed to a 0.5 M NaCl solution. The 0.5 M NaCl solution maintained electrolytic contact between the anode and the concrete, while cotton wool saturated with 0.3 M NaOH solution bridged the 3 mm gap between the concrete and the cathode. The experiment was conducted in a temperature controlled room (22 °C). The cell was pressure tested to ensure that the specimen maintained intimate contact with the side of the tube.

2.2 Testing

A constant current density was applied for 30 min in every 12 hr between the anode (activated titanium strip) and the cathode (stainless steel plate) of the test cell using a galvanostat and timer for a period of 6 weeks. More than 99.5% of the tidal zone would be under water for 30 minutes if tidal behaviour was sinusoidal. The work was undertaken using

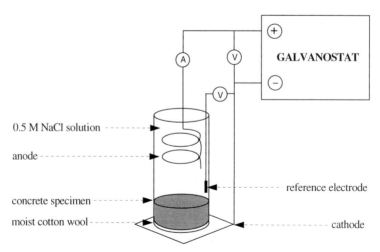

Figure 1 *Experimental arrangement*

a current density of 100 mA/m^2 (cathode surface area). This is a typical design current density used to protect uncoated steel in sea water and therefore installation of such a system should be practical. A data logger recorded the current, the drive voltage and potential of the cathode relative to a saturated calomel reference electrode (SCE) attached to a Luggin probe terminating in the NaCl solution (Figure 1). Periodically the IR voltage drop across the concrete was determined on interrupting the current while recording the potential of the cathode through the concrete at 25 readings per second using a multimeter.

Chloride profiles were determined on a control specimen and the specimen subjected to the intermittent current. Precision grinding was used to produce powdered samples from the central portion of the specimens after they had been split. The depth increment was 2 mm and 20 powdered samples representing the full depth of each specimen were obtained. The total chloride content was determined on each sample by acid soluble extraction using concentrated nitric acid followed by potentiometric titration against silver nitrate (AgNO$_3$).

3 EXPERIMENTAL RESULTS

3.1 Electrochemical Changes

The rest potential of the cathode prior to the application of current was -216 mV (SCE). The potential measured 3 hours before each application of current (i.e. 9 hours after current was last applied) is plotted in Figure 2. After approximately 20 days there was a sharp negative shift in the potential to a value close to -1000 mV (SCE). In the 9 hours following the current pulse the potential only rose by approximately 60 mV.

Figure 3 gives the typical cathode potential changes monitored over one current cycle after 20 days. The rapid potential change of approximate value of 300 mV following the application or interruption of the current results from the current induced voltage drop across the concrete specimen. The slower changes following this result from cathodic polarisation or depolarisation.

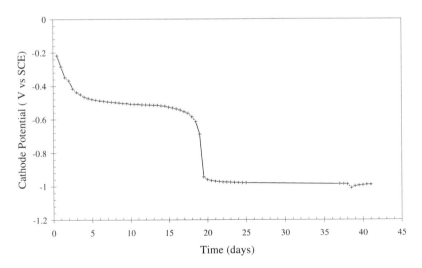

Figure 2 *The cathode potential 9 hrs after each current pulse as a function of time*

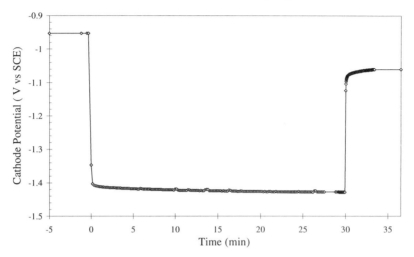

Figure 3 *Cathode potential over one current cycle monitored after 20 days*

The current related voltage drop across the concrete specimen (the IR drop) is plotted as a function of time in Figure 4. This increased from 165 to 300 mV over the 6 week exposure period. A similar trend was also noted in the drive voltage measured after the current had been applied for 15 minutes. This increased from 1880 to 2120 mV over the same period.

3.2 Chloride Profiles

The chloride profiles determined on the control and exposed specimens are given in Figure 5. The chloride profile of the control rises slightly at the cast surfaces due to increased

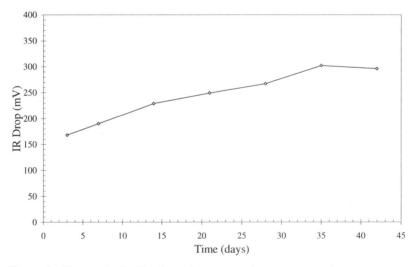

Figure 4 *Changes in the IR voltage drop across the concrete specimen*

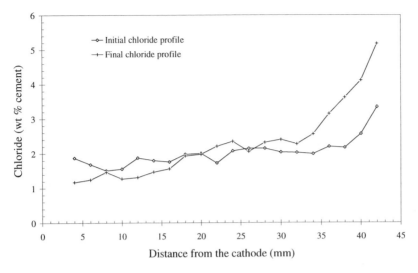

Figure 5 *Chloride profiles determined on control and exposed specimens*

cement content at these locations. An increase in the chloride content occurs at the concrete surface exposed to the salt solution probably as the result of the concentration gradient occurring across this interface. Furthermore in the electrolyte at the cathode the chloride content of the 0.018 litres of solution increased from 0 to 0.22 mole/l as the result of back diffusion of chloride from the concrete. The chloride to hydroxyl concentration ratio in this solution was 1.2 at the end of the test.

The determination of the effective transference number (electrical efficiency) of chloride ions was complicated by the back diffusion of chloride from the 0.5 M NaCl solution into the concrete and from the concrete into the electrolyte at the cathode. The transference number through a surface a given distance from the cathode may be calculated using the amount of the chloride removed by the current from the segment enclosed by that surface and the cathode. At a distance of approximately 20 mm from the cathode, a maximum transference number of 0.63, equivalent to the removal of 3% chloride from the enclosed segment, was determined (cf. Figure 5). Thus the intermittent current has the overall effect of removing chloride from the cathodic half of the specimen.

4 DISCUSSION

4.1 Chloride Removal

The corrosion problems of interest in the present work arise from chloride contamination, thus the basis for protection when applying intermittent CP could be to reduce the chloride level below a threshold value. In practice corrosion threshold values vary from 0.2% to more than 2% chloride ion by weight of cement.[10,11] While lower values are usually incorporated into codes of practice, the maintenance of a high hydroxyl concentration at the cathode and the water-logged nature of the concrete in a tidal zone may allow the use of a higher value when intermittent CP is applied.

The electrical efficiency of chloride removal (the chloride transference number - t_{Cl^-}) is given by the amount of DC current carried by the chloride ions (I_{Cl^-}) relative to the total current (I_{tot}):[12]

$$t_{Cl^-} = \frac{I_{Cl^-}}{I_{tot}} = \frac{C_{Cl^-} \upsilon_{Cl^-}}{\Sigma C_i \upsilon_i z_i} \qquad (1)$$

where the mobility, activity, and electronic charge for each ion in the solution denoted by the variable i are given by υ_i, C_i and z_i respectively with υ_{Cl^-} and C_{Cl^-} being the mobility and activity of chloride ions with an electronic charge of 1. This suggests that, in addition to the total charge passed and further chloride contamination via diffusion, chloride removal is affected by its mobility and activity relative to other ions in the solution. Thus chloride binding and its concentration relative to more mobile ions such as OH^- are also important.[13]

The chloride transference number through any cross-section of the specimen can also be calculated by comparing the equivalent charge of chloride ions removed from the segment of concrete and electrolyte at the cathode enclosed by this cross-section (Q_{Cl^-}) with the total charge that passed through it (Q_{tot}).[12]

$$t_{Cl^-} = \frac{Q_{Cl^-}}{Q_{tot}} = \frac{Cl_{ex} \times F}{I \times t} \qquad (2)$$

where Cl_{ex} is the amount of chloride removed in moles, F is the faraday constant in coulombs/mole, I is the current in Amps and t is the time in seconds. The chloride extracted through a cross-section approximately 20 mm from the surface of the steel resulted in an effective transference number of 0.63. A rise in the chloride content occurred at a greater distance from the steel partly as the result of back diffusion of the chloride from the 0.5 M NaCl electrolyte (Figure 5).

Chloride transference numbers above 0.5 are considered to be relatively high.[8,12] Such values were determined in the initial phase of studies on the removal of chloride from concrete. However it was noted that as the removal process proceeds, the extraction efficiency falls probably as the result of a falling free chloride content. The rate of removal is restricted by the release of bound chloride into the concrete pore solution and in one work, passing a charge greater than 10^7 C/m^2 proved to be very inefficient.[8] In another study chloride removal treatment was applied for a second time to local areas 9 months after the initial treatment to remove chloride which had been subsequently released into the pore solution.[13] The slow extraction process employed when applying intermittent CP (100 mA/m^2 applied for 30 minutes in every 12 hours for 6 weeks amounted to a charge of only 1.5×10^4 C/m^2 in the present work) may render the process more efficient by allowing the bound-free equilibrium to re-establish while maintaining a low free chloride content in the vicinity of the cathode.

The present work was complicated by the problems of back diffusion of chloride from the NaCl solution into the concrete and from the concrete into the electrolyte at the cathode. In practice a lower chloride gradient will exist at the boundaries which will minimise the effects of back diffusion and may render the chloride removal process more efficient. By contrast a reduction in chloride ion mobility resulting from a lower w/c and better curing may hinder the removal process. Further work is needed to assess these effects.

4.2 Oxygen Starvation

Dissolved oxygen is an important if not essential factor affecting the corrosion of steel in concrete. In its absence the corrosion rate is generally considered to be negligble and a lower corrosion rate and more uniform corrosion attack can be obtained by a sufficient reduction in the oxygen concentration at the steel surface.

As noted in the introduction a cathodic current removes oxygen via the cathodic reduction reaction:

$$O_2 + 2H_2O + 4e^- \rightarrow 4OH^- \tag{3}$$

and therefore reduces the amount available to support an anodic dissolution reaction on the surface of the protected steel. A reduction in the concentration may therefore provide a basis for the intermittent cathodic protection of steel in concrete in appropriate circumstances.

Evidence that oxygen starved conditions have been reached in the present work is given by the very negative potential of the cathode which fails to rise to values above -900 mV (SCE) after 20 days of intermittent current (Figure 2). This occurs despite the cathode being stainless steel which normally requires a very small level of cathodic activity to maintain a positive potential indicating stable passivity. This suggests that oxygen starved conditions were easy to achieve with the applied level of current.

It has been noted that, if the potential of steel in concrete is reduced below a value of -750 mV (SCE) by cathodic protection, further corrosion activity is unlikely.[14] This is particularly true if a high pH, which would minimise the rate of hydrogen evolution and its contribution to the cathodic reaction, is maintained at the steel surface. The achievement of cathode potentials significantly more negative than this value suggests that oxygen removal may provide a viable basis for the intermittent protection of steel in concrete.

In the present work the specimens were poorly cured prior to current application and then continuously exposed to the sodium chloride solution. Periodic drying would promote oxygen access while improved curing may restrict it, thus further work is necessary to assess the influence of these parameters on the achievement of oxygen starved conditions in practice.

4.3 Specimen Changes

The increase in concrete resistivity, resulting in a rise in the drive voltage and IR voltage drop across the specimen of the concrete, suggests that changes have occurred within the specimen. This may result from the removal of conductive ions such as chloride as well as a more refined pore structure due to continued curing. Similar observations have been reported by others.[15] In sea water the precipitation of calcareous deposits will also block the pore structure although this would not occur in the present work.[16] Further work is necessary to rank the importance of the various factors affecting resistivity. It is interesting to note that resistivity is sometimes considered to exert a controlling influence of the rate of corrosion of the reinforcement in concrete, and an increase in the resistivity therefore reduces the corrosion risk.[17]

5 CONCLUSIONS

This preliminary investigation into a basis for applying intermittent cathodic protection to reinforced concrete structures suggests that persistent protection may be achieved via either chloride removal or oxygen starvation. Despite the relatively small charge passed, a significant quantity of chloride was removed from the concrete close to the cathode. The transference number suggests that chloride removal in these circumstances was relatively efficient. Furthermore oxygen starved conditions were readily achieved in the present experimental set-up.

In the next phase of the work the effect of wet/dry cycles on chloride removal and oxygen starvation will be examined. In addition the effect of better curing, lower w/c, test duration and current magnitude will also be studied, as will be the implications of a change in the resistance of the concrete during current flow on the protection achieved.

Acknowledgements

This research was supported at Imperial College by the Engineering and Physical Sciences Research Council (grant no. GR/J 85554).

6 REFERENCES

1. G. K. Glass and J. R. Chadwick, *Corros. Sci.*, 1994, **36**, 2193.
2. R. E. West and W. G. Hime, *Mater. Perf.*, 1985, **24**(7), 29.
3. J. P. Broomfield, *Mater. Perf.*, 1992, **31**(9), 28.
4. T. K. Ross and H. J. Wood, *Corros. Sci.*, 1972, **12**, 383.
5. G. K. Glass, *Corros. Sci*, 1986, **26**, 441.
6. G. K. Glass, W. K. Green and J. R. Chadwick, 'Long term performance of cathodic protection systems on reinforced concrete structures', UK Corrosion 91 Conf. Proc., Vol. 3, Institute of Corrosion, Leighton Buzzard, 1991.
7. P. F. G. Banfill, 'Features of the mechanism of realkalisation and desalination treatments for reinforced concrete' in 'Corrosion and corrosion protection of steel in concrete', (edited by R. N. Swamy), Vol.2, 1489, Sheffield Academic Press, Sheffield, 1994.
8. J. E. Bennet, T. J. Schue, K. C. Clear, D. R. Lankard, W. H. Hartt and W. J. Swiat, 'Electrochemical chloride removal and protection of concrete bridge components', Strategic Highway Research Program report SHRP-S-657, 1993.
9. M. I. Jaffer, J. L. Dawson, and D. G. John, 'Cathodic protection studies on steel in concrete' in 'Corrosion of Reinforcement in Concrete', (edited by C. L. Page, K. W. J. Treadaway and P. B. Bamforth), 525, Elsevier Applied Science, London, 1990.
10. W. Lukas, *Betonwerk und Fertigteil-Technik*, 1985, **51**(11), 730.
11. P. Vassie, *Proc. Inst. Civil Engrs.*, 1984, Part 1 **76**, 713.
12. R. B. Polder, R. Walker and C. L. Page, 'Electrochemical chloride removal tests of concrete cores from a coastal structure' in 'Corrosion and corrosion protection of steel in concrete', (edited by R. N. Swamy), Vol.2, 1463, Sheffield Academic Press, Sheffield, 1994.
13. B. Elsener and H. Böhni, 'Electrochemical chloride removal field test' in 'Corrosion and corrosion protection of steel in concrete', (edited by R. N. Swamy), Vol.2, 1451, Sheffield Academic Press, Sheffield, 1994.
14. Concrete Society Technical Report No 36, *Cathodic Protection of Reinforced Concrete*, The Concrete Society, Slough, 1991.
15. N. R. Buenfeld and J. P. Broomfield, 'Effect of chloride removal on rebar bond strength and concrete properties' in 'Corrosion and corrosion protection of steel in concrete', (edited by R. N. Swamy), Vol.2, 1438, Sheffield Academic Press, Sheffield, 1994.
16. N. R. Buenfeld and J. B. Newman, *Cement and Concrete Research*, 1986, **16**, 721.
17. P. G. Cavalier and P. R. Vassie, *Proc. Inst. Civil Engrs.*, 1981, Part 1 **70**, 461.

PASSIVATION MECHANISM OF STEEL UNDER CATHODIC PROTECTION IN MEDIUM SIMULATING CONCRETE

S. Joiret[1], A. Hugot-Le Goff[1], J. P. Guilbaud[2] and A. Raharinaivo[2]

1 - Laboratoire "Physique des liquides et Electrochimie"
UPR 15 du CNRS, UPMC, 4 place Jussieu
F-75252 PARIS Cedex 05, FRANCE
2 - Laboratoire Central des Ponts et Chaussees,
58 Boulevard Lefebvre, F-75732 PARIS Cedex 15, FRANCE

1 INTRODUCTION

1.1 Scope

In sound concrete, with a pH of about 12, embedded steels are protected by a passivation film. But, sometimes, reinforcing steel corrodes after chloride ions or carbon dioxide penetrate the concrete cover. In such carbonated structures, the concrete pH is lowered to about 9. Cathodic protection is sometimes applied to rehabilitate these deteriorated structures and to protect the reinforcing steels. The object of this investigation is to show in what cases a corroded rebar can be effectively re-passivated by cathodic protection.

1.2 Previous Results

A previous study[1] has been carried out on reinforced concrete slabs, under cathodic protection lasting four years. The results have shown that, in some cases, magnetite (Fe_3O_4), akaganeite (ß-FeOOH) and ferrihydrite (δ-FeOOH) are formed on the steel surface.

Another investigation[2] has examined steel polarised in a medium simulating sound and carbonated concrete at potentials similar to that of cathodic protection. The visual examination of steel specimens showed that, at some potentials, when chloride was present in the electrolyte, steel could corrode and the corrosion products were greenish corresponding to "green rust", an unstable corrosion product of iron.

1.3 Outline of this Study

The objectives of this investigation were to validate the previous results concerning the presence of green rust on steel under cathodic protection, when concrete is carbonated and polluted with chlorides, and to determine whether a reinforcing steel previously corroded can be passivated under cathodic protection.

2 EXPERIMENTAL

2.1 Steel Specimens

Specimens tested were plain carbon steel (C: 0.06%; Si: 0.172%; Mn: 0.495%; P: 0.012%; S: 0.019%) wires, diameter 6 mm. They were cut on the longitudinal axis to give the flat surface needed for Raman studies. The entire specimen was insulated with a lacquer deposited by electrophoresis, except for the 2 cm^2 active area. To simulate the porosity of concrete, as described in reference 3, a porous sheet filter was wrapped around the electrode and secured at both ends by adhesive tape. Before being wrapped with the filter, the active steel surface was polished (with 600 SC emery paper) and degreased with acetone. Filters made from borosilicate glass microfibers without organic binding (type AP40, Millipore), suitable for Raman spectroscopy, were used rather than the polyvinyldiene fluoride filters used in previous works[3] which present a large fluorescence under laser illumination.

2.2 Solutions Simulating Concrete

Carbonated concrete has a pH of about 9.5. Two solutions were used to simulate the ageing of concrete under carbonation. Their compositions were as follows:
- Solution A: 0.025 M Na_2CO_3 +0.125 M K_2CO_3 + 0.05 M $NaHCO_3$ + 0.25 M $KHCO_3$ (pH = 9.7)
- Solution B: 0.1 M $NaHCO_3$ + 0.5 M $KHCO_3$ (pH = 8.3)

Chloride ions were introduced via calcium chloride dissolved in water, so that the concentration ratio $[OH^-]/[Cl^-] = 2$. (It is to be noted that green rust can form if the ratio $[OH^-]/[Cl^-] \geq 1$[4])

These solutions were made with analytical grade chemicals and high purity water. In some cases the solutions were deaerated thoughout the experiment using argon bubbling.

2.3 Polarisation for Cathodic Protection

2.3.1. Experimental Equipment. A three electrodes cell was used for polarising steel specimens in solution simulating concrete into which the electrolyte had been previously introduced. The steel specimen (working electrode) was placed with the active side up in this cell into which the electrolyte had been previously introduced. The counter-electrode, a platinum gauze, was placed near of the specimen active area. Mercury/mercurous oxide was employed as the reference electrode (0mV Hg/HgO = -110mV/SCE). This was chosen in order to avoid any chloride contamination. All the potentials are presented against this reference. Constant potential was applied to the steel through a potentiostat (Tacussel Corrovit) for one week, unless otherwise stated.

2.3.2. Polarisation for Obtaining Steel Corrosion. The values of the applied potential correspond to the green-rust domain in the Pourbaix diagram.[5] It means that the steel can corrode in the solution A, when the steel potential ranges between -800 mV and -500 mV, and in the solution B when this potential ranges between -700 mV and -400mV.

These results were confirmed by plotting potentiodynamic curves for steel in the solutions. For potentials more negative than the lower limit of the green-rust domain, the steel was cathodically protected. For potentials less negative than the upper limit of the green-rust domain, steel was covered with a thin layer of Fe_3O_4 and γ- Fe_2O_3.[6,7]

2.3.3 Polarisation of Steel Peviously Corroded. Steel specimens which had been corroded after a cathodic polarisation for one week were stored for three weeks, without polarisation. They were then polarised at -900 mV, simulating steel under cathodic protection.

2.4 Raman Spectroscopy and X-ray Diffraction

Raman spectroscopy was performed using a DILOR OMARS 89, equipped with a multichannel detector and an Argon ion laser. All the studies were performed with the 514.5 nm laser line. The power at the sample was 1 mW, the laser beam being collimated through an 80X microscope objective onto a 1.2 μm^2 surface. The Raman light emitted was back scattered through the same objective. To obtain a Raman spectrum from the steel surface, the filter was carefully removed after the end of the polarisation, the electrode being still immersed in 1 mm of solution. When the solution was deaerated, argon bubbling was maintained throughout the acquisition of the Raman spectra.

When corrosion products layers were very thick, an adhesive tape was attached to the surface to remove the outer layers of product, so that the inner layers could be examined. The performance of the microscope (focusing) allowed analysis of the deepest parts of the corrosion layer .

The X-ray diffraction patterns were recorded directly on the electrode surface using a diffractometer (model Philips PW. 10/30) and Mo $K\alpha$ radiation.

3 RESULTS

3.1 Visual Examinations and Electron Microscopy

3.1.1. Products on Initially Clean Steel. After polarisation for one week, general corrosion occurred on steel polarised in the domain of green-rust. This corrosion had a specific and reproducible appearance, as illustrated by the scanning electron microscope picture (Figure 1). Dark green, hexagonally shaped crystals lay directly on the whole surface of the steel specimen, their sizes were up to 4 or 5 μm. Around the crystals an homogeneous zone covered with thick corrosion products was observed; this layer was orange in aerated and black in deaerated solution.

If steel was polarised for at least three weeks in aerated solution, or if it was exposed to atmosphere for one week, the initial "dark green" crystals turned orange - brown, but retained their shape.

The existence of these crystals was determined by the experimental arrangement. Without the filter or if the electrode was turned face down during the polarisation, the corrosion in aerated solution was homogeneous .

3.1.2. Products on Previously Corroded Steel. On steel specimens which had previously been corroded under cathodic polarisation, blackish products were formed at a potential of -900 mV.

3.2 Raman Spectroscopy

3.2.1. Products on Initially Clean Steel. The corrosion products were analysed with Raman spectroscopy. The Raman spectrum of the dark green crystals, given in Figure 2a, corresponds to green rust 1,[8] and the Raman spectrum of the brown-orange oxidised form of the crystals, shown in Figure 2b, is typical of akaganeite (β-FeOOH).[8,9] In argon atmosphere the size and number of these crystals grew with polarisation time.

Figure 1 *SEM examination of iron electrode after one week at V = -600 mV in aerated solution* A

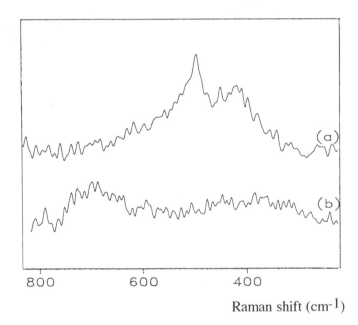

Raman shift (cm^{-1})

Figure 2 *Raman spectra of the crystals shown in Figure 1: (a) in situ spectrum , (b) after three weeks in air*

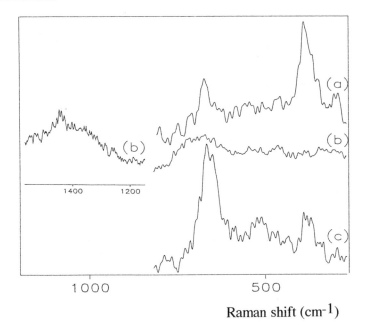

Figure 3 *Raman spectra of an electrode polarised at -600 mV in aerated solution A:*
(a) external layer , (b) intermediate layer , (c) internal layer

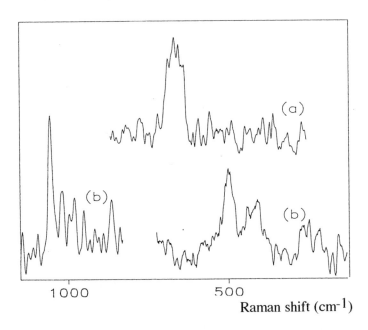

Figure 4 *Raman spectra of an electrode polarised at V=-700mV in deaerated solution A:*
(a) homogeneous corrosion layer, (b) spectrum near the crystalline zone

The Raman spectra of the products obtained in aerated solution A, in a zone where the corrosion is homogeneous are given in Figure 3. Close to the steel surface, spectrum 3c is typical of magnetite (black spinel Fe_3O_4). For the intermediate layer, spectrum 3b is typical of brown γ- Fe_2O_3. For the top of the layer, spectrum 3a corresponds to orange goethite (α-FeOOH). The same types and features of spectra were found for steel corrosion products in aerated solution B, around the green or brown crystals on steel as for the whole electrode when the polarisation has been performed without a filter or when the electrode has been turned face down .

In solution A deaerated with argon, this layer was greenish and gave the spectrum of Fe_3O_4 in Figure 4a together with spectrum 4b in a zone very close to the crystals. It is to be noted that this spectrum is the only one which shows the CO_3^{2-} ν_1 vibration peak as in $FeCO_3$.[10] In solution B only Fe_3O_4 is found .

3.2.2. Products on Previously Corroded Steel. The black layer formed on steel which had been previously corroded was composed of magnetite (Fe_3O_4).

3.3. X-Ray Diffraction

In order to confirm the structure of the green crystals formed on steel, X-ray diffraction studies were made of these products. Our results were then compared with data published by other authors (Table 1).

Table 1 *Interplanar Spacings and Relative Intensities of XRD Reflections for Green Rust*
I = intensity , vs=very strong , m = medium , vw = very weak

Present work			Mc Gill [11]			Sagoe-Crentsil [12]			Rezel [13]		
hkl	d, Å	I	hkl	d, Å	I	hkl	d, Å	I	hkl	d, Å	I
003	7.56	vs	003	7.50	100	003	7.75	100	0003	8.0	100
006	3.78	m	006	3.755	25	006	3.86	42	0006	4.402	50
			101	2.718	3	101	2.74	7	1010	2.77	20
									1012	2.71	5
102	2.67	vw	102	2.666	20	102	2.68	25	0009	2.66	10

Mc Gill and Sagoe-Crentsil have studied green rusts in carbonate solution and agreed on a formula: $[(Fe^{2+})_4 (Fe^{3+})_2 (OH^-)_{12}]^{2+} (CO_3)^{2-}$, $2H_2O$ with a rhombohedral symmetry. Rezel studied green rust in chloride media and gives: $2 Fe(OH)_2$, $Fe(OH)Cl$, $Fe(OH)_2Cl$ with a trigonal structure. Our results are in better agreement with the carbonated formula in which carbonate ions and water molecules are inserted between two layers of iron hydroxides, the carbonate ion forming a double hydrogen bond with hydroxyl ions from adjacent layers.

4 CONCLUSION

This study explains two results obtained regarding cathodic protection of steel in concrete which is carbonated and polluted with chloride.

If the applied potential is not lower than -800 mV/SCE, depending on the oxygen content in the medium, green rust may be formed on steel. It can be transformed into other corrosion products (magnetite, akaganeite, etc.) which are not commonly formed in rusts, but were obtained on reinforcing steel in slabs .

If a rebar which has been previously corroded is put under cathodic protection, it can be re-passivated only if the applied potential is lower than -800 mV/ SCE .

References

1 L. Delorme, P. Refait, J. M. Génin, G. Grimaldi and A. Raharinaivo, Proc. Int. Conf. Corrosion and Protection of Steel in Concrete, Sheffield, U. K., July 1994, **2**, 1402.
2 J. P. Guilbaud, G. Chahbazian, F. Derrien and A. Raharinaivo, Proc. Int. Conf. Corrosion and Protection of Steel in Concrete, Sheffield, U. K., July 1994, **2**, 1382.
3 A. Raharinaivo, J. P. Guilbaud, G. Chahbazian and F. Derrien, *Corr. Sci.,* 1992, **33**, 1607.
4 A. Raharinaivo and J. M. Génin, *Materiales de Construcción*, 1986, **36**, 5.
5 P. Refait and J. M. Génin, *Corr. Sci.*, 1993, **34**, 797.
6 M. Nagayama and M. Cohen, *J. Electrochem. Soc.,* 1962, **109**, 781.
7 A. Hugot-Le Goff, J. Flis, N. Boucherit, P. Delichere, S. Joiret and J. Wilinski, *J. Electrochem. Soc.,* 1990, **137**, 2684.
8 N. Boucherit, A. Hugot-Le Goff and S. Joiret, *Corr. Sci.*, 1991, **32**, 497.
9 G. Nauer, P. Strecha, N. Brinda-Konopik and G. Lipstay, *J. Therm. Anal.,* 1985, 813.
10 R. G. Herman, C. E. Bogdan, A. J. Sommer and D. R. Simpson, *Applied Spectroscopy,* 1987, **41**, 437.
11 J. R. Mc Gill, B. Mc Enaney and D. C. Smith, *Nature ,* 1976, **259**, 200.
12 K. K. Sagoe-Crentsil and F. P. Glasser, *Corrosion ,* 1993, **49**, 457.
13 D. R. Rezel, Thesis, University of Nancy I, 1988.

CATHODIC PROTECTION OF THE NORTH SEATON BRIDGE

Graham Gedge and Tony Sheehan

Ove Arup + Partners
13 Fitzroy Street
London W1P 6BQ
UK

1 INTRODUCTION

The North Seaton bridge was designed by Ove Arup + Partners (OAP) between 1970 and 1973, to carry the A189 trunk road across the River Wansbeck estuary in Northumberland, Figure 1. The bridge was built in 1974 as part of the South East Northumberland spine road. The bridge columns are classified as a location of extreme exposure according to BS5400,[1] being exposed to a tidal estuarine environment.

In 1990 a principal bridge inspection was undertaken and this revealed that, in general, the structure was in good condition. However, it also revealed that five of the six central columns had hollow sounding areas and/or vertical cracks near the high tide level.

In 1991 OAP were commissioned to determine the cause of these defects and to recommend the optimum remedial measures to ensure the integrity of the structure. The resulting survey concluded that the cracking was due to corrosion, arising from the penetration of chloride to the depth of the reinforcement.

After considering a number of repair options it was decided, by the County Council, to extend the life of the structure by installing an impressed current cathodic protection system. OAP were retained by the client to design the remedial works and to provide specialist assistance during construction.

2 STRUCTURAL DESIGN

The concrete was specified to be class 37.5/20 using sulphate resisting Portland cement (srpc), which is now known to have inferior resistance to chloride ion ingress compared to conventional Portland cement concrete. The sulphate resisting cement content was between 350-475 kg/m³ and the maximum water cement ratio was 0.5. The cover to the main reinforcement was 70 mm, in the lower part of the columns 20 Y32 bars were located and the links were sets of R8 and R10 bars at 300 mm pitch. Cover to the links was therefore 60 mm, Figure 2.

When checked against current design codes it was found that the columns still had adequate load carrying capacity, despite being designed to the less onerous standards prevailing in the early 1970s. Nonetheless, if remedial works were not carried out then the condition of the bridge would continue to deteriorate and could, in the future, be unable to carry its design loads.

Figure 1 *The North Seaton Bridge*

3 BRIDGE SURVEY

3.1 Visual Survey

The original bridge inspection in 1990 revealed vertical cracking in the vicinity of the high water level. This cracking was recorded on five of the six central columns. The cause of cracking was considered to be the same in all cases, the most likely reason being corrosion of the outer layers of reinforcement.

As the cause of cracking appeared to be the same in all cases, only one pair of columns was further investigated in detail. The chosen pair were located at pier 4, as these were readily accessible at low tide and showed typical signs of deterioration. The investigation concentrated on the east face of the east column, where cracking was worst. However, a number of comparative tests were also conducted on the west faces. Removal of the cracked concrete on the column corner showed that, although the main steel was only very lightly corroded, the links were badly pitted.

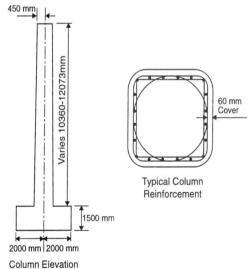

Typical Column Reinforcement

Figure 2 *Typical column section*

3.2 Survey Techniques

In addition to the visual assessment a number of other tests were undertaken to provide a clearer understanding of the problem and possible remedial works. These tests were:

measurement of the concrete cover, petrographic analysis via cores, carbonation, and surface electro-potential surveys as well as chemical analyses to determine chloride and sulphate concentrations.

The carbonation measurements and sulphate tests showed that neither were likely to have contributed to the cracking. Petrographic analysis indicated the concrete to be of good quality with a dense, compact, matrix free of microcracks. No signs of any deleterious reactions (e.g. ASR) were present in the concrete.

3.3 Test Results

The cover survey revealed cover on the east faces of the columns to be in the range 35-60 mm. On other faces it was in the range 50-100 mm.

The electrical potential survey was carried out to investigate whether corrosion was likely to be occurring in areas that had not cracked. However, meaningful results were only obtained in the vicinity of cracks.

Chloride analysis was carried out by obtaining powdered drilling samples from various depths. Concentrations varied between 1.3% and 0.1% by weight of cement. In general the concentration decreased with distance into the concrete and height up the column. In the initial survey sampling for chloride analysis was carried out to a height of 5 m above the foundation.

For chlorides present in the original concrete mix, BRE digest 264[2] and Concrete Society Technical Report No. 26[3] suggest that chloride ion concentration by weight of cement may be placed into three categories with respect to the risk of corrosion of embedded reinforcement:

Low	-	with content up to 0.4%
Medium	-	content 0.4% to 1%
High	-	above 1%

Where chlorides have penetrated from external sources, the risk of corrosion may be further enhanced.

Concentrations in excess of 0.4% were found in all samples taken at depths of 25-50 mm and in half the samples taken at 50-75 mm, Table 1. Cracking of the concrete had occurred where the cover was less than 50 mm and where the chloride concentrations were greater than 0.4%. The specified cover was 60 mm and the average was found to be in the range 60-70 mm. At this depth a significant number of samples were found to exceed 0.4% chloride. Furthermore, since the concentration of chlorides will increase with time, widespread corrosion and associated cracking would have been expected unless remedial measures were taken.

From these studies it was concluded that the corrosion to date had been caused by a combination of significant chloride ingress and low concrete cover. The high chloride concentrations measured in good quality concrete, of low permeability, were attributed to the exposure conditions of the columns. The columns were exposed to continual wetting and drying on a twice daily basis as they are located in the tidal zone of a river estuary.

Recently, Bamforth[4] has suggested that such significant chloride concentrations may be typical of modern normal portland cement concretes. The resistance of srpc concretes to chloride ion ingress is generally expected to be lower than normal portland cement concretes. In this case, however, the level of ingress into the srpc concrete is similar to that found by Bamforth for modern normal portland cement concretes.

Table 1: *Chloride Analysis Results*

Height above base (m)	Depth (mm)	Chloride Content (% by weight cement)	BRE Classification
1m	5 - 30	1.25	High
	30 - 60	0.71	Medium
	60 - 90	0.42	Medium
	90 - 120	0.12	Low
1½m	5 - 30	1.01	High
	30 - 60	0.6	Medium
	5 - 20	1.9	High
	20 - 40	0.95	Medium
	40 - 60	0.71	Medium
	60 - 80	0.54	Medium
	80 - 100	0.3	Low
2m	5 - 25	0.95	Medium
	25 - 50	0.54	Medium
	50 - 75	0.3	Low
	75 - 100	0.24	Low
	100 - 125	0.12	Low
2m	5 - 25	1.31	High
	25 - 50	0.89	Medium
	50 - 75	0.48	Medium
	75 - 100	0.17	Low
2½m	5 - 30	0.83	Medium
	30 - 60	0.83	Medium
	60 - 90	0.77	Medium
	90 - 120	0.65	Medium
3½m	5 - 25	1.01	High
	25 - 50	0.77	Medium
	50 - 75	0.42	Medium
	75 - 100	0.12	Low
4½m	5 - 25	0.83	Medium
	25 - 50	0.6	Medium
	50 - 75	0.3	Low
	75 - 100	0.18	Low
4½m	5 - 25	0.71	Medium
	25 - 50	0.48	Medium
	50 - 75	0.24	Low
	75 - 100	0.18	Low

4 REMEDIAL OPTIONS

The survey had demonstrated that if no action were taken, the bridge columns would continue to deteriorate due to corrosion of the reinforcement. In time this could have led to a reduction in load carrying capacity. A number of remedial options were therefore considered. These options were: local repair, local repair+coating, prop and breakout, staged repair and cathodic protection.

The first two options were discounted because they could not address the main concern, the future corrosion of areas unaffected at the time of surveying. Both the next two options were feasible and have a track record. However, they are not without problems and it has been found that, even when substantial repairs are undertaken, these methods have a finite life.

In 1991 the use of cathodic protection (CP) represented an uncertain option. The technique had not been used in the situation under consideration. The nearest parallel was the Tay bridge in Scotland where CP had been installed in 1986 to protect columns above the high water level. This system was apparently operating successfully.[5]

Despite these doubts, installation of a CP system represented a cost effective solution that could achieve the minimum life extension of 10 years. It was decided to adopt this option. The system would differ from the Tay bridge in that protection would include the permanently submerged parts of the structure. In addition the client required the load carrying capacity to be increased. This required an extra layer of reinforcement and concrete.

5 CATHODIC PROTECTION DESIGN

The design of the CP system on the North Seaton bridge had to recognise the physical constraints the site presented. These were primarily related to its tidal nature and the difficult access to the central columns. These constraints required provision of substantial temporary civil engineering works to allow installation of the CP system.

5.1 Temporary Works

A major part of the contract was the provision of access by the main contractor. These works involved the construction and removal of a temporary causeway. This ran from the north shore access road to the central columns. This causeway was essential for the construction of coffer dams around each column, to protect them during the construction period.

On completion of these works, each column in turn had areas of damaged concrete removed and repaired. The CP system anode mesh, cabling and reference electrodes were installed and new reinforcement and concrete placed. Once this had been satisfactorily completed the shuttering and dams were removed and work began on the next pair of columns.

5.2 CP Design

5.2.1 Specification.   The detailed CP design was let as a specialist sub-contract to the main civils contract, awarded to Sir Robert McAlpine, the chosen sub- Contractor being The Cathodic Protection Company (CPC). CPC's design had to meet performance specifications prepared by OAP.

This specification generally followed the recommendations of The Concrete Society Technical Report (CSTR) and Model Specification.[6,7] The criteria specified by OAP were:

-The CP system should be impressed current and be capable of delivering 20 mA/m² of reinforcement area.

-The CP system should provide sufficient current to polarise the structure. The acceptance criteria being a 100 mV decay, with a limit of -1100 mV Ag/AgCl/KCl instant-off potential.

- Monitoring was to be by a combination of embedded and surface mounted reference electrodes.

- The anode material was specified as being an expanded Titanium mesh.

5.2.2 Detail Design. The detail design was undertaken by CPC and adopted a solution in which the anode mesh was located between the new and old reinforcement, Figure 3.

Due to the differing exposure conditions with column height CPC decided to split each column into two zones, the upper and lower. In general the upper zone provided protection to atmospherically exposed concrete and the lower zone

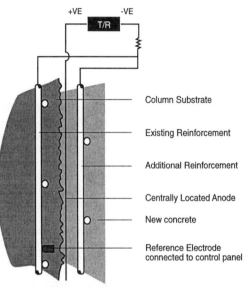

Figure 3 *Schematic of CP system*

protected the tidal and submerged exposure conditions.

In order to reduce the length of DC cabling CPC located the transformers in housings at each end of the bridge span. Each housing containing the transformer/rectifiers for the two zones of five individual columns.

5.3 CP Installation

The anode mesh was installed after repairs to the damaged concrete were completed, with the mesh held 10 mm from the existing surface using plastic spacers. In order to achieve accurate control of current to each zone, a 100-200 mm gap was left between the mesh zones. The mesh within each zone was overlapped at each joint by a minimum of 75 mm. At each joint in a zone titanium distribution bars were welded to the mesh.

Permanent reference electrodes Ag/AgCl were placed at predetermined locations on each column, in general these were located in areas of repair, with two reference electrodes per column.

5.4 System Commissioning

The system was energised in December 1994 and commissioned, in accordance with CSTR Nos. 36[7] and 37[8], during January 1995. During commissioning the contractor encountered a number of difficulties in meeting the stated acceptance criterion. In all cases it was found impossible to achieve the 100 mV shift within four hours of measuring the instant-off potential.

For the reasons given below the acceptance criteria were therefore changed to reflect the particular conditions of this structure. For the upper zone the 100 mV shift was maintained, but the decay time was extended to eight hours. For the lower zone the criteria was changed to an absolute value of instant-off in the range of -720 to -1100 mV Ag/AgCl/KCl.

These changes allowed the satisfactory commissioning of all zones, except those known to have either faulty connections or defective reference electrodes.

6 DISCUSSION

In general the design and installation of the CP system went well, despite the physical constraints of a difficult tidal site. However, it was not fully appreciated at the outset that the acceptance criteria originally specified would be difficult to achieve. These difficulties arose because the acceptance criteria were adopted from systems with different forms of construction and exposure conditions.

In the UK most applications of CP have been in motorway bridge structures which are exposed to a more clearly defined and constant environment. In essence the environment is that of atmospheric exposure. In addition these applications have not involved the placement of additional concrete and reinforcement over the anode mesh.

At North Seaton the condition was very different. Not only was there additional concrete but also the environment varied with height up the column. The effect of these features, allied with the high quality of the concrete, has a significant effect on the operation of a CP system.

When steel is cathodically protected the anodic, metal dissolution, reaction is stopped. The cathodic reduction reaction however, continues at a rate controlled by the current supplied by the transformer. For constant current devices this rate is fixed. In the case of reinforced concrete the species reduced will be oxygen. As the rate at which this is reduced is constant over the whole structure, assuming a common current density, then it is possible that the oxygen concentrations at different locations will vary with exposure.

The oxygen concentration at the steel surface will determine the potential of the steel, as measured by the embedded reference electrode at the surface. These differences will arise because of both macro and micro changes in the permeability of the concrete.

In the present context, macro variations are particularly important. In the atmospheric zones permeability will be relatively high with oxygen freely available at the steel surface. In the submerged and tidal zones the concrete is permanently saturated with water, and under these conditions oxygen availability at the steel surface will be restricted. Data from a number of authors confirms that the oxygen permeability of saturated concrete is very much lower than non saturated material.[8,9]

The oxygen concentration at the steel surface will also be affected by the concrete cover to the bars. In this case the cover at corroding bars was increased from approximately 50 mm to 175 mm, thus significantly increasing the diffusion distance.

The resulting differences in oxygen concentration would be expected to affect both the natural and protection potentials. In areas of higher oxygen availability (i.e. areas of low cover), potentials would be expected to be less negative than areas of low oxygen availability and vice versa. This was found to be the case for all the columns. Prior to energising the natural potentials on the upper zones were all between +200 and +500 mV compared to the lower (saturated) zones.

Once the CP system was energised the potentials of all zones changed only very slowly with time away from the natural values. After approximately one month none of the zones could meet the acceptance criteria.

However, most of the upper zones were showing a four hour shift of between 60 and 80 mV. It was therefore decided to increase the permitted decay time to approximately 24 hours and to continuously record the potential. This decision was based on the recognition of the role of oxygen in the corrosion process. The potential of the steel will change positively as the oxygen concentration increases at the steel surface.

It was concluded, in this instance, that both the concrete cover and quality were restricting oxygen access to the steel. Increasing the decay time would, therefore, negate this problem by increasing the time available for oxygen to reach the steel. It was found that the 100 mV shift criterion could be met if the decay period was set at 10 hours, Figure 4.

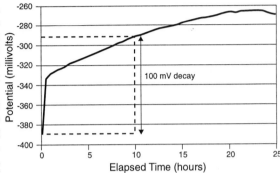

Figure 4 *Column 1 East Upper-R1 Off potential decay*

An alternative proposal to retain the four hour period but increase the current was also investigated. This did not prove successful, as would be expected if oxygen availability were controlling potentials.

A similar exercise on the lower zones was less successful. Even when left to decay for 70 hours the 100 mV shift was not achieved, Figure 5. Again this can be accounted for by oxygen availability to the steel surface. In the lower zones this would be even more restricted than in the upper zones because of the reduced permeability of saturated concrete.

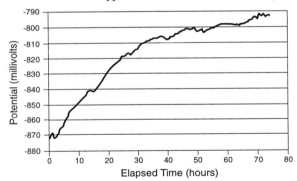

Under such conditions it is unlikely that the 100 mV requirement could ever be met.

Under such conditions a more appropriate criterion is the absolute potential criterion recommended by Linder[8] and others[§]. This criterion requires an instant off potential of between -720 and -1100 mV Ag/AgCl/KCl.

Figure 5 *Column 2 East Lower- Decay to natural potential*

7 CONCLUSIONS

(1)   The use of CP has been shown to be practically possible on a difficult tidal structure. It has also been shown that the provision of temporary works for such contracts should not be underestimated.
(2)   The survey results have shown that the steel bars in reinforced concrete, exposed to a tidal environment, are susceptible to corrosion particularly where cover to bars is reduced. The survey also highlighted the unreliability of using potential surveying as the sole method of assessing the risk of corrosion.
(3)   In specifying acceptance criteria caution should be exercised. From the experience of this project it seems unlikely that a single universal acceptance criterion is possible. This is often the case with other CP applications, such as, for example, in the case of buried pipelines.
(4)   The criteria specified by designers need to reflect both the form of construction and the operating environment of the structure. On structures where the environment varies it is possible that different criteria need to be adopted for different operating zones. Where a designer is unsure of the most appropriate criteria for a given zone, provision should be made in the specification for the development and testing of acceptance criteria.

References

1.   British Standards Institution, BS5400, 'Code of practice for the design of steel, concrete and composite bridges', BSI, 1980.
2.   Building Research Establishment, 'The durability of steel in concrete: Part 2: Diagnosis and assessment of corrosion-cracked concrete', BRE Digest 264, August 1982. HMSO, UK.
3.   Concrete Society, 'Repair of concrete damaged by reinforcement corrosion. Report of a working party. October 1984', Concrete Society Technical Report 26, 1985.
4.   P Bamforth, *Concrete*, 1994, **28**, 18.
5.   A Watters, *Ind. Corr.*, 1991, April/May, 4.
6.   Concrete Society, 'Cathodic protection of reinforced concrete. Report of a joint Concrete Society and Corrosion Engineering Association Working Party', Concrete Society Technical Report 36, 1989, The Concrete Society, UK.
7.   Concrete Society, 'Model specification for cathodic protection of reinforced concrete. Based on the ICE Conditions of contract. Report of a joint Working party of the Concrete Society and Corrosion Engineering Division of the Institute of Corrosion', Concrete Society Technical Report 37, 1991, The Concrete Society, UK.
8.   B Sederholm, and B Linder, 'UK Corrosion and Eurocorr 94' Chameleon Press, London, 1994, **3**, 189.
9.   CL Page, and P Lambert, *J. Materials Science.* 1987, **22**, 6.

ELECTROCHEMICAL CHLORIDE REMOVAL

ELECTROCHEMICAL CHLORIDE REMOVAL - A CASE STUDY AND LABORATORY TESTS

J. Tritthart

Graz, University of Technology
Stremayrgasse 11
A-8010 Graz
Austria

1 INTRODUCTION

The principle of electrochemical chloride removal (ECR) is to apply a direct current between the reinforcement acting as cathode and an anode that is placed temporarily onto the outside of the concrete surface. The anode is preferably a titanium wire mesh which must stay in contact with an electrolyte to ensure the electro-conductive contact between anode and cathode. Since negatively charged ions, such as chloride ions, migrate in the direct-current field from the reinforcement to the positively charged outer electrode, relatively large amounts of chloride can be removed from the concrete within a relatively short time (the usual period of application is 6-10 weeks). The method is similar to cathodic protection, however, the current density applied is about 100 times higher (often 1-2 A/m^2), and the anode is removed together with the pulp afterwards. In contrast to cathodic protection, chloride extraction is not a permanent matter, so that there is no need for continuous and elaborate monitoring.

The possibility of removing chloride from concrete by an electro-chemical process was first reported in the USA in the 1970s,[1-3] and it involved the application of very high direct voltages (up to 220 V). These investigations had been carried out on horizontally positioned test objects like slabs or bridge decks, using a liquid outer electrolyte. In Europe, ECR was patented in 1986 by the Norwegian company "Noteby", especially for vertical concrete surfaces, using a material around the anode with good properties regarding adherence to the concrete surface and water retention, such as retarded gunite.[4] Considerably lower voltages, preferably less than 30 V are used in this method, which is known as the NORCURE™ method. The studies preceding the patent had not been published.[5] It was only at the beginning of the 1990s that reports on this method became more frequent.[6-27] At present, electrochemical chloride removal is studied within the framework of the EU COST 509 program by cooperating research institutions in several European countries. The studies presented are part of the research project "A 1" carried out by the author within COST 509.

In order to improve the method or, in case of failure, to identify the causes, it is crucial to understand the mechanisms underlying the method. As the electric current can be transported in the pore solution only by ions dissolved therein, the key to understanding the mechanism of chloride removal is to know what changes occur in the composition of the pore solution of concrete during the flow of electricity. However, such tests have only been included recently.[11,13,21,24,27] In the following, I describe the tests and results obtained prior to and during electrochemical chloride removal in a reinforced-concrete hall with extremely high chloride concentrations in the concrete. Then the laboratory tests that have been performed to arrive at a better understanding of some effects are described.

2 ELECTROCHEMICAL CHLORIDE REMOVAL FROM A HALL MADE OF REINFORCED CONCRETE

2.1 General

The investigated hall had been used for more than 10 years as a depot for loose storage of deicing salt. Numerous spots of corrosion products had appeared at the concrete surface. The hall was made available for the study by the Styrian Government. Chloride removal was performed in an area of about 36 m^2 by the company System Bau (Salzburg) at no cost. A commercially available titanium mesh was used as an anode which was embedded in paper pulp. Throughout the test, the pulp was sprinkled with tap water almost daily to keep it wet and thus to ensure electrical contact with the concrete acting as electrolyte. The average current density applied to the concrete surface was 1 A/m^2. The necessary voltage was continually readjusted by hand.

2.2 Tests

Five cores of 50 mm and one core of 200 mm diameter were extracted at different spots within the area to be repaired, before the anode was attached. In the laboratory, slices of 10 mm thickness were cut from the cores parallel to the base. As the saw blade had a thickness of 5 mm, the slices came from depths of 0-10 mm, 15-25 mm, 30-40 mm, etc. Determinations were made of the contents of total chloride and cement.[27] The pore solution had been expressed from the slices of the 200 mm core and was examined for chloride and hydroxide concentrations.[27]

After 40 days of migration, a first control core of 50 mm diameter was extracted from near the point where core 4 had been extracted before migration, which was located almost in the center of the area to be repaired, and after 83 days of migration, another core of 200 mm diameter was taken from close to the extraction point of the first core of the same diameter. Results were available after 96 days of migration and showed a high Cl$^-$-content from the reinforcement downwards. The flow of current was interrupted now for 98 days. Before the current was restored, another core of 200 mm diameter was taken to analyse the liquid phase and the solids. Chloride removal was finished after a total of 139 days of current flow. Five cores of 50 mm diameter and one core of 200 mm diameter were again extracted from close to the sampling points from which the cores had been taken prior to the start of chloride removal.

2.3 Results and Discussion

2.3.1 Total Chloride Distribution Before, During and After Chloride Removal. Figure 1 shows the distribution of chloride before Cl$^-$ removal. It shows a group of curves with similar patterns, whose chloride content in the 0-10 mm zone is between about 12% and 15% by weight of cement. At depths of 40 to 60 mm (it was approximately in this area of cover that overlapping bars were situated), the concrete still contained chloride in very high concentrations. Even at a depth of 170 mm, chloride was still detectable. Core 2 showed significantly lower values in the outer zone as well as a lower concrete penetration depth. This sample came from an area where the paint still adhered to the concrete surface and had inhibited chloride penetration.

Figure 2 depicts the results of chloride distribution in the cores of 50 mm diameter obtained after the end of current flow. They show that the chloride content was very low in the outermost zone (0-10 mm) compared to the initial values, and that it was below 1% of the cement content in all cases. As the depth of extraction increased, the chloride values also rose, reaching, at various depths between 60 and 100 mm, values similar to those before chloride removal. The only exception was core 1. The further history of chloride values was comparable to the one prior to chloride removal, in some instances the values were even higher. The unequal efficiency of chloride removal as well as the unequal depth down to which chloride could be extracted can be explained mainly by differences in the distribution of current density in the concrete. Chloride is removed, for example, much more rapidly from the zones directly above the reinforcement than from the zones between the reinforcement bars as current density decreases with distance from the reinforcement. In case of differing cover thickness, chloride removal happens faster in zones where the cover is thin because the electric resistance of the concrete layer above the reinforcement is lower, and thus the current density is higher than in places with a thick cover. The fact that, after chloride removal, the Cl⁻-content of the concrete reached its initial values only behind the reinforcement indicates that chloride was not only removed from within the cover. Why the chloride values of core 1 in Figure 2 were far below the initial values after chloride removal is unknown.

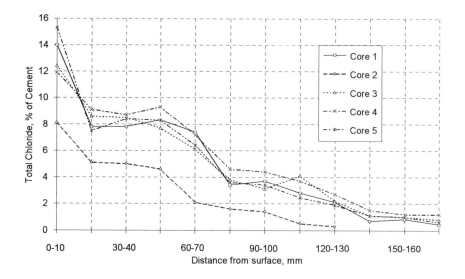

Figure 1 *Chloride distribution before chloride removal*

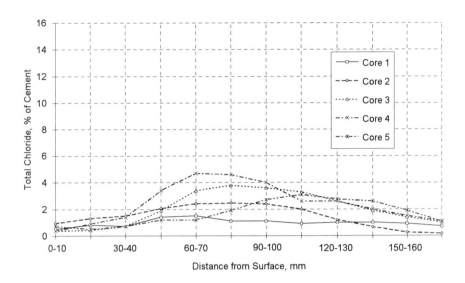

Figure 2 *Chloride distribution after chloride removal*

However, there are some indications that the core had not been removed after the end of ECR at the point from which the core had been extracted before Cl⁻-removal, but from a more distant point with lower Cl⁻-contamination.

Figure 3 shows the results which were obtained on cores 4 and 6 before, during and after chloride removal. As the cores with the same number as in Figure 3 were drilled out from the same area, statements regarding the speed of chloride extraction can be made from the results. As the values of cores 4, 4a and 4b show, the majority of the chloride was extracted during the first 40 days. In extraction zone 6 (cores 6, 6a, 6b and 6c), the changes were qualitatively the same as at extraction site 4 (cores 4, 4a and 4b). The differences between curve 6 and the curves 6a, 6b and 6c are less pronounced due to the "low" initial chloride contents and because the first control core was extracted only after 83 days of migration. With $>1\%$ Cl⁻ in the reinforcement area after the end of current flow, the Cl⁻ content was still above the Cl⁻ limit acceptable for reinforced concrete.

As expected, it was not possible to extract chloride from far behind the reinforcement. Although chloride also moves away from behind the reinforcement due to the negative potential of the reinforcement, it cannot be removed to a greater extent due to the relatively low current density and the long transport distances.[15] However, the differences in the amount of chloride removed from within the cover were surprising. As it was assumed that the chloride which was located at the end of the migration path and moved from the concrete into the pulp would be replenished by chloride ions from lower depths, it was expected that chloride removal would be slowest in the area close to the surface. This behaviour, which is due to the migration direction of chloride in the cement matrix, had been observed in model tests with cement paste

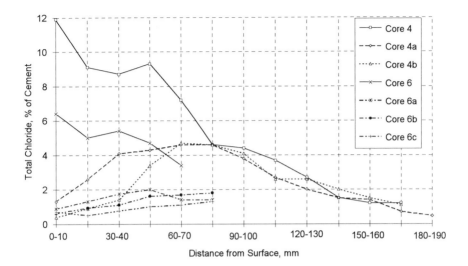

Figure 3 *Chloride content (i) before chloride removal (cores 4 and 6), (ii) after 40 days of migration (core 4a), (iii) after 83 days of migration (core 6a), (iv) after a break of 96 days (core 6b), and (v) at the end of migration (cores 4b and 6c).*

samples in the laboratory.[13,14] This was not the case in concrete, due to changes in the liquid phase of concrete, which will be discussed later.

2.3.2 Changes in Hydroxide Concentration in the Pore Solution of Concrete during Chloride Removal. Figure 4 shows that there was a gradient in initial OH⁻ concentration, with the values rising from the outside to the inside. At a depth of 0-10 mm, for example, 0.024 mole OH⁻/L were measured, and 0.26 mole OH⁻/L at 60-70 mm. The respective calculated pH-values were 12.38 (0-10 mm) and 13.41 (60-70 mm). The latter value lies in the pH-range typical of a pore solution in non-carbonated portland cement concrete. As indicated by the height of the bars after 83 days and the end of migration, the OH⁻ concentration decreased in the two outermost zones in the course of electrochemical chloride removal, while it increased very strongly, from a depth of 30-40 mm downwards.

This phenomenon is caused by reactions that occur at the electrodes:

Cathode (reinforcement):

$$H_2O + 1/2O_2 + 2e^- \rightarrow 2OH^- \qquad\qquad (1)$$

$$H_2O + e^- \rightarrow OH^- + H_{(ads)}; \qquad 2H_{(ads)} \rightarrow H_2\!\Uparrow \qquad (2)$$

Existing oxygen is used up at the cathode while forming OH⁻-ions (1). As only a little oxygen is present in the concrete and oxygen diffuses from outside into the concrete very slowly, this reaction can be neglected. Here, reaction (2) predominates, i.e. the decomposition of water with the simultaneous formation of OH⁻-ions and

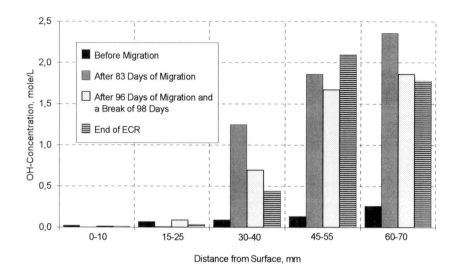

Figure 4 *Changes in hydroxide concentration of the pore solution*

gaseous hydrogen. Positively charged, easily soluble alkali-ions such as Na^+ and K^+ move in the electrical field to the cathode and form, with the OH^--ions, sodium-hydroxide and potassium-hydroxide, respectively. So, a continuous production of OH^- ions takes place at the cathode. The OH^- concentration in the pore solution rises in the vicinity of the reinforcement although OH^- ions migrate continuously away from the cathode because for each OH^--ion and Cl^--ion leaving the concrete new OH^--ions are formed. The increase of the OH^- concentration of the pore solution around the reinforcement is therefore greater the more sodium chloride is present in the concrete.

Anode (titanium mesh):

$$H_2O \rightarrow 1/2O_2 + 2H^+ + 2e^- \tag{3}$$

$$2Cl^- \rightarrow Cl_2 + 2e^- \tag{4}$$

According to (3), electrolysis of water occurs at the anode. Provided that the water used to wet the pulp does not contain any salts, this is the only reaction that occurs until anions have moved in from the concrete and are discharged. The H^+ ions that have formed migrate in the direct voltage field towards the cathode and meet OH^-- and Cl^--ions moving in the opposite direction. The OH^- ions are neutralized forming water, and hydrochloric acid is formed with the Cl^- ions. As a consequence, the pH-value of the initially neutral water shifts into the acidic range in the outer electrolyte (pH-values between 1.5 and 3 were measured); in the concrete zones close to the surface, the OH^- concentration falls due to direct contact with the acid electrolyte on the outside. According to (4), the Cl^- ions arriving at the anode are discharged by forming chlorine gas.

2.3.3 Changes in Chloride Concentration in the Pore Solution of Concrete during Chloride Removal. Figure 5 shows the changes in chloride concentration of the pore solution. Like the total chloride content, the decrease in Cl⁻ concentration of the pore solution within the cover was strongest in the 0-10 mm zone and weakest in the reinforcement zone (45-55 and 60-70 mm sections). The reason for the difference in chloride removal capacity at different depths is the shift in OH⁻ concentration of the pore solution.

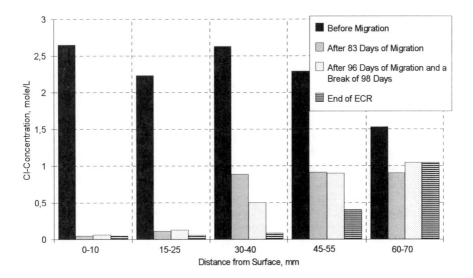

Figure 5 *Changes in chloride concentration of the pore solution*

The proportion of the total current flowing through the electrolyte that is transported by the chloride ions - the Chloride Transference Number - is larger the fewer other negatively charged ions are present and the lower their mobility in the direct voltage field when compared to the mobility of the Cl⁻ ions. The anions contained in Cl⁻-free, non-carbonated concrete are mainly OH⁻ ions, and other anions are normally present in such a low concentration - compared to that of OH⁻-ions - that they can be neglected. However, Page *et al.*[24] observed that the strong increase in OH⁻ concentration at the reinforcement was accompanied by a rise in the SO_4^{2-} concentration of the pore solution from about 10-30 mmole SO_4^{2-} (pH ≈ 13.5) to more than 100 mmole SO_4^{2-} (pH ≈ 14). In the area close to the surface, the OH⁻ concentration was initially low and the current had to be carried primarily by the Cl⁻ ions. Due to the continuous decrease of the OH⁻ concentration, the Chloride Transference Number increased constantly in this area, therefore here the chloride removal was particularly efficient. In the reinforcement area, where the OH⁻ concentration had been higher than in the peripheral zone, it rose during migration to a value about five times higher than the initial value. As the period of migration grew longer, the current was transported here more and more by the OH⁻ ions, and there was a steady and significant decrease of the Chloride Transference

Number, so that chloride removal became less and less efficient. This was visibly demonstrated by the fact that, after 83 days of current flow, about 0.9 mole Cl⁻/L (about 32000 ppm) were still present in the pore solution of the section extracted from 40-50 mm depth, whereas the pore solution of the 0-10 mm zone contained no more than 0.05 mole Cl⁻/L (about 1800 ppm).

The Cl⁻ concentrations obtained after interrupting the flow of current for 98 days differ slightly from those after 83 days of migration and are higher at 60-70 mm and 75-85 mm depth than after 83 days of migration (Figure 5). This is related to the local fluctuation of the total chloride content (Figure 3). In the period between the end of the interruption and the end of the migration test, the Cl⁻-concentration also decreased significantly in the 30-40 mm zone and even at a depth of 45-55 mm, a considerable drop in Cl⁻-concentration was observed while no change occurred at 60-70 mm depth.

The reason for the interruption was to determine whether more firmly bound chloride (e.g. strongly adsorbed Cl⁻) would dissolve during such a break. It has been reported[12] that Cl⁻ removal becomes inefficient with time, because the dissolved and loosely bound chloride has already been removed, and the firmly bound chloride dissolves very slowly, therefore, Cl⁻ removal can be continued more efficiently after a pause during which chloride has been dissolved. In the case of extremely high chloride contamination, no bound chloride was dissolved either in the zones near the surface or near the reinforcement, which follows from the fact that the Cl⁻ concentrations of the liquid phase had not changed appreciably during the interruption of current flow. These as well as earlier results[13,14] indicate that chloride binding is not the primary factor that governs the efficiency of chloride removal. According to the results shown in Figures 4 and 5, it appears to be the rising OH⁻ concentration at the reinforcement and its influence on the Chloride Transference Number that are more important.

3 LABORATORY TESTS

3.1 Test Method

To verify the interactions that had been derived, some additional tests were performed at the laboratory. The migration cells that were used are outlined in Figure 6 and described in more detail in reference 13. The test specimens were cylindrical, chloride-free hardened cement paste samples of 50 mm length and 50 mm diameter. The chloride had been dissolved in the liquid filling of the cathode chamber and was transported from it in a direct voltage field through the sample to the anode chamber. The current applied was 10 mA, which corresponds to a current density through the cement paste of about 5 A/m². This is significantly higher than the current density applied for chloride removal from the concrete (1-2 A/m²). The current intensity was kept constant throughout the test by automatic voltage control.

First, the influence of the OH⁻-concentration of the chloride-containing solution in the cathode chamber on chloride transport was investigated, and then the migration-time-dependent changes. At the time of testing, the cement paste samples were divided into four pieces by cutting first a 10 mm slice off each of the two sample ends parallel to the base and then by dividing the remainder of the sample in the middle. The sections were marked with the letters A, B, C, and D, with A (D) specifying the 10 mm-thick

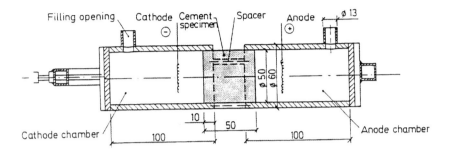

Figure 6 *Sketch of the migration cell (dimensions in mm)*

end zone extending into the cathode (anode) chamber. The section marked B (C) is the part from the middle of the sample adjacent to A (D). For details on sample preparation, installation in the migration cell, etc., see reference 13.

3.2 Cl⁻-Transport at Different OH⁻-Concentrations

To study the influence of the OH⁻ concentration of the chloride-containing electrolyte on the speed of chloride transport, several samples were prepared of Austrian Portland cement, type "375", with a w/c ratio of 0.55. They were placed into the migration cells at the age of 28 days. From another sample, the OH⁻ concentration of the pore solution was determined before migration. The anode chambers of the migration cells were filled with a (chloride-free) saturated $Ca(OH)_2$-solution. Solutions with different OH⁻ concentrations were used to fill the cathode chambers, namely distilled water; 0.1 N NaOH; 0.2 N NaOH; 0.4 N NaOH; 0.6 N NaOH; 0.8 N NaOH and 1.0 N NaOH. The chloride concentrations of these solutions were all equal to 0.5 mole/L (NaCl).

The duration of current flow was 10 days. Then the chloride and hydroxide concentrations in the cathode chambers and in the pore solutions of the four sections of each sample as well as the total chloride content of the sections were determined.

The values contained in Table 1 show that the Cl⁻ concentration of the solution in the cathode chamber decreased and the OH⁻ concentration increased, as was to be expected, during the 10 days of migration. The change was greater the lower the initial concentration of OH⁻-ions, i.e. when a larger proportion of current was carried by the Cl⁻-ions. The Cl⁻ concentration of the initially neutral solution, for example, decreased by 0.14 mole/L, while the decrease in the 1 N NaOH solution was as small as 0.04 mole/L.

Figure 7 shows the Cl⁻-concentrations of the pore solutions within the samples. As can be seen, there was always a strong concentration gradient from zone A to zone D, which can be explained by the migration direction of the ions. When the neutral NaCl

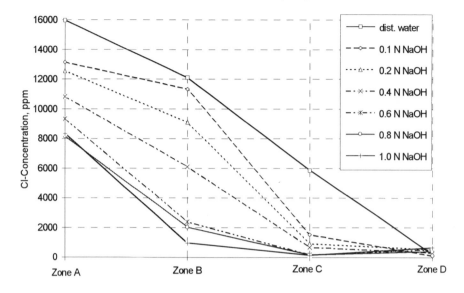

Figure 7 *Chloride concentration in the pore solution of cement pastes*

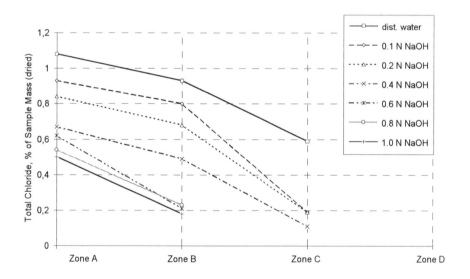

Figure 8 *Total chloride content of cement pastes*

Table 1 *Test results obtained from cathode chamber solutions*

before migration		after 10 days of migration			
Cl⁻	OH⁻	Cl⁻	Δ Cl⁻	OH⁻	Δ OH⁻
(mole/L)	(mole/L)	(mole/L)	(mole/L)	(mole/L)	(mole/L)
0.50	--	0.36	- 0.14	0.17	+ 0.17
0.50	0.10	0.40	- 0.10	0.24	+ 0.14
0.50	0.20	0.42	- 0.08	0.31	+ 0.11
0.50	0.40	0.44	- 0.06	0.49	+ 0.09
0.50	0.60	0.45	- 0.05	0.65	+ 0.05
0.50	0.80	0.46	- 0.04	0.84	+ 0.04
0.50	1.00	0.46	- 0.04	1.04	+ 0.04

solution was used to fill the cathode chamber, the Cl⁻-concentration in zone A rose to 16000 ppm (0.45 mole/L). The values measured in the pore solutions of zones A of the remaining specimens decreased as the OH⁻-concentration of the electrolyte in the cathode chamber increased. In the samples containing 0.8 and 1.0 N NaOH in the cathode chamber, the Cl⁻ concentrations were only slightly above 8000 ppm Cl⁻ (0.22 mole/L) although the Cl⁻ concentration was still about twice as high outside the sample.

Figure 8 illustrates the history of total chloride content of the samples. The relations obtained correspond more or less to those of the pore solution. However, it was somewhat surprising that most of the chloride was present in bound form which could be concluded from the total chloride content, the Cl⁻ concentration of the pore solution and the content of free water in the sample. The fact that chloride was partly bound in current-carrying hardened cement paste indicates that under direct current, too, Cl⁻-binding occurs according to an equilibrium between solid and liquid phase. This is in agreement with evidence from other research.[24] Whether chloride is taken up from the pore solution by the solids or whether it is given off to the pore solution depends on the ratio between the chloride concentration of the pore solution and the chloride content of the solids.

3.3 Changes of pore solution composition

As regards the influence of OH⁻ concentration of the pore solution on chloride-transport, the most interesting results were obtained from a test performed as described in section 3.1, but for a period of 35 days. The experimental set-up also differed inasmuch as the two electrode chambers were filled with a saturated Ca(OH)₂ solution, and the Cl⁻-concentration of the solution in the cathode chamber was 1 mole/L.

Figure 9 shows the changes in Cl⁻-concentration of the pore solutions of the four sample sections. First, the Cl⁻-concentration rose, but decreased again as the migration time grew longer; the changes were strongest in zone A, which was closest to the cathode chamber. Surprisingly, the Cl⁻-concentration in the pore solution of zone A dropped to about 8000 ppm (about 0.22 mole/L) after 35 days although the adjacent solution in the cathode chamber, from which the chloride was transported into the sample, still contained about 24000 ppm Cl⁻ (about 0.68 mole/L). These changes can be

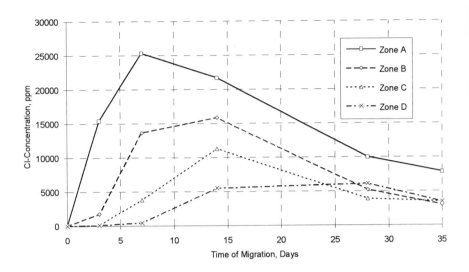

Figure 9 *Changes of Cl-concentration of the pore solution versus migration time*

explained by shifts in the Cl⁻- and OH⁻-concentrations of the pore solution and the electrolyte of the cathode chamber.

Table 2 contains a summary of the Cl⁻- and OH⁻-concentrations in sample zone A and in the electrolyte of the cathode chamber as well as the calculated values of the Cl⁻-content as a percentage of the sum total of Cl⁻- and OH⁻-concentrations.

What was striking was that the OH⁻-concentration of the pore solution decreased rapidly and strongly. This is primarily due to the fact that, in the pore solution of the samples prepared without chloride, initially almost the only anions available for the transport of current were negatively charged OH⁻ ions, which were transported in the direct current field towards the anode. In the filling solution of the cathode chamber, on the other hand, the situation was reversed; very few OH⁻-ions were present but many Cl⁻-ions, so that in the beginning no virtually other ions than Cl⁻-ions were transported into the sample. This explains why, especially in the pore solution of zone A, the Cl⁻-concentration rose so quickly at first, while the OH⁻-concentration decreased. After 7 days, the Cl⁻-concentration in the pore solution of this zone was no longer far below that in the cathode chamber. However, the OH⁻ concentration of zone A was already somewhat lower than in the electrolyte of the cathode chamber, where it rose due to cathode reaction. The chloride content in relation to the sum total of OH⁻- and Cl⁻-ions was already a little higher in the pore solution than in the solution of the cathode chamber.

As, to a first approximation, the concentration of the other anions can be neglected, this means that, after 7 days, the Cl⁻-ions were more strongly involved in the transport of current in the pore solution than in the solution of the cathode chamber, i.e. that more Cl⁻-ions were transported away from zone A per unit of time than were transported into it, and therefore the Cl⁻-concentration of the pore solution began to

Table 2 *Changes in Cl⁻-and OH⁻-concentrations*

Days of migration	Pore solution of sample section A			Electrolyte in cathode chamber		
	Cl⁻ (mole/L)	OH⁻ (mole/L)	Cl⁻ % of OH⁻ + Cl⁻ (mole/L)	Cl⁻ (mole/L)	OH⁻ (mole/L)	Cl⁻ % of OH⁻ + Cl⁻ (mole/L)
0	0	0.60	0	1.00	0.04	ca. 96
3	0.43	0.15	ca. 74	0.92	0.11	ca. 89
7	0.71	0.12	ca. 86	0.86	0.18	ca. 83
14	0.61	0.14	ca. 81	0.81	0.27	ca. 75
28	0.28	0.12	ca. 70	0.72	0.44	ca. 62
35	0.22	0.09	ca. 71	0.68	0.55	ca. 55

drop. As can be seen, the difference in chloride content in relation to the sum total of OH⁻- and Cl⁻-concentrations between the pore solution and the solution in the cathode chamber increased, which means that the quantity of chloride removed from the pore solution per unit of time rose steadily compared to the chloride carried into the sample. This explains why after 35 days the Cl⁻-concentration in zone A could drop to about one third of the Cl⁻-concentration existing at that time in the cathode chamber.

Altogether, the laboratory tests confirm the strong influence of the OH⁻ concentration of the liquid phase on electrochemical chloride removal and makes clear why, in the case of the salt storage hall, the chloride concentration decreased much more slowly and to a much lesser extent in the area of the negatively charged reinforcement than in the more distant concrete zones close to the surface.

4 CONCLUSIONS

It has been confirmed that a considerable amount of chloride can be extracted by non-destructive electrochemical removal. The chloride that has penetrated behind the reinforcement can be removed to a limited extent only. When tap water is used as the outer electrolyte, chloride removal becomes most efficient in the area of most severe chloride contamination, i.e. close to the concrete surface, and it becomes less and less efficient at greater depths and as the time of migration lengthens due to the continuous rise in OH⁻ concentration at the reinforcement and the concomitant drop in the Chloride Transference Number. As, however, there is normally a pronounced Cl⁻ concentration gradient in concrete and usually much less chloride is present at the reinforcement than in the deicing-salt storage hall reported here, the chloride content will, in practice, drop to "safe" levels in many cases before the OH⁻ concentration of the pore solution will rise to values that make chloride removal inefficient. However, if this is not possible within the normal period of application, as was the case in this study, the extension of migration time does not seem to be desirable because the efficiency of chloride removal declines in the vicinity of the reinforcement.

An increase in the OH⁻ concentration of the pore solution around the reinforcement has a positive effect in terms of corrosion protection. But in case of concrete that contains reactive siliceous aggregate particles, this can be dangerous because it can accelerate the damage due to alkali-silica reaction.[23,24] Moreover, the fact that the

OH⁻ concentration is raised to levels that never occur in normal concrete, adds another element of uncertainty. It is conceivable that, apart from other possible and hitherto unknown and perhaps detrimental influences, aggregates that have been considered harmless so far may become reactive under such extreme conditions.

Acknowledgements

The author wishes to thank the Austrian Federal Ministry of Economic Affairs for the financial support given to this project and the Styrian Government and the Company Systembau for their interest and cooperation.

References

1. D. R. Lankard, J. E. Slater, W. A. Hedden and D. E. Niesz, 'Neutralization of Chloride in Concrete', Report No. FHWA-RD-76-60, 1975, 1-143.
2. J. E. Slater, D. R. Lankard and P. L. Moreland, 'Electrochemical Removal of Chlorides from Concrete Bridge Decks', Transportation Research Record No. 604, 1976, 6-15.
3. G. L. Morrison, Y. P. Virmani, F. W. Statton and W. J. Gilliland, 'Chloride Removal and Monomer Impregnation of Bridge Deck Concrete by Electro-Osmosis', Report No. FHWA-KS-RD-74-1, 1976, 1-41.
4. European Patent of the Company 'Noteby', for 'Removal of Chlorides from Concrete': Application No.86 30 2888.2, 1986.
5. Personal information of the patent holder, 1994.
6. J. B. Miller, 'Chloride Removal and Corrosion Protection of Reinforced Concrete', Proceedings of the SHRP-Conference 1989, VTI-rapport 352 A of the Swedish Road and Traffic Research Institute, Göteborg, Sweden, 1990, 117-119.
7. D. G. Manning, 'Electrochemical Removal of Chloride Ions from Concrete', Proceedings of the Symposium 'Elektrochemische Schutzverfahren für Stahlbetonbauwerke', SIA Documentation D 065, Schweizer Ingenieur- und Architektenverein, Zürich, 1990, 61-68.
8. M. Molina, 'Erfahrungen mit der elektrochemischen Chloridentfernung an einem Stahlbetonbauwerk: Wirkungsweise und Beurteilung', Proceedings of the Symposium 'Elektrochemische Schutzverfahren für Stahlbetonbauwerke', SIA Documentation D 065, Schweizer Ingenieur- und Architektenverein, Zürich, 1990, 77-82.
9. H. R. Eichert, B. Wittke and K. Rose, *Beton*, 1992, **42**, 209.
10. R. Polder and H. van den Hondel, 'Electrochemical Realkalisation and Chloride Removal of Concrete - State of the Art, Laboratory and Field Experiments', Proceedings of RILEM International Conference on 'Rehabilitation of Concrete Structures', Melbourne, Australia, 1992, 135-147.
11. R. Polder and R. Walker, 'Chloride Removal from a Reinforced Concrete Quay Wall, - Laboratory tests', COST 509 annual report 1992 of project NL-2; TNO Report 93-BT-R1114, Delft, The Netherlands, 1992, 21.
12. B. Elsener, M. Molina and H. Böhni, *Corrosion Science*, 1993, **35**, Nos 5-8, 1563.
13. J. Tritthart, K. Petterson and B. Sørensen, *Cement and Concrete Research*, 1993, **23**, 1095.

14. J. Tritthart, 'Korrosionsrelevante Mechanismen und Beurteilungskriterien von Chloridverseuchtem Stahlbeton', Schriftenreihe Straßenforschung des Bundesministeriums für wirtschaftliche Angelegenheiten, Vienna, 1993, 78-92.

15. I. L. H. Hansson and C. M. Hansson, *Cement and Concrete Research*, 1993, **23**, 1141.

16. J. Bennett, K. F. Fong and T. J. Schue, 'Electrochemical Chloride Removal and Protection of Concrete Bridge Components, Volume 1: Laboratory Studies' Strategic Highway Research Program (SHRP), Report SHRP-S-657, 1993.

17. J. Bennett, K. F. Fong and T. J. Schue, 'Electrochemical Chloride Removal and Protection of Concrete Bridge Components, Volume 2: Field Trials', Strategic Highway Research Program (SHRP), Report SHRP -S-669, 1993.

18. J. Bennett and T. J. Schue, 'Evaluation of NORCURE Process for Electrochemical Chloride Removal from Steel-Reinforced Concrete Bridge Components', Strategic Highway Research Program (SHRP), Report SHRP -C-620, 1993.

19. J. Bennett and T. J. Schue, 'Chloride Removal Implementation Guide', Strategic Highway Research Program (SHRP), Report SHRP-S-347, 1993.

20. R. Polder, 'Chloride Removal of Reinforced Concrete Prisms after 16 Years Sea Water Exposure', COST 509 annual report 1993 of project NL-2; TNO Report 94-BT-RO462, Delft, The Netherlands, 1993, 22.

21. R. Polder, R. Walker and C. L. Page, *Magazine of Concrete Research*, in press.

22. W. K. Kaltenegger and G. Martischnig, 'New Gentle Method of Concrete Repair', Proceedings of 1st Slovak Conference on Concrete Structures, Bratislava, 1994, 353.

23. C. L. Page and S. W. Yu, *Magazine of Concrete Research*, 1995, **47**, No. 170, 23.

24. C. L. Page, S. W. Yu and L. Bertolini, 'Some Potential Side-Effects of Electrochemical Chloride Removal from Reinforced Concrete', Proceedings of the International Conference 'UK Corrosion & Eurocorr 94', Bournemouth, UK, 1994, 228.

25. R. Polder, 'Electrochemical Chloride Removal of Reinforced Concrete Prisms Containing Chloride from Sea Water Exposure', Proceedings of the International Conference 'UK Corrosion & Eurocorr 94', Bournemouth, UK, 1994, 239.

26. S. Chatterji, *Cement and Concrete Research*, 1994, **24**, 1051.

27. J. Tritthart, 'Changes in the Composition of Pore Solution and Solids during Electrochemical Chloride Removal', Proceedings of the second CANMET/ACI-International Symposium on Advances in Concrete Technology, Las Vegas, USA, June 11-14, ACI SP-154, 1995, 127.

ELECTROCHEMICAL REMOVAL OF CHLORIDES FROM CONCRETE - EFFECT ON BOND STRENGTH AND REMOVAL EFFICIENCY

Øystein Vennesland and Egil Peder Humstad
Norwegian Institute of Technology, 7034 Trondheim, Norway

Olav Gautefall
SINTEF Structures and Concrete, 7034 Trondheim, Norway

Guri Nustad
Norwegian Concrete Technologies A/S, P.O. Box 6626, 0502 Oslo, Norway

1 INTRODUCTION

One technique for dealing with corrosion of embedded steel due to the action of chlorides is electrochemical removal of the chlorides from the concrete. Investigations carried out in the USA in the 1970s proved that it was possible to remove chloride ions from concrete by using an electric current.[1] During the last 10 years this technique has been improved and a commercially available repair technique developed.[2] Many reports on practical experiences, laboratory investigations and general information about the process are to be found in the literature.[3-13]

During the electrochemical removal of chloride from concrete a strong DC current is applied between an external anode and the embedded steel which acts as an internal cathode. This causes chloride ions to move within the electrical field from the interior cathode towards the anode at the concrete surface. Not only the chloride ions move in the electrical field. Other ions present in the pore solution like sodium, potassium and hydroxyl ions will move in the electrical field and this high electrical charge on the reinforcement might affect the bonding between the steel and the concrete.[5,7]

Other possible harmful effects on the concrete structure have been suggested, including water accumulation,[6,7] alkali aggregate reaction (AAR) and cracking by some undefined mechanism.[13]

In the extensive SHRP program in the USA,[6] the concrete to rebar bond strength was evaluated for treated specimens by measuring the ultimate bond stress (maximum load just before failure), the bond stress at 0.01 inch loaded-end slip, and the bond stress at 0.001 inch free-end slip. The ultimate bond stress was not affected by any treatment process, up to maximum current density of 50 A/m^2 and a charge of 2000 Ah/m^2 of steel. Loaded-end slip, which is probably the most important parameter, was reduced by about 40% for the specimen subjected to both the highest current and charge, but not when either current or charge was applied independently. Free-end slip, considered less important, was reduced significantly for specimens subjected to either the highest current density or the highest charge. In summary, the chloride removal process as recommended in the SHRP program, less than 5 A/m^2 and about 1500 Ah/m^2 of steel, can be expected to have little or no effect on rebar bond strength. In view of these results, higher current densities and higher charges are not recommended.

2 EXPERIMENTAL

2.1 Specimens

Details of the specimens are presented elsewhere.[3] The specimens were concrete prisms with a central steel reinforcement insulated against electrical contact with the electrolyte by an epoxy coating. The water-cement ratio of the concrete was 0.70, the mean 28 day compressive strength was 30.8 MPa and the concrete contained 2% chlorides by weight of cement, added as sodium chloride. Effective steel surface area was 0.0503 m^2.

The prisms were de-moulded one to two days after casting, wrapped in ordinary plastic bags and allowed to cure outdoors for at least three months.

2.2 Procedure

After curing, the protruding rebar ends were electrically insulated with a silicone compound. The prisms were then placed vertically into cylindrical plastic containers lined with titanium anode mesh and filled with potable water as an electrolyte.

The current applied to each prism was controlled and recorded once a day. The temperature was normal room temperature. The current densities applied were 1.6, 4 and 8 A/m^2 rebar surface and the durations of the chloride removal tests were 7, 14, 26 and 56 days. After 26 days some specimens were decoupled and left to be tested for bond stress. The testing was done 1, 2, 7, 17 and 27 days after decoupling of the current.

After treatment the test specimens involved were cut into small segments,[3] different distances from the steel. The 15 small prisms measuring about 17 x 17 x 50 mm were crushed to powder and then analysed for chlorides.

2.3 Pull-out Tests

The prisms were loaded progressively until a maximum value for the bond stress was obtained. The loading rate was kept constant manually. According to RILEM/CEB/FIB[14] a loading target rate determined by Equation (1) can be:

$$v_p = 0.5d^2 \qquad (1)$$

where: v_p = Loading rate (N/s)
          d = Bar diameter (mm)

The bar diameter was 32 mm and the loading rate was kept constant at 512 N/s. However, due to the manually controlled equipment used, some variations in the loading rate were unavoidable.

The bond stress is defined according to Equation (2)[15]:

$$\tau = F/A = F/\pi DL \qquad (2)$$

where: τ = Bond stress (MPa)
       F = Force (N)
       A = Area (m^2)
       D = Diameter (m)
       L = Length (m)

3 RESULTS AND DISCUSSION

3.1 Chloride Removal Efficiency

The current applied to the specimens selected for study of chloride removal efficiency was 0.4 A, i.e. a current density of 8 A/m^2 rebar steel. Remaining chlorides and removal efficiency are presented in Table 1. After four weeks of treatment almost 50% of the chloride was removed and after eight weeks this had risen to almost 60%. This is in good agreement with other investigations.[4, 6, 7]

The chloride removal efficiency decreases with reduced chloride content and increased content of other ions, in this case the increased concentration of hydroxide ions produced at the embedded steel surface. Therefore the effectiveness of the treatment is reduced when the treatment time is extended from 26 to 56 days. The chloride removal efficiency of the process is about 10%, which is in agreement with other works.[6]

Table 1 *Remaining chlorides and removal efficiency*

Charge (Ah/m^2)	Days of treatment	Remaining chlorides (%)	Removal efficiency (%)
1361	7	92.5	6.8
2721	14	72.0	12.7
5053	26	52.3	11.6
10063	56	41.2	7.2

3.2 Pull-out tests

In Figures 1 to 3 changes in bond stress as a percentage of the average bond stress of the reference specimens (not subjected to current treatment) are shown against the time of treatment.

The results presented are liable to a certain degree of uncertainty. This is due to the manual control of the pressure loading rate and the fact that the results are based on single measurements only. The results are summarised in Figure 4 where changes in bond stress are shown plotted against total charge applied. There is a tendency toward a significant reduction in bond stress within the charge range 600 to 5000 Ah/m^2 of rebar steel. At the highest charge, however, there is a considerable increase in bond stress. This result is somewhat contradictory, but it should be kept in mind that the results are based on only a single measurement.

The change in bond stress with time after treatment was determined for the specimens treated for four weeks and the results are presented in Figures 5 to 7. The results show that there is a significant reduction in bond stress until between 7 and 17 days after treatment. After this time the bond stress increases until it is almost equal to that of the reference after 28 days.

Examination of the specimens revealed blackish deposits at the concrete-bar interface. The greasy consistency of these deposits was most pronounced in the prism treated with approximately 5000 Ah/m^2 of rebar surface. Part of this layer, and the underlying concrete, was removed to a depth of 1 mm from the rebar-concrete interface for further investigation. Chemical analyses revealed that the contents of calcium, sodium, potassium and iron given as their respective oxides were 5.6, 6.1, 9.4 and 16.0 % by weight of sample, respectively.

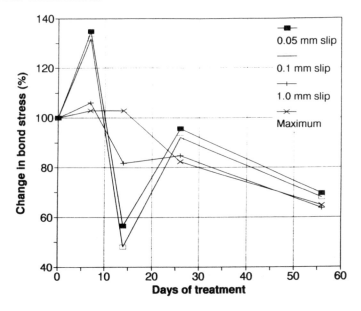

Figure 1 *Change in bond stress at a current density of 1.6 A/m² rebar surface*

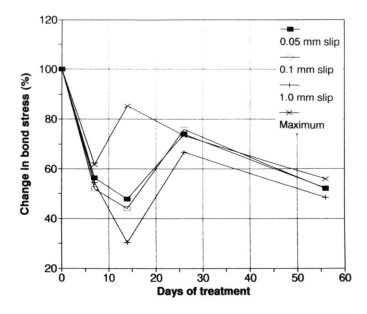

Figure 2 *Change in bond stress at a current density of 4.0 A/m² rebar surface*

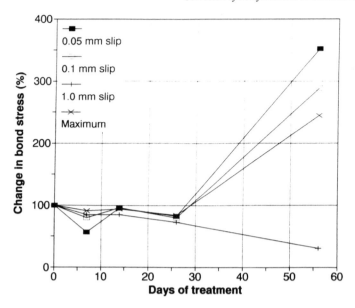

Figure 3 *Change in bond stress at a current density of 8.0 A/m² rebar surface*

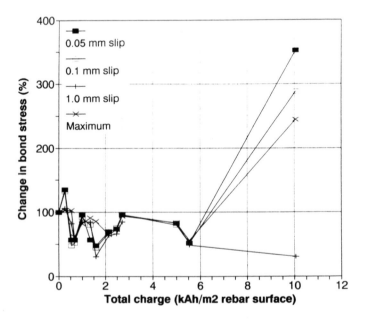

Figure 4 *Change in bond stress vs. total charge applied*

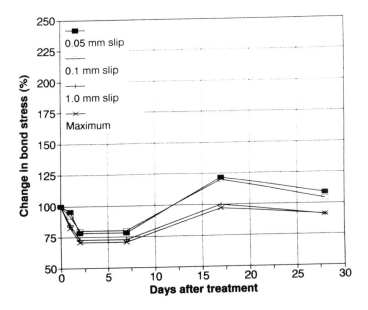

Figure 5 *Change in bond stress with time after treatment at a current density of 1.6 A/m² of rebar surface*

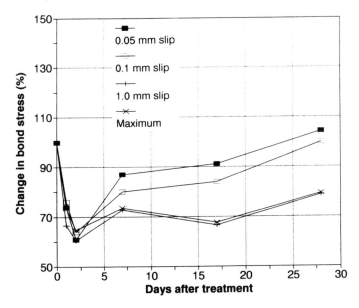

Figure 6 *Change in bond stress with time after treatment at a current density of 4.0 A/m² of rebar surface*

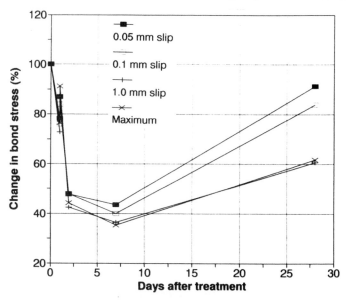

Figure 7 *Change in bond stress with time after treatment at a current density of 8.0 A/m² of rebar surface*

These values, except that for calcium, exceed the amounts normally found in concrete with ordinary portland cement and ordinary aggregate. This shows that there has been a migration of cations towards the steel surface. Thorough examination did not reveal any cracks or microcracks in the specimens.

4 CONCLUSIONS

There was a significant reduction in bond stress within the charge range 600 to 5000 Ah/m² of rebar steel. At higher charges, however, there was an extreme increase in the bond stress.

After the end of treatment the bond stress continued to decrease when the treatment period was between 7 and 17 days, and after that increased to the level of the reference specimens after 28 days.

The total chloride removed from the specimens by the treatment described above was equivalent to 40-50% of that originally present. Most of the chlorides remaining were located in the outer region of the specimens.

A small part of the total applied current contributes to the electromigration of chloride ions. The current efficiency during the first four weeks was about 11-13%, but after eight weeks of treatment the total current efficiency of chloride had been reduced to 7%.

References

1. J. E. Slater, D. R. Lankard and P. J. Moreland, *Materials Performance*, 1976, **15**, (11), 21.
2. US Patent 4832803.
3. J. B. Miller and G. E. Nustad, NACE International Conference on Engineering solution to Industrial Corrosion Problems, 1993, Sandefjord, Paper No. 49.
4. B. Elsener, M. Molina and H. Bohni, *Corrosion Science*, 1993, **35**, 1563.
5. F. G. Collins and G. A. Kirby, Proceedings of the RILEM International Conference on Rehabilitation of concrete structures, Melbourne, Australia, 1992, 171.
6. J. Bennett, T. J. Schue, K. C. Clear, D. L. Lankard, W. H. Hartt and W. J. Swiat, Electrochemical Chloride Removal and Protection of Concrete Bridge Components: Laboratory Studies, Strategic Highway Research Program, SHRP-S-657, Washington DC, 1993.
7. P.-T. Støen, Graduate work, The Norwegian Institute of Technology, Division of Structural Engineering, Trondheim, 1990, (in Norwegian).
8. J. Tritthart, K. Pettersson and B. Sørensen, *Cement and Concrete Research*, 1993, **23**, 1095.
9. I. L. M. Hansson and C. M. Hansson, *Cement and Concrete Research*, 1993, **23**, 1141.
10. W. K. Green, S. B. Lyon and J. D. Scantlebury, *Corrosion Science*, 1993, **35**, 1627.
11. F. G. Collins and P. A. Farinha, *Transactions of the Institution of Engineers, Austra lia: Civil Engineering*, 1991, **CE33**, 43.
12. A. J. van den Hondel and R. B. Polder, *Construction Repair*, 1992, **6**, 19.
13. F. Karlsson, *Nordisk Betong*, 1990, **4**, 20 (in Swedish).
14. RILEM/CEB/FIP Recommendation RC 6, Bond Test Reinforcing Steel. 2. Pull-out Test, 1st Edition, 1978.
15. CEB-FIP Model Code 1990, 83.

REDISTRIBUTION OF CHLORIDE AFTER ELECTROCHEMICAL CHLORIDE REMOVAL FROM REINFORCED CONCRETE PRISMS

Ben T.J. Stoop and Rob B. Polder

TNO Building and Construction Research
PO Box 49
NL 2600 AA Delft

1 INTRODUCTION

Penetration of chloride is the most common initiation mechanism for corrosion of reinforcement causing damage to concrete in civil engineering structures. Conventional repair methods are based on the removal of chloride laden concrete followed by application of fresh, chloride-free, concrete. As an alternative to this laborious approach electrochemical chloride removal (ECR) may be used to remove chloride from the concrete and bring the reinforcement back into a passive condition.[1] This paper gives an overview of the results obtained with ECR one year after the treatment. The work was part of the European collaboration project COST 509. It was supported by Rijkswaterstaat (Dutch ministry of public works).

2 SPECIMEN HISTORY

The specimen history is outlined schematically in Table 1.

2.1 Production (1975)

In these investigations specimens of three concrete compositions were made: OPC and w/c ratio of 0.4 (420 kg cement per m^3); OPC and w/c 0.54 (300 kg/m^3) and BFSC (portland blast furnace slag cement, with about 70% slag) and w/c 0.4 (420 kg/m^3). Originally, the concrete contained a negligible quantity of chloride. Siliceous river sand and gravel (32 mm) was used as aggregate. The specimens were prisms with dimensions of 500 x 100 x 100 mm·mm·mm, with three plain 8 mm rebars at 15, 30 and 46 mm cover depth.

2.2 North Sea Exposure (1976-1993)

The characteristics of the concrete after 16 years submersion in the North Sea can be summarised as follows. Except for strong chloride penetration, the concrete showed relatively few changes as a result of the exposure. Chemical analyses showed that penetration of sulphate and magnesium was limited to the outer few millimeters. Compressive strengths at 16 years were found to be between 40 and 80 N/mm^2. Chloride penetration was strong: OPC concrete contained a high chloride content throughout the whole cross-section, BSFC contained a high chloride content in the outer 30 mm (see Table 2 for average chloride content). More details have been given elsewhere.[2]

Table 1 *Specimen history*

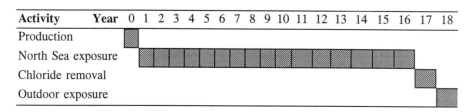

Activity	Year	0	1	2	3	4	5	6	7	8	9	10	11	12	13	14	15	16	17	18
Production																				
North Sea exposure																				
Chloride removal																				
Outdoor exposure																				

Table 2 *Main experimental details for Electrochemical Chloride Removal*

Specimen code	Composition			Treatment					Average chloride content (by mass % of cement)	
	OPC 0.4	BFSC 0.4	OPC 0.54	1 A/m² lime	4 A/m² lime	1 A/m² water	0 A/m² lime	20 °C 80 % RH	before treatment	after treatment
1	x			x					4.1	2.3
2		x		x					1.0	0.7
3			x	x					4.4	2.6
4	x				x				4.1	1.2
5		x			x				2.8	0.9
6			x		x				2.7	0.8
7	x					x			2.7	1.8
8		x				x			1.9	1.0
9			x			x			3.2	1.8
10	x						x		2.5	-
11		x					x		1.0	-
12			x				x		2.7	-
13	x							x	2.5	-
14		x						x	1.9	-
15			x					x	3.2	-

- not determined

2.3 Chloride Removal (1993)

After cutting off the top 100 mm of the prisms, electrical contacts were fixed to each of the rebars and the top and bottom surfaces were epoxy coated. Twelve prisms were placed in PVC cylinders containing an activated titanium anode mesh and an electrolyte solution. Three prisms were kept in a climate room at 20°C and 80% RH as 'dry controls'. A constant current was passed, from galvanostats, between the reinforcement bars and the anode of 9 prisms for 39 days. The applied current densities were either 1 or 4 A/m^2 steel surface in saturated calcium hydroxide solution or 1 A/m^2 in tap water. Three prisms placed in saturated lime but receiving no current served as 'wet controls'. Table 2 shows the main experimental details. After treatment chloride contents were found to have been reduced to between 40 and 70 % of the initial contents depending on the amount of charge passed. More details have been given elsewhere.[2]

2.4   Outdoor Exposure (1993-1994)

The remaining parts of each prism, with dimensions of about 100 x 100 x 200 mm·mm·mm, were used for the corrosion investigations. Electrical contacts were fixed to the rebars. The specimens were then exposed outside the TNO laboratory for a period of one year. During this period measurements of steel potential and resistivity were made on several occasions.

3 ELECTROCHEMICAL MEASUREMENTS

Steel potentials were measured shortly after ECR (November 1993) and several times during the outdoor exposure (until November 1994) using a Ag/AgCl saturated KCl reference electrode. The concrete resistivity was measured with a Wenner four electrode probe (spacing 30 mm) using 108 Hz AC. The results are given in Table 3.

4 CHLORIDE REDISTRIBUTION

During the 16 year period of North Sea exposure chloride was able to penetrate into the concrete. The chloride penetration profile after ECR was determined by analysing samples taken from the locations shown in Figure 1. It was found that the local chloride contents had been altered by the chloride removal treatment, in that part of the chloride had been removed from the concrete and part of the chloride had been displaced. As a result, significant chloride concentration gradients were present after removal. In particular, the chloride content adjacent to the rebars was low. This non-equilibrium situation was expected to change during the one year outdoor exposure period due to rediffusion of chloride. This could lead to an increase in chloride content at the rebars, which would influence the corrosion state of the rebar. During the project the local chloride content was measured three times:
   (a) after 16 years of North Sea exposure;
   (b) after chloride removal;
   (c) after one year of outdoor exposure.
For redistribution (b) and (c) are compared. Since measurement of the chloride level is inherently destructive, measurements need to be carried out on slices taken from different cross-sections of the same specimen. In the discussion of these results it is assumed that the initial chloride profile and material properties of the three cross-sections taken from the same specimen are identical.

Table 3 *Steel potential vs. Ag/AgCl reference electrode and resistivity of reinforced concrete prisms after chloride removal treatment Specimens 1-9: treated, specimens 10-15: controls*

Specimen	November 1993			November 1994			
	E_{steel} (mV)		ρ ($\Omega \cdot m$)	E_{steel} (mV)			ρ ($\Omega \cdot m$)
	Rebar cover			Rebar Cover			
	30 mm	46 mm		15 mm	30 mm	46 mm	
1	+35	-130	110	+ 60	- 35	+ 50	120
2	+10	-5	990	+ 60	- 10	+ 10	1095
3	0	-25	90	- 175	0	- 110	105
4	+25	+80	110	+ 80	+ 30	+ 40	95
5	+75	+5	430	+ 115	+ 45	- 45	545
6	+145	+140	160	+ 150	+ 165	+ 190	400
7	+75	+80	155	+ 140	+ 110	+ 150	170
8	-70	-125	610	+ 10	- 15	- 55	1005
9	+55	+25	120	+ 45	+ 105	+ 75	125
10	-345	-315	200	- 465	- 435	- 225	155
11	-330	-330	990	- 350	- 375	- 230	990
12	-295	-420	160	- 455	- 470	- 560	155
13	-205	-250	200	- 140	- 30	- 165	385
14	-110	-330	610	- 240	- 55	- 195	1130
15	-335	-310	120	- 230	- 305	- 315	160

5 LOCAL CHLORIDE CONTENT DETERMINED EXPERIMENTALLY

After one year of outdoor exposure, samples were taken for local chloride determination as was done after ECR. In Figure 2 for one prism the ten local samples in the cross section are shown in each of three situations: before ECR, immediately after ECR and after one year of outdoor exposure. The figure shows the general decrease in chloride as a result of ECR (cf. dark grey bars vs. black bars); furthermore during one year of outdoor exposure rediffusion of chloride is seen to take place (cf. light grey vs. dark grey bars). Chloride redistribution during outdoor exposure will influence the corrosion state of the reinforcement. The change in local chloride content at the three rebars with respect to the local chloride content immediately after chloride removal treatment is summarised in Table 4.

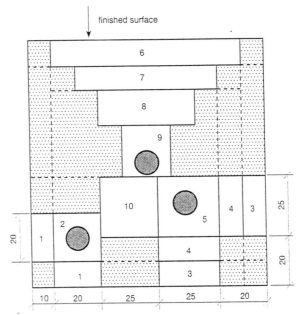

Figure 1   *Layout of sampling for local chloride analysis (cross-section). Dimensions are given in mm.*

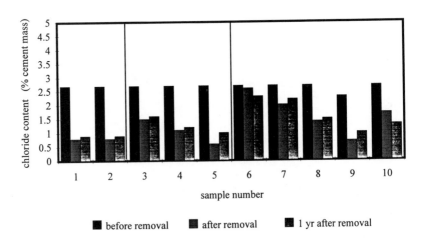

Figure 2   *Local concrete chloride contents in Specimen 7, before, after and one year after ECR; rebars were located at positions 2, 5 and 9.*

Table 4 *Chloride redistribution during one year outdoor exposure, measured at the location of the rebars.*

Specimen	Apparent increase in local chloride content (% by mass of cement) at rebar with cover of		
	15 mm	30 mm	46 mm
1	0.1	0.1	-0.6
2	0	0	0.9
3	0.3	0.4	0.5
4	0.2	0.2	-0.1
5	0	0	0
6	0.1	0.2	0.1
7	0.1	0.4	0.3
8	0.2	-0.1	-0.2
9	0.1	0.1	0.2

6 CALCULATED CHLORIDE REDISTRIBUTION

In order to predict the pattern of chloride redistribution over a long term period from the local chloride content in the specimen after chloride removal treatment the redistribution of chloride was analyzed using the finite element package DIANA.

The concrete was modeled by the grid shown in Figure 3, which consists of 10 x 10 elements having four nodes each. It was assumed that the chloride transport could be described by diffusion with a constant diffusion coefficient as determined after North Sea exposure:[2]

BSFC 0.4 $D = 0.3 \cdot 10^{-12}$ m^2/s
OPC 0.4 $D = 2.0 \cdot 10^{-12}$ m^2/s
OPC 0.54 $D = 3.0 \cdot 10^{-12}$ m^2/s

In order to check the sensitivity of the model to an incorrect value of the diffusion coefficient, control calculations were carried out with diffusion coefficients twice as large and twice as small as the values specified above. These resulted in a poor matches with the actual chloride distribution.

For the calculations the following initial and boundary conditions were used:

(1) all external fluxes are 0;
(2) at t = 0 an initial chloride concentration is present; this chloride concentration is dependent on the location in the grid and is derived from the local chloride concentrations determined experimentally in 1993;
(3) the presence of rebars in the concrete is not taken into account (i.e. in the calculations the specimen was assumed to consist only of concrete to which homogeneous transport properties were assigned).

Based on the initial chloride distribution, a transient analysis was carried out with time steps of $3.15 \cdot 10^6$ s (i.e. 1/10th of a year). Calculated chloride levels were obtained by taking 10 steps for a one year, or 100 steps for a ten year redistribution period. The calculated change in chloride distribution after one year is given for specimen N° 2

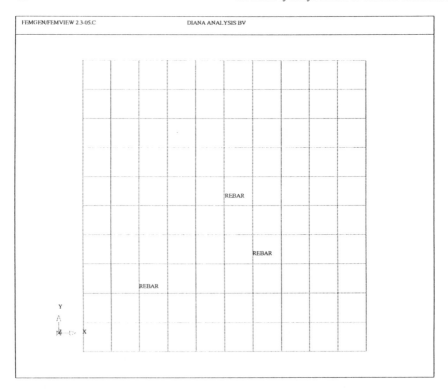

Figure 3   *Finite element grid used for calculation of chloride redistribution.*

(BSFC 0.4) in Figure 4 (1 to 3) and for specimen N° 3 (OPC 0.54) in Figure 4 (2 to 4).

7 DISCUSSION

7.1   Corrosion Related Measurements

From the steel potential measurements it is clear that chloride removal treatment is an effective means of shifting the corrosion potential to more positive values. Both dry and wet control specimens retain their negative corrosion potentials. For example in specimen 2, which had an chloride content of 1.0 % before and 0.7 % after treatment, the steel potential shifted from around -300 mV vs. Ag/AgCl range to close to 0 mV (cf. control specimen 11). Even larger shifts were recorded in the case of specimen 6 in which the average chloride content was reduced from 2.7 to 0.8 % and potentials were shifted from between -500 and -300 mV to between +130 and +190 mV (cf. control specimen 12).

A potential more negative than -260 mV vs. Ag/AgCl is taken to mean that the probability of corrosion is > 90 %. Potentials more positive than -110 mV indicate that there is a probability of > 90 % that the steel is not corroding. Within the intervening interval the corrosion activity is uncertain. These results have been confirmed elsewhere.[3]

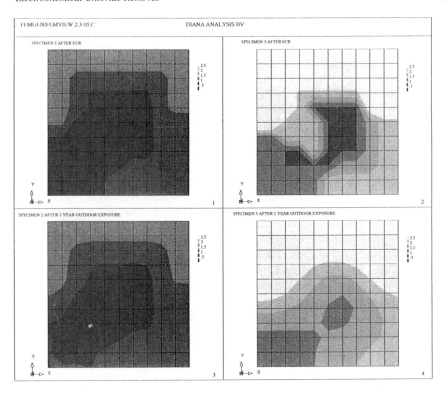

Figure 4   *Calculated chloride redistribution one year after ECR*

7.2   Measured Chloride Redistribution

The three different types of concrete used in the experiments differ with respect to the mobility of the chloride. The mobility of the chloride can be characterized by the diffusion coefficient.[2] The mobility of chloride in blast furnace slag cement concrete is small ($D = 0.3 \cdot 10^{-12}$ m^2/s). Chloride is more mobile in ordinary portland cement concrete ($D = 2.0 \cdot 10^{-12}$ m^2/s for OPC 0.4 and $D = 3.0 \cdot 10^{-12}$ m^2/s for OPC 0.54): the OPC 0.54 was fully saturated with chloride.

The difference in chloride mobility becomes apparent during the outdoor exposure period after chloride removal treatment. From Figure 2, it can be seen that the chloride removal treatment caused the largest changes in chloride content to occur in the vicinity of the reinforcement (locations 2, 5 and 9). In the more distant parts of the cross-section and in the area between the rebars (location 10), a larger fraction of the initial chloride remained. Chloride redistribution causes of the chloride to migrate back towards the reinforcement. As the corrosion state of the reinforcement is partly determined by the local chloride content, redistribution of chloride after the chloride removal treatment influences the long-term effectiveness of the treatment.

From Table 4 it can be seen that at critical locations within the concrete during the one year period subtle changes in the local chloride content were observed. In the BFSC 0.4 specimens (2, 5, 8) negligible chloride redistribution occurred. In one OPC 0.54 specimen (3) rediffusion caused the local chloride level at the rebars to increase to a level at which active corrosion could occur. This was confirmed by the potential measurements.

7.3   Calculated Chloride Redistribution

The chloride redistribution process will eventually reduce the beneficial effect of chloride removal treatment, but this effect will be limited since some of the chloride was removed from the specimens. The remaining chloride tends to spread out over the whole cross-section. From the outdoor exposure experiments it was found that within a period of one year, redistribution of the chloride did not have a large influence on the corrosion state of the steel. In order to investigate the effect of chloride redistribution on the long-term effectiveness of the chloride removal treatment, finite element calculations were carried out to assess the chloride redistribution. Results were given after one year and ten years redistribution. The best correlation between calculated and measured local chloride levels at the rebar surface after one year outdoor exposure was found in the OPC 0.54 concrete.

The calculations illustrate that after the chloride removal treatment, differences in the local chloride content across the specimen level off gradually. In specimens characterized by a large overall chloride content this process leads to loss of rebar passivation after 10 years.

During exposure over a ten year period, further redistribution of chloride takes place. The increase in the chloride level at the rebars depends on the chloride mobility and the amount of chloride stored in the concrete. BSFC 0.4 concrete is characterized by a low chloride mobility and a low chloride content, hence it has the lowest value of 0.08 % increase in the chloride level at the rebars compared to the chloride level after one year of outdoor exposure. In OPC 0.54 the chloride mobility is higher, more chloride is present. This induces a calculated increase in the chloride level of 1.03 % compared to the level after one year at the rebar surface.

The calculated chloride distributions for a ten year exposure period show that in the blast furnace slag cement concrete, chloride levels at the rebars of 0.6-0.9 % by mass of cement could be expected. In the ordinary portland cement concrete specimens, in which a large amount of chloride remained after treatment, chloride levels at the rebars of up to 2.2 % were calculated.

8 CONCLUSIONS

In the previous phase of this work it was shown that electrochemical removal of chloride is an effective means of shifting the steel potential towards more positive values. The technique was demonstrated to work on three types of concrete in which a considerable amount of chloride was present. The treatment removed 1/3 to 2/3 of the chloride present.

The present phase of this work has shown that after a one year outdoor exposure period, the steel potentials of the treated specimens have retained their quite positive values, whereas in the control specimens negative steel potentials were still recorded, indicating that the rebars are actively corroding.

After the removal treatment, some chloride remained in the concrete: at the rebars a low chloride content was found, whereas near the surface the chloride level was much higher. From these data it is evident that 2-dimensional chloride concentration gradients are present after the chloride removal treatment.

In the one year outdoor exposure period chloride redistribution occurs, small changes in the chloride level at the rebars. These changes were most noticeable in the case of OPC 0.54 concrete. It was assumed that these changes could be attributed to transport of chloride by diffusion.

Based on the experimentally determined chloride distribution and on the chloride diffusion coefficients of the concrete mixes involved, a reasonable fit was obtained between observed and calculated chloride redistribution patterns after one year. Subsequently the long term effect of the chloride removal process was evaluated by calculating the chloride redistribution over a ten year period.

From the calculations performed it became clear that in BFSC (w/c) 0.4 concrete, characterized by a low chloride mobility and a small amount of total chloride, redistibution of chloride will not give rise to loss of passivation. On the other hand in the OPC concrete corrosion may be reactivated due to back diffusion of chloride over a ten year period.

References

1. R. B. Polder, 'Electrochemical chloride removal from reinforced concrete prisms containing chloride from sea water exposure', in 'UK Corrosion and Eurocorr 94', Institute of Materials, London, 1994, 239.

2. R. B. Polder and J. A. Larbi, 'Investigation of concrete exposed to North Sea water submersion for 16 years', *HERON*, 1995, **40**, N° 1, 31.

3. S. Fiore, R. B. Polder and R. Cigna, this book.

THE TRIAL REPAIR OF VICTORIA PIER, ST. HELIER, JERSEY USING ELECTROCHEMICAL DESALINATION

K Armstrong M. G. Grantham B. McFarland,

K. Armstrong	M.G. Grantham	B. McFarland
States of Jersey	MG Associates	Martech Services Ltd.
Public Services Dept	15, County Gate	Midas House
PO Box 412, South Hill	Barnet	21, Church Street
St. Helier	Herts.	Sawtry, Huntingdon
Jersey JE2 4UY	EN5 1EH	Cambs. PE17 5SZ

1 INTRODUCTION

Victoria Pier is located within the main harbour of St. Helier, Jersey, Channel Islands. Situated within the Bay of St. Malo on the north coast of France, it is subjected to a tidal range of 12.2 m (40 ft) reported to be the fourth highest in the world. The pier has housed, until recently, the main transit warehouse for a shipping company with docking facilities for fishing, naval and coastal freighters.

The structure, was built in 1950, in reinforced concrete on steel box piles with concrete infill (later encased with concrete). Reinforced columns support the deck above the piles. The pier is built over an older masonry breakwater and berthing area and the void formed under the structure is subjected to tidal action (Fig 1).

Figure 1 *Victoria Pier, St. Helier Jersey*

A visual concrete condition survey highlighted considerable spalling and loss of section to the reinforcement in the supporting beams. Half cell, covermeter and concrete dust sample surveys were commissioned to assess the extent of the degradation. This confirmed that all the concrete elements were in the high risk category and undergoing intense electrochemical corrosion activity, caused by the ingress of chlorides.

After evaluating various repair options, the electrochemical desalination process was adopted and the condition of the concrete before and after the desalination process was assessed to assist in the interpretation of the results.

Desalination trials were carried out by a specialist contractor on typical concrete elements of the structure and the concrete analysed before and after desalination to assess the full effects of using the process. A post-desalination half cell potential survey was also undertaken.

2 INVESTIGATIONS PRIOR TO DESALINATION

2.1 **Visual Survey.**

The visual survey of the structure revealed spalling of the concrete on the soffit and sides of the beams supporting the deck and on the deck soffit. The majority of the spalls were located to the rear of the structure adjacent to the original breakwater wall. Considerable loss of section had occurred in the tension reinforcement at the soffit of the beams and in areas of deck slab soffit. The supporting columns were heavily rust stained with many corners cracked although not spalled. The staining was later discovered to be due to corrosion of the original steel box piles causing expansion and resulting in cracking in the encasing concrete. Some areas of deck soffit had delaminated, especially adjacent to the rear wall.

It was clear from this preliminary investigation that additional information was required concerning the extent of degradation, the nature of the concrete and the location of the reinforcement.

2.2 **Preliminary Corrosion Survey.**

The preliminary survey comprised the following site testing; visual survey, covermeter survey, half-cell potential survey and dust sampling to establish cement and chloride content. The survey area consisted of all four faces of four columns and two soffit areas, with one beam surveyed running between a column and an abutment wall. The area surveyed totalled 55m².

The results of the half cell potential survey indicated very negative potentials (<-500 mV Ag/AgCl/0.5 M KCl) at the base of the columns which became less negative towards the top. High potentials and closely spaced contour lines indicating intense electrochemical activity were seen at the top of one column and the sides and soffit of the adjacent beam. Moderately intense contour packing and moderately high potential values in two other columns indicated the likelihood of continuing corrosion. The results for the soffits indicated a low risk of active corrosion at the time of the survey, but the visual inspection showed that some of the reinforcing bars were so corroded that electrochemical contact had been lost in the areas of delamination.

Chloride contents by weight of cement results gave readings in the high risk category (> 1.0% by mass of cement) for 0 mm to 50 mm depth and medium to high risk

(0.4 to 1.0%) for 50 mm to 100 mm depth. A cement content of 13.8% was determined for the columns and 18.3% for the soffits.

The covermeter readings for the columns ranged from 33 mm to 91 mm over the four columns tested and 51 mm to 81 mm for the beam. The estimated depth of reinforcement to the soffits varied between 21 mm to 68 mm for four test areas and 41 mm to 89 mm for the remaining four. Carbonation readings taken on freshly exposed concrete revealed no carbonation.

The survey indicated high levels of chloride penetration and a moderate to high risk of ongoing corrosion. In some areas where corrosion had spalled the concrete, less corrosion activity was indicated by the half cell potentials. This may have resulted from drying out of the surface, while at the depth of the rebars, or inside incipient spalls, moisture may have been entrapped for some time.

The concrete quality appeared to be consistent with the period of construction. The original drawings reported the concrete mixes to be 1:11/2:3 (approximately 45 N/mm^2) for slender columns, 1:3:6 (approximately 17.5 N/mm^2) for the stocky columns and 1:2:4 (approx.' 30N/mm^2) for the slabs.

The survey data was evaluated and a comparative analysis of the repair options available was undertaken, as in Table 1.

The client required the continued use of the transit building and no loss of berthing space during the repairs, and after considering the cost-effectiveness of various options it was decided to investigate the suitability of electrochemical desalination, followed by the application of a coating, as a method of repair.

Table 1 *Repair Options*

Repair	Advantage	Disadvantage
No Remedial Action	Defers expenditure	Possible failure of Structure
Patch Repairs	Inexpensive?	Corrosion will continue
Coating	Could slow rate of corrosion	Expensive and corrosion will still continue
Cathodic Protection	Stops further corrosion	Difficult environment for CP and requires further maintenance
Desalination + coatings	Stops further corrosion?	Untried in marine environment
Infill Void Area	Removes problem	Loss of transit building and disruptive

2.3 Pre-Desalination Testing

The concrete elements examined in the preliminary survey were selected for trials, recognising that the Channel Island's aggregates have a potential for alkali aggregate reaction (AAR),[1] and that the electrochemical treatment will increase the alkalinity at the steel.[2] Cores were taken and analysed petrographically. The results indicated no evidence of AAR although the aggregate type, a greywacke/argillite, was potentially reactive.

Before commencing trials, therefore, an accelerated mortar bar test using the ASTM P1260 procedure[3] and gel pat tests[4] using 1M and 10M alkali solutions were used to increase alkalinity and observe the effect on the aggregate. The results of the mortar bar test concluded that the aggregate does expand, albeit slowly, and that alkali silica gel is formed, concluding that the aggregate is potentially reactive. This test is, however, conducted at much higher temperatures than that which could be experienced on site during the desalination process. The gel pat test was considered more relevant in determining the effects of an increase of alkalinity within the concrete. The gel pat tests showed no evidence of reaction. The risk of development of AAR in the trial areas was considered acceptable and it was recommended that further tests be carried out post-desalination to establish whether any deleterious reaction had occurred. It was decided that the desalination trials should go ahead.

3 DESALINATION TRIALS

Trials started on site in the Autumn of 1993 with standard repairs to the spalled areas before commencing the desalination process on the 5 May 1994 and completing on the 20 July 1994. The system used comprised a steel anode inserted inside a shallow cassette shutter filled with electrolyte (Figs 2 and 3).

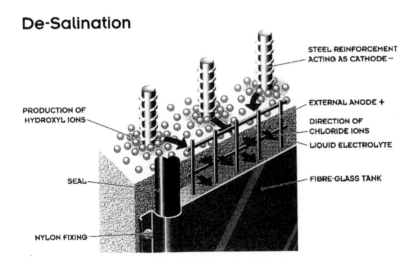

Figure 2 *Schematic Arrangement Showing Desalination System*

The cassette shutter contained a compressible gasket around the perimeter to provide a seal. Portable power supplies with data logging capabilities were employed to provide power for the desalination process, which operates by attracting negatively charged chloride ions from within the concrete into an electrolyte solution external to the concrete, by virtue of the positive charge on an external anode. Simultaneously, large amounts of hydroxyl ion are generated at the steel reinforcement, restoring passivity and raising the pH around the bar. A detailed explanation of the process is given in a report by the Strategic Highways Research Program (SHRP).[5] Dust samples were taken from the concrete before, during and after the process to be tested for chloride content. The chloride ion content of the water used as electrolyte was also determined. A core was taken post-desalination and analysed for chloride ion content. The effectiveness of chloride ion removal relates to the applied current and the time of application. Columns D4 and E4 achieved better results than columns G3 and F3 due to a longer period of application (see Table 2)

Figure 3 *The Cassette Shutter System in use on a Column*

From these results, it was recommended that a total charge of 1200 Amp hours per square metre should be applied to attain a level of chloride ions of 0.2% by mass of cement at the steel depth.

Location	Total Charge Passed	Area Treated
Soffits	187 hrs at 1A/Sq m	24m²
Columns G3, F3	293 hrs at 1A/Sq m	17m²
Columns D4, E4	748 hrs at 1A/Sq m	13m²

Table 2 *Treatment Times and Areas Treated*

3.1 **Problems and Solutions**

Various difficulties were encountered in carrying out the trial:

3.1.1 Repairs. Prior to desalination, it is necessary to use conventional repair methods to cut out and replace spalled concrete, leaving the desalination process to deal with incipient damage. Sprayed concrete repairs were chosen, but were difficult to effect on the column areas within the time and tide limitations. This meant that cassette shutter bolts were difficult to fix and compression of the seal was hard to achieve. Loss of electrolyte was also apparent through cracked repairs. In essence, the cassette shutters provided a check on the quality of the concrete repairs. Concrete repairs required better specification before a full contract could be carried out.

3.1.2 Shutters For The Temporary Anode. Whilst achieving excellent results, the fibreglass shutters, used on deck soffits, proved difficult to maintain. The opaque nature of the fibreglass meant that electrical resistance of the circuit had to be relied upon to detect loss of electrolyte or build up of gas. The tidal range and salt water also caused problems with the shutters and anode connections. This prompted a redesign of the shutters from glass fibre to clear perspex sandwiched between a neopene gasket and a rigid metal frame. With the exception of these, the process was undertaken without undue problems or disruption to the operation of the quay.

Anode connections, exterior to cassette shutters, corroded extremely quickly under tidal conditions. Indeed the movements of the tide and the corrosion of the connections were well recorded on the electrical histograms. These conditions have since been corrected using heat-shrink and mastic materials to encapsulate the connections.

3.1.3 Anode. Mild steel fabric anode used with the fibreglass shutters is sacrificial and corrodes, often needing to be replaced during the treatment. Activated Titanium mesh, on the other hand, is inert and can be re-used. In addition, when using Titanium the concrete need not be cleaned after treatment to remove rust stains. This system has been used for later work on other structures.

3.1.4 Electrolyte. Calcium hydroxide proved unable to buffer the pH for any substantial period thus necessitating frequent replacement of the buffer. During 1995 a suite of electrolytes have been developed for a variety of treatment parameters including the use of lithium based electrolytes to mitigate the effects of any alkali aggregate reactions. Surfactants and inhibitors have also been used as aids to treatment.

3.1.5 Current Density. Current densities of 1.0 A/m^2 of concrete were delivered to the deck soffits and column elements. Treatment densities of 1-2.5 A/m^2 of reinforcement and sometimes as high as 5 A/m^2 of reinforcement are more common in more recent work. Differences in current densities have significant effects on the treatment and thus affect the contract duration.

3.1.6 Surface Treatment. In order to protect the concrete from further chloride ingress, the surfaces of the desalinated areas were gunited with a low chloride permeability mortar to a thickness of 5 mm. This was less than the 15 mm thickness that had been intended. This product could be applied and could reach initial set before being covered by the tide. The sheltered nature of the environment precluded the need for temporary protection of the gunite from washout. The product chosen was used on the basis of previous experience, and on information on chloride permeability provided by the supplier. Although in these trials the gunite finish was as sprayed, the product is capable of being trowel finished. It would have lessened the extent of marine growth and water borne detritus attached to the surface, which developed within about a year, if the finish was less rough.

4 POST-DESALINATION TESTING

4.1 Initial Tests.

The initial core results indicated a successful reduction of chloride ion concentration in the vicinity of the reinforcement of the columns as shown in Table 3, and further cores and dust samples were taken to include the soffit areas. A half cell potential survey was also scheduled for a later date allowing for a period of stabilisation of the electrical potentials affecting the reinforcement.

The results for Column F3 and to some extent D4 caused some concern as the chloride ion concentration levels were considerably higher in the case of F3, or almost as high in the case of D4, than before desalination. This phenomenon is to a certain extent corroborated by the chloride analysis of the paste in the concrete cores. Earlier tests did not show this rise in chloride ion levels and showed lesser levels to Column D4. The difference in results may reflect a variation in chloride ion levels within the concrete but such a variation is unlikely. These anomalies require explanation from an engineering point of view as they greatly affect the perception of the effectiveness of the desalination process or indeed of the testing regime. It should be noted that two different testing companies were used for sampling and laboratory tests. Recent work has shown very poor reproducibility in chloride testing between laboratories despite NAMAS accreditation.[6] The key to good data is the use of control samples of known composition, run alongside the samples under test. Further and more comprehensive sampling is planned, together with corrosion rate measurements, early in 1996.

The results from the soffit areas show that almost all the chlorides have been removed which is probably due to the greater volume of reinforcement within this element.

Table 3 *Chloride Ion Content By Mass of Cement At Reinforcement Depth*

Location	Chloride Ion % Of Cement	
	Before	**After**
Column F3	1.87	0.73
Column G3	1.4	0.64
Column E4	1.04	0.35
Column D4 (1)	0.87	0.43
Column D4 (2)	0.87	0.21
Soffit F6, F7, G6, G7	n/a	n/a
Soffit F3, F4, G3, G4	n/a	n/a

All results assume a cement content of 15%

Table 4 *Results Of Post Desalination Tests*

Location	Chloride Ion % Of Cement	
	Before	**After**
Column F3 (over bar)	n/a	0.91
Column F3 (between bars)	0.29	1.46
Column D4 (over bar)	n/a	0.98
Column D4 (between bars)	1.04	0.96
Soffit F6, F7, G6, G7 (over bar)	n/a	0.08
Soffit F6, F7, G6, G7 (between bars)	1.17	0.06
Soffit F3, F4, G3, G4 (over bar)	n/a	0.06
Soffit F3, F4, G3, G4 (between bars)	n/a	0.12

4.2 Analysis of Concrete Cores

The concrete cores taken post-desalination were analysed with respect to potential Alkali Aggregate Reaction and effects of desalination on the reinforcement and surrounding concrete.

The cement paste in all five samples was based on portland cement. The paste showed varied but minor amounts of microcracking as seen in thin section with some of the microcracking passing through the coarse aggregate. These cracks had no gel or crystal growth within them, and there was no crystal growth developing in the voids seen in the thin sections.

Studies using the electron microscope showed that the aggregate had remained stable at the time of examination, with no evidence being found of reaction or of entry of alkalis into the microcracks in the aggregate. However the analyses showed that the concentration of alkalis had become exceptionally high in the cement paste within some 30 mm of the steel (Figure 4).

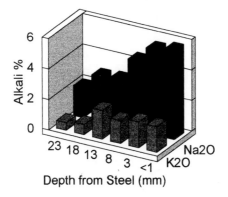

Figure 4. *Graph Showing Alkali Content around the Steel Reinforcement*

Levels of combined alkalies were up to 20 kg/m³ which is some four times the level needed for significant reaction. Since the enrichment was confined to a particular zone of concrete around the steel, the overall expansion developed if reaction took place would probably be low. It was considered that there was the potential for expansion of the order of 0.5% around the reinforcement over some years. Over time, it would be expected that the alkalis and the accompanying hydroxyl ions would migrate away from the steel, reducing the risk of AAR. Research is currently being carried out into the development and subsequent migration of hydroxyl ions and methods to measure hydroxyl concentration in concrete are being evaluated.

The core results confirmed that chloride ions had been effectively depleted from the soffit areas but only partially removed from some of the columns. Results from one of the columns for electron microprobe analysis are given in Table 5 and Figure 5. It can be seen that the chloride has been depleted around the steel, but not entirely removed.

The petrographic inspection of the concrete showed no deterioration and chloride ion levels in the immediate vicinity of the reinforcement have been found to be low.

Table 5. *Analyses of cement paste in a section 0 to 30 mm from reinforcement towards surface.*

Depth (mm)	23	18	13	8	3	<1
SiO_2	22.88	22.11	27.3	23.01	22.93	20.73
TiO_2	0.34	0.27	0.28	0.6	0.2	0.22
Al_2O_3	6.42	6.09	6.57	5.93	6.7	5.56
Fe_2O_3	6.34	2.83	2.68	3.67	3.62	2.77
MnO	0.06	0.12	0.04	0.02	0.06	0.01
MgO	0.93	0.88	1.34	1.28	1.23	1.28
CaO	58.7	61.62	54.4	57.02	57.09	61.26
Na_2O	1.98	2.96	3.05	4.57	5.58	5.8
K_2O	0.5	0.66	1.86	1.48	1.69	1.65
SO_3	1.35	1.8	1.92	1.96	0.78	0.63
Cl	0.5	0.66	0.57	0.45	0.14	0.1

(Depth = distance from reinforcement in mm. All analyses are as a percentage of the anhydrous cement paste)

4.3   Remigration of Chloride Ions

Consideration was then given to the rate at which chloride ions would migrate within the concrete. It was hoped that this data would provide a guide to the service life of the desalination process as applied to the Victoria Pier. An estimation of the Diffusion Coefficient was made from gradient chloride data, and this information was used to calculate time to remigration of chloride back to the steel, based upon measurements made with the electron microprobe. It was calculated that the chloride content would take approximately 25 years for the concentration to double at the depth of the reinforcement. In conclusion, if the steel could be passivated to a chloride ion content of 0.2% by mass of cement by the desalination process, then a service life of maybe another 25 years could be expected. If the concrete surface could be effectively coated against chloride ingress, then this period may be enhanced considerably as the driving mechanism for the chloride ingress would be reduced as redistribution occurred

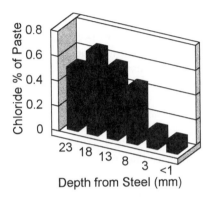

Figure 5 *Graph Showing Chloride Distribution From the Steel Reinforcement Towards the Concrete Surface*

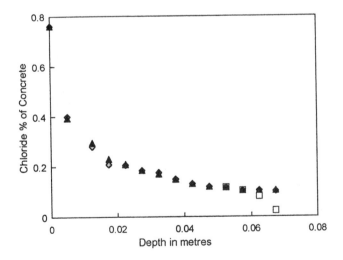

Figure 6 *Graph Showing Predicted Vs Actual Chloride Profiles*

Figure 6 is a plot of the measured chloride mass percentage against the depth in metres in one of the column sample sets, using data obtained from interpolation of the average chloride profiles given in the report (diamonds).

The amounts of chloride given by the calculated empirical formula $Cl = (DT/d)^{0.6}$ are given by the triangles. (D = Diffusion Coefficient., T is time in seconds and d is the penetration depth in metres). The squares to the right of the diagram are the compositions measured using the electron microprobe on these samples after treatment. Removal of chloride around the bar can clearly be seen.

5 POST DESALINATION SURVEY

A repeat survey was carried out in May 1995, to evaluate chloride levels and half cell potentials. The authors were not entirely satisfied with the results of this work and it is planned to carry out a comprehensive examination, to include half cell potentials, chloride tests, and linear polarisation corrosion rate measurements, early in 1996.

6 CONCLUSIONS

- The desalination process reduced the chloride ion levels within the deck slab soffit areas.
- There was only limited removal of chloride ions from columns.
- No AAR developed but high levels of alkalinity were produced in the immediate vicinity of the reinforcement.
- Before using electrochemical desalination process ensure that a full analysis of the concrete is undertaken and includes petrographic analysis of aggregates for potential alkali aggregate reaction.
- The success of desalination is proportional to the density of reinforcement. Closely spaced reinforcement gives better chloride removal than more widely spaced bars, although in the latter case, chlorides are removed from in front of the bar and for some distance around the steel.
- Prior to starting desalination, decide on the target chloride ion level required at reinforcement depth. This will provide the contractor with an aim and a figure to which the successfulness of the process can be measured.
- These trials were commissioned to establish the suitability or non-suitability of the desalination process as a viable repair method to a tidal zone structure. At this moment in time the trials cannot be said to have been successfully concluded. There are questions unresolved especially relating to the column chloride ion profiles and unexpected post desalination half cell results. It is intended to carry out a further survey including corrosion rate measurements using linear polarisation techniques early in 1996 to resolve these questions.

7 ACKNOWLEDGEMENTS

Public Services Committee and Officers for permission to produce this work. The Harbour Office, St. Helier, Jersey, also for permission to produce this paper.

8 REFERENCES

1. R. G. Cole, P. Horswill, Alkali-silica reaction: Val de La Mare dam, Jersey, Case History, Proceedings Institution of Civil Engineers: Part 1, 84, 1988, 1237.

2. C. L. Page, S. W. Yu, L. Bertolini, Potential Effects Of Electrochemical Desalination Of Concrete On Alkali Silica Reaction, Magazine Of Concrete Research, 47, No.170, March, 95, 23.

3. ASTM C1260-94, Standard Test Method For the Potential Alkali Reactivity of Aggregates (Mortar Bar Method), ASTM, Philadelphia, Pennsylvania 19103.

4. F.C. Jones, and R.D. Tarleton, Reaction Between Aggregates and Cement, National Building Studies Research Paper No 25, Part 5, Alkali Aggregate Interaction, 1958.

5. Electrochemical Chloride Removal and Protection of Concrete Bridge Components: Laboratory Studies, SHRP-S-657, Strategic Highway Research Program, National Academy of Sciences, Washington D.C., 1993.

6. M. Grantham , and R. van-Es, Admitting That Chlorides Are Variable, *Construction Repair*, 1995, Nov/Dec, 7.

REALKALISATION

DURABILITY OF ELECTROCHEMICALLY REALKALISED CONCRETE STRUCTURES

J. S. Mattila, M. J. Pentti and T. A. Raiski

Tampere University of Technology
Structural Engineering
P.O.Box 600
FIN-33101 TAMPERE
FINLAND

1 INTRODUCTION

Electrochemical realkalisation is a repair method for concrete structures suffering from carbonation induced rebar corrosion. The method was developed in the late 1980s and its use is increasing. The principle of realkalisation is to increase permanently the alkalinity of carbonated concrete by transferring artificial alkaline solution into the concrete pore system by means of a weak electric current.[1] The solution used is typically one molar sodium carbonate solution with a pH of 11.5. The aim of the treatment is to reduce significantly the corrosion rate of reinforcement in the carbonated zone and so prolong the service life of the structure.

Electrochemical realkalisation is a relatively new repair method without a long track record at least as far as the durability of the treatment is concerned. However, the short term effects of the method have been examined widely in several research laboratories, which have provided plenty of essential facts for understanding the behaviour of realkalisation.

Trials have been carried out using the electrochemical realkalisation method in Finland since 1991. During these trials it has been noticed that structures which are potential for realkalisation in Finland may differ essentially from the structures in countries where realkalisation has been developed and examined.[2] This is why it was seen as important to investigate points affecting corrosion protection and long-term behaviour of realkalisation in typical Finnish structures and climatic conditions.

The most common structures to be realkalised in Finland are concrete panel facades, which are widely used in large residential and office buildings. The structural system in these facades is a sandwich panel system (Figure 1), in which two thin concrete panels are connected together with metallic ties through a mineral wool layer. The outer panel is usually very thin (40 to 80 mm), and there is effective thermal insulation in the wall (typically 90 to 145 mm mineral wool). Thus, the climatic exposure of the outer panel is often severe and there will be intense rainfall exposure and moisture movement in the concrete. Therefore, the leaching of alkalinity may endanger the corrosion protection in realkalised concrete.

The concrete facades to be realkalised are usually coated for aesthetical reasons. However, the artificial alkalinity may weaken the adhesion of coating films to concrete and shorten the maintenance periods of the facade coatings. This is why it was important to know whether realkalisation would affect the durability of different kinds of coatings.

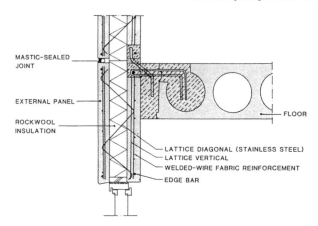

Figure 1 *Vertical cross-section of a sandwich-type exterior wall*

The sandwich panel structure may be problematic as far as electrochemical realkalisation is concerned because carbonation can also affect the inside surface of the outer panel against the insulation. In many cases a major part of the reinforcement is near the interior surface. In addition, the lattice verticals which are a structurally significant part of the lattice are made of unalloyed steel (only the diagonals which go through the insulation layer are of stainless steel).

On the basis of the considerations mentioned above the aims of the research project were stated as follows:

1) to investigate whether the alkalis are prone to leach out of thin concrete panels due to severe climatic exposure,

2) to evaluate the effect of realkalisation on the durability of facade coatings,

3) to evaluate the corrosion protection in sandwich wall structures, in which carbonation and corrosion can occur on the inside surfaces of the outer panels against insulation,

4) to evaluate the effect of further carbonation on the corrosion protection in a realkalised structure.

2 EXPERIMENTAL

This research project is based on experiments which were mostly carried out as accelerated laboratory tests. The principle of the laboratory tests described here was to simulate the behaviour of a realkalised concrete facade in severe climatic exposure. The specimens were normally realkalised (in the laboratory) concrete panels obtained from the facades of existing old buildings.

2.1 Specimen Preparation

2.1.1 Concrete Surfaces to be Realkalised. The concretes used for the specimens were pieces from two different old concrete facades. Despite having the same nominal strength the quality of the concrete was remarkably different between these two facades. Therefore the concretes from the different origins were divided into two categories respectively (C1P = porous, C2D = dense). The basic properties of the concretes are listed in Table 1.

Table 1 *The basic properties of the concretes used for specimens*

Mark	Nominal Strength [MPa]	Age [y]	Average Cover [mm]	Average Carbonation Depth [mm]
C1P	25	24	40	22
C2D	25	25	21	9

The concrete panels were sawn to dimensions of 450 by 600 mm^2. They had also been originally coated with organic paint which was removed with careful sand blasting.

The electrical connections required for realkalisation were made by chiseling one end of a rebar out of the concrete on each specimen and soldering a copper cable to it. The connections as well as cracks on the surfaces and the ends of the other bars were sealed with self-adhesive bitumen strips to avoid short circuits.

2.1.2 Realkalisation Treatment. The specimens were realkalised in the laboratory by the experts of BAC-Engineering Oy (license holder). The original Norwegian cellulose fibre method was used.

For the realkalisation treatment the specimens were laid out horizontally on a wooden platform. Cellulose fibre bulk was installed manually on top of the specimens, and the bulk was wetted with 1 M Na_2CO_3-water solution. Anodes, which were pieces of normal steel reinforcement mesh, were buried in the fibre bulk layer.

The realkalisation current was adjusted with a galvanostat to 1 A/m^2-concrete. The current could not be adjusted for each specimen separately, but it was monitored specimen by specimen several times a day. The fibre bulk was also rewetted daily.

The treatment was completed after 3 days except in the case of one specimen (C2D concrete) which was realkalised for a further 3 days to investigate the effect of longer treatment on the penetration depth of alkalis.

The current density during realkalisation was much greater than 1 A/m^2-reinforcement, which is the normative value,[1] because of the small amount of reinforcement in the specimens, only 0.13 m^2-steel/m^2-concrete. Too high a realkalisation current was selected by mistake, and it was realised only after the treatment had been completed. The average amount of current passed through specimens was 630 Ah/m^2-steel. However, in this case, since the major object was to investigate the leaching of alkalis and the durability of coatings, the high intensity of the treatment should not be very harmful.

2.1.3 Coating. After the realkalisation had been completed the specimen surfaces were cleaned carefully by high pressure water jetting and the specimens were allowed to dry in conditions T ≈ 20 to 25 °C and RH ≈ 65 % for three weeks. During drying minimal salt formations occured on the concrete surfaces. The salt formations were removed before coating by brushing.

The specimens were coated with four different types of facade coating typically used in Finland. The coatings are identified in Table 2.

The coating work was carried out according to manufacturers' instructions. Special primers were used with coatings Ke and De (Ke: polysiloxane-based and De: modified epoxy resin based primer). After coating, the specimens were stored in conditions T ≈ 20 to 25 °C and RH ≈ 65 % for three weeks.

Table 2 *The coatings used in tests*

Mark	Trade Name	Type of Binder	Description
Ke	Kenitex EH	alkyd	organic facade coating
De	Decadex	acryl	organic facade coating
CePa	Optiroc Cement Paint	white cement	cement paint
RmCp	Vetonit Repair Mortar + Optiroc Cement Paint	OPC + white cement	cement-based finishing mortar + cement paint

2.2 Tests

2.2.1 Accelerated Weathering Test. An accelerated weathering test in a weathering chamber was used to investigate leaching of alkalis and durability of coatings in weather exposure. The specimens (panels 450 by 600 mm^2, all made of concrete C2D) were fixed to a special 6 m^2 wall frame to form an exterior wall between outdoor and indoor chambers. In the wall there were 12 specimens altogether: one half realkalised and the other half untreated. Eight specimens were coated with four different coatings and in addition to this, there were four uncoated specimens (two realkalised and two untreated) in the weathering test. The weathering exposure program is described in Table 3.

The purpose of phase I was to cause intense moisture transportation through the concrete which may cause alkalis to move towards the surface and to leach out of the concrete. On the other hand, in these conditions moisture and alkalinity will accumulate behind the coating films, which may weaken the adhesion of the coatings.

In phase II very intense wetting-drying periods were achieved by cyclic spraying and drying, which probably leads to a maximal leaching rate.

At the end of the weathering test a normal freeze-thaw weathering program was executed (phase III). The purpose of this was to continue the leaching and to test the freeze-thaw resistance of the coatings.

Table 3 *The exposure program in the accelerated weathering test*

Phase	Conditions		Dura-tion [d]	Remarks
	Outdoor Chamber (realkalised & coated side)	Indoor Chamber (back side)		
I	constant: 0 °C, 45 % RH	25°C, 75% RH	30	heavy condensation on back surfaces
II	cyclic: water spray 2 h + 32 °C, 40 % RH 22 h		17	intense wetting and drying
III	cyclic: freezing-thawing-spraying in 8 h periods (5 h -30 °C & 3 h +20 °C 0.75 h water spraying)		33	100 mm mineral wool without vapour barrier on back surface

2.2.2 Accelerated Carbonation. The aim of the accelerated carbonation test was to investigate whether continued carbonation could create a zone between realkalised and uncarbonated concrete where corrosion protection may be insufficient. For this reason un-coated specimens (concrete C1P and C2D) were carbonated in 4 % CO_2-concentration and at 20 °C RH 75 % for 90 days. On the other hand, accelerated carbonation was carried out to confirm the effect of realkalisation on the carbonation rate of concrete.

2.3 Measurements and Other Observations

2.3.1 Leaching. Leaching of alkalis was examined by measuring at intervals the sodium concentration of the concrete at a depth range of 0 to 15 mm in the specimens during the accelerated weathering test. Leaching was examined only in uncoated speci-mens because leaching will be more likely in uncoated concrete structures.

For the sodium concentration measurement φ45 mm cylinders were drilled out of the specimens and 15 mm discs were sawn from the surface end. The discs were crushed and ground with ball mill to the maximum particle size of 0.5 mm. The sodium concentration was determined by the Florent method using an atomic absorption spectrometer.

The sampling for the sodium concentration measurement was performed four times during the weathering test: before the weathering test, after phase II and after 30 as well as 100 freeze-thaw cycles.

2.3.2 Durability of Coatings. The behaviour of the coatings was observed by bond strength measurements and by monitoring visual changes in the coating films.

For the bond strength measurements circular grooves were drilled through the coating film and 3 φ 75 mm test pieces were glued to the coating at each testing using solvent-free epoxy resin. The testing was carried out with the manual pull off tester Proceq Dyna 25. The bond strength measurements were performed four times during the weathering test: before the test, after phase I and after 30 as well as 100 freeze-thaw cycles.

It is known from experience that the moisture content in the concrete will affect the bond strength of all kinds of coatings. For this reason the moisture content in the concrete was approximately equalised by adjusting the conditions to 30 °C and RH 40 % in the weathering chamber for a week before each test.

2.3.3 Penetration of Alkalis and Effect on Carbonation. Penetration of the realka-lisation treatment was examined by means of the sodium concentration measurements as well as pH-indicator tests during accelerated carbonation.

For measuring the penetration of realkalisation φ45 mm cylinders were drilled out of the realkalised specimens and three successive 15 mm discs were sawn starting from the surface end of the cylinder. Sodium concentration was determined as described earlier.

The sampling during a 90 day accelerated carbonation test was performed by sawing slices from the edges of the specimens. The fresh surfaces were analysed with the standard phenolphtalein test to determine the pH profile in the concrete. Simultaneously it was possible to observe whether realkalisation affected the carbonation rate.

3 RESULTS AND DISCUSSION

3.1 Leaching of Alkalis

The changes in sodium concentration (kg/m^3) in the depth range of 0 to 15 mm in un-coated specimens measured during the accelerated weathering test are shown in Figure 2.

It can be noticed that the sodium concentration decreased from about 10 kg/m³ to about 6.5 kg/m³ due to exposure. In untreated specimens the sodium concentration has, naturally, remained almost unchanged.

When evaluating the amount of leaching it must be kept in mind that the measurement has been made from a depth of 0 to 15 mm, and the sodium concentration (at different depths) has not been even. Probably there have been much higher concentrations of sodium near the surface straight after realkalisation than the average value. It is also probable that leaching has been most intense near the surface. Therefore the measurements, which will show only the average value in the layer of 0 to 15 mm, may give us a slightly too negative impression of the volume and significance of leaching.

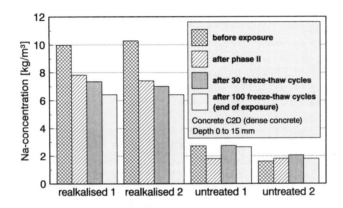

Figure 2 *Sodium concentrations in uncoated specimens (concrete C2D, depth range 0 to 15 mm) measured during the accelerated weathering test*

3.2 Durability of Coatings

In a visual examination, no defects in the paint films could be seen. Despite heavy condensation on the back surfaces of the specimens and intense moisture transportation through the concrete there were no blisters or other defects which would indicate, for example, any harmful osmosis. On the uncoated specimens there were no significant salt formations either.

The bond strength values measured during the accelerated weathering test are shown in Figure 3.

At the beginning of the weathering test (during phases I and II) the bond strengths increased in all the coatings in spite of the high alkalinity and moisture content of the concrete. One reason for this could be the age of the coatings. The first bond strength measurement was carried out as early as three weeks after coating, when the coatings had not yet reached their final strengths. The increase in bond strengths does, however, indicate that the adhesion of the coatings tested is not very sensitive to alkaline moisture in the concrete.

Later in the weathering test the bond strengths started to decrease slowly. The decrease was, however, similar in the realkalised and in the untreated specimens. In the case of the coating named Keni the bond strength was systematically somewhat weaker in the realkalised specimen compared with the untreated one and it also decreased more rapidly. The reason for this might be in the binder type (alkyd) which is known not to be totally alkali resistant, or in the primer, which should be alkali resistant, but which probably can not fully isolate the actual coating film from the alkaline substrate.

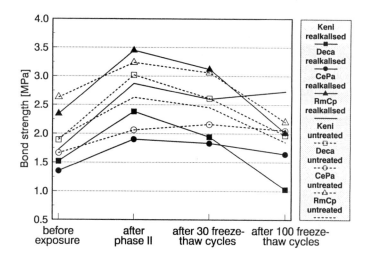

Figure 3 *The average bond strength values measured from the coated specimens during the accelerated weathering test*

At the end of the weathering test the bond strengths were more than 1.0 MPa in each specimen, when the average value was about 2.0 MPa. The weakening of the adhesion can be said to be reasonable especially when the severity of the weathering test is taken into account.

3.3 Penetration of Realkalisation and its Effect on Carbonation

The penetration of realkalisation into uncarbonated concrete was examined by measuring the sodium concentration of samples taken from different depths in the concrete and by investigating changes of the pH-profile in the specimens due to further carbonation.

The sodium concentrations measured straight after realkalisation from three depths are shown in Figure 4.

On the basis of the measurements it can be seen that in porous concrete (C1P), where carbonation has penetrated fairly deeply, realkalisation can also penetrate fairly deeply in three days treatment.

In good quality concrete (C2D), where the carbonation depth was much smaller, the sodium concentration in the zone of 0 to 15 mm remained much lower after three days of treatment. Sodium concentration could be increased by continuing the treatment for up to six days. However, behind the carbonation front, the sodium concentration did not increase at all in three days and it could not be raised significantly even by lengthening the treatment time up to six days.

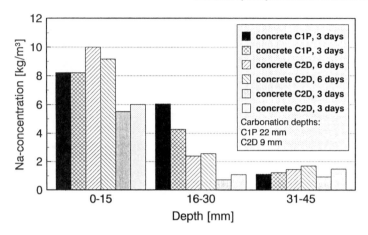

Figure 4 *Sodium concentrations at different depths straight after realkalisation*

On the basis of the pH profile changes in realkalised and untreated specimens due to accelerated carbonation, it can be stated that realkalisation does not accelerate nor reduce the carbonation rate in realkalised concrete in comparison with similar untreated concrete. This is to be expected because the amount of solids transferred into the pore system during realkalisation is quite small.

In the same context it could be noticed that because of further carbonation of 8 to 9 mm in C1P concrete, a zone had developed between the realkalised and the uncarbonated concrete where the pH was clearly lower than in the realkalised zone. This means that the penetration of realkalisation into uncarbonated concrete is very slow even in very porous concrete.

4 CONCLUSIONS

4.1 Leaching of Alkalis

The leaching of alkalis was examined by means of sodium concentration measurements performed during the weathering test. On the basis of these measurements it can be stated that alkali concentration may be lowered by intense weather exposure (cyclic wetting and drying). However, the effect of leaching will probably be limited to near the surface. Deeper in the concrete the alkali concentration may even increase due to moisture transportation, although this could not be confirmed by measurements (Figure 5).

The exposure in the accelerated test was much more intense but also of shorter duration than in normal facade structures. However, it is probable that the alkali concentration could be lowered due to leaching in normal climate exposure, although the leaching will affect primarily the zone near the surface. It can be concluded that corrosion protection in realkalised concrete may be endangered due to leaching in low cover areas (cover < 5 to 10 mm) in structures where weather exposure is intense.

In structures which are sheltered from free rainfall leaching will probably be insignificant. In structures exposed to rainfall the effect of leaching can probably be decreased by applying a hydrophobic treatment or coating.

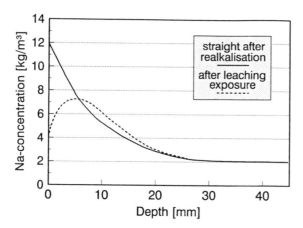

Figure 5 *Changes in sodium concentration due to leaching (hypothesis)*

4.2 Durability of Coatings

At the beginning of the accelerated weathering exposure test the bond strengths increased in all the coatings. Therefore it may be concluded that realkalisation would be at least not disasterous for coating adhesion. Later in the test the bond strengths started to decrease, but this was not clearly stronger in realkalised specimens than in untreated ones. Only the coating with a non-alkali resistant binder performed less well because the bond strength in the realkalised specimen were systematically weaker than in the untreated one and, in addition to this, the bond strength decreased more seriously in the realkalised than in the untreated one.

On the basis of the accelerated weathering test it may be concluded that realkalisation would not adversely affect the durability of alkali resistant organic or inorganic coatings. As far as other coatings are concerned, realkalisation might restrict the service life of coatings on realkalised surfaces in comparison with normal concrete surfaces.

4.3 Penetration of Realkalisation and its Effect on Carbonation

On the basis of the measurements carried out it can be stated that the effect of realkalisation will be fairly limited to the carbonated zone. It can also be confirmed that realkalisation will neither accelerate nor reduce the carbonation rate.

The fact that realkalisation does not really penetrate into uncarbonated concrete and does not retard the carbonation rate means that further carbonation may endanger the corrosion protection of the rebars behind the carbonation front. On the basis of this it may be further concluded that realkalisation can not be used as a preventive measure too early, when the carbonation front is still proceeding quite rapidly, at least not when the present treatment durations are followed. This means that realkalisation should not be recommended for relatively new structures where carbonation is still progressing fairly rapidly.

The restricted penetration of alkalis will affect the suitability of realkalisation for use in sandwich-type exterior wall structures, where rebar corrosion can be induced by carbonation from the inside surface of the outer slab. As far as these rebars are concerned the corrosion protection gained by realkalisation will be limited. The protection of this part of the reinforcement is based only on the cathodic reactions at the surface of reinforcement.

However, this kind of protection may not be very durable because the hydroxyl ions produced can be neutralised by carbonation. On the basis of this it may be further concluded that corrosion protection of reinforcement and the unalloyed part of the ties in sandwich type exterior walls may be insufficient.

Acknowledgements

The project described here was carried out as an experimental research in Tampere University of Technology. The project was financed by BAC-Engineering Oy. (licence holder in Finland), Rak.tsto A. Puolimatka Oy. (contractor), Technology Development Center of Finland and TUT Laboratory of Structural Engineering.

References

1.   R. B. Polder and H. van den Hondel, 'Electrochemical Realkalisation and Chloride Removal of Concrete', TNO Report, Delft, 1992.
2.   M. J. Pentti and J. S. Mattila, Case-study of Facade Panels Realkalised in Yläkiven-tie 5C (a residential building in Helsinki), unpublished, 1993.

MECHANISMS AND CRITERIA FOR THE REALKALISATION OF CONCRETE

G. Sergi, R. J. Walker and C. L. Page

Department of Civil Engineering,
Aston University,
Aston Triangle,
Birmingham B4 7ET, UK.

1 INTRODUCTION

Carbonation of concrete is the neutralisation of the normally basic cementitious phases of the material by acidic gases such as CO_2 present in the atmosphere. This can lead to corrosion of the steel reinforcement and a reduction in the serviceability of the structure concerned.[1] Electrochemical realkalisation has recently been developed as a rehabilitation technique for reinforced concrete where carbonation has penetrated beyond the depth of cover. It is a temporary treatment which re-establishes high alkalinity around the steel reinforcement by promoting the production of hydroxyl ions at the steel cathode and inward migration of alkali ions from an external electrolyte. The alkalinity of the cover concrete is further enhanced by absorption and diffusion of the alkaline electrolyte.[2]

Assessment of the effectiveness of the realkalisation procedure normally involves spraying a sample of the realkalised concrete with phenolphthalein indicator. Such a test can only indicate whether the pore electrolyte of the sample has exceeded pH 10, but cannot determine the degree of pH enhancement or establish whether passivation of the reinforcement has occurred. Carrying out a potential survey of the steel reinforcement, a well established test for determining the corrosion activity of the reinforcement in concrete,[3] is open to misinterpretation as it is likely that complete depolarisation of the steel, following such large polarising currents (1-2 A/m2), may require considerable periods of time. There is a need, therefore, for a better and more accurate assessment procedure, which can be applied to realkalised concrete soon after treatment. This paper describes such a procedure based on the use of a range of acid/base indicators and an anodic polarisation technique. It was developed to assess the relative performance of a number of lithium-based electrolytes considered to be suitable for realkalisation of concrete. These were of interest because lithium compounds are inhibitors of the alkali-silica reaction, which can cause expansion and cracking of concretes where high levels of alkalinity and susceptible aggregates are present.

2 METHODS FOR ASSESSING THE DEGREE OF REALKALISATION

2.1 Testing with Acid/Base Indicators

Phenolphthalein indicator is normally used to measure the depth of carbonation in concrete. The test can detect the carbonation front quite accurately because the pH of the concrete at the front usually drops fairly sharply from over 13 to less than 9. It was anticipated that, during the realkalisation process, a shallower pH gradient could develop around the reinforcement. A technique that could distinguish values in the pH range 10 to 13 should be able to detect this gradient and help elucidate the mechanisms of realkalisation. It should also be able to determine the effectiveness of the realkalisation process more accurately.

Table 1 *Acid/base indicators used for detecting changes of pH in the mortars*

	Colour change	Approximate pH change
Phenolphthalein	colourless/pink	<10
Thymolphthalein	colourless/blue	11
Titan Yellow	yellow/brown	12.5
Indigo Carmine	blue/green/yellow	13

A number of acid/base indicators were, therefore, considered and tested in a range of solutions and on cement pastes which were equilibrated in solutions of variable pH levels. The four identified as the most useful were as shown in Table 1.

2.2 Anodic Polarisation Techniques

Mietz and Isecke[2,4] adopted a polarisation technique to assess the degree of realkalisation of carbonated concrete. Their procedure involved anodically polarising a known area of the steel embedded in the concrete either potentiostatically or galvanostatically. The resultant shape of the anodic current/time or potential/time curve indicated the corrosion activity of the steel at the time and the stability of any protective film that may have formed on the surface.

The galvanostatic version of the test lends itself well for use in any laboratory equipped with simple electrochemical instrumentation so, after trial tests on laboratory concrete specimens carbonated to various degrees, it was decided to use this version throughout the programme.

3 EXPERIMENTAL PROCEDURE AND RESULTS

Mortars of 0.8 water to cement ratio and a cement content of 300 kg/m^3 were cured in water at 38° C for 4 days, dried at the same temperature for 3 days and exposed to a 100 % CO$_2$ atmosphere at 65 % RH for 7 days. The specimens were then re-immersed in water for a further 3 days, dried for 4 days and exposed to CO$_2$ for between 7 and 14 days until complete carbonation was established on replicate mortars containing no steel.

Each specimen had dimensions of 40 x 80 x 100 mm and had a steel bar, 8 mm in diameter centrally positioned along the 80 mm length and protruding at both ends at a depth of 25 mm from the 40 x 80 mm face. The steel bars had one end drilled and tapped to facilitate electrical connections. A small length of plastic tube containing fully cured flowable mortar was inserted in the top cast surface of every mortar above the centre of each steel bar. This tube was to act as an extension to the reference electrode during monitoring of the potential of the steel bar. All the faces of each specimen except the 40 x 80 mm face closest to the steel were masked with a waterproof epoxy resin. The specimens were then conditioned in a 65 % RH environment.

3.1 Assessment of Electrolytes at Constant Current Density and Time

Duplicate specimens were positioned with their exposed surfaces resting on two plastic dividers which were in turn placed on a strip of activated titanium mesh. The mesh and the first 3 mm of each specimen were immersed in 80 ml of the appropriate electrolyte inside a shallow dish (Figure 1). The titanium mesh as the anode and the steel as the cathode were connected to a galvanostat which maintained the current at the appropriate level. The solutions tested were those shown in Table 2. The current density and duration of realkalisation were kept constant at 1 A/m^2 of steel area and 5 days respectively. This condition was deliberately chosen to provide an incomplete level of realkalisation for these specimens so that differences in the behaviour of the electrolytes would be distinguishable.

Table 2 *Solutions tested in part 1 of the programme*

1. 0.5 molar LiOH	5. 0.5 molar LiNO$_3$	9. Water
2. 2 molar LiOH	6. 2 molar LiNO$_3$	10. 0.5 molar Na$_2$CO$_3$
3. 0.5 molar LiNO$_2$	7. 0.17 molar Li$_2$CO$_3$	11. 2 molar Na$_2$CO$_3$
4. 2 molar LiNO$_2$	8. 0.2 molar LiBO$_2$	

During realkalisation, the driving voltage and the anode and cathode potentials of each specimen were monitored daily to ensure the correct polarisation conditions. At termination, the instantaneous-off potential of each steel electrode was determined using a small datalogger capable of recording potentials at one second intervals. One specimen was then stored at 100 % RH for 24 hours after which it was subjected to galvanostatic anodic polarisation at a constant current density of 20 µA/cm^2 for 20 hours. Immediately before and five hours after polarisation, the corrosion rate of the embedded steel (Icorr) was determined by linear polarisation[5] and the rest potential (Ecorr) was recorded.

The duplicate specimen was sliced in half with a single cut perpendicular to the steel in such a way that two 40 mm^2 sections, 100 mm long, were produced. One section was subjected to profile cutting at 1 mm intervals from the exposed surface and the other was split in half along the plane of the steel axis so that two freshly exposed surfaces were produced. These surfaces were quickly sprayed with each of the four acid/base indicators and photographed. The zones of apparent realkalisation were measured for each indicator as a maximum and a minimum length from the surface of the steel. Where a realkalisation zone developed from the exposed surface of the specimen inwards because of the natural alkalinity of the electrolyte, this was also recorded in the same way. Figure 2 depicts diagrammatically typical realkalised zones achieved for the 2 molar LiOH solution, as shown by the four different indicators. Figure 3 represents the same zones in graphical form. All the non-alkaline solutions produced the same alkali enhancement around the steel but no realkalised band from the surface. The result obtained for 2 molar Na$_2$CO$_3$ is depicted in Figure 4 and shows a continuous band at around pH 11, the pH level associated with a buffered Na$_2$CO$_3$ solution, extending along the cover depth and joining the realkalised bands around the steel.

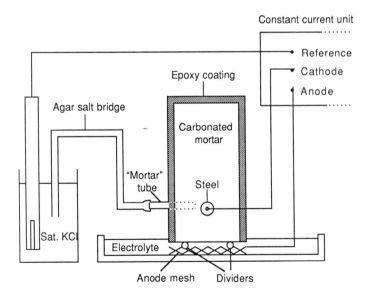

Figure 1 *Experimental set-up for laboratory tests*

Figure 5a-c shows three separate galvanostatic polarisation scans obtained for 2 molar LiOH, 2 molar Na_2CO_3 and 2 molar $LiNO_2$ respectively. In the case of LiOH (Figure 5a), the resultant curve showed that any oxide film produced during realkalisation was unable to sustain passive behaviour of the steel. The potential rose to a maximum of around 200 mV, which is much lower than the 600mV that would normally indicate complete passivity of the steel, and was followed by a gradual reduction. The shape of the curve obtained for 2 molar Na_2CO_3 was slightly different (Figure 5b) but it too showed lack of steel passivity. The polarisation curves produced for all the other conditions except for the specimens exposed to 2 molar $LiNO_2$ also showed the inability of the electrolyte and current density conditions to sustain passive behaviour. The passivation mechanism involved in the case of lithium nitrite (Figure 5c) was likely to be different as it was probably influenced by the inhibitive action of the nitrite ions.[6]

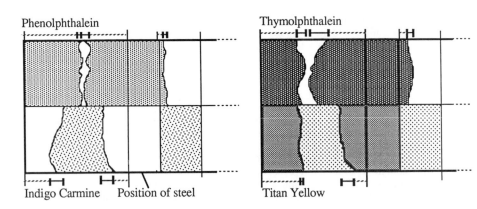

Figure 2 *Realkalisation bands shown by four different acid/ base indicators achieved during realkalisation of the mortar in 2 molar LiOH over a period of 5 days at a current density of 1 A/m²*

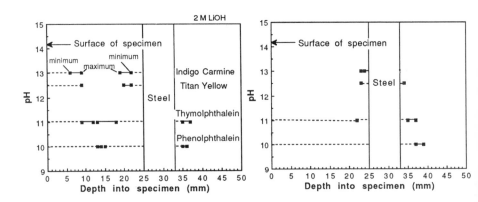

Figure 3 *Realkalisation bands of Figure 2 shown in graphical form*

Figure 4 *Realkalisation bands achieved for a mortar realkalised in 2 molar Na₂CO₃ for 5 days at 1 A/m²*

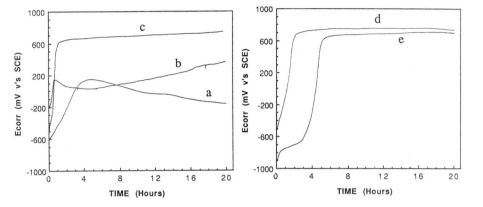

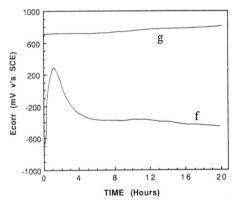

Figure 5 *Galvanostatic anodic polarisation scans of mortars 1 day after realkalisation (except g at 7 days) for the conditions shown*
a 2 molar LiOH, 5 days, 1 A/m²
b 2 molar Na₂CO₃, 5 days, 1 A/m²
c 2 molar LiNO₂, 5 days, 1 A/m²
d 2 molar LiOH, 5 days, 2 A/m², 25 mm steel cover
e 2 molar LiOH, 10 days, 2 A/m², 10 mm steel cover
f 2 molar LiOH, 10 days, no current, 10 mm steel cover, 1 day after removal from solution
g 2 molar LiOH, 10 days, no current, 10 mm steel cover, 7 days after removal from solution

A decision was made to use 2 molar LiOH as an appropriate electrolyte for part 2 of the programme, based on engineering judgment, as electrochemical data were not entirely conclusive. LiOH was at least comparable to Na_2CO_3 in terms of the pH enhancement near the steel cathode and was thought to offer extra benefits as it provided a high pH zone extending 10-15 mm from the mortar surface whilst introducing an inhibitor of alkali-silica reaction into the material (Figure 3). $LiNO_2$ was considered to present a possible risk of promoting localised corrosion which would require further investigation beyond the timescale of the present research.

3.2 Optimisation of Current Density and Time for the Realkalisation Process

The test conditions for part 2 of the programme are summarised in Table 3. Sodium carbonate electrolyte was used as well as lithium hydroxide to enable comparison. The current densities were 0, 1 or 2 A/m² of steel area and the duration of realkalisation was set at either 5 days or 10 days. A depth of cover to the steel of 10 mm was also investigated as it was thought that such shallow depths of cover would not be unusual in reinforced concrete suffering corrosion as a result of carbonation. In order to study the effects of mortar quality, some specimens were made at 0.65 water to cement ratio. This necessitated a further period of exposure in the CO_2 cabinet of about three weeks to achieve full carbonation. In one case (condition 5) the two duplicate specimens were wetted and conditioned in a 100 % relative humidity environment prior to realkalisation. Otherwise the test procedure was as for part 1.

All the resultant anodic polarisation scans, apart from those of the unrealkalised controls, showed typical passive behaviour where the potential reached around 600 mV (v's SCE), the potential associated with oxygen evolution at these pH levels (Figure 5d & e). The results showed that the realkalisation conditions tested in part 2 were all sufficient to repassivate the steel reinforcement within the initial 24 hours after realkalisation. The unrealkalised control specimen of condition 11, which was exposed to 2 molar LiOH for 10 days, was subjected to a second anodic polarisation scan, 7 days after the first. Even though the original scan showed the embedded steel to be corroding, during the repeat scan it exhibited passive behaviour as a sufficient concentration of lithium hydroxide had reached the steel by diffusion (Figure 5f & g). In the intervening period the specimen was kept in the laboratory wrapped in PVC 'clingfilm'. Anodic polarisation of the specimens of conditions 3 and 7 (2 molar LiOH and 2 molar Na_2CO_3 respectively at 2 A/m^2 for 10 days) was repeated after around 80 and 250 days. Both were shown to offer continuing protection over these periods.

All the electrochemical results of part 2 are summarised in Table 4. Each of the realkalised specimens showed a relatively high corrosion rate immediately prior to anodic polarisation. As the very negative potentials also imply (<-600 mV v's SCE), the embedded steel exhibited a low potential active condition often associated with the low availability of oxygen. This is supported by the shape of the anodic polarisation scans (Figure 5e) which first reached a potential plateau associated with the formation of an Fe^{2+}-containing oxide film formed when there is a restriction in the supply of oxygen.[7] Their corrosion rates after anodic polarisation when a 'passive' (Fe^{3+}) oxide film had formed, were lowered significantly in all cases. Their recorded potentials, although becoming more noble with time, were still low even after 80 days. It is significant that the most noble potentials at 80 days of the specimens with high cover and high water to cement ratio were achieved when the polarising conditions from realkalisation were the least severe (-163 mV and -189 mV v's SCE for conditions 2 and 6 respectively). This offers clear evidence that a potential survey carried out after realkalisation could be misleading if it is used on its own to assess the effectiveness of the treatment.

Table 3 *Realkalisation conditions chosen for part 2 of the programme*

Test Condition	Current density (A/m^2)	Period of exposure (Days)	Internal RH of mortar (%)	Electrolyte
0.8 w/c, 25 mm depth of cover				
1	1	10	65	2M LiOH
2	2	5	65	2M LiOH
3	2	10	65	2M LiOH
4	0	10	65	2M LiOH
5	1	10	100	2M LiOH
6	1	10	65	2M Na_2CO_3
7	2	10	65	2M Na_2CO_3
0.8 w/c, 10 mm depth of cover				
8	1	10	65	2M LiOH
9	2	5	65	2M LiOH
10	2	10	65	2M LiOH
11	0	10	65	2M LiOH
12	1	10	65	2M Na_2CO_3
0.65 w/c, 25 mm depth of cover				
13	1	10	65	2M LiOH
14	2	5	65	2M LiOH
15	2	10	65	2M LiOH
16	0	10	65	2M LiOH
17	1	10	65	2M Na_2CO_3

The pH profiles determined by spraying the various indicators on the freshly exposed surfaces of the realkalised specimens confirmed the increase in alkalinity around the steel and in a zone extending from the exposed surface. In this respect, lithium hydroxide was again shown to be comparable to sodium carbonate but with the added advantage of providing a reservoir of high pH solution which is able to migrate several millimetres into the specimen by absorption and diffusion. It was significant that considerable penetration of the lithium hydroxide was seen even in condition 5 (2 molar LiOH, 1 A/m^2, 10 days, 100 % RH), in which the specimens were initially wetted (Figure 6).

Significantly greater penetration of the lithium hydroxide was achieved when the depth of cover was only 10 mm (Figure 7). As the same phenomenon was seen in the control specimens at zero current, a process that was subsequently successful on its own in passivating the embedded steel (Figure 5g), transport may have increased simply because the quality of the mortar was lower within the cover zone owing to problems with compaction.

3.3 Realkalisation Applied to Cores from Existing Carbonated Structures

A number of cores were taken from an internal ceiling of a demolished block of flats in the UK. Each core contained a number of steel sections, some of which were lying in carbonated concrete. Six cores were chosen for realkalisation, four of which had a cover depth to the reinforcement of between 10 and 12 mm. The other two had a cover depth of 15-16 mm. All had carbonation depths significantly greater than the depth of cover. The procedure was the same as for the manufactured specimens apart from the exposure to CO_2. Duplicate specimens of the two cover depths were then exposed to 2 molar LiOH at a current density of 2 A/m^2 for 5 days, conditions shown to be adequate for realkalisation in part 2. The other two specimens were exposed to 2 molar Na_2CO_3 at the same current density and over the same period. Anodic polarisation scans showed that all three sets of cores allowed passivation of the steel reinforcement following the realkalisation process.

Table 4 *Summary of the electrochemical results for the part 2 tests*

ID	Current density (A/m^2)	Period of realk. (Days)	% RH before realk.	Electrolyte	Before anodic polarisation Ecorr (mV)	Icorr (μA/cm^2)	After anodic polarisation Ecorr (mV)	Icorr (μA/cm^2)	Ecorr 80 days after realk. (mV)	Anodic polaris. passed (P) or failed (F)
0.8 w/c, 25 mm cover										
1	1	10	65	2M LiOH	-1006	3.34	-456	1.06	-433	P
2	2	5	65	2M LiOH	-695	3.85	-550	1.69	-163	P
3	2	10	65	2M LiOH	-740	5.41	-501	0.73	-514	P
4	0	10	65	2M LiOH	-600	2.52	-580	3.51	-524	F
5	1	10	100	2M LiOH	-988	2.08	-534	1.23	-495	P
6	1	10	65	2M Na$_2$CO$_3$	-830	8.19	-500	0.62	-189	P
7	2	10	65	2M Na$_2$CO$_3$	-1048	2.52	-510	1.28	-358	P
0.8 w/c, 10 mm cover										
8	1	10	65	2M LiOH	-1004	3.21	-597	2.46	-383	P
9	2	5	65	2M LiOH	-747	6.83	-392	0.53	-331	P
10	2	10	65	2M LiOH	-724	7.26	-392	0.43	-303	P
11	0	10	65	2M LiOH	-246	0.12	-254	0.45	-258	F then P
12	1	10	65	2M Na$_2$CO$_3$	-925	4.02	-419	0.52	-249	P
0.65 w/c, 25 mm cover										
13	1	10	65	2M LiOH	-998	2.78	-371	0.56	-437	P
14	2	5	65	2M LiOH	-326	0.50	-280	0.23	-397	P
15	2	10	65	2M LiOH	-517	7.66	-340	0.40	-372	P
16	0	10	65	2M LiOH	-480	1.79	-520	2.89	-488	F
17	1	10	65	2M Na$_2$CO$_3$	-538	6.70	-327	0.59	-333	P

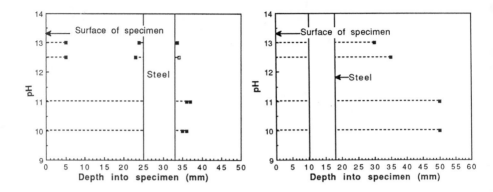

Figure 6 *Realkalisation bands achieved for a mortar realkalised in 2 molar LiOH for 10 days at 1 A/m² (mortar preconditioned in a 100 % RH environment)*

Figure 7 *Realkalisation bands achieved for a mortar realkalised in 2 molar LiOH for 5 days at 2 A/m²*

The pH profiles of the duplicate cores were all very similar and showed them to be totally realkalised to a pH level of at least 11. At a depth of cover of 10 mm, the 2 molar LiOH solution appeared to achieve almost complete realkalisation to at least pH 12.5 (Figure 8). Unfortunately, the pH changes detected by the indigo carmine indicator proved very difficult to assess on the concrete cores.

3.4 Migration of Lithium Ions During the Realkalisation Process

Powdered samples from a selection of specimens used in part 1 and part 2 were collected by profile cutting from the exposed surface to the steel. The samples were dissolved in nitric acid and the solution was filtered before being analysed with a flame photometer.

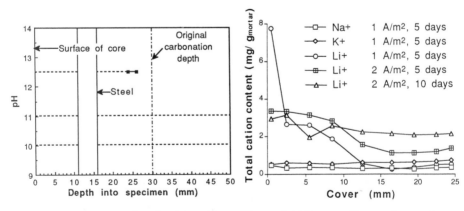

Figure 8 *Realkalisation bands achieved for a concrete core realkalised in 2 molar LiOH for 5 days at 2 A/m²*

Figure 9 *Lithium, sodium and potassium concentration profiles for mortars realkalised in 2 molar LiOH for variable periods and current densities*

Figure 9 shows typical lithium concentration profiles immediately after realkalisation. Most showed a significant concentration of lithium throughout the length of the specimen with higher concentrations near the steel and particularly near the surface. The penetration depth of the lithium ions from the surface did not appear to be greatly influenced by either the current density or the time of exposure. There was, however, some evidence to suggest that concentrations of lithium near the steel increased as the current density and time of realkalisation were increased. It was shown by others[8,9] that lithium ions reduce the likelihood of alkali-silica reaction in susceptible concretes. It is encouraging, therefore, that in most cases, measured concentrations of lithium ions were significantly greater than those of sodium and potassium ions (Figure 9 shows typical sodium and potassium concentration profiles achieved during realkalisation in 2 molar LiOH at 1 A/m^2 for 5 days).

4 DISCUSSION

During this investigation it was possible to establish a two-stage technique for testing the effectiveness of the realkalisation treatment. The first stage involves the application of several pH indicators, covering a range between pH 10 and pH 13, to freshly broken, recently realkalised mortar or concrete. This establishes a pH profile around the steel and within the cover; it represents an improvement over the phenolphthalein test which simply determines a band of concrete achieving a pH higher than about 10. The second stage involves subjecting the steel reinforcement to galvanostatic anodic polarisation whilst monitoring its potential in order to determine the degree of repassivation of the steel. Trials on steel-containing cored reinforced concrete samples indicated that it would be possible to adopt these techniques as performance criteria for concrete structures during different stages of the realkalisation process.

Lithium hydroxide, a base known to inhibit alkali-silica reaction in susceptible concretes, was shown to be a suitable electrolyte for the realkalisation process on small mortar specimens, carbonated in a laboratory. It appeared to be as successful as the more generally used sodium carbonate. A combination of 2 A/m^2 current density by area of steel for 5 days or 1 A/m^2 for 10 days appeared to repassivate the metal for periods of only a few weeks. A higher polarisation level (2 A/m^2 for 10 days) was more effective for longer periods and when tested 250 days after its application, was able to maintain passivation of the steel. Establishing a reservoir of high alkalinity within the cover concrete by promoting ingress of LiOH could potentially be very beneficial as the behaviour of the control specimen with a 10 mm cover suggested (see section 3.2), so it would appear worthwhile to optimise drying and application procedures to achieve this. Research along these lines is in hand.

It must be emphasised that the results of this investigation apply to well-characterised carbonated laboratory samples and further work will need to be undertaken to assess their applicability to real structures where non-uniform current densities could be experienced.

5 CONCLUSIONS

The main conclusions of the investigation can be summarised as follows:

1. A two-stage method involving the use of acid/base indicators and galvanostatic polarisation was established for determining the effectiveness of realkalisation conditions. The technique could be applied to cored samples taken from concrete structures which have undergone realkalisation.

2. Under controlled laboratory conditions, 2 molar LiOH was shown to be successful in realkalising fully carbonated mortar specimens and its effectiveness was comparable to that of 2 molar Na_2CO_3. The process might be improved by developing ways of increasing the depth of the reservoir of LiOH solution absorbed.

3. Concentrations of lithium ions that penetrated the cover zones of mortar specimens during realkalisation with 2 molar LiOH were substantially higher than the concentrations of sodium and potassium ions in the material. It is likely, therefore, that the risk of inducing

alkali-silica reaction in concretes with susceptible aggregates will be correspondingly reduced.

Acknowledgement

We are grateful for support of this research from Tarmac Global.

References

1. G. Sergi, S. E. Lattey and C. L. Page, 'Influence of Surface Treatments on Corrosion Rates of Steel in Carbonated Concrete', 'Corrosion of Reinforcement in Concrete', (Eds. C. L. Page, K. W. J. Treadaway and P. B. Bamforth), Elsevier Applied Science, London, 1990, 409.
2. J. Mietz and B. Isecke, 'Investigations on Electrochemical Realkalisation for Carbonated Concrete', Proc. Corrosion 94, NACE, Houston, 1994, Paper no. 297.
3. 'ASTM Standard Test Method for Half-Cell Potentials of Uncoated Reinforcing Steel in Concrete', C 876-87, American Society for Testing and Materials, Philadelphia, PA, 1989.
4. J. Mietz and B. Isecke, 'Mechanisms of Realkalisation of Concrete', Proc. UK Corrosion & Eurocorr 94, The Institute of Materials, London, 1994, **3**, 216.
5. C. Andrade and J. A. Gonzales, 'Quantitative Measurements of Corrosion Rate of Reinforcing Steels Embedded in Concrete Using Polarization Resistance Measurements'*Werkst. Korros.*, 1978, **29**, 515.
6. B. El-Jazairi and N. S. Berke, 'The Use of Calcium Nitrite as a Corrosion Inhibiting Admixture to Steel Reinforcement in Concrete', 'Corrosion of Reinforcement in Concrete', (Eds. C. L. Page, K. W. J. Treadaway and P. B. Bamforth), Elsevier Applied Science, London, 1990, 571.
7. M. Pourbaix, 'Atlas of Electrochemical Equilibria in Aqueous Solutions', Pergamon Press, Oxford, 1966, 307.
8. S. Diamond and S. Ong, 'The Mechanisms of Lithium Effects on ASR', Proc. 9th Int. Conf. on Alkali-Aggregate Reaction in Concrete, The Concrete Society, London, 1992, 269.
9. D. C. Stark, 'Lithium Salt Admixtures- An Alternative Method to Prevent Expansive Alkali-Silica Reactivity', Proc. 9th Int. Conf. on Alkali-Aggregate Reaction in Concrete, The Concrete Society, London, 1992, 1017.

AN EXPERIMENTAL INVESTIGATION INTO THE EFFECTS OF ELECTROCHEMICAL RE-ALKALISATION ON CONCRETE

T.K.H. Al-Kadhimi**, P.F.G. Banfill*, S.G. Millard§, J.H. Bungey§

*Department of Building Engineering and Surveying, Heriot-Watt University, Riccarton, Edinburgh, EH14 4AS, UK.
**School of Architecture and Building Engineering, and §Department of Civil Engineering, University of Liverpool, PO Box 147, Liverpool, L69 3BX, UK.

1 INTRODUCTION

Previous publicly available literature on re-alkalisation has been concerned with reporting development work and case studies or with raising awareness of the issues. Little experimental work has been reported. This paper is the first detailed report of an experimental investigation into the effect of the treatment on the properties of concrete. The main features and overall conclusions have been summarised elsewhere but without the supporting experimental detail which is provided for the first time by this paper.

2 PREVIOUS WORK

Electrochemical re-alkalisation is a commercial process which has been used to rehabilitate tens of thousands of square metres of concrete which is suffering from carbonation to such an extent that the steel is at risk of corrosion. A low voltage direct current is applied between the reinforcing steel (cathode) and an external temporary anode, which in turn is immersed in an electrolyte solution held against the external face of the concrete. During treatment the alkaline electrolyte solution (commonly sodium carbonate) migrates into the carbonated concrete, raising its pH, while simultaneous production of hydroxyl ions by reduction of water at the cathodic steel re-establishes the passive layer.

Faced with buildings and structures in which the concrete is showing signs of reinforcement corrosion consulting engineers advising building owners have traditionally had a single option - hack out and patch repair. In such a case unrepaired concrete remains at risk and it is not uncommon for further patch repairs to be needed. The advent of electrochemical methods presents the possibility of lasting rehabilitation for concrete through tackling the causes instead of symptoms. Re-alkalisation sits alongside chloride removal and cathodic protection in the engineer's armoury of electrochemical methods. Being longest established, standards and codes of practice exist for the latter but re-alkalisation and its sister technique, chloride removal, have been introduced relatively recently to the UK and despite a record of success, engineers are understandably wary of potential side effects.

The main and desirable effect of re-alkalisation is to re-establish passivity of the reinforcing steel but would-be users are concerned about the effect the treatment has on other properties of the concrete. Miller[1] reviewed evidence on the structural effects of both chloride removal and re-alkalisation, dealing with such matters as cracking, effects on concrete-steel bond strength, and durability.

The heating effect of passing electric current at densities of about 1 A/m^2 of concrete surface, corresponding to 1-5 A/m^2 of steel surface, could adversely affect the integrity of the concrete and possibly cause a system of cracks to develop. Such cracks could weaken the concrete and could also allow aggressive agents in the environment to penetrate and cause durability problems. Miller concludes that microcracking and microstructural damage due to re-alkalisation treatment are insignificant and much less than what is produced on concrete as a result of chiselling during patch repairs.

It might also be expected that the steel-concrete interface could be weakened by softening of the concrete immediately around the steel due to electrolytic production of hydroxyl ions. Softening was reported by Bennett and Schue,[2] but in fact, the reductions in bond strength that have been observed in chloride removal occurred at levels of total charge passed which are far in excess of those used in re-alkalisation.[3] At low levels the effects would be expected to be insignificant although no evidence is available to confirm this.

There is a further concern that because alkali metal ions migrate into the concrete during the treatment and the pH of the concrete rises, this may promote conditions conducive to the alkali-silica reaction (ASR). Recognising that ASR requires the simultaneous presence of suitable quantites of alkali reactive aggregates, moisture and alkali metal ions at temperatures above freezing and at pH above about 12,[4] it seems more likely that ASR will be a potential problem in chloride contaminated concrete than in the lower pH of carbonated or re-alkalised concrete. Page and Yu[5] confirmed the expansion due to ASR in chloride free concrete specimens, which had been electrochemically treated at the current densities typical of chloride removal, and found 'pessimum' effects of the total charge passed. However, Miller argued that because the equilibrium pH of the sodium carbonate solution introduced into the concrete is about 10.7 the pH will not be high enough to activate ASR in the bulk of the concrete even though a higher pH will be generated by electrolysis around the reinforcement.

Potential users are also rightly concerned about the long term durability or lifetime of the treatment after re-alkalisation. They need to be assured that the high alkalinity will remain around the reinforcement to preserve its passivity and that continuing atmospheric exposure will not recarbonate or reacidify the concrete. Such durability will be influenced partly by chemical factors related to the ability of the treatment to buffer the environmental acidity and partly by physical factors such as porosity and permeability which govern the ease of both ingress of aggressive environmental agents and leaching of the treatment out of the concrete. Changes in the pore structure as a result of electrochemical re-alkalisation would not be unexpected. Indeed, chemical re-alkalisation achieved by application of a cementitious material to carbonated concrete was found to change the pore size distribution significantly and the re-alkalised material was denser than the original carbonated paste.[6]

Despite this, most of the literature on re-alkalisation published to date has been concerned with reporting development work and case studies or to raising awareness of the issues and little experimental work has been reported. The research described here sets out to provide independent information on the material and mechanical properties of concrete before and after re-alkalisation. Since the objective of the experimental work was to produce strictly comparable results between uncarbonated, carbonated and re-alkalised concrete, great care was taken to ensure that possible differences due to variations in curing, age, moisture content and condition at test were minimised.

3 EXPERIMENTAL WORK

3.1 Carbonation of Concrete

In order to carry out any experimental work on re-alkalisation it is first necessary to have well characterised and reproducible carbonated concrete. After scrutiny of the alternative methods available it was felt that methods based on exposure of concrete to high concentrations of carbon dioxide at atmospheric pressure would not yield results in a reasonable time. Accordingly a hyperbaric method was developed and the final procedure consisted of two stages. First, in a preconditioning period of typically 6-8 weeks the concrete specimens were dried in air until their internal relative humidity had dropped to 70%. This was measured using an electronic humidity probe (Concretemaster II) inserted in a 10 mm diameter hole drilled 40 mm deep into a dummy concrete specimen. Second, the specimens were exposed to a pure carbon dioxide atmosphere in an enclosed cylindrical chamber (Figure 1) at 15 bar pressure for two weeks. By this means it was possible to produce fully carbonated concrete with a minimum dimension of 100 mm in as little as 10 weeks from casting. Petrographic and microstructural examination of thin sections of the carbonated concrete so produced confirmed that it showed all the features of naturally carbonated concrete[7]. These features are the migration of calcium within the microstructure, leading to precipitation of the sparingly soluble calcium carbonate wherever there is space, for example in pores. Carbon dioxide also reacts with calcium silicate hydrate and again calcium carbonate appears in pores leaving the calcium silicate hydrate as a silica rich residue. These microstructural changes account for the gross densification that occurs in carbonated concrete. By contrast, it should be noted that methods operating at atmospheric pressure but increased carbon dioxide concentration would have taken at least a year to achieve the same result.

3.2 Specimen Design and Testing

The experimental approach was to carbonate and re-alkalise specimens of an appropriate geometry for the experimental parameter under investigation, for instance, to enable effects on compressive strength to be determined by use of 100 mm cubes. This approach was adopted in preference to the alternative of carbonating and re-alkalising large concrete specimens followed by cutting cores because of the ease of ensuring full carbonation of small specimens and the reproducibility of the mechanical property measurements which result. The hyperbaric carbonation chamber had an internal diameter of 150 mm and was one metre long. These dimensions restricted the size of specimens which could be tested and the experimental programme concentrated on 100 x 100 x 500 mm and 75 x 75 x 270 mm prisms, 100 mm cubes and 100 mm diameter cylinders. The cubes were used directly for tests of compressive strength and ultrasonic pulse velocity and the larger prisms were used for tests of flexural strength and dynamic modulus of elasticity, all in accordance with appropriate standard methods. The 75 x 75 x 270 mm prisms were used in tests of alkali-silica expansion, which will be reported elsewhere.[8] Pullout bond strength was determined using 16 mm diameter smooth bars cast axially in 100mm diameter cylinders 150 mm long. No secondary helical restraining reinforcement was used because of the need to ensure that the re-alkalisation current flowed to the test bar without interruption. The set up is shown in Figure 2.

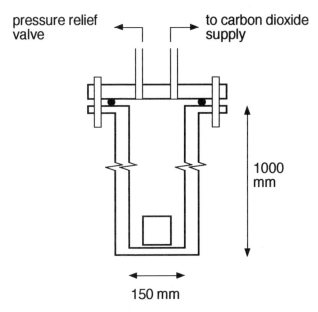

pressure relief valve

to carbon dioxide supply

1000 mm

150 mm

Figure 1 *Hyperbaric chamber for accelerated carbonation of concrete (schematic)*

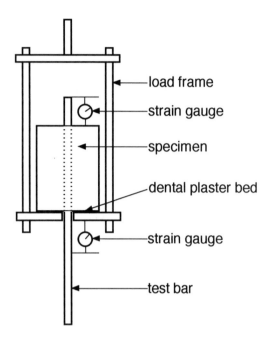

Figure 2 *Bond strength pullout test rig used in a tensile testing machine (schematic)*

Before load testing half cell potential measurements using a copper/copper sulphate reference electrode were carried out on the specimens to show the effect of re-alkalisation on the state of corrosion activity or passivity of the steel.

In re-alkalisation the current flows between anode and cathode through the minimum path length of concrete, causing migration of electrolyte.The model for electro-osmotic flow[9] suggests that both current flow and electrolyte penetration occur in the direction of the electric field gradient. In an incorrectly designed specimen this could bypass some parts of the material so it was necessary to design specimens to ensure that the whole volume was subjected to the treatment current. Cubes and prisms were cast with one face against sheet steel plates, which were perforated to add some degree of mechanical adhesion to the specific adhesion developed by interfacial contact (Figure 3). During electrochemical treatment the steel plate became the cathode, with an electrical connection to a self tapping screw, and the specimens were immersed in a tank of electrolyte solution (1M sodium carbonate solution) and surrounded by the inert metal mesh anode (platinised titanium mesh). In the pullout specimens the axial projecting steel rod became the cathode in the same way. In each case care was taken to ensure that sodium carbonate solution could only penetrate in one direction towards the cathode (rectilinear from the face opposite the cathode for cubes and prisms, radial from the circumference for pullout cylinders) by coating the other faces of the specimens with an impermeable silicone based sealant. The metal plates were removed from the cubes and prisms after treatment and before testing, so that the mechanical properties measurements were not affected by the presence of the steel. This experimental system worked very successfully.

In addition total, capillary and initial surface water absorption measurements on 100 mm cubes, and scanning electron microscopy and mercury intrusion porosimetry on samples from different parts of specimens were all performed. Total water absorption was tested using the following modification to the standard method. After drying, all specimens were coated with silicone sealant on four faces leaving the anode and cathode faces bare. Care was

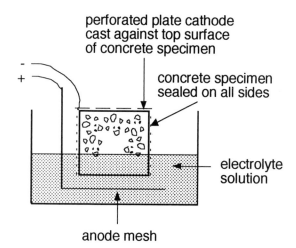

Figure 3 *Re-alkalisation set up for a cube*

taken to ensure that the equivalent faces (according to the orientation of casting) of the uncarbonated and carbonated specimens were also left bare. The cubes were then placed in water 125 mm deep for 24 hours (i.e. 25 mm clear depth over the top surface). This ensured that both anode and cathode faces were subjected to equal hydrostatic head during the test, and that any changes in absorptivity due to the re-alkalisation treatment would be shown with maximum sensitivity. Capillary water absorption was determined by standing the appropriate face of the cube in a 25 mm depth of water and initial surface absorption was determined according to British Standard.

3.3 Materials

All concrete was produced from a standard mix containing Grade 42.5 portland cement (Castle cement, Padeswood) and siliceous aggregate with a maximum size of 20 mm (Croxden Gravel, Staffordshire). Concrete for alkali-silica reactivity testing contained different proportions of calcined flint cristobalite as part of the fine aggregate (Blue Circle Industries, Greenhithe). The mix was designed to be representative of structural concrete, containing 300 kg/m^3 cement with a 28 day compressive strength of 35 N/mm^2, but the specimens were cured in their moulds for only 24 hours before exposure to a dry atmosphere for the preconditioning period prior to high pressure carbonation. Because of restrictions imposed by the size of the available equipment, the specimens for different tests were cast from different batches of nominally identical concrete and each batch was given a sample series designation.

3.4 Re-alkalisation Procedure

After carbonation the specimens were fully saturated using a vacuum saturation procedure, in which specimens were placed in a vacuum chamber which was evacuated to a pressure of 50 mbar prior to filling with water. Specimens were left immersed for 24 hours before they were connected to the rectifier. A multichannel dc rectifier allowed 10 specimens to be re-alkalised simultaneously. Each channel was individually controlled by a manually variable resistance to ensure that the treatment current passing through each specimen was maintained at 1 A/m^2 of concrete surface (1 A/m^2 of steel surface for some pullout specimens). To keep the current constant the treatment voltage was reduced in line with the drop in electrical resistance as electrolyte solution penetrated the concrete. Measurement of the electrical resistance gave a simple guide to the progress of the treatment. Re-alkalisation was continued until the resistance had reached a constant value and with the specimens being typically 100 mm thick the time to complete penetration was about 2-3 weeks. This is somewhat longer than is used in industrial practice, where the carbonation zone is normally thinner. The total charge passed was 500 ± 50 Ah/m^2.

After re-alkalisation the specimens to be tested for ultrasonic pulse velocity and the various strength tests were wrapped in polythene sheet and kept wet until they were tested wet. The uncarbonated and carbonated specimens were also saturated and kept wet at the same time to ensure that the results are directly comparable. Similarly, the specimens (uncarbonated, carbonated and re-alkalised) to be tested for total water absorption, capillary absorption and initial surface absorption were all dried at 40°C to constant weight, cooled and tested.

4 RESULTS

For every test uncarbonated (control) specimens, specimens which were carbonated only, and specimens which were both carbonated and re-alkalised were prepared. Tests were carried out on three specimens in each of the three conditions at the same time. Except where stated, results in the tables are the mean of three specimens. The different sample series designations denote different batches of nominally identical concrete. Samples F were pullout bond strength specimens, samples P were 100 x 100 x 500 mm prisms, samples S, T, U and V were 100mm cubes. The samples denoted AR/4, AR/10 and AR/H10 were all 75 x 75 x 270 mm prisms for testing of alkali-silica expansion. These latter, to be reported in detail elsewhere, contained respectively 4% and 10% calcined flint cristobalite as the reactive ingredient, and 10% with additional alkali.

Table 1 shows the variation in ultrasonic pulse velocity (PUNDIT with 54 kHz transducers) and cube compressive strength (100 mm cubes). Table 2 shows the variation in flexural strength (100 x 100 x 500 mm prisms) and dynamic modulus of elasticity and fundamental frequency (100 x 100 x 500 and 75 x 75 x 270 mm prisms). Table 3 shows the variation in pullout strength. Table 4 shows the effect on total water absorption, capillary water absorption and initial surface absorption rate (100 mm cubes). Table 5 shows half cell potentials measured on pullout specimens before and after treatment. Figure 4 shows cumulative pore size distribution data obtained by mercury intrusion porosimetry on subsamples taken from 100 mm cubes.

5 DISCUSSION

As already mentioned, the experimental procedure aimed to ensure that treated specimens were directly comparable to the controls (both uncarbonated and carbonated). It is possible that the limited initial hydration, essential to facilitate the carbonation of the specimens in a reasonable time, leaves scope for the development of hardened properties to restart when the specimens are rewetted by the re-alkalisation treatment and that this further hydration might mask the effects of the treatment. However, comparability was maximised by soaking control specimens alongside test specimens for the same time, by leaving control specimens immersed in clean water while the test specimens were re-alkalised and ensuring that both control and test specimens for every test were exposed to the same environmental conditions. Thus differences in properties between re-alkalised, carbonated and uncarbonated concretes can be attributed reliably to the treatment and not to differences in the degree of hydration.

Ultrasonic pulse velocity, compressive strength and flexural strength all increased as a result of carbonation and increased further as a result of re-alkalisation. The differences in pulse velocity are small enough to be within experimental error but since great care was taken to ensure that the specimens were all saturated at the time of testing, the differences are probably real. Taken in conjunction with the compressive and flexural strengths, the results indicate a genuine increase in strength. The dynamic modulus of elasticity increased as a result of carbonation and re-alkalisation. The fundamental frequency is higher in the smaller

Table 1 *Effect of re-alkalisation on average ultrasonic pulse velocity (PUNDIT) and average cube strength*

Sample series	Path length mm	UPV (km/sec)			Cube strength (N/mm^2)		
		Uncarb-onated	Carbonated	Carbonated +Re-alkalised	Uncarb-onated	Carbonated	Carbonated +Re-alkalised
P	500	4.45	4.61	4.64			
S	100	4.55	4.69	4.74			
T	100	4.83	4.88	4.90	22.4	33.6	34.9
U	100	4.57	4.65	4.67			
AR/H10	270	4.58	4.70	4.75			
AR/10	270	4.58	4.74	4.75			
AR/4	270	4.60	4.72	4.73			

Table 2 *Effect of re-alkalisation on average fundamental frequency, f, average dynamic modulus of elasticity, E, and average flexural strength*

Sample series	Size mm	f (Hz) (and E (kN/mm^2))			Flexural strength - N/mm^2		
		Uncarb-onated	Carbonated	Carbonated +Re-alkalised	Uncarb-onated	Carbonated	Carbonated +Re-alkalised
P	500 x 100 x 100	4140 (41.7)	4060 (41.1)	4210 (44.4)	3.57	4.68	5.72
AR/H10	270 x 75 x 75	7750 (43.8)	7540 (42.2)	7610 (43.2)			
AR/10	270 x 75 x 75	7820 (44.6)	7770 (45.3)	7800 (45.9)			
AR/4	270 x 75 x 75	7850 (45.0)	7720 (44.4)	7810 (45.7)			

Table 3 *Effect of re-alkalisation on pullout strength - (N/mm^2) (individual values, 16mm smooth bars)*

Sample series	Uncarbonated	Carbonated	Carbonated+Re-alkalised at 1 A/m^2	
			steel surface	concrete surface
F	14, 15	27, 33	51, 56, 61	57

Table 4 *Effect of re-alkalisation on water absorption (average of three specimens each)*

Series	Uncarbonated	Carbonated	Carbonated + Re-alkalised
		24 hour total water absorption (%)	
U	5.41	4.18	3.83
		24 hour capillary water absorption (%)	
V Anode face	3.02	1.74	1.40
V Cathode face	1.46	1.13	0.84
		ISA (10 mins) - ml/mm^2/sec	
U Anode face	0.53	0.29	0.36
U Cathode face	0.36	0.24	0.15

Table 5 *Half cell potentials (volts relative to Cu/CuSO4)*

Sample series	Uncarbonated	Carbonated	Carbonated+Re-alkalised at 1 A/m²	
			steel surface	concrete surface
F	-0.27, -0.26	-0.54, -0.66	-0.55, -0.44, -0.62	-0.49

specimens, as would be expected, and shows a trend of reduction as a result of carbonation followed by an increase upon re-alkalisation back to the uncarbonated value or slightly higher. The pullout strength of plain bars doubled upon carbonation and doubled again upon re-alkalisation, with a suggestion from the limited available results that calculating the current density at 1 A/m² based on the steel surface area (7.5 mA/specimen) or the concrete surface area (47 mA/specimen) has little effect on pullout strength.

Half cell potential measurements confirm the loss of passivity of the steel by carbonation. Although there was some shift towards more positive (passive) potentials after re-alkalisation, this was not as significant as might be expected. However, it must be noted that the measurements were carried out immediately after completion of the re-alkalisation and before the specimens were load tested. It was not possible to leave the specimens long enough for passivity to fully rebuild. In the long term the half cell potential values would be expected to return to those shown by the uncarbonated specimens.

Total water absorption, capillary absorption and initial surface absorption rate were all reduced in those specimens which were carbonated but not re-alkalised compared to those which were uncarbonated. Absorption was further reduced in those specimens which had been subsequently re-alkalised. The effects were greater at the cathode than at the anode, i.e. when absorption or capillary rise was measured through the surface of the concrete which had been exposed to the treatment solution (anode face) it was higher than when measured through the surface which had been in contact with the steel plate cathode (cathode face). The

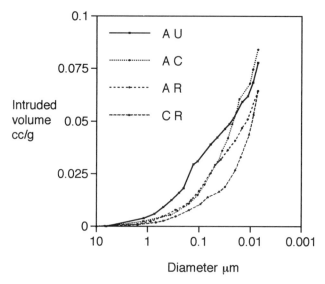

Figure 4 *Mercury intrusion porosimetry data for uncarbonated concrete (AU), carbonated concrete (AC), carbonated and re-alkalised concrete in the anode zone (AR) and carbonated and re-alkalised concrete in the cathode zone (CR)*

mercury intrusion porosimetry results confirm this effect and show that the pore sizes diminish progressively from the uncarbonated concrete through the carbonated concrete to the re-alkalised concrete and that, while the total porosity in the latter is the same in both zones, the sample from the cathode zone has a significantly larger proportion of finer pores than that from the anode zone. The pullout results give no evidence of softening in the cathode zone around the steel and these observations of differences in the pore size distribution in the anode and cathode zones suggest a reason for this: the concrete is denser around the steel resulting in increased bond strength.

These results are all consistent with a hypothesis that re-alkalisation densifies the concrete. This is confirmed by the scanning electron micrographs (Figures 5 and 6) which show the anode zone near the external surface of a 100 mm cube and the zone adjacent to the steel plate cathode. The texture in the latter is noticeably denser and more compact and the anode zone has significant areas of high porosity with many voids. Both regions exhibit the large areas of crystalline calcite and of decalcified calcium silicate hydrate which are characteristic of fully carbonated concrete, together with small amounts of the hydrated ferrite phase which is unaffected by carbonation or subsequent re-alkalisation. As noted already, such effects are not likely to be due to differences in the extent of hydration because of the care taken to match the exposure history of both control and re-alkalised specimens.

The water absorption and capillary absorption results, combined with visual observations of the cut specimens from which other tests were carried out, provide no evidence in support of the assertion that re-alkalisation produces a network of fine cracks throughout the treated concrete. Significant cracking, however fine, would lead to increased water absorption, not the significant reduction observed.

The observed changes in pore size distribution also have implications for the susceptibility of re-alkalised concrete to alkali-silica reaction. It is possible that the densification and resulting reductions in water absorption and permeability may outweigh the increased alkali content produced by re-alkalisation and prevent the ingress of water needed to fuel the expansive reactions. As a result ASR expansion may not be a problem in re-alkalised concrete, a point to be confirmed by the parallel studies on ASR.

6 CONCLUSIONS

This work on a limited number of concretes has shown no detrimental effects of electrochemical re-alkalisation. The properties change in a manner consistent with the deposition of material in the pores of the concrete leading to densification. The pore size distribution changes in the direction of smaller pores. The total water absorption, capillary absorption and initial surface absorption all decrease. The compressive strength, flexural strength, pullout strength, dynamic modulus of elasticity and ultrasonic pulse velocity all increase.

Hyperbaric treatment with carbon dioxide at 15 bar enables full size specimens of concrete, preconditioned by drying in air until the internal relative humidity has dropped to 70% or less, to be fully carbonated in typically two weeks. This is considerably quicker than previously available methods of producing artificially carbonated concrete.

While all these results are encouraging and reassuring to engineers and their clients, it must be recognised that it is not possible to predict the long term durability of the treated concrete from them and further work is needed to confirm the longevity of the treatment.

7 ACKNOWLEDGEMENTS

We are grateful to the Engineering and Physical Sciences Research Council for financial support under grant number GR J/10501 and to the members of the COST 509 Action on corrosion and protection of steel in concrete for helpful discussions. We are grateful to Dr D E Macphee for mercury intrusion porosimetry, to Dr K Scrivener for scanning electron microscopy and to D Clark for laboratory assistance.

Figure 5 *Scanning electron micrograph of re-alkalised concrete in the anode zone showing the relatively porous texture*

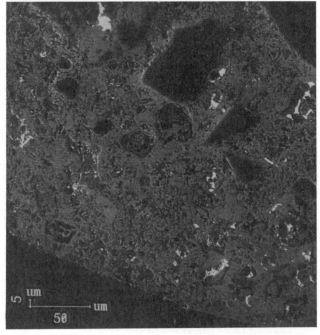

Figure 6 *Scanning electron micrograph of re-alkalised concrete in the cathode zone*

8 REFERENCES

1. J. B. Miller, 'Structural aspects of high powered electro-chemical treatment of re-inforced concrete', in 'Corrosion and Corrosion Protection of Steel in Concrete' (Ed. R. N. Swamy), Sheffield Academic Press, Sheffield, 1994, 1499.
2. J. E. Bennett and T. J. Schue, 'Electrochemical chloride removal from concrete: a SHRP status report', in Corrosion 90, NACE International, Las Vegas, USA, 1990, paper 316.
3. G. E. Nustad and J. B. Miller, 'Effect of electrochemical treatment on steel to concrete bond strength', in 'Engineering solutions to industrial corrosion problems', NACE International, Sandefjord, Norway, 1993, paper 49.
4. D. W. Hobbs, 'Alkali-silica reaction in concrete', Thomas Telford, London, 1988.
5. C. L. Page and S. W. Yu, 'Potential effects of electrochemical desalination of concrete on alkali-silica reaction', *Mag. Concr. Res.*, 1995, **47**, 23.
6. Th. A. Bier, J. Kropp and H. K. Hilsdorf, 'Realkalisation of carbonated concrete surfaces using cementitious materials', DFG Colloquium on Durability of Nonmetallic Inorganic Building Materials, Karlsruhe, 1988, 125.
7. T. K. H. Al-Kadhimi, P. F. G. Banfill, S. G. Millard and J. H. Bungey, 'An accelerated carbonation procedure for studies on concrete', *Adv. Cem. Res.*, in press.
8. T. K. H. Al-Kadhimi and P. F. G. Banfill, 'The effect of electrochemical re-alkalisation on alkali-silica expansion in concrete', 10th Int. Conf. on alkali-aggregate reaction in concrete, Melbourne, 1996, in press.
9. P. F. G. Banfill, 'Features of the mechanism of re-alkalisation and desalination treatments for reinforced concrete', in 'Corrosion and Corrosion Protection of Steel in Concrete' (Ed. R. N. Swamy), Sheffield Academic Press, Sheffield, 1994, 1489.

RE-ALKALISATION OF CARBONATED CONCRETE BY CEMENT-BASED COATINGS

J. S. Mattila and M. J. Pentti

Tampere University of Technology
Structural Engineering
P.O.Box 600
FIN-33101 TAMPERE
FINLAND

1 INTRODUCTION

Carbonation induced corrosion is a common problem affecting concrete structures in Finland. The main reasons for this are poor quality concrete, and, even more often, small cover depths due to thin structures, relatively heavy reinforcement and bad workmanship. In many cases corrosion damage is extensive, but is still often repaired with patch repair techniques. However, by using these techniques the damage can not be prevented or retarded elsewhere, just in the repaired spots.

When the corrosion damage is extensive it is often necessary to apply a finishing mortar all over the surface for aesthetic reasons. By applying a cement-based coating on the concrete surface it may be possible to retard the extension of corrosion damage. Actually, the new alkaline layer may be beneficial in several ways. The alkaline layer may prevent further carbonation and the carbonation can not penetrate deeper to affect any greater part of the reinforcement. The corrosion rate of reinforcement in carbonated concrete can also be retarded because of the re-alkalisation effect, which is the transfer of alkalinity from the alkaline surface layer into the carbonated concrete.[1] The surface layer may also affect the corrosion rate by controlling the moisture content of the carbonated concrete surrounding the reinforcement.

The effect of cement-based coatings on rebar corrosion has not been investigated widely. Cement-based coatings on concrete surfaces have become more popular in Finland, but their use only goes back to the early 1980s, and therefore there is little long-time experience available.

The aim of the research project described here was to look into the effect of different kinds of cement-based coatings on the reinforcement corrosion in carbonated concrete. The practical aim of the project was to evaluate whether it is possible to prevent or reduce the corrosion damage by using cement-based coatings.

Because of the complexity of the problems involved in the subject they were divided into six sub-questions as follows:
1. Possibility of re-alkalisation: Is it possible to increase the alkalinity of carbonated concrete by applying a cement-based coating?
2. Moisture conditions: What type of moisture conditions are needed for re-alkalisation?
3. The coatings suitable for re-alkalisation: What kinds of coatings are suitable for re-alkalisation purposes?
4. The depth of re-alkalisation: What is the effective depth of re-alkalisation?

5. Durability of re-alkalisation: Is the re-alkalisation so durable that it can be used as a preventive and/or repair method in practice?
6. Effect of further carbonation: Is it possible that a cement-based surface layer can reduce the carbonation rate significantly at the original carbonation front after the surface layer has carbonated?

Due to the complexity of the research problem numerous test series were carried out during the project. Therefore all the details and results of individual measurements are not presented here. The test methods and the main results are described concentrating on the most essential results, observations, and details.

2 EXPERIMENTAL

The aim of the experiments was to treat different concrete substrates with different kinds of commercial cement-based coatings and to monitor the changes in pH and the reinforcement corrosion due to the alkaline coatings under different conditions. The majority of the experiments were carried out in the laboratory, but sampling from one existing concrete structure coated with a cement-based coating was included.

2.1 Specimen Materials

The materials used for the specimens were old carbonated concrete panels obtained from a prefabrication plant. In addition, new concrete panels were cast in the laboratory. A part of the new panels (concrete of poor quality) was carbonated in normal atmosphere in the laboratory and the other part was exposed to accelerated carbonation in 4 % CO_2. After casting the new panels were immersed in water for one month to prevent hydration from continuing strongly during the re-alkalisation, which could give a false impression of the re-alkalisation. The basic properties of the concrete panels used in the tests are listed in Table 1.

The concrete panels used for specimens were mainly normal concrete slabs except for the specimens which were made for corrosion measurements. These specimens were provided with graphite electrodes serving as cathodes and lead electrodes serving as reference electrodes in corrosion monitoring, and they were exposed to accelerated carbonation in 4 % CO_2 (Figure 1).

Table 1 *The basic properties of the concrete panels*

Group (see Tables 2 & 3)	Mark	Age [y]	Nominal Strength [MPa]	Type of Cement	w/c	Carb. Depth [mm]
1	V	26	40	OPC	?	0 to13
	Us	2	30	OPC	0.55	0 to 5
	O	0.5	22	OPC	0.80	3 to 5
2	K	accelerated carbonation	25	OPC + 30 % BFS	0.58	≈ 13
	R	65	20	OPC	?	30 to 40

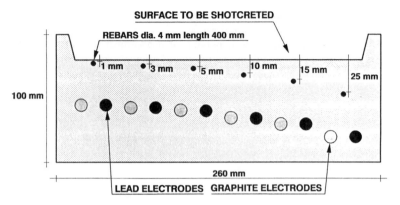

Figure 1 *Schematic diagram of the specimens made for the corrosion measurements*

The concrete panels were coated in the laboratory with several kinds of cement-based coatings. The panels of group 1 were coated with commercial cement-based facade coatings (coatings in the upper part of Table 2). In addition, a couple of organic coatings were included to look at their effect on corrosion. The panels of group 2 were shotcreted using a dry method (coatings in the lower part of Table 2).

Table 2 *The coatings tested*

Group	Mark	Description	Thickness [mm]
1	1-Al	white cement-based mortar coating	3
	2-Kui+Ter	OPC-based rendering mortar + white cement-based finishing mortar (sprayed)	5
	3-Cap	water-dilutable acrylic paint	0.5
	4-Ter	cement-based finishing mortar (sprayed)	1.5
	5-Rap	traditional lime-cement rendering	13
	6-Julk	cement-based finishing mortar	1
	7-Sem	cement paint	0.5
	8-Cons	cement-based polymer modified finishing mortar + oligomer-siloxane/acryl-based CO_2-resistant paint	1
	9-Ken	solvent-dilutable alkyd paint	0.5
	10-Puhd	mark for uncoated reference specimens	0
2	Sc1	shotcrete (dry), OPC 500 kg/m^3, w/c ≈ 0.40	20
	Sc2	shotcrete (dry), OPC 500 kg/m^3, w/c ≈ 0.45	20

2.2 Storing Conditions after Coating

After the specimens were coated they were exposed to different moisture conditions. The variations in moisture conditions are listed in Table 3.

Table 3 *The storing conditions*

Group (see Table 2)	Mark	Description
1 (facade coatings)	RH70	70 % RH in laboratory
	RH90	90 % RH in laboratory
	SS	in testing field exposed to rain
	SK	in testing field sheltered from rain
2 (shotcretes)	V	water immersed 30 d, after that 95 % RH in lab
	RH95	95 % RH in laboratory
	SS	in testing field exposed to rain
	SK	in testing field sheltered from rain

2.3 Measurements

The re-alkalisation in the specimens was evaluated by using pH-indicators, by monitoring potentials and by measuring the corrosion rates of the rebars in carbonated concrete from time to time.

2.3.1 pH-measurements. The sampling for the pH-measurements was performed by drilling cores out of the concrete. The cores were split immediately prior to analysing with indicators. The indicators used were phenolphthalein (range pH 8.2 to 9.8), thymolphthalein (range pH 9.3 to 10.5) and sometimes also a wide-range indicator (Merck 9176, range pH 9 to 13). The pH measurements were carried out at two weeks and one, two and six months after the coating.

The pH-measurements proved to be problematical because of the difficulties in evaluating the degree of the colour change and in documenting it. It was noticed that the behaviour of the indicators was somewhat different in the re-alkalised concrete compared with normal carbonated or uncarbonated concrete. In the re-alkalised concrete the colour change occurred slowly, and the colour also disappeared slowly after a couple of minutes. It seems to be possible that the pH-range of an indicator does not reflect the exact pH in the re-alkalised concrete. The pH-level may be underestimated due to the diluting effect of the indicator solution which might be essential in the case of re-alkalised concrete because there seems to be alkalinity only in the pore water, as will be discussed later.

2.3.2 Corrosion Measurements. The degree of re-alkalisation was also evaluated by measuring the rebar potentials and the corrosion rates in the specimens with embedded monitoring electrodes. These specimens were coated with the shotcretes.

The potential monitoring was started immediately after the shotcreting. The potentials were measured by means of lead/lead oxide electrodes, which will give potentials + 720 mV higher than normal Cu/CuSO$_4$-electrodes.[2] In addition to monitoring potentials, the corrosion rates were measured from time to time. Two methods were used to evaluate the corrosion rates of the rebars. Firstly, the galvanic current between a rebar and a noble graphite electrode was measured. This method was used because it is rapid and easy to apply. Secondly, polarization measurement using the Tafel method was used. The polarization measurements were carried out mainly to check the accuracy of the galvanic current method.

On the basis of the corrosion measurements carried out it can be stated that the galvanic current reflects the exact corrosion current very well. Also the values from the galvanic current measurements and the corrosion potentials matched fairly well (Figure 2). This means that the corrosion state of the reinforcement can be evaluated with reasonable accuracy on the basis of the potential monitoring.

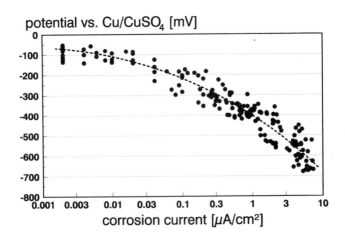

Figure 2 *The relationship between measured potentials and galvanic currents*

2.4 Field Tests

In addition to the laboratory tests, sampling from an existing concrete structure treated with a cement-based coating was included in the research. The structures examined were exterior and supporting walls of a 20 to 25 year old parking garage. The structures were shotcreted 3.5 years before the sampling.

The original carbonation depth of the concrete had been about 10 to 20 mm depending on the degree of the rainfall exposure. These values lead to carbonation coefficients of about 2.0 to 4.1 mm/√year, which represents normal carbonation rates and concrete qualities in Finland.[3]

The sampling was performed by drilling ϕ 45 mm cores out of the concrete. All together seven cores were drilled out of both the surfaces exposed to rain and the sheltered surfaces. The thickness of the shotcrete layer was about 15 to 25 mm. The measured carbonation depths of the shotcrete reached up to 10 mm during 3.5 years which means that the quality of the shotcrete was not very good.

The cores were split and analysed by means of the indicators (phenolphtalein and thymolpthtalein) to evaluate the re-alkalisation in the concrete.

3 RESULTS AND DISCUSSION

Numerous separate tests and measurements have been carried out during the research. The individual values can not be presented in this paper, but they can be found in the original report.[2] This is why the results are described as summaries using the division into sub-questions, as in section 1.

3.1 Possibility of Re-alkalisation

Re-alkalisation was examined by the means of pH-indicators and corrosion measurements. Clear evidence of re-alkalisation due to cement-based coatings was obtained from all these methods in numerous specimens and test series, as was expected.[1] By using indicators, clear colour changes indicating re-alkalisation were noticed even as early as two weeks after coating in the poor quality concrete. Re-alkalisation was somewhat slower in the specimens of good quality concrete.

The effect of re-alkalisation was also recognized by corrosion measurements (Figure 3). Figure 3 shows the potentials measured from shotcreted and untreated specimens which were both stored under the storing condition V (see Table 3). The potentials in the shotcreted specimen (solid lines) were continuously rising during the monitoring period while the potentials in the untreated specimen (dotted lines) were constantly and clearly at the corrosion level.

From the pH-measurements it could be noticed that re-alkalisation occured not only from the direction of the cement-based coating but also from the opposite direction i.e. from the uncarbonated concrete. The evaluation of the uncarbonated side re-alkalisation was difficult because no clear reference level was available, but it seemed that re-alkalisation from the uncarbonated concrete was at least as intense as it was from the cement-based coating.

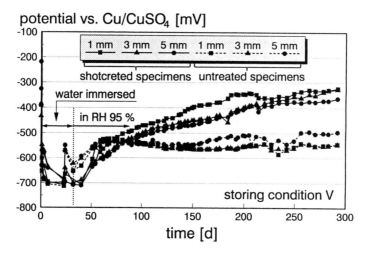

Figure 3 *The potentials in the shotcreted specimen and the untreated specimen stored under the condition V (numbers indicate the cover depths of the rebars)*

The cores taken from a real shotcreted concrete structure showed strong re-alkalisation (indicator test). The whole carbonated zone was thoroughly re-alkalised (10 to 20 mm) on the surfaces exposed to rain as well as on the surfaces totally sheltered from rain.

All the test series proved that re-alkalisation is a slow phenomenon. However, the re-alkalisation front seems to proceed into the concrete at at least the same speed as the carbonation front had proceeded when the structure was new (the square root model).[4]

3.2 Moisture Conditions

Re-alkalisation probably occurs mainly by diffusion and sometimes through capillary forces in the pore solution of the concrete. Therefore it follows that there has to be a threshold moisture content in concrete which is required for re-alkalisation.

In the test series re-alkalisation was noticed to occur at a relative humidity of 90 % or more (also in 95 % RH and outdoors, Table 3). At a relative humidity of 70 % no re-alkalisation was seen. Re-alkalisation was most rapid in the specimens stored outdoors and exposed to rainfall. In the specimens stored in the constant RHs or sheltered from rain re-alkalisation was some 30 to 50 % slower.

The specimens with the corrosion monitoring electrodes and stored exposed to rain were an exception. In these specimens no re-alkalisation was found during the first year of storage. The potentials were more or less constant at a level of about -600 mV vs. $Cu/CuSO_4$ in the shotcreted as well as in the untreated specimens despite the fact that the conditions for re-alkalisation were favourable (moist). The reason for this may, however, be found in the blast furnace slag in the concrete substrate. During the 28 day water curing only part of the slag had reacted, and a lot of unreacted slag had remained in the concrete. This unreacted part may be able to absorb the alkalinity transferred into the carbonated concrete by re-alkalisation.

The threshold moisture content for the reinforcement corrosion in carbonated concrete is usually considered to be 85 % RH.[4] This is very near to the value found critical for re-alkalisation in this research. It seems that re-alkalisation is possible in conditions where the rebar corrosion can also proceed.

3.3 Coatings Suitable for Re-alkalisation

In the tests the cement-based coatings did not prove equal as far as re-alkalisation is concerned. It was noticed that re-alkalisation mainly occurred in samples with relatively thick coatings, such as 2-Kui+Ter, 5-Rap and shotcretes. Re-alkalisation was seen only in RH 90 % or more. No re-alkalisation was found in drier conditions. Actually, in dry conditions the carbonation front proceeded into the concrete despite the new alkaline cement-based facade coatings (group 1 in Table 2). This means that CO_2 is able to penetrate through an alkaline coating into the concrete substrate and, consequently, no re-alkalisation can occur.

The thinnest facade coatings like 4-Ter, 6-Julk and 7-Sem, did not cause any re-alkalisation even in the most humid conditions. Therefore it is obvious that CO_2 can penetrate through these thin and permeable coatings even in very moist conditions.

On the basis of the tests it seems that CO_2-permeability is the most important property of a cement-coating as far as re-alkalisation is concerned. The coating has to act as a CO_2-barrier so that CO_2 can not carbonate the re-alkalisation which has occurred during moist periods. The only coatings to give CO_2 protection were shotcretes, which can also preserve the re-alkalisation over dry periods.

3.4 Depth of Re-alkalisation

The effective depth of re-alkalisation is a complicated question because there are numerous factors to consider, such as the duration of favourable moisture periods, the density of the concrete substrate, the acceptable amount of further corrosion, etc. However, the question proved to be irrelevant to a certain extent because re-alkalisation can also occur from the direction of the uncarbonated concrete. This improves the efficiency of re-alkalisation because the majority of the rebars to be protected are usually near the carbonation front.

As stated earlier, re-alkalisation is a fairly slow phenomenon. However, the re-alkalisation rate is not very important because the rebars should be protected before the structural capacity is significantly lowered or further corrosion leads to cracking or spalling of the concrete cover. If the surface to be re-alkalised is cleaned with, for example, an ultra high pressure water jetting or sand blasting the rebars which are about to cause cracking will be revealed and there will probably be a period of at least a couple of years during which re-alkalisation can occur. The rate of re-alkalisation is also dependent on the moisture content of the concrete, i.e. the re-alkalisation rate will be high during the same periods that the corrosion rate is high.

3.5 Durability of Re-alkalisation

During the research it was noticed that re-alkalisation is not always very durable. In the specimens stored outdoors re-alkalisation following a moist period disappeared later despite the fact that the coatings were still clearly alkaline. It is probable that this happened because the CO_2-permeability of the coatings increased due to drying.

Later it was also noticed that re-alkalisation disappears instantly after the coating carbonates through its whole thickness. It was also noticed that in the cores drilled out of re-alkalised specimens carbonation proceeds several millimetres in a couple of days from the unprotected surface of the core into the re-alkalised concrete. These observations indicate that there are very small amounts of alkalinity in the re-alkalised concrete, which can carbonate very rapidly if CO_2 can penetrate the coating.

During long term monitoring it was noticed that the cement-based facade coatings (group 1 in Table 2) carbonate quite rapidly. In most of the coatings the whole thickness of the coating was carbonated in less than two years, especially on the surfaces sheltered from rain. This means that re-alkalisation induced by cement-based facade coating will be of relatively short duration.

The shotcretes, however, proved to be very slowly carbonating. The carbonation depths were only 1 to 3 mm after a two year storage sheltered from rain.

3.6 Effect of Further Carbonation

All the coatings possess a certain diffusion resistance which may retard the CO_2-diffusion into concrete even after the coating has been thoroughly carbonated. This effect may be beneficial as regards the expansion of the corrosion damage.

The long term pH-measurements carried out in the specimens stored in the test field produced information about the effect of the coatings on the carbonation rate of the concrete substrate. The carbonation depths measured at the age of three years are presented in Figure 4.

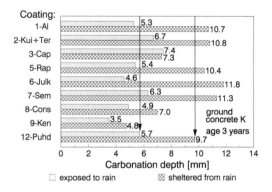

Figure 4 *The carbonation depths of the specimens (concrete substrate) at the age of three years*

Figure 4 shows that in the specimens sheltered from rain the cement-based coatings show no decrease in the carbonation depth at all. This is a little bit surprising because the coatings will inevitably increase the diffusion resistance of the carbonation zone at least to some extent. This finding may, however, be explained by the effect of the coating on the moisture behaviour at the surface. Porous cement-based coatings may absorb the capillary water condensing on the surface from time to time which would otherwise block the smaller pores of the uncoated concrete and reduce the carbonation rate. The porous coatings may let CO_2 through even when they are moist, and the carbonation reducing effect of the moisture becomes less efficient. On the other hand, this effect would not necessarily be harmful because the corrosion rate of the rebars will be reduced due to the same mechanism.

In the specimens exposed to rain no clear differences were noticed. It can, however, be stated that carbonated permeable cement-based coatings will not significantly affect the carbonation rate of the concrete surfaces exposed to rainfall.

The effect of carbonated shotcrete on the carbonation rate of the concrete substrate could not be examined due to the extremely low carbonation rates of the shotcretes. It is, however, probable that the carbonated shotcrete will function in the same way as at least the same thickness of normal good quality carbonated concrete cover. This means that the original carbonation front is deep in the concrete when the shotcrete has carbonated and further carbonation will be very slow.

4 CONCLUSIONS

The aim of the research was to evaluate theoretically the serviceability of cement-based coating induced re-alkalisation for the prevention and repair of corrosion damage in the reinforced concrete structures. Because there is not a great deal of previous research available on this subject the research problem was quite complex. This is why the research program and the tests were more general than specific, and the results and the conclusions are mainly suggestive, serving as a basis for possible further work. However, some conclusions may be drawn.

1. When coating a carbonated concrete surface with a cement-based coating of certain thickness and permeability, the carbonated concrete will be re-alkalised in favourable

moisture conditions. The carbonated concrete will be re-alkalised not only from the direction of the alkaline coating but also from the uncarbonated concrete. The re-alkalisation is usually more efficient from the uncarbonated concrete than from the coating. It should be noticed that this signifigantly improves the efficiency of the re-alkalisation because the major part of the corroding reinforcement is usually near the carbonation front.

2. The re-alkalisation rate is slow and depends on the porosity of the concrete so that re-alkalisation is faster in more porous concrete. However, it seems that the re-alkalisation front proceeds at at least the same rate as the carbonation front proceeded into the concrete when it was new (square root model).

3. Re-alkalisation disappears rapidly if CO_2 can penetrate the coating. This can happen due to insufficient "CO_2-tightness" of the coating or because the coating has thoroughly carbonated. It is apparent that because re-alkalisation occurs through the pore solution in concrete any alkalinity reserves can not easily develop in the re-alkalised concrete which could serve as a buffer against CO_2. Therefore it is essential that the coating used will remain constantly "CO_2-tight" for the required service life, and any coating used for re-alkalisation purposes has to be relatively thick, preferably more than 10 mm, and as dense as possible to ensure "CO_2-tightness" and slow carbonation. Naturally, the coating layer must also be uniform, without visible cracks, and tightly bonded to the surface. The best coatings for re-alkalisation are normal good quality concrete either cast or shotcreted. Cement-based mortars are not suitable for re-alkalisation because they do not give sufficient CO_2-protection in dry conditions. In addition to this, their carbonation rate is fairly high which usually restricts the service life of the re-alkalisation to a couple of years.

4. The threshold moisture for reasonably rapid re-alkalisation in the concrete is about 90 % RH. In natural outdoor exposure the humidity will vary all the time. By monitoring the specimens stored in the test field as well as the results from the actual shotcreted structure it may be concluded that re-alkalisation can occur in cases of natural outdoor exposure even without the effect of rainfall. It may be further concluded that, as far as the moisture requirements are concerned, re-alkalisation is able to occur when the reinforcement corrosion is also likely to proceed.

5. Overall it may be concluded that it seems possible to use re-alkalisation at least as a beneficial side-effect when large-scale corrosion damage is to be repaired.

Acknowledgements

The research project was financed by Technology Development Center of Finland, Lohja Co. and Partek Co.

References

1. T. A. Bier, 'Karbonatisierung und Realkalisierung von Zementstein und Beton', PhD Thesis, Universität Karlsruhe, 1988.
2. J. S. Mattila, 'Re-alkalisation of Carbonated Concrete by Cement-based Coatings', Licentiates Thesis, Tampere University of Technology, Tampere, 1995 (in Finnish).
3. L. Mehto, M. Pentti and H. Käkönen, 'Carbonation of Concrete Facades', Report 41, Tampere University of Technology, Tampere, 1990 (in Finnish).
4. K. Tuutti, 'Corrosion of Steel in Concrete', Swedish Cement and Concrete Research Institute, Stockholm, 1982.

PATCH REPAIR AND HYDROPHOBIC TREATMENT

LOCAL REPAIR MEASURES AT CONCRETE STRUCTURES DAMAGED BY REINFORCEMENT CORROSION - ASPECTS OF DURABILITY

P. Schießl and W. Breit

Institute for Building Materials Research
RWTH Aachen - University of Technology
Schinkelstraße 3
D-52056 Aachen (Germany)

1 INTRODUCTION

Corrosion of steel in concrete is one of the major problems with respect to the durability of reinforced concrete structures. The enormous amount of money required for repair measures results in a strong need to improve the durability of new structures as well as to guarantee the durability of existing structures after repair or strengthening. To avoid poor performance of remedial work and recurrence of damage, repair procedures must be based on sound design and good workmanship. Design and workmanship need to be based on knowledge of the critical processes which have caused the damage and which need to be excluded after the repair. In most practical cases a number of different approaches to repair or intervention is possible. To ensure that a sound decision is made on how to proceed, it is essential to be aware of the different options for repair and the basic requirements to be fulfilled for each of them. These different *Principles of Repair* are defined in the RILEM Draft Recommendation - 124-SRC Repair Strategies for Concrete.[1]

The studies carried out in this research project should demonstrate the consequences of bad workmanship in which local damage due to reinforcement corrosion, e.g. spalls and cracks, is repaired solely in the area of visible surface damage. When damage is dealt with in this way, the carbonated or chloride-contaminated concrete is often not completely removed. In consequence, even after the repairs there are areas of the reinforcement where there is no guarantee of sufficient protection against corrosion and a high corrosion risk remains. The reinforcement in the repaired area may even accelerate corrosion in unrepaired areas adjacent to it.

2 AIM OF THE INVESTIGATIONS

In the past, corrosion damage has been observed repeatedly in areas adjoining those which have been repaired. The occurrence and extent of such corrosion is essentially dependent on the quality of the concrete and repair mortar, the concrete cover and on the temperature and moisture content within the concrete structure.

The studies in this paper were intended to resolve the following questions:

- How does reinforcing steel behave in the transition zone between the repaired and adjacent unrepaired regions? Is it possible to prevent the formation of macrocells through an appropriate coating of the steel?
- To what extent are coatings on the concrete surface able to influence the corrosion process effectively by reducing the water content of the concrete?

3 PREPARATION OF TEST SPECIMENS

For the tests,[2] concrete corrosion cells were devised in which macrocell corrosion due to carbonation or chloride attack could occur. Plastic containers were used as shuttering and for storage (600 × 130 × 150 mm). To simulate realistic construction practice where local repairs are made in areas with visible concrete damage (see Figure 1, top), three areas were differentiated within the specimens (Figure 1, bottom):

- **Area A** - Repaired area (visible damage)
  one steel embedded (BSt 500 S, $\varnothing$ = 16 mm, l = 100 mm)
- **Area B** - Unrepaired area (with remaining chloride contamination or carbonation)
  one steel (not ribbed) embedded each side (St 37, $\varnothing$ = 10 mm, l = 80 mm)
- **Area C** - Area of good quality concrete
  two steels embedded each side (BSt 500 S, $\varnothing$ = 16 mm, l = 100 mm)

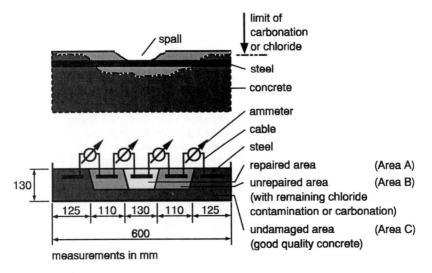

Figure 1 *Diagram showing local damage in the concrete and the resulting specimen structure*

To obtain defined separation between the anodic and cathodic surface areas of the steel, a single piece of reinforcement was embedded in each area, instead of a continuous reinforcing bar. A cable was fixed to each single piece of steel. The connection between steel and wire was protected against corrosion by using a thermal shrinkage tube, filled with epoxy resin. The macrocell current was measured not directly but by voltage measurement with an inserted resistor (10 Ω), allowing the current to be calculated via Ohm's law. The current intensity of each cell can then be determined by differentiation of the currents measured between the individual pieces of reinforcing steel.

4 TEST-PARAMETERS AND REPAIR STRATEGIES

The tests were performed on typical concretes and repair systems used in practice. The specimens varied in terms of concrete composition (water/cement ratio, type of cement), concrete cover and the kind of attack to which they were exposed (moisture, chloride and carbonation). In the test series, corrosion was initiated by addition of chloride to the mixing water and by external chloride exposure as well as by means of accelerated carbonation with 3 vol.% carbon dioxide. Various repair measures then followed.

The studies relate to the repair principle based on the use of alkaline repair mortar for local repassivation (method R3 rsp. R1-Cl).[1] The following additional protection measures were chosen:

- no additional protection,
- coating the steel to prevent a cathodic reaction,
- coating the concrete surface to reduce water content,
- coating of both the steel and the concrete surface.

After repair, the specimens were wetted to varying extents in order to examine the corrosion behaviour of the reinforcement in different environments.

5 SIMULATION OF CONDITIONS PRIOR TO REPAIR

The outer areas (Area C, good-quality concrete) were designed to permit exclusively cathodic reactions. This entails guaranteeing a supply of oxygen and excluding chloride wetting or rapid carbonation. Area C simulated the sound areas of a structure. As these areas are much larger on the jobsite, the amount of reinforcement used was four times that in the area intended for repair (Area A).

The centre (Area A, intended for repair) represents the local damage, initiated by addition of chloride or accelerated carbonation. A high water/cement ratio (w/c = 2.0) and low cement content ($c = 50$ kg/m^3) in relation to the aggregate size of 2-4 mm were selected for this area. This concrete has high internal porosity and can easily be removed to simulate the repair job. Until the repairing of the reinforcement, the anodic corrosion reaction should take place in the porous Area A.

The concrete in the adjacent area was either carbonated or wetted with chloride. Contrary to the guidelines[1,3] Area B was not repaired at the same time as Area A, since the intention was to investigate the consequences of such a procedure.

6 RESULTS OF TESTS WITH ADDITION OF CHLORIDE TO THE FRESH CONCRETE

Where chloride was added directly to the mixing water, the chloride concentrations were between 0.5 % and 2.5 % in relation to the cement content of Area B (Table 1). This method ensures uniform distribution of the chloride throughout the concrete, creating

Table 1 *Test-Parameters With Addition of Chloride to the Fresh Concrete*

Cement Type	w/c Ratio	Concrete Cover in mm	Chloride by Cement Content in % Added in Area B
	0.6	25	0.5 / 1.0 / 1.5 / 2.0 / 2.5
OPC 35	0.5 / 0.7	25	1.5
	0.6	15 / 35	1.5
	0.6	25	1.0 / 1.5 / 2.0
BFSC 35	0.5	25	1.5
	0.6	15 / 35	1.5
FAC 35	0.6	25	1.0 / 1.5 / 2.0

- Area C without chloride, Area A was made with porous concrete (concrete mixture c:w:a 1:2:25) and a chloride content of 2.5 % by cement weight.

more rigorously-defined experimental conditions than can be obtained by applying chloride to the concrete surface. The area intended for repair (Area A) was made with porous concrete on all specimens and a chloride content of 2.5 % by cement weight. When investigating specimens with added chloride application of water to the concrete surface was allowed to accelerate the corrosion process.

Figure 2 shows the results for a specimen made with ordinary portland cement and a chloride content of 0.5 % by cement weight in Area B. The figure indicates the macrocell current versus time for the Areas A, B and C. Due to the high chloride content of 2.5 %, corrosion occurred in Area A prior to repair. The presence of a macrocell is confirmed by a clear anodic (positive) cell current in Area A. Changing the humidity by applying water led to a rapid increase in cell current. On drying, the current rapidly decreased again. This was also observed in a previous study.[4] No anodic reactions were observed in Area B, containing 0.5 % chloride.

After local repair at an age of 300 days, no cell current is measurable once initial passivating currents have ceased. Even wetting after repair failed to produce any significant corrosion activity. This means that under these conditions a chloride concentration of 0.5 % by cement weight is below the critical limit for corrosion to occur.

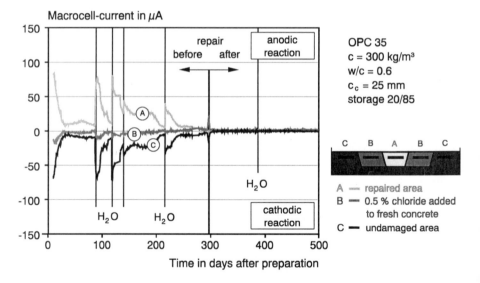

Figure 2 *Results of macrocell current measurements on a specimen with 0.5 % chloride related to the cement content added to fresh concrete in the unrepaired Area B*

Figure 3 illustrates the results from a specimen made of blast-furnace slag cement, with a chloride content of 2.0 % by cement weight, in Area B. As described above Area A, intended for repair, was made with porous concrete and a chloride content of 2.5 % by cement weight. Unlike the example in Figure 2, this specimen exhibits a clear anodic reaction in Area B prior to repair. This reaction develops faster in Area B than in Area A. Area A can be more strongly activated by applying water. This changes the previously anodic reaction in the adjacent Area B into a largely cathodic reaction. Area A dries out more quickly, owing to its higher porosity, and the cell current falls to negligible values. The cathodic protection effect for Area B is reduced when Area A dries out and this area reverts to an anodic (positive) cell current. The results show that the critical limit in Area B has already been exceeded prior to the repair. In very humid conditions, however, surface defects initially occur only in Area A.

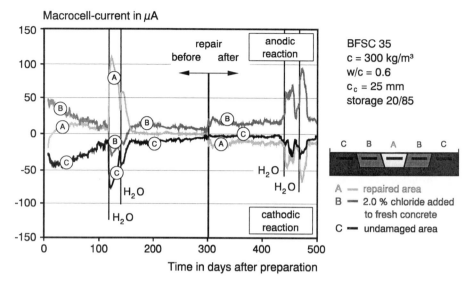

Figure 3 *Results of macrocell current measurements on a specimen with 2.0 % chloride related to the cement content added to fresh concrete in the unrepaired Area B*

As expected, repair creates an active cathode in the central Area A, contributing to anodic polarisation of the adjacent Area B. On application of water after repair the cell current increased rapidly, but without changing the sign of the electrochemical reaction, as happened before the repair measure.

6.1 Influence of Cement Type

Under the described test conditions, the critical, corrosion-initiating, chloride content for a uniform concrete cover of 25 mm and ordinary portland cement was found to be 0.5-1.0 % chloride content by cement weight when added to fresh concrete (Table 2).

As the blast-furnace slag content of the cement increases, the pore structure becomes denser, and the diffusion resistance of the concrete is increased.[5] This is confirmed by measurements of the electrolytic resistances, which are significantly higher than those for ordinary portland cement. In consequence, all transport phenomena in the concrete are inhibited. As expected, the blast-furnace slag cement with 50 % slag content had a higher critical limit at 1.0-1.5 % chloride by cement weight.

Table 2 *Results of Specimens With Addition of Chloride to the Fresh Concrete*

Cement Type	w/c Ratio	Chloride by Cement Content in % Added in Area B			
		0.5	1.0	1.5	2.0
OPC 35	0.5	×	×	●	×
	0.6	O	●	●	●
	0.7	×	×	●	×
BFSC 35	0.5	×	×	●	×
	0.6	×	O	●	●
FAC 35	0.6	×	O	●	●

O = no corrosion   ● = corrosion   × = no specimen

The use of fly ash in concrete produces a denser cement matrix with fewer capillary pores, due to the pozzolanic reaction.[6] This positive effect is particularly apparent in a slower reaction mechanism, especially at older ages compared to normal concrete. Moreover, supplementation of the particle size distribution in the finest range for the other concrete constituents (filler effect) increases diffusion resistance. As a result, there is a fall in the electrolytic conductivity of the concrete, similar to that with blast-furnace slag cement, and a resulting fall in the reinforcement corrosion rate. The critical limit of chloride content for fly ash cement in the tests (fly ash content 24 wt.%) was the same as that for the blast-furnace slag cement, 1.0-1.5 % chloride by cement weight.

After the critical chloride content had been exceeded, the corrosion rate rose with increasing chloride content, confirming previous results.[4,5]

6.2 Influence of Concrete Cover

The concrete covers (15, 25 and 35 mm) were tested using the same type of specimen at a uniform chloride concentration of 1.5 % by cement weight for ordinary portland and blast-furnace slag cement. The specimens showed anodic corrosion reaction in Area B, however an influence of the concrete cover depth was not discernible in this test-procedure.

7 RESULTS OF STUDIES ON CHLORIDE ATTACK FROM THE CONCRETE SURFACE

7.1 Situation Before Repair

Studies of macrocell formation due to chloride penetration covered repair variants with and without coating of the steel. Figures 4 and 5 illustrate the results of two specimens (Table 3) exposed to cyclic surface wetting with a 1.0 and 2.0 % chloride solution respectively, in Areas A and B, where Area A, intended for repair, was made with porous concrete and an additional chloride content of 2.5 % by cement weight to cause accelerated corrosion. The figures show the macrocell current versus time for the three areas. Due to the high chloride content, corrosion occurred in Area A, while a cathodic cell current was observed for Areas B and C only. The wetting periods prior to repair are reflected by the sudden increases in cell current. As expected, the current returns to significantly lower levels in the ensuing drying phases.

7.2 Situation After Repair Without Coating of the Steel

After repair at an age of about 270 days, virtually no cell current is measurable once the initial passivation currents[4] have ceased and the subsequent drying phase is completed. When the concrete is wetted again at an age of roughly 370 days, the expected shift of the anodic element currents to the adjacent unrepaired areas takes place (Figure 4). This implies that the chloride content in Area B is above the critical corrosion-initiating limit under the test conditions. Repair creates an active cathode in the central Area A, contributing substantially to anodic polarization of the adjacent Area B.

Table 3 *Test-Parameters For Chloride Attack From the Concrete Surface*

Cement Type	w/c Ratio	Concrete Cover in mm	Chloride Solution in % Applied to Areas A and B
OPC 35 BFSC 35	0.6	15	1.0 / 2.0

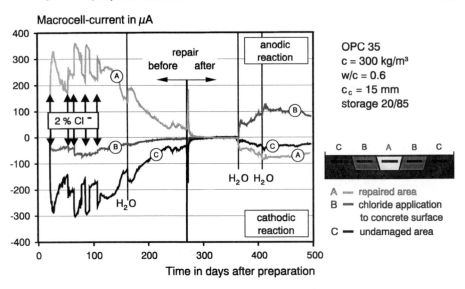

Figure 4 *Results of macrocell current measurements on a specimen of uncoated steel in the repaired Area A*

7.3 Situation After Repair With Coating of the Steel

In order to eliminate the cathodic effect of the repaired Area A the reinforcing steels of some specimens were treated with steel coating systems as an additional repair measure. In this case, a solvent-free, cementitious, thixotropic two-component system based on cold-

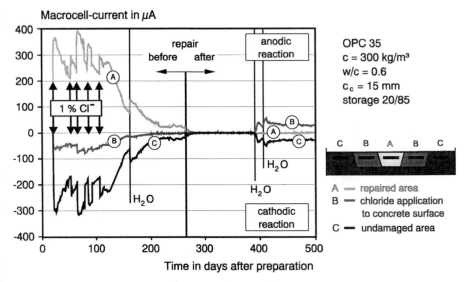

Figure 5 *Results of macrocell current measurements on a specimen of coated steel in the repaired Area A*

curing epoxy resin was employed. This prevented cell formation in the repaired area after wetting at an age of about 390 days (Figure 5). The steel coating systems are being investigated in an additional test series paralleling the macrocell studies.[2]

Coating the steel screened the corrosion process from the cathodic effect of the repaired Area A, nevertheless corrosion was able to take place in the adjacent chloride contaminated Area B due to large regions of passivated steel surfaces in the remaining Area C. Coating the steel with repair mortar is therefore likely to have only a subordinate effect in terms of reducing the cathodic reaction under real conditions, i. e. there will be only a slight reduction in the total cell current, since other areas also act as cathodes.

8 RESULTS OF INVESTIGATIONS OF CORROSION DUE TO CARBONATION

For the corrosion tests initiated due to carbonation (Table 4), the concrete cover was 10 mm in all cases, to ensure that the concrete was always carbonated to below the reinforcement (10 mm steel diameter). The carbonation depths measured on reference specimens were between 20 and 25 mm. Due to the differing airing conditions in the different areas, the anodic reaction could only be triggered by application of water (similar to Figure 3) to simulate the situation prior to repair in Area A. Due to rapid drying of the aggregate-porous concrete, corrosion in this area fell to negligible cell current values a few days after wetting.

Table 4   *Test-Parameters For Accelerated Carbonation With 3 Vol.% of CO_2*

Cement Type	w/c Ratio	Concrete Cover in mm
OPC 35	0.5 / 0.6 / 0.7	10
BFSC 35	0.5 / 0.6	10
FAC 35	0.6	10

- Area C was protected against carbonation, Area A was made with porous concrete (concrete mixture c:w:a 1:2:25)

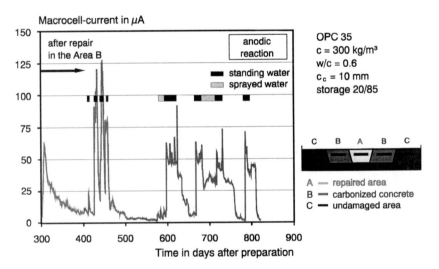

Figure 6   *Results of macrocell current measurements on a specimen with carbonation in the unrepaired Area B*

Figure 6 illustrates the cell currents in the carbonated Area B of an ordinary portland cement specimen after repair. During storage in a climate of 20 °C and 85 % relative atmospheric humidity, the cell currents fell to negligible values. Only after wetting with standing or sprayed water there was a sharp rise in cell currents. This means that unless all carbonated concrete in the vicinity of the reinforcement is removed completely after local repairs, macrocell corrosion will occur if the concrete is not kept constantly dry.

9 RESULTS OF INVESTIGATIONS USING CONCRETE COATINGS

In the ongoing tests, some specimens have been treated with different surface protection systems, in order to determine the efficacy of *Repair Principle W* (reduction of water content) according to the guidelines.[1,3] Previous research projects[7,8] have already addressed the topic of surface protection systems for concrete. No significant change in the specific electrolytic resistance of concrete as a result of wetting was observed following application of a coating system.[7]

Three among the twelve existing surface protection systems (SPS), according to the guidelines,[3] were chosen for laboratory experiments. To check the efficiency of the coating, the test specimens were wetted with 30 mm water applied to the (horizontal) concrete surface after about one month. After surface coating the depth dependent moisture distribution was checked using the Multi-Ring-Electrode method.[2,9] A percentage distribution relative to the resistance value before wetting is the most useful means of analyzing changes in the distribution of resistance due to such wetting. As Figure 7 shows, significant changes in resistance due to wetting were measured on the specimen, coated with SPS 2. This behaviour is similar to an uncoated specimen.[2,9] A distinctly improved protective effect was achieved with SPS 5, but a slow decrease in resistance, however, was evident at a depth of 6.2 mm. With SPS 11, no significant changes in resistance for the different areas due to wetting were found after an initial decrease of 10 %, which means that no water penetrated the coating. The initial decrease, also found with SPS 5, was caused by temperature difference between specimen and water applied to the concrete.

Provided that the coating has a sufficiently high penetration resistance against water and if no water enters the concrete from other sources, carbonation-induced corrosion can be reduced significantly. In the case of chloride-induced corrosion a maximum decrease of 50 % in the macrocell current was achieved after coating.

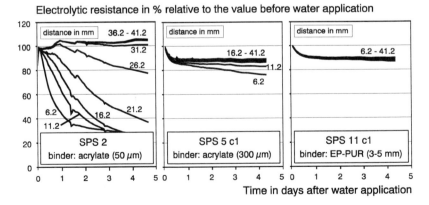

Figure 7 *Influence of wetting on relative electrolytic resistance of concrete specimens coated with different types of surface protection system (SPS 2, 5, 11)*

10   CONCLUDING REMARKS

Laboratory tests were carried out on reinforced concrete specimens to determine the corrosion behaviour of steel in concrete after local repair of damage caused by corrosion of the reinforcement in carbonated or chloride contaminated areas. The results of the investigations are summarised below.

Where carbonated concrete was not removed completely in the vicinity of the reinforcement, macrocell corrosion occurred after repair in all specimens when water was applied. This means that *Repair Principle R* (method R3 - local repassivation)[1] is applicable only after all carbonated concrete at the level of the reinforcement has been removed or if a sufficiently low water content of the concrete can be assured.

The corrosion rate is essentially dependent on moisture content and is less influenced by specific characteristics of the concrete. With specimens stored in a defined environment (20 °C, 85 % relative atmospheric humidity), macrocell currents of carbonated, chloride-free specimens declined to negligible values. The implication is that a relative atmospheric humidity of 85 % close to the surface of the concrete results in such low water contents that carbonation-induced corrosion virtually ceases. Even if carbonation extends behind the reinforcement, no damage is to be expected under such dry environmental conditions. Further investigations showed that the critical limit for corrosion is between 90 and 95 % relative atmospheric humidity.

In the case of chloride-induced macrocell corrosion, all areas in which the critical chloride content is exceeded must be removed, irrespective of whether damage is visible in such areas or not. Under the test conditions, a critical chloride content of 0.5-1.0 % was found for concrete containing ordinary portland cement and a critical chloride content of 1.0-1.5 % for blast-furnace slag cement and fly ash cement concretes.

Cathodic reaction of the reinforcement within the repaired areas was reduced to negligible values by epoxy-coating the steel. Under real conditions, there are usually large steel surfaces outside the repaired area of reinforced concrete structures which act as cathodes, as demonstrated by macrocell current measurements.

Provided that water absorption can be prevented by means of a suitable concrete coating *(Repair Principle W)*, carbonation-induced corrosion can be reduced significantly. In the case of chloride-induced corrosion an insufficient decrease of macrocell current was achieved after coating.

References

1.   P. Schießl, 'Draft Recommendation for Repair Strategies for Concrete Structures (RILEM TC 124-SRC)', *Materials and Structures*, 1994, **27**, 415.
2.   P. Schießl and W. Breit, Institut für Bauforschung F 332, Aachen, 1994.
3.   German Committee on Reinforced Concrete, Guidelines for Protection and Repair of Concrete Components, Beuth, Berlin, 1990-1992.
4.   M. Raupach, 'Zur chloridinduzierten Makroelementkorrosion von Stahl in Beton', Beuth, Berlin/Köln, *DAfStB*, 1991, **433**.
5.   P. Schießl and M. Raupach, *Beton-Informationen*, 1988, **3/4**, 33.
6.   P. Schießl, 'Corrosion of Steel in Concrete (Rilem TC 60-CSC)', Chapman and Hall, London/New York, 1988.
7.   P. Schießl and M. Raupach, Institut für Bauforschung F 310, Aachen, 1991.
8.   P. Schießl and M. Raupach, Institut für Bauforschung F 290, Aachen, 1991.
9.   P. Schießl and W. Breit, 'Monitoring of the depth-dependent moisture content of concrete using Multi-Ring-Electrodes', International Conference on Concrete Under Severer Conditions in Sapporo, Japan, August 2-4, 1995.

NATURAL EXPOSURE IN UK OF REPAIRED SPECIMENS IN BRITE PROJECT 3291

K. Hollinshead, D. J. Bigland and R. J. Pettifer

Building Research Establishment
Garston
Watford WD2 7JR

1 INTRODUCTION

The aim of BRITE II Project 3291 was to develop standardised performance tests and criteria for concrete repair systems. Seven partners from the UK (Taywood Engineering Ltd. and BRE), Spain (LABEIN, Fosroc SA and Intemac), Germany (BAM) and Italy (Institute Ricerche Breda) were involved in the project, which ran from September 1990 to August 1995. The project was split into eight tasks, each dealing with one phase of the project, including an initial review, assessment of real repaired structures, natural weathering, accelerated weathering, critical materials property testing and large scale testing. The data from the natural and accelerated weathering tasks are being compared and used in the preparation of the final report which will make recommendations on performance criteria for concrete repair materials and systems, goals for product development, the prediction of repair performance and durability and accelerated tests for evaluation of repair materials.

In the natural weathering task specimens were exposed for three years in UK, Spain and Italy by BRE, LABEIN and Breda. This paper describes the natural exposure programme undertaken in the UK and presents the main findings.

2 SPECIMEN DESIGN

The specimens were upright panels of dimensions 340 x 300 x 100 mm, reinforced with four vertical and two horizontal 8 mm diameter deformed plain carbon steel bars. Figure 1 shows a panel before repair. Four stainless steel tubes were also cast in each specimen with the purpose of accelerating any corrosion initiated on the carbon steel bars. Each bar and tube had an electrical connection, allowing them to be linked. The vertical reinforcement was designated 'a' to 'd' from left to right.

Simulated spalls were formed at two corners and in the centre of the front face of the panel by casting in polystyrene inserts, which were removed after curing. The inserts were designed so that, once the panels were repaired, bars a and c had 20 mm cover and bars b and d 10 mm cover. The central patch repair had a ninety degree overhang and feathered edges of 35° and 45°. The concrete mix was as outlined in Table 1, with a free water : cement ratio of 0.5. The mean 28 day compressive strength was 49.0 MPa and the mean slump was 75 mm.

Figure 1 *Unrepaired natural exposure panel*

Table 1 *Concrete mix proportions*

Component	Mass/kg
OPC	36.9
<5 mm aggregate	74.8
10 - 5 mm aggregate	110.8
Total water	22.7

3 REPAIR AND EXPOSURE OF PANELS

Three exposure conditions were used in the UK: tidal, sea spray during rough weather and adjacent to a motorway. Forty-eight specimens were pre-exposed at each site for 6 months before repair. Six repair systems were chosen which were typical of those available in Europe. These were designated Types 1 to 6 as shown in Table 2.

Table 2 *Designation of repair types*

Type 1	OPC mortar
Type 2	Polymer modified mortar
Type 3	Epoxy resin mortar
Type 4	Free flowing micro concrete
Type 5	Polymer modified mortar
Type 6	Polymer modified mortar

Bar primers were part of all systems except Type 1. The repair systems were chosen to be representative of those available and to give a range of performances under the different exposure conditions. The repair mortars were applied by a specialist repair contractor. The bars were grit blasted before application of the repair systems. Where possible the panels were repaired in the upright position. However, it was impossible to apply the Type 3 mortar in this way and so the panels were repaired in a horizontal position. A letterbox shutter was used as formwork for Type 4. After repair the specimens were returned to their exposure sites.

4 MONITORING PROGRAMME

The panels were monitored visually and electrochemically at one and four months after repair and then every four months up to three years of exposure. During each monitoring session any significant visual features, such as rust staining, cracking or surface crazing were recorded on a proforma. Linear polarisation measurements, with iR compensation, were made on each of the four vertical bars and the corrosion current, corrosion potential and ohmic drop between the reference and working electrodes recorded. After each 12 month period of exposure, two specimens of each repair type were removed from each exposure site for destructive examination. This involved making a photographic record of each specimen, measuring the pull off strength of the repair in the central patch, collecting dust samples from the concrete and repairs for chloride analysis and finally dismantling the specimens to inspect the condition of the rebars and any other significant features.

5 RESULTS

5.1 Electrochemical Data

Figure 2 shows typical corrosion current data from the three exposure conditions collected over the three years of exposure. The data shown are from one bar 'b' from each type of repair. The data from the panels subjected to motorway exposure provided a baseline from which to develop a criterion for classifying active and passive behaviour. A value of 16 μA was chosen and at currents greater than this the bars were assumed to be likely to be actively corroding. The area of the bars in these panels was nominally 80 cm^2 and so 16 μA was equivalent to a current density of 0.2 μAcm^{-2} if corrosion is assumed to be uniform. This is in broad agreement with Andrade *et al*[1] who proposed that corrosion currents greater than 0.1 to 0.2 μAcm^{-2} indicated active corrosion. It was subsequently found that even bars at the

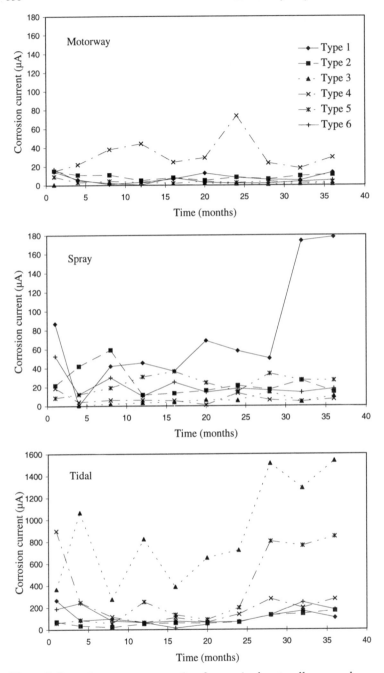

Figure 2 *Corrosion current versus time for repaired, naturally exposed specimens*

motorway site had some small areas of corrosion and so the 16 µA criterion may not be strictly accurate. Higher currents were measured on bars in panels repaired with the Type 4 material. This system included a zinc rich bar primer which may account for this observation. These data were not included in the development of the corrosion criterion.

The data from panels subjected to spray indicate that bars in Types 2 to 6 had broadly similar currents of 10 to 20 µA but Type 1 panels had higher currents indicating active corrosion. The data from the tidal site show that all bars were actively corroding with Types 3 and 5 having currents significantly higher than the other panels after three years of exposure.

Figure 3 shows the relationship between corrosion current and potentials using data collected after three years of exposure. As expected there is a roughly linear relationship between log current and potential, although the plot tends to flatten at potentials more positive than 0 mV. The data from the three sites tend to fall into separate groups reflecting the relative severity of exposure.

5.2 Chloride Ingress

The chloride content of the repair materials was determined at three depth ranges: 0 to 5 mm, 5 to 10 mm and 10 to 20 mm. Figure 4 shows the chloride profiles of each repair type after three years of exposure to spray and the tidal zone. The chloride content of the panels at the motorway site did not increase over the exposure period and remained at background levels. Figure 5 shows the relationship between average corrosion current and the chloride content in the 10 to 20 mm depth range.

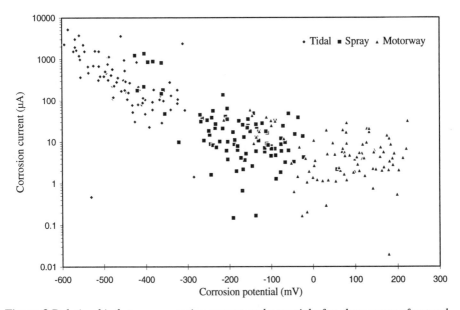

Figure 3 *Relationship between corrosion current and potential after three years of natural exposure*

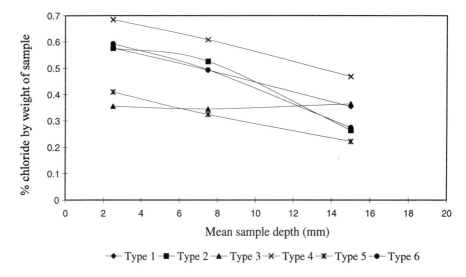

Figure 4a *Chloride profiles in repairs after three years of exposure to spray*

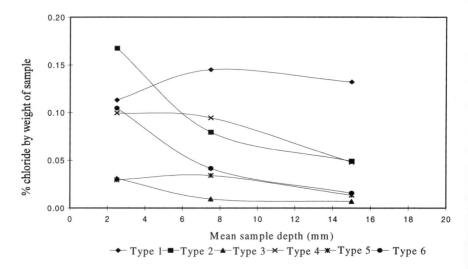

Figure 4b *Chloride profiles in repairs after three years of exposure in tidal zone*

5.3 Destructive Examination

During destructive examination the performance of the panels was assessed in terms of the compaction of the repair materials and the amount of corrosion of the reinforcement. Generally the compaction of the repair mortars was poor behind the reinforcement and around the cross pieces, and these were the areas where corrosion was most severe. Figure 6 shows typical voids behind the reinforcement. The quality of compaction was rated on a scale of 1 to 5 as follows:

1. Near perfect compaction at all points in the repair and little or no voiding at the top and bottom repair concrete interfaces.
2. Compaction of a good standard but small to medium sized voids (<5 mm) at the top and bottom concrete interfaces.
3. Compaction obviously deficient at some points behind the reinforcement and large voids at the repair concrete interface.
4. Compaction of a poor quality along the entire length of the rebar and large voids evident at the repair concrete interface.
5. Very little repair material at any point behind the rebar.

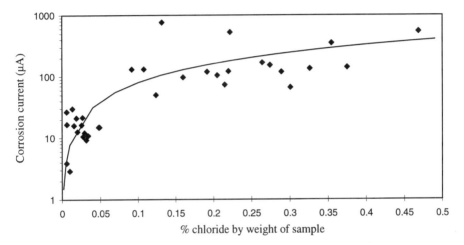

Figure 5 *Relationship between corrosion current and chloride content after three years of natural exposure*

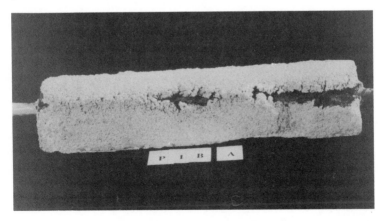

Figure 6 *Poor compaction behind reinforcement (dimensions of repair approximately 220 x 40 x 40 mm)*

Figure 7 shows the compaction rating of the left, middle and right patches in each of the six materials. Type 4, the micro concrete, showed the best compaction, although it tended to suffer from air voids at the 90° overhang. Type 2 was the next best in terms of compaction, while Types 1, 3, 5 and 6 were broadly similar. Compaction was generally better on the right hand corner patch, where the clearance behind the bar was about 20 mm, and worst on the left hand corner patch, where the clearance was only 10 mm.

Figure 8 shows the percentage areas of the bars corroded. Even under the least severe exposure conditions, where electrochemical data indicated that the reinforcement was passive, significant areas of corrosion were seen. Corrosion tended to be present at the interface between the repair and the concrete, at voids and on the reverse of the bars where grit blasting was difficult.

Carbonation of the bulk materials was very variable and had not generally reached the depth of the reinforcement. However, carbonation had progressed down the repair/concrete interface. It is probable that chloride ions also entered in this manner. Once carbonation or chloride reached the reinforcement the corrosion tended to propagate along the reinforcement via voids in the repair. Some pitting of the reinforcement was observed in specimens subjected to tidal exposure, particularly Type 3. This was the only repair type where corrosion was seen over the entire surface of the reinforcement. Figure 9 shows the relationship between area of bar corroded and corrosion current using data collected.

6 DISCUSSION

In order to assess the influence of factors affecting the performance of the repairs in protecting the reinforcement from corrosion, the systems were ranked with regard to compaction, chloride content, corrosion current, area of bar corroded and corrosion at the interface between the repair and the concrete. Table 3 shows the rankings after three years of

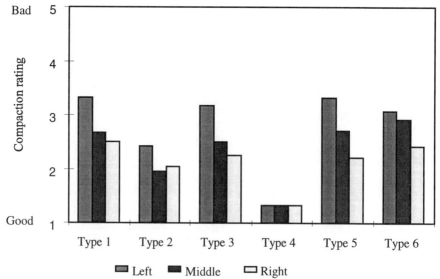

Figure 7 *Average compaction rating of repairs*

Tidal exposure. This highlights that overall performance is influenced by a number of factors.

Type 5 had the lowest chloride content after three years of tidal exposure but was ranked fifth in terms of the area of bar corroded and fourth in terms of compaction, corrosion current and interface corrosion. Type 6 is ranked lower for compaction (fifth) and chloride content (second) but higher for corrosion current (first), area of bar corroded (second) and interface corrosion (second). This supports the observations made during the destructive examination that corrosion can initiate at the interface and then propagate along the rebar through voids in the repair.

Type 4 performed well in terms of compaction (first), poorly for chloride content (sixth) and moderately (third) for interface corrosion. In this case the overall performance seems to have been determined by the chloride content of the bulk repair. Type 2 performed fairly well in all aspects and had the lowest area of bar corroded. Type 1 and Type 3 performed fairly poorly in all aspects.

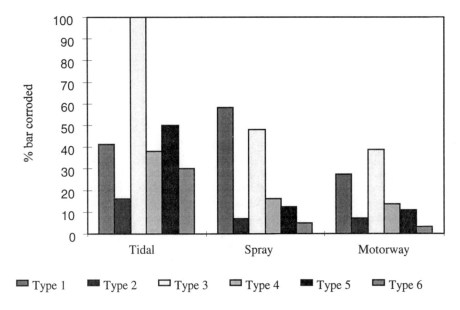

Figure 8 *Average percentage of bar corroded after three years of exposure*

Table 3 *Ranking of performance of repair systems after three years of Tidal exposure*

	Compaction	Chloride content	Corrosion current	% Bar corroded	Interface corrosion
Type 1	6	4	3	4	5
Type 2	2	2	2	1	1
Type 3	3	4	6	6	6
Type 4	1	6	4	3	3
Type 5	4	1	4	5	4
Type 6	5	2	1	2	2

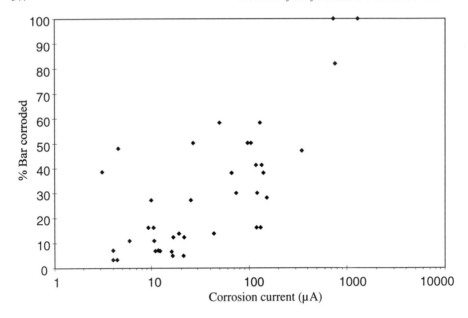

Figure 9 *Relationship between corrosion current and area of bar corroded*

The rankings of materials at the spray and motorway sites was broadly similar but the influence of the interface seemed to be slightly less critical in determining overall performance. The problems with assessing non-cementitious materials, such as Type 3, were also highlighted. Type 3 was still ranked low in terms of area of the bars corroded but this was not reflected in the corrosion currents measured or chloride content. Any protection given by non-cementitious materials tends to be by acting as a barrier to moisture and corrosive ions, and by the high resistivity of the material when dry, and not by promoting passivation. This highlights the need to consider the mechanism of protection of the reinforcement when specifying performance characteristics and tests. Electrochemical measurements were generally not suitable for Type 3 due to its high resistivity when dry and relatively low levels of chloride and moisture were apparently required to cause corrosion.

7 CONCLUSIONS

The natural exposure tests highlighted several factors important in the corrosion protection of reinforcement by repair materials. The interface between the repair and concrete was found to be vulnerable to carbonation and chloride ingress, which could then spread along the reinforcement if voids were present, causing corrosion. All the repair systems used in the tests suffered from this to some extent. Compaction was best where there was more clearance behind the reinforcement. Chloride ingress through the bulk material became more important when the interface was less vulnerable and compaction was good.

The electrochemical measurements were useful in predicting the corrosion protection given by the cementitious materials but are not suitable for high resistivity, non-cementitious materials. Where sacrificial bar primers (e.g. zinc rich) are used the normal electrochemical criteria for predicting corrosion may not be applicable.

These findings are being correlated with those from the accelerated testing before final recommendations are made. It may also be necessary to determine whether real repairs perform in a similar fashion, in particular with regard to compaction. However, with regards to natural exposure testing the present form of specimen is probably too complex, with two different depths of cover, reinforcement in corner and central patches and three different angles in the central repair patch. Manual handling is also a problem.

Acknowledgement

The authors would like to thank the many people who have been involved in this project for their assistance over the last five years.

References

1. C. Andrade, M.C. Alonso and J.A. Gonzalez, 'An Initial Effort to Use the Corrosion Rate Measurements for Estimating Rebar Durability', *Corrosion Rates of Steel in Concrete, ASTM STP 1065,* (N. S. Berke, V. Chaker and D. Whiting, Eds.), American Society for Testing and Materials, Philadelphia, 1990, 29.

HYDROPHOBIC TREATMENT OF CONCRETE AGAINST CHLORIDE PENETRATION

Rob B. Polder, Huibert Borsje

TNO Building and Construction Research,
P.O. Box 49, 2600 AA, Delft, The Netherlands

Hans de Vries

Ministry of Transport, Civil Engineering Division,
P.O. Box 20 000, 3502 LA Utrecht, The Netherlands

1 INTRODUCTION

As part of their maintenance policy, Rijkswaterstaat decided in 1994 to apply hydrophobic treatment to all newly constructed concrete bridge decks as an additional protective measure against penetration of deicing salts. A typical cross section is given in Figure 1. This measure was believed necessary because of the increasing use of porous asphalt, which requires more deicing salt and allows easier penetration. Hydrophobic treatment was thought to be a protective measure with a good performance/cost ratio. The decision was supported by a research programme into the hydrophobic treatment of concrete, consisting of a literature study, the development of test methods and requirements for commercial products and additional research into the performance of hydrophobised concrete and application oriented tests.

The test setup, the results for nine hydrophobic products and the first results of the research have been published elsewhere.[1] This paper reports the main results of the research programme.

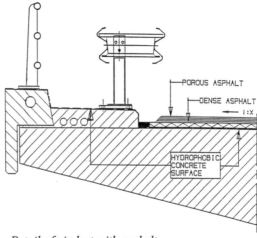

Figure 1   *Detail of viaduct with asphalt*

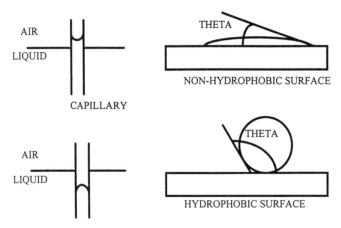

Figure 2 *Interaction between water and either hydrophobic or non-hydrophobic concrete surface*

2 BACKGROUND

2.1 Theory

When a porous building material such as concrete comes into contact with (liquid) water, the water is absorbed into the pores of the material by capillary forces. The capillary forces are determined by: the surface tension of the liquid, the contact angle between the liquid and the pore walls, and the radius of the pores. Narrow pores attract moisture more strongly than do wide pores. A viscous liquid is less quickly absorbed than a liquid with lower viscosity. In the case of a deicing salt loaded concrete bridge deck, the pore structure and the viscosity are fixed and only the contact angle (Θ) can be influenced. A small contact angle (<90°) indicates molecular attraction between the liquid and the substrate. A drop wets the surface, tends to spread and is absorbed; the level inside a capillary lies above the level outside. If this attraction between solid and liquid molecules is absent, the contact angle is >90°, a water drop remains spherical and the liquid inside the capillary is depressed below that outside. The principles are illustrated in Figure 2.

The molecular attraction between water and concrete can be weakened by hydrophobic treatment: the concrete becomes water repellent after impregnating with hydrophobic agents, such as silicones. Silicones form a group of related compounds, in which silanes and siloxanes are the most important for concrete. Silanes are small molecules having one silicon atom; siloxanes are short chains of a few silicon atoms. Their molecules contain organic alkoxy groups linked to the silicon atoms, which can react with the silicates in the concrete to form a stable bond. In addition, silanes and siloxanes contain organic alkyl groups (CH_3-) which have a fatty character. After reaction of the alkoxy groups, the alkyl groups protrude from the pore walls into the pores, as shown in Figure 3. As a result, water molecules will be repelled and will no longer be able to wet the surface: the contact angle is greater than 90° and water is no longer absorbed by capillary suction.

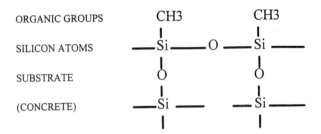

Figure 3   *Chemical bond between the hydrophobic agent and concrete*

In commercial hydrophobic agents for concrete, silanes are dissolved in alcohol or hydrocarbons (10 to 40% silane) or consist of 100% silane. Silane contents must be high because a substantial amount is lost due to evaporation. Siloxanes are often dissolved in alcohol or hydrocarbon solvents (about 10 to 20% siloxane). With siloxanes less is lost by evaporation. Both types are also used in the form of water-borne systems ("micro-emulsion").

2.2 Literature survey

From the literature the following preliminary conclusions were drawn:
Hydrophobic treatment of concrete can reduce its water absorption by 70 to 90% and also reduces chloride penetration. The best results are obtained with hydrophobic agents based on silanes and oligomeric siloxanes. Applying a hydrophobic agent does not adversely affect the adhesion of coatings or asphalt bond aids. Due to reduced water absorption, hydrophobised concrete will reach its equilibrium moisture content more quickly than untreated concrete; this may slow down ongoing reinforcement corrosion.[2] Most research concerned ordinary portland cement concrete; no information is available on blast furnace slag cement concrete, which is used in many Dutch concrete structures.

3 TEST METHODS AND REQUIREMENTS FOR HYDROPHOBIC AGENTS

Test methods and requirements were designed and nine commercially available products were tested. The composition of the concrete used as test material was specified precisely. Two cement types were used to cast slabs of which both formwork and finished surfaces were tested. Further details are given in.[1] The concrete composition, the specimen size and the application are summarized in Table 1 and the tested products are described in Table 2.
   The main requirements are:
- water absorption less than 20% of control concrete;
- water absorption after heating at asphalt application temperature less than 30% of control;
- penetration depth at least 2 mm;
- water absorption less than 20% of control when applied to strongly alkaline portland cement-sand mortar;
- evaporation of water through hydrophobic concrete at least 60% of control.

Table 1   Standard concrete composition, curing, specimen preparation and application for testing hydrophobic products[3]

Cement Type	portland CEM I 32,5 R blast furnace slag CEM III/B 42,5 (>65% slag)
Cement Content	340 kg/m^3
Water-cement Ratio	0.50 (strength class B35)
Maximum Grain Size	32 mm, distribution precisely defined (NEN 3532)
Curing	3 days covered in formwork, then in air at 20°C 65% RH
Specimen Test Surface	100 x 100 mm^2 (height about 75 mm)
Age at Application	≥4 weeks
Application	dip in liquid 5 seconds, wait 10 minutes, dip again 5 seconds
Curing before Testing	at least 4 weeks in 20°C 65% RH

Table 2   Overview of the tested hydrophobic agents and main results[1]

Code	type	content	solvent	density (kg/m^3)	overall result form	fin
A	silane	99%	none	878	-	+
B	silane	100%	none	915	-	+
C	silane/siloxane	12%	water	997	-	-
D	silane/siloxane	12%	water	998	-	-
E	silane	20%	water	969	-	+
F	oligom.siloxane	>9%	hydrocarbon	814	-	x
G	silane	40%	ethanol	825	-	-
H	silane	20%	hydrocarbon	813	-	x
I	silane	10%	hydrocarbon	799	-	-

form formwork surface   fin finished surface
+ complies   - does not comply   x complies if not heated

3.1 Test Methods and Results

The water absorption was determined by placing the specimens' treated face in a layer of water and weighing after 1, 2, 3 and 24 hours. The results varied considerably: some products had a very low water absorption (10% of controls or less), for others it was reduced to about 30% of the controls.

The heat resistance was measured by heating the treated specimens at 160°C for 30 minutes and determining the water absorption as above. Only three products remained water repellent; six lost most of the water repellency.

The penetration depth of the hydrophobic agent was measured by splitting the specimens and wetting the broken surface, which reveals the extent of the hydrophobic zone as it appears much lighter than the non-hydrophobic concrete. In particular on formwork surfaces most products were unable to attain the required 2 mm penetration; finished surfaces were more deeply penetrated and blast furnace slag cement finished surfaces were penetrated more deeply than OPC finished surfaces.

The resistance to alkali was generally sufficient. The evaporation showed no major changes with respect to control values. For more details refer to.[1]

To identify the tested products, the liquid density was determined and the infrared spectrum was recorded. Measuring the density can be used as a simple on site test for identity verification. The infrared spectrum is a more accurate laboratory method of identification.

The overall results are given in Table 2. The performance of the treatment on formwork surfaces and the penetration in particular proved insufficient in many cases. On finished surfaces, three products proved satisfactory (A, B and E). After evaluation of the methods and results, some changes were made to the test setup and the requirements. The final test procedure has been laid down in a recommendation.[3]

4 EFFECT ON CHLORIDE PENETRATION

The effect of hydrophobic treatment on the penetration of chloride has been investigated by exposing specimens to wetting and drying cycles with a salt solution. These tests were intended to simulate deicing salt application. The test may be particularly significant to non-protected parts of concrete structures like sidewalks and vehicle barriers (without asphalt overlay).

The specimens were 100 x 100 x 75 mm^3 blocks (only finished surfaces) made with both portland cement and blast furnace slag cement (standard concrete, see Table 1). They were treated with a hydrophobic agent containing 100% silane (product B) or a 20% silane dispersed in water (product E).

During each cycle, the treated surfaces of the specimens were allowed to absorb a 10% NaCl (by mass) solution for 24 hours and then dry for 6 days in air at 20°C and 50% RH. The chloride penetration profiles were determined after 12 months by dry grinding of layers 4 mm thick and analysing for total (acid soluble) chloride.

After 52 weekly cycles the chloride content at a depth of about 20 mm in hydrophobised concrete was less than 0.5% by mass of cement; in non-treated concrete the equivalent value was close to 3%, as shown in Figures 4 and 5. This means that the hydrophobic treatment has reduced the chloride penetration by about a factor of 5. Similarly, the chloride content of the outermost layer was reduced by about a factor of 3. This corresponds to the reduction in (pure) water absorption.

It may be concluded that hydrophobic treatment of concrete, made with either portland cement or blast furnace slag cement, strongly reduces the penetration of chloride under salt application-drying cycles.

5 EFFECT ON CORROSION

The effect of hydrophobic treatment on ongoing corrosion caused by chloride penetration was investigated using macrocell specimens. Specimens of 300 x 150 x 150

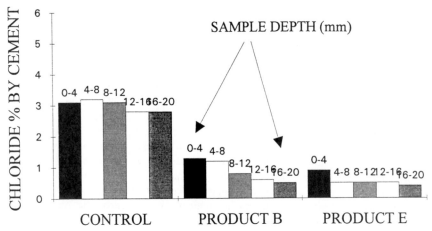

Figure 4   *Chloride penetration in control and hydrophobic portland cement concrete after 12 months weekly ponding*

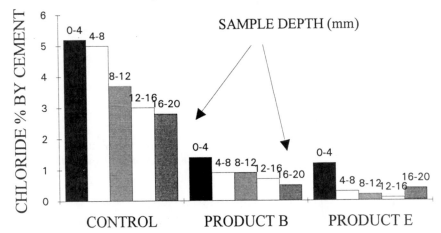

Figure 5   *Chloride penetration in control and hydrophobic blastfurnace slag cement concrete after 12 months weekly ponding*

mm^3 with two reinforcing bars at 25 mm and two bars at 110 mm from the upper surface were exposed to salt solution ponding on this surface for 24 hours followed by drying at 20°C and 50% RH. During the major part of the ponding period, the salt solution contained 10% NaCl, and the cycle lasted 14 days. The upper bars were expected to start corroding, which would be indicated by the macrocell current in a 10 Ω resistance connecting one upper bar and the two lower bars. The surface ratio of anodic and cathodic bars was 1:3. Six specimens were made with OPC and six with BFSC, all having a standard composition as given in Table 1.

After about one year no corrosion had been initiated; this was demonstrated by negligible macrocell currents and passive steel potentials (see Table 3). In week 59 after starting the salt applications, chloride penetration was stimulated by polarising the steel positively for one week, applying 1 A/m^2 (steel surface), with an activated titanium mesh placed in the ponding solution as a counter electrode.

After this treatment, macrocell currents became significant and steel potentials became negative, suggesting initiation of corrosion had taken place. In week 70, three specimens of each cement type were treated with a hydrophobic agent (product E) and all were transferred to a climate room at 20°C and 80% RH. This climate was thought to represent the outside conditions more realistically than 50% RH. After that, the ponding cycles were continued for about one year. During this period, steel potentials and macrocell currents in all OPC specimens were quite stable at levels which indicated active corrosion. In the BFSC specimens steel potentials gradually became more positive and macrocell currents decreased to negligible values, suggesting repassivation, probably because of the relatively low chloride content (see next paragraph). These results were confirmed by polarisation resistance measurements. It appeared that the hydrophobic treatment had no effect on corrosion rates. Results are given in Table 3.

After a total period of about two years, four specimens were destructively investigated. The chloride penetration profiles of hydrophobised and control specimens were very similar. Apparently the profiles were established in the first 70 weeks, before the hydrophobic treatment. The chloride content at the depth of the rebars in the two OPC specimens was about 1% (by mass of cement), which apparently caused and sustained active corrosion. In the two BFSC specimens the chloride content at the bar depth was between 0.2 and 0.6%, which apparently was insufficient to sustain active corrosion. It was concluded that ongoing corrosion due to high chloride content is not slowed down significantly by hydrophobic treatment, at least not on the time scale of one year.

Table 3   *Corrosion potentials and macrocell corrosion currents of upper bars in ponded macrocell specimens (average values of three specimens)*

Cement Type	E_1 (-mV Ag/AgCl)			I_{macro} ($\mu A/cm^2$)
	week 44	week 62	week 120	week 120
OPC	40	310	240 H	0.8 H
	40	270	240 N	0.9 N
BFSC	90	350	100 H	0.01 H
	100	310	110 N	0.01 N
Note	1	2	3	4

Notes: H hydrophobised in week 70; N non-hydrophobised
1   representative for first year of ponding
2   shortly after forced chloride penetration
3   at the end of the exposure period
4   macrocell current between upper and lower bar at the end of the exposure period

6 DURABILITY OF WATER REPELLENT EFFECT

In order to study the durability of the water repellent effect, specimens were hydrophobised and exposed on the roof of the TNO laboratory. Twenty four (including non-treated control, formwork and finished surfaces) specimens were exposed for 21 months; 40 more specimens (only finished surfaces) were exposed for 14.5 months. Specimens were made with OPC and BFSC (concrete composition as in Table 1). Three hydrophobic products were applied (codes A, B and E). Occasionally the specimens were taken inside and were allowed to equilibrate at 20°C and 65% RH. Then their water absorption was tested, whereafter the outside exposure was continued.

During these tests it was found that the water absorption coefficient of the controls decreased over time. Direct comparison of hydrophobised samples with the controls was no longer considered useful. Consequently the water absorption coefficient (WAC) over 24 hours was determined. The WAC of the OPC controls showed a strong reduction, whilst that of the BFSC controls showed only a small reduction. The WAC was found to be fairly constant for all hydrophobised specimens. Some results are given in Table 4.

It was concluded that the water repellency of concrete treated with each of the three products, had not deteriorated during 21 months exposure to outdoor climate. In non-treated OPC concrete, the water absorption of the control concrete decreased quite markedly over time, so the **relative** effect of hydrophobic treatment decreased. Clearly this is not due to the degradation of the hydrophobic effect. For non-treated BFSC concrete, the water absorption of control specimens decreased only slightly. Hydrophobised BFSC concrete retained its much lower water absorption compared to non-treated concrete. The reduction in the water absorption of untreated OPC concrete is probably due to carbonation of the surface layer, which makes it denser. BFSC concrete does not show densification upon carbonation, so the water absorption decreases only as a result of further hydration.

Table 4   *Water absorption coefficients of hydrophobised and control specimens after outside exposure*

Cement Type Surface	Product	Water absorption coefficient over 24 hours ($g/m^2\sqrt{s}$) after n months exposure			
		Start	11 months	16 months	21 months
OPC Finished	Control	4.5	1.5	1.1	1.2
	B	0.5	0.3	0.3	0.3
	A	0.6	0.4	0.4	0.4
	E	0.7	0.3	0.2	0.2
BFSC Finished	Control	3.8	3.7	3.1	3.3
	B	0.7	0.3	0.2	0.3
	A	0.7	0.5	0.4	0.5
	E	0.9	0.3	0.2	0.3

7 APPLICATION VARIATIONS

The results of application oriented tests are only summarised here.

It was found that heating to asphalt temperature (160°C) between 1 and 28 days after application of the hydrophobic agent, did not decrease the water repellent effect. Consequently, asphalt may be applied from 24 hours after the application of the hydrophobic treatment.

A low substrate temperature of 5°C had a negative effect on the treatment in that the penetration depth was unacceptably small. This was probably related to the decrease of the consumption of the products. The water absorption, however, had the usual value. Application should therefore take place at temperatures above 10°C.

The waiting period between application of two consecutive coats was varied from 1 to 28 days (compared with ten minutes in the standard application procedure). In all cases the penetration depth was less than after standard application and in a few cases the water absorption was higher than usual. It was concluded that two consecutive coats applied wet-in-wet gives the best result.

In other tests, the substrate was wetted (simulating a shower of rain) and then allowed to dry for 1 to 9 days at 20°C and 80% RH before application of the agent. The penetration depths did not depend on the drying time after wetting; they were lower (between 1.2 and 2.4 mm) than after application on substrates which had equilibrated in standard (20°C and 65% RH) conditions (usually 3 to 4 mm). It was concluded that waiting for one day after rain would result in acceptable penetration of the hydrophobic product; waiting longer was concluded not to be useful.

Finally, the influence of applying a curing compound to the concrete after casting was investigated. It appeared that the application of an acrylic based curing compound did not have a negative effect on the subsequent application of a hydrophobic treatment to OPC concrete. However it did have a negative effect on the penetration depth and water absorption of hydrophobised BFSC concrete. Using a curing compound is not advised for BFSC concrete that is intended to be hydrophobised.

8 CONCLUSIONS

The following conclusions were drawn from the investigation.

The proposed test setup for hydrophobic agents is satisfactory and the requirements are reasonable.

Laboratory tests indicate that several products on the Dutch market show a good performance on concrete typically used in The Netherlands for bridge construction. Various other products, however, did not perform satisfactorily. Proper identification of a hydrophobic agent is therefore necessary.

The performance of the hydrophobic agents was different on different types of concrete in terms of cement type and finished or formwork surface. In particular the penetration of the hydrophobic agent is better in finished surfaces than in formwork surfaces.

Hydrophobic treatment of concrete strongly reduces the penetration of chloride under salt/drying cycles. This is the case for concrete made with portland cement and with (high slag) blast furnace slag cement.

The effect on ongoing corrosion which was caused by penetration of chloride was negligible. This means that hydrophobic treatment of concrete is suitable only as a preventive measure; it is not effective as a corrective measure in the case of active chloride induced corrosion (under wet/dry cycles).

During outdoor exposure for up to two years, the hydrophobic protection remained intact. Apparently the durability of the products was good over that time scale.

Useful recommendations can be made as a result of investigation of variations in the application conditions and procedures.

References

1. J. de Vries, R.B. Polder, 'Hydrophobic treatment of concrete', Construction Repair, September/October 1995, 42-47; Structural Faults and Repair, London, 1995, Vol II, 289-295.
2. G. Sergi, S.E. Lattey, C.L. Page, 'Influence of surface treatments on corrosion rates of steel in carbonated concrete', Proc. 'Corrosion of Reinforcement in Concrete', Eds. C.L. Page, K.W.J. Treadaway, P.B. Bamforth, 1990, SCI/Elsevier, 409-419.
3. Bouwdienst Rijkswaterstaat, 'Aanbeveling voor de keuring van hydrofobeermiddelen voor beton volgens de eisen van Bouwdienst Rijkswaterstaat', 1993, report BSW 93-26 ('Recommendation for testing hydrophobic agents for concrete according to the requirements of Ministry of Transport, Civil Engineering Division'), in Dutch, pp. 24.

A PERFORMANCE SPECIFICATION FOR HYDROPHOBIC MATERIALS FOR USE ON CONCRETE BRIDGES

A.J.J. Calder and Z.S. Chowdhury

Transport Research Laboratory
Crowthorne
Berkshire RG45 6AU

1 INTRODUCTION

The use of deicing salts for the winter maintenance of highways in the UK has caused major problems with existing reinforced concrete bridges. Extensive reinforcement corrosion can occur where the chloride ion concentration at the reinforcement exceeds a threshold value. One solution is to protect the surface of the concrete from the ingress of salts using a hydrophobic surface treatment.

In 1990, the Department of Transport issued a Departmental Standard[1] for the use of surface impregnation to prevent chloride induced reinforcement corrosion from de-icing salts and from marine environments. The Standard applies to reinforced and prestressed concrete members in new constructions and in some existing structures. The hydrophobic impregnating material, a silane, is specified by material type rather than by a performance specification.

This paper describes the development of a performance specification so that impregnants other than those currently specified by the Department of Transport can be considered for use on UK bridges. The specification was then used to assess the performance of a range of products and the results have been used to set pass/fail criteria.

2 DEVELOPMENT OF A PERFORMANCE SPECIFICATION

2.1 Approach

The specification was to be based on a series of performance tests with pass-fail criteria. These performance tests were selected and evaluated initially against materials meeting the current Departmental Standard and then relative to other types of hydrophobic materials so that pass-fail criteria could be considered in that context.

A review of the literature showed that a wide variety of methods have been used to test the performance of hydrophobic materials. They fall into three broad classes;

(1)   assessment of resistance to chloride penetration,
(2)   assessment of water vapour transmission,
(3)   assessment of the durability of the impregnant.

The moisture content of the concrete is likely to affect the penetration and performance of the treatment and the rate at which the untreated concrete loses vapour and absorbs water. The review showed that none of the currently used tests satisfactorily control the moisture content of the concrete specimens at the time of treatment or at the start of the water vapour transmission or chloride absorption tests.

2.2 Test Methods

2.2.1 Preparation of Specimens. All the tests described in this paper were carried out on 100 mm cubes which were cast in the laboratory from 40 Grade concrete (water cement ratio: 0.50), and cured under water at 20 °C for 3-5 days. Three cubes from each batch cast were then dried in an oven at 105 °C to constant weight (normally 7 days) and then discarded. The oven dried weights of these cubes were used to estimate the oven dry weights of the remaining cubes from the batch. The test cubes were dried from their saturated surface dry condition either on the bench or in an oven at 30 °C until they reached a required moisture content and were then stored in air tight boxes containing an appropriate saturated salt solution which maintained constant relative humidity until required for treatment and testing.

For each test method, the performance of three treated cubes was compared with the performance of three untreated control cubes from the same batch of concrete.

Cubes from each batch were treated with the appropriate product by dipping each face in 60 ml of the material for 2 minutes. After 48 hours storage in a fume cupboard, the treated cubes were either replaced in the air tight boxes until required for testing or placed in an environmental cabinet for the start of the water vapour transmission test. Each set of treated cubes was kept separate from all other cubes to prevent cross contamination by silane vapour.

2.2.2 Water Vapour Transmission Test. One advantage of the use of hydrophobic materials compared with other surface protection systems is that the concrete is able dry out by water vapour transmission through the treated surface. A test has been devised to assess this effect by comparing drying rates of treated and untreated cubes.

After treatment, the cubes were kept in a fume cupboard for 48 hours and then placed in an environmental cabinet running at a temperature of 30 °C and a relative humidity of 40 %. The drying rates were measured over a 24 hour period which started after a minimum of 24 hours or when the weight of the cubes was less than their pretreatment weight.

Figure 1 shows typical changes in weight that occur as a result of treatment and subsequent drying; there was an initial increase in weight due to the uptake of silane followed by a reduction in weight due to evaporation of silane vapour. During the first 24 hours that the cube treated with Product A was in the environmental cabinet, its weight reduced to slightly less than its pretreatment weight. It was assumed that once this point was reached, the evaporation of silane vapour was complete and further weight losses were due to water vapour transmission across the treated surfaces. The drying rate was measured over the next 24 hours. In the case of Product B, the rate of weight loss after treatment was much slower (Figure 1). After 24 hours in the environmental cabinet, the cube still weighed about 100 g/m² more than its pretreatment weight and it was evident that silane vapour was still evaporating from it. In order to ensure that the test measured

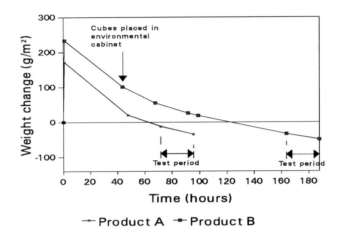

Figure 1 *Changes in weight of cubes during treatment and subsequent drying*

the rate of water vapour transmission only, it was necessary to leave the cube in the environmental cabinet until its weight was less than its pretreatment weight. In this case, the drying rate was measured over a 24 hour period approximately 160 hours after treatment.

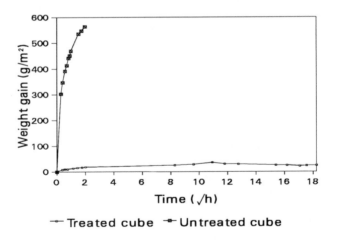

Figure 2 *Weight gain of treated and untreated cubes during immersion tests*

Table 1 *Comparison of Total Immersion and Single Sided Capillary Absorption Tests*

Test method	Absorption Coefficient (%)			
	Silane A	Silane B	Silane C	Silane D
Total immersion	0.70	1.79	0.57	1.62
Single side capillary absorption	1.12	1.69	0.59	1.00

To avoid cross contamination of silane vapour, treated and untreated cubes were kept separate during testing. The drying rate of the untreated cubes was measured between 6 and 24 hours after being placed in the environmental cabinet. The water vapour transmission coefficient was defined as the ratio of the drying rates of treated and untreated cubes expressed as a percentage.

2.2.3 Resistance to Chloride Ion Penetration. The resistance to chloride ion penetration can be determined by measuring absorption either by total immersion in salt solution or by single sided capillary absorption tests. Figure 2 shows the weight gain of a treated cube and an untreated cube plotted against square root of time during total immersion. There was a gradual increase in the weight of treated cubes whereas the weight gain of the untreated control cubes was rapid and became non-linear as they became saturated. The absorption coefficient expressed as a percentage has been defined as the ratio of the weight gain per unit root time of treated and untreated cubes measured over 24 and 1 hour periods respectively. Similarly shaped curves were obtained from single sided capillary absorption tests.

Total immersion and single sided capillary absorption tests were compared using cubes with a moisture content of 4 % treated with four different silanes. The results (Table 1) indicate that performance of silanes B and C as measured by the absorption coefficient was very similar whether measured by total immersion or by single sided capillary tests. Although there were differences in coefficients for silanes A and D, it was considered from these results that a reasonable estimate of the capillary absorption of a treated surface could be obtained from total immersion tests on treated cubes.

The test procedure can be further simplified if the uptake of chloride ions is estimated from the increase in weight during total immersion. Tests were carried out on sets of cubes with moisture contents of 4 % 5 % and 6 % which had been treated with two different silanes. Treated and untreated cubes were totally immersed in 15 % salt solution for 10 days and 4 hours respectively. The total chloride ion content of each cube was then estimated from chloride profiles obtained from one face by "profile grinding". Figure 3 compares the mean weight of chloride ions found in each set of three cubes from the chemical analysis of the dust samples with the weight of chloride ions absorbed during the immersion tests. The results are scattered about the line of equality, especially for the treated cubes. The scatter may be attributed to the fact that the total weight of chloride ions absorbed into each cube has been calculated from a single profile from one face. A better estimate of the weight of chlorides absorbed by each cube would have been obtained by obtaining chloride profiles for each of the six faces. The weight gain slightly overestimated the uptake and suggested that the performance test can be based on immersion in water.

The results of the tests described above suggest that the resistance to the ingress of

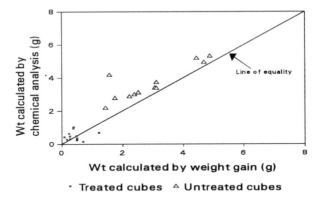

Figure 3 *Comparison of estimates of chloride ion uptake based on chemical analysis and weight gain*

chloride ions by capillary absorption provided by treatment of concrete with silane can be estimated by total immersion tests based on 100 mm treated and untreated concrete cubes.

2.2.4 Silane Penetration. The depth of penetration of silane was measured on three cubes from each batch. The cubes were first sawn in half and points marked along the four edges of one the sawn faces at 10 mm intervals. The samples were washed with tap water for 10-15 minutes. The depth of penetration of silane was then estimated by measuring the distance from the edge of the cube to the boundary between the wet and dry areas at each point using a stereo microscope with twin fibre optics lights. The mean depth of penetration for each cube was calculated.

2.2.5 Resistance to Alkali. The resistance to alkali was measured using a method similar to that specified by the Department of Transportation Utilities, Alberta.[2] Immediately after the immersion test the three treated cubes were placed in beakers containing 3 litres of 0.1 M potassium hydroxide solution. After 21 days, the cubes were removed and dried on the bench until they reached their weight prior to the start of the immersion tests. A further immersion test was then performed on each cube.

2.3 Effect of Moisture Content

The effect of moisture content on the absorption coefficient and the depth of silane penetration was investigated by treating and testing cubes from the same batch of concrete with different moisture contents. The results (Table 2) indicate that for both silanes tested, the mean absolute weight gain of both treated and untreated cubes during the immersion tests and the depth of silane penetration were significantly greater for cubes with a moisture content of 4 % compared with cubes with a moisture content of 6 %. However the effect on the absorption coefficient was less marked.

Table 2 *Effect of Moisture Content on Absorption Coefficient and Silane Penetration*

Moisture content (%)	Total immersion tests		Absorption coefficient (%)	Silane penetration (mm)
	Mean weight gain (g/m²/√h)			
	Treated cubes	Untreated cubes		
	Silane A			
4	6.1	658	0.93	3.5
5	4.5	477	0.94	2.8
6	3.1	290	1.07	2.2
	Silane B			
4	18.1	753	2.40	1.3
5	11.5	455	2.52	0.7
6	9.8	352	2.87	0.3

These results confirm the importance of controlling the moisture content of cubes at the time of treatment and testing.

3 DRAFT SPECIFICATION

The development tests described above resulted in a draft specification for test methods for hydrophobic materials. The specification was evaluated by using it to assess the performance of a range of hydrophobic products. The results were also used to assess the repeatability of each of the tests and to set pass/fail criteria.

The draft specification requires treated cubes from a single batch of concrete to be tested and compared with untreated cubes. Each test cube was first conditioned to a moisture content of 5 % before treatment or testing. Water vapour transmission and absorption coefficients before and after immersion in alkali solution, and silane penetration were measured. The procedure outlined in the draft specification was followed using each product on two different batches of concrete and this generated the pairs of results which were used to calculate repeatability.

3.1 Water Vapour Transmission and Absorption Coefficients

The water vapour transmission and absorption coefficients before and after immersion in alkali solution have been calculated and the mean result for each product is shown in Figures 4-6. The results for each of the tests have been arranged in order of performance with the best performers in each test plotted on the left of the chart.

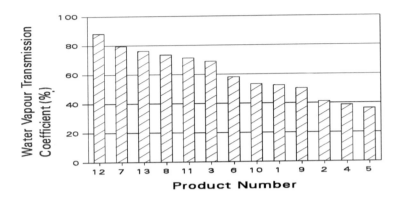

Figure 4 *Results of water vapour transmission tests*

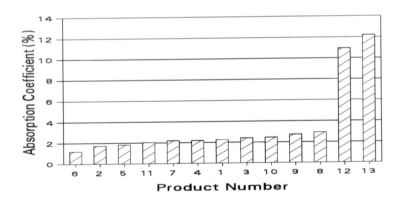

Figure 5 *Results of immersion tests*

The water vapour transmission coefficient (Figure 4) for all the products tested varied between about 40 and 90 %. The values for products currently approved by the Department of Transport (products 1 and 2) were 52 % and 41 %. This indicates that concrete treated with these products will dry out at about half the rate of untreated concrete. The coefficient of most of the other products tested was somewhat higher.

The mean absorption coefficient for each product, prior to the immersion in alkali

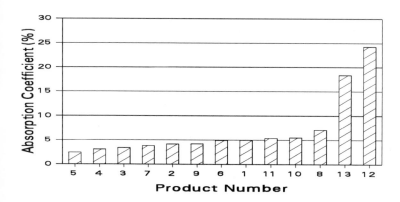

Figure 6 *Results of immersion tests after exposure to alkali*

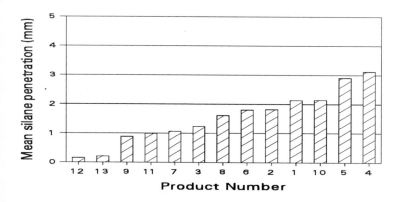

Figure 7 *Mean silane penetration of each product*

solution, is shown in Figure 5. The values for all the silanes, siloxanes and silane/siloxane blends tested (products 1-11) were less than 3 % indicating a very good performance. An analysis of variance of the results from each batch showed no statistically significant differences between the absorption coefficients measured on any of these products (1-11). However, the performance of products 12 and 13 (which were not specifically recommended for use on concrete) was significantly worse with absorption coefficients greater than 10 %.

Figure 6 shows the mean absorption coefficients for each of the products after immersion in alkali solution. The coefficients ranged from 2.41 to 7.06 %, and for each product was greater than the value measured prior to the alkali immersion. This indicated that immersion in potassium hydroxide solution had adversely affected the performance of all the treatments. It also changed the ranking of the products. The performance of products 12 and 13 was again significantly worse than the rest. An analysis of variance of the results from products 1-11 showed that the differences between the products were significant at the 2.5 % level. However, in practice the absolute differences in performance were very small.

3.2 Silane Penetration

The average penetration for each product is shown in Figure 7. For most of the products tested, the variability within each batch was small and there was good agreement between the results from duplicate batches. The average penetration of the products recommended for the treatment of bridges varied between 0.9 mm and 3.1 mm. There was virtually no penetration of products 12 and 13. The penetration of the water based products (3 and 7) was approximately 1 mm.

3.3 Repeatability

British Standard BS 5532[3] defines repeatability as the value below which the absolute difference between two single test results may be expected to lie, with a 95 % probability. The repeatability of each of the test methods was calculated from the pairs of results from each product using the method outlined by Sym.[4] These are given in Table 3. These repeatability values are associated with pairs of results from a single laboratory (TRL). It is likely that the values would be different for different laboratories. If these test procedures are to be adopted as standard, it would first be necessary to estimate the repeatability at different laboratories and the reproducibility between laboratories. This would require comparison of results from pairs of tests on identical materials from a number of different laboratories. This work is currently in progress.

Table 3 *Calculated Repeatability Values for each of the Tests*

Test Method	Repeatability Value
Silane Penetration	1.5 mm
Water Vapour Transmission Coefficient	16.4 %
Absorption Coefficient	1.0 %
Absorption Coefficient (After Alkali test)	3.6 %

3.4 Pass/Fail Criteria

The acceptance level could be set such that new products are required to perform as well in each of the tests as the products that are currently accepted by the DOT. For normal testing of new products, a single test result will be obtained. In order to establish suitable pass/fail criteria, the one sided 95 % confidence limits for individual results were first calculated from the tests on products 1-11. For the absorption coefficient before and after alkali immersion, the pass/fail value was then calculated as the highest mean value obtained from products 1 and 2 plus the 95 % confidence limit calculated above. For the water vapour transmission coefficient and silane penetration, the pass/fail value was calculated as the lowest mean value obtained from the tests on products 1 and 2 minus the 95 % confidence limits. From this analysis, the following criteria were proposed:

Water vapour transmission coefficient	> 25 %
Absorption coefficient	< 5 %
Absorption coefficient (after immersion in alkali solution)	< 10 %
Silane penetration	> 1 mm

The pass/fail criteria for absorption coefficients represent improvements of 95 % and 90 % before and after immersion in alkali solution respectively compared with untreated concrete.

4 CONCLUSIONS

This paper described the development of a performance specification for hydrophobic materials for the protection of concrete highway structures. The specification is based on tests on 100 mm treated and untreated cubes. The tests measure water vapour transmission, resistance to chloride ion penetration, resistance to alkali and depth of silane penetration. Results from the development tests confirmed that it is very important to control the moisture content of the cubes during treatment and testing.

The draft specification was evaluated by testing eleven products which were recommended as being suitable for use on concrete highway structures. Two further products which were considered not suitable for use on bridges were also tested. The following conclusions were drawn:

1. The water vapour transmission coefficients for the different products varied between 40 and 90 %.
2. The absorption coefficients for the products which had been supplied as suitable for the protection of concrete in bridges ranged from 1 % to 3 % prior to immersion in alkali solution. The differences between the products were not statistically significant. The absorption coefficients were higher after immersion in alkali solution (2 to 7 %) and the differences between the products were found to be statistically significant; although in absolute terms these differences were small.
3. The products which were recommended for the protection of concrete bridges penetrated the concrete between 1 mm and 3 mm; whereas there was virtually no penetration of the products which were specifically specified as not being suitable for

the protection of bridges.
4. The performance of the products which were considered unsuitable for use on bridges as measured by the immersion tests were significantly worse both before and after immersion in alkali solution.

The repeatability of each of the test methods has been calculated and pass/fail criteria proposed.

References

1. Department of Transport, Criteria and material for the impregnation of concrete highway structures, Departmental Standard BD43/90, 1990.
2. Department of Transportation, Alberta test procedure for alkaline resistance of penetrating resistance for bridge concrete, Test Procedure BT-002, Province of Alberta, Canada, Undated.
3. British Standard Institution, "Statistical terminology, Part 1, Glossary of terms relating to probability and general terms relating to statistics", London, 1978.
4. R. Sym, "Using repeatability values in an aggregate testing laboratory", *Concrete*, 1987, July, 9.

EFFECT OF SURFACE TREATMENT OF CONCRETE ON REINFORCEMENT CORROSION

A. M. G. Seneviratne[*], G. Sergi[+], M. T. Maleki[+], M. Sadegzadeh[+] and C. L. Page[*]

* Department of Civil Engineering,
Aston University, Aston Triangle, Birmingham B4 7ET, UK.

+ Aston Material Services Limited,
Aston Science Park, Love Lane, Birmingham B7 4BJ, UK.

1 INTRODUCTION

The two main processes responsible for the initiation of reinforcement corrosion are (a) carbonation of the concrete cover to the level of the reinforcing steel causing loss of alkalinity of the concrete and general depassivation of the steel and (b) contamination of the concrete with significant concentrations of chloride leading to breakdown of the protective oxide layer on the steel reinforcement and to pitting corrosion.

The basic requirements for reinforcement corrosion are anodic and cathodic sites connected through the concrete pore electrolyte and a supply of oxygen to fuel the cathodic reaction ($1/2O_2+H_2O+2e^-\rightarrow2OH^-$). As the exclusion of oxygen is very difficult in practice, a more promising way of limiting reinforcement corrosion is to reduce the availability of pore water which in turn will increase the resistivity of the concrete. Research carried out at Aston University has shown that, under laboratory conditions, some concrete surface treatment systems are capable of controlling carbonation-induced corrosion of steel reinforcement exposed to wetting and drying environments, by simply excluding moisture from the concrete.[1] Indeed the ability of many surface treatments to reduce chloride, carbon dioxide and moisture ingress and thereby to control corrosion of reinforcement has been shown by other researchers to be good under laboratory conditions.[2-4] Very little published data is available, however, on the performance of these treatments when applied to structures under field conditions and exposed to natural weathering.

The research reported in this paper forms part of a collaborative project involving Aston University and a number of industrial and science-based partners, supported by the LINK Construction Maintenance and Refurbishment Programme. The aims of the project are to assess whether or not four commercially-available surface treatment systems are likely to be useful in extending the service-lives of concrete structures in which reinforcement corrosion has been initiated as a consequence of carbonation or chloride contamination and to determine whether their performance is affected by exposure to natural weathering.

2 EXPERIMENTAL PROGRAMME

The main scope of the experimental programme was to identify structures with reinforcement corrosion which had been initiated as a result of carbonation and/or chloride contamination. Parts of the identified structures where the corrosion damage had not reached destructive levels were chosen for the investigation. The degrees of corrosion of reinforcements in selected elements were investigated before application of the four selected surface treatment systems and the corrosion activity of the reinforcement and the internal RH of the concrete were monitored regularly. Small specimens were recovered from these structural elements and monitored for corrosion activity of the embedded steel under controlled environmental conditions in order to study the mechanisms of action of the surface treatments in providing protection to the reinforcements. The details of the two structures included in the investigation are described below.

2.1 Reinforced Concrete Structure Suffering from Carbonation

A 40-year-old two-storey building was identified as being suitable for inclusion in the programme. It included eight identical reinforced concrete columns exposed at first floor level and all highly carbonated with carbonation depths varying between 15 and 57 mm. Some columns also contained significant concentrations of chloride. Concrete cover to the reinforcement varied between 15 and 28 mm. The exposed parts of the columns measured 280 x 120 x 2520 mm.

Based on the results of a half-cell survey, four columns were chosen to receive one each of the four selected surface treatment systems. Two columns were left as reference elements. Two silver/silver chloride embeddable reference electrodes were installed in the concrete to enable the measurement of the half-cell potential of the reinforcing steel, one electrode being near the top and another close to the first floor level. Two additional holes of 20 mm ϕ x 35 mm were drilled close to the reference electrodes to accommodate stainless steel inserts through which probes could be introduced to measure the internal relative humidity of the concrete.[5] Electrical connections to the reinforcing cage were made at two points. All exposed steel faces were masked with SBR-modified cement slurry and covered with epoxy adhesive in order to reduce the possibility of crevice corrosion. The details of these installations are illustrated in Figure 1.

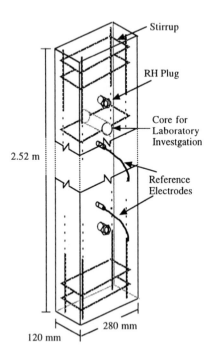

Figure 1 *Details of a typical carbonated column*

2.2 Reinforced Concrete Structure Suffering from Chloride Contamination

Two eight-storey high-rise buildings about 35 years old were selected for the programme since they included a large number of identical balcony railings which were known to be contaminated with calcium chloride. These balcony railings measured 290 x 100 x 5000 mm and many were suffering from cracking and spalling due to reinforcement corrosion. Preliminary investigations showed that chloride levels in the concrete were between 0.18% and 0.27% by weight of sample. Carbonation depths ranged between 5 and 15 mm whilst the cover to the reinforcement varied between 21 and 52 mm. Since the balcony railings were being replaced with new ones, six railings were selected and transported to Aston University to be included in the investigation.

The balcony railings were cut to a manageable size of 400 mm length ('slabs'). Based on half-cell potential and visual surveys when in a wet condition, ten such 'slabs' were selected for the investigation. Five had moderately negative potentials (-250 to -300 mV vs SCE) and no significant surface cracks, and were allocated to group 1. The other five had considerably lower potentials (-450 to -500 mV vs SCE) and had developed large cracks particularly along the top surface. These were allocated to group 2. Two silver/silver chloride electrodes and a relative humidity plug were installed in each slab as shown in Figure 2. Electrical connections were made to the reinforcing steel to facilitate corrosion measurements. All steel faces exposed due to cutting were masked with SBR-modified cement slurry and the cut faces of the concrete were covered with epoxy adhesive so that only the original exposed faces of the railings which were to receive the surface treatments were left uncovered.

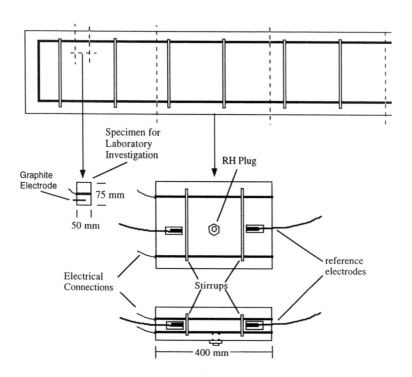

Figure 2 *Details of a typical balcony 'slab'*

2.3 Application of Surface Treatments

The selection of the four surface treatment systems used in this investigation was made by the project steering committee with specialist advice from the three manufacturers of surface coatings, who collaborated in the project. This was done in an attempt to specify the most suitable commercially-available materials for the work, a formidable task in view of the diverse range of products on the market. The details of the four surface treatment systems selected for the investigation are listed in Table 1. For the carbonated columns, the application of surface treatments was carried out by the material manufacturers in-situ, complying with their specifications. In the case of the chloride- contaminated 'slabs', treatments were applied by the manufacturers indoors at Aston University. Four pairs of 'slabs', each pair containing one slab from group 1 and one from group 2, received one each of the four different surface treatments and one pair was left as controls. All ten 'slabs' were then exposed on the flat roof of a four-storey building at Aston University in such a way as to simulate as closely as possible their original exposure conditions. The details of the structural elements and the allocation of the elements for each surface treatment are given in Table 2.

Table 1 *Details of the surface treatment systems used in the investigation*

Code	Description
A	Acrylic modified siloxane primer/water based elastomeric acrylic coating
B	Low-viscosity water based acrylic primer/water based acrylic elastomeric coating
C	Water based epoxy primer/water based acrylic elastomeric coating
D	Solvent diluted oligomeric alkyl alkoxy siloxane

Table 2 *Details of the allocated structural elements for each surface treatment system*

Carbonated Columns from the Two Storey Building						
Specimen	Control 1	A	B	C	Control 2	D
Average Carbonation Depth (mm)	24	28	21	27	52	38
Average Chloride Content (% wt. of sample)	0.02	0.03	0.01*	0.01	0.22	0.29

Chloride Contaminated Balcony Rails from the High Rise Building					
Specimen Type	Controls	A	B	C	D
Average Chloride Content (% wt. of sample) Group 1	0.21	0.24	0.18	0.19	0.23
Group 2	0.27	0.26	0.27	0.26	0.26

* The top third of this column which was cast separately had a chloride content of 0.12%.

2.4 Removal of Specimens for Laboratory Investigation

After the surface treatments were fully cured, small specimens were removed from the structural elements for study in the laboratory. The specimens from the carbonated columns were removed by taking 45 mm ϕ cores through the narrow section of the column, cutting through the links twice (see Figure 1), in order to produce a core specimen with two sections of steel reinforcement embedded in the concrete.

The specimens from the chloride-contaminated balcony railings were removed by cutting with a diamond saw. Each specimen contained two sections of the main reinforcing steel from the top part of the railing and measured approximately 100 x 75 x 50 mm (see Figure 2).

The laboratory specimens were adapted to enable corrosion monitoring by making an electrical connection to the steel and installing a 4 mm ϕ graphite counter electrode and a 3 mm ϕ gel salt-bridge (5% Agar gel + 10% KNO_3) close to the steel reinforcement sections using a fine cement slurry. After masking the exposed steel surfaces with SBR-modified cement slurry, the cut faces of the concrete were masked with epoxy adhesive leaving only the treated faces exposed.

The completed specimens were immersed in water for a week. This tested the ability of the treatments to resist water ingress under full immersion. They were then placed in custom-made environmental cabinets where the relative humidity (RH) could be maintained at a constant level over a period of time with the use of saturated salt solutions.[6] The RH of the cabinet was then changed in approximately 20% steps between 100% and 40%. The RH of the cabinet was monitored with a combined RH/Temperature probe (Vaisala 31UT) and a data logger (Squirrel SQ32).

The corrosion rate of the embedded steel was determined periodically by linear polarisation using a Potentiostat/Galvanostat (Amel, Model 551) with positive feedback IR compensation. The potential of the steel was shifted 20 mV above its rest potential (Ecorr) and then 20 mV below Ecorr. At each potential shift the resulting current was measured after a period of one minute. The polarisation resistance (Rp) was determined from the ratio of the applied potential shift and the resultant current density. The corrosion intensity (Icorr) was calculated assuming B=26 mV in the Stern Geary equation.[7,8]

$$Icorr = B/Rp \qquad (1)$$

3 RESULTS AND DISCUSSION

3.1 Carbonated Reinforced Concrete Columns

Prior to the application of the surface treatments, the columns were wetted thoroughly over a period of a fortnight by spraying water with a hose. Early internal RH values of the columns were, therefore, high (>90%) and appeared to reach a maximum about 40 to 100 days after the initial wetting period. These early high RH values were accompanied by potentials more negative than -300 mV (SCE scale) indicating a significant level of corrosion activity of the reinforcement. With time, the concrete became drier (RH <75%) and the potentials of the steel reached noble values (>-150 mV vs SCE). From around 450 days, all the columns were wetted for 20 minutes with a hose pipe twice a week and, from around 650 days, the chloride-free columns were wetted daily for two hours. The elastomeric coatings were resistant to the uptake of moisture during the wetting period whilst the untreated control columns showed moisture penetration and internal RH increasing to a value close to 90% (Figure 3a). This was sufficient to increase the corrosion activity of the reinforcement and bring potentials down to around -250 mV vs SCE (Figure 3b). The behaviour of the two chloride-contaminated columns was significantly different even during moderately wet conditions. This behaviour, which is depicted in Figures 3c and 3d, shows that the potentials of the steel reinforcement of both the treated and the untreated columns fell quickly at around 450 days, following a small increase in the RH to no more than about 75%. The appearance of significant cracks on

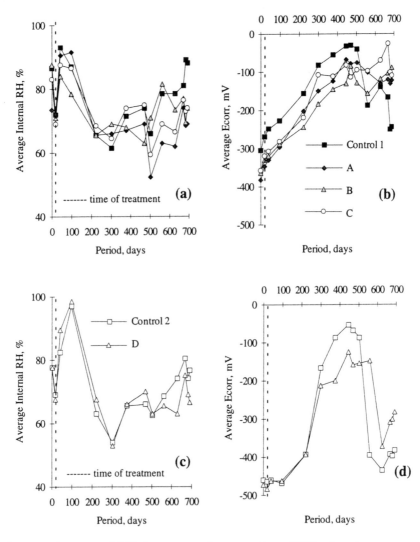

Figure 3 *The internal RH of concrete and average Ecorr (SCE) of steel in carbonated/chloride contaminated columns*

these two columns suggests that the corrosion of steel lying in carbonated and chloride-contaminated concretes may be difficult to control by the simple application of a non-crack-bridging surface treatment as even moderate levels of moisture can penetrate through the growing cracks down to the reinforcement and promote significant corrosion.

The laboratory experiments supported the above findings and provided some information on the fundamental behaviour of the treated specimens. Figure 4 represents the variation of the RH of the environmental cabinet with time. Figure 5 shows how the potentials of the embedded steel varied over the same period. Whilst the behaviour of the chloride-free specimens treated with elastomeric surface treatments (A, B and C) remained fairly constant throughout the period, the potential of the equivalent control specimen responded to the 'wet/dry' cycles and resulted in significantly more negative values during periods of high RH (Figure 5a). The resultant average, maximum and minimum corrosion rates of the steel reinforcement over the exposure period are shown in Figure 6. Of the four chloride-free specimens (control 1 and specimens A, B and C) only the control specimen exhibited corrosion rates that exceeded 0.1 $\mu A/cm^2$ during wet periods. This value is normally accepted as a threshold below which the corrosion rate of steel in

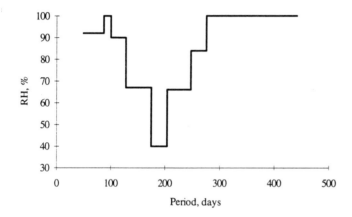

Figure 4 *Relative humidity of the environmental cabinet*

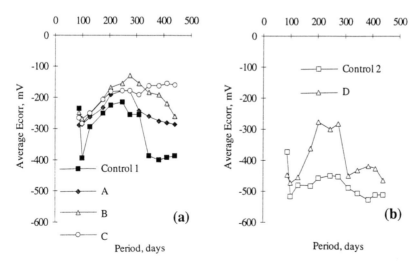

Figure 5 *Average Ecorr (SCE) of the steel in cores from carbonated columns*

concrete is considered to be acceptably low.[9] The elastomeric surface treatments tested in this programme appear, therefore, to be successful for a period of just over a year in excluding moisture, both as surface water and as external vapour; in doing so, they appear to restrict the corrosion of the reinforcement in the carbonated concrete to an acceptable rate.

The effect of chloride contamination of carbonated concrete is seen very clearly in Figures 5b and 6. The steel reinforcement of control 2 did not reach potentials less negative than - 400 mV even during the drying cycle (Figure 5b) and the corrosion rate was an order of magnitude higher than for the chloride-free control (Figure 6). Treatment of the chloride-contaminated concrete with a siloxane-based material reduced the overall corrosion rate significantly (Figure 6) and allowed the potential of the steel reinforcement to reach more noble values (~-300 mV vs SCE) during the drier period   (Figure 5b). It appeared, however, to allow some penetration of external vapour during periods of high RH. Knowing that the siloxane-based material does not have crack-bridging properties and taking into account the in-situ results for the columns, it is reasonable to assume that treatment of reinforced concrete, suffering from both carbonation and chloride infestation, with a hydrophobic type treatment is unlikely to be successful if corrosion initiation has already occurred, owing to the cracks that can develop from reinforcement corrosion and which cannot be accommodated by the treatment.

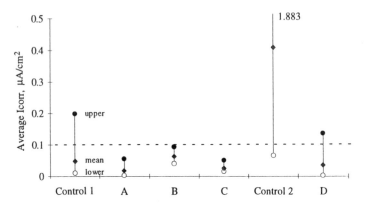

Figure 6 *Variation of the average Icorr of steel in cores from carbonated columns*

3.2 Chloride Contaminated Balcony Rails

The behaviour of the two groups of 'slabs' continued to be very different after the application of the treatments (at 30 days) and their subsequent exposure on the roof of the university building. The internal RH of the group 1 treated 'slabs' generally remained below 80% throughout the 650 day exposure period whilst that of the untreated control 'slab' was frequently as high as 100% (Figure 7a). Evidently, the treatments applied to this group of slabs were able to reduce the penetration of water significantly. The resultant steel potentials were very noble for the treated slabs (> -100 mV, SCE) whilst those of the control were considerably lower, about -200 mV vs SCE (Figure 7b). It is significant to note, however, that corrosion rates of the sections of the reinforcement in smaller specimens exposed in a controlled environment were normally higher than 0.1 $\mu A/cm^2$ even though potentials appeared to be similarly noble. Although some reduction in the corrosion rate may be possible, the overall rate is likely to remain significant in the presence of chloride.

The early behaviour of the group 2 slabs (not illustrated in this paper) was essentially the same for a period of about 100 days but then gradually, hairline cracks started to appear on the elastomeric treatments above the corrosion-induced cracks (in the concrete underneath the treatments) which appeared to become wider with time. These small cracks appeared to be sufficiently large to allow moisture penetration and increase the internal RH considerably so that all the 'slabs', treated or untreated, behaved in the same way as the untreated control of group-1 (Figure 7). Laboratory experiments showed that such levels of potential (-200 to -300 mV vs SCE) coincided with corrosion rates for these types of samples of considerably higher than 0.1 $\mu A/cm^2$. Even though the elastomeric surface coatings were chosen specifically for their crack-bridging capabilities, early results suggest that where corrosion due to chloride contamination of the concrete is ongoing, such surface treatments may not be successful in controlling reinforcement corrosion. The large cracks that can develop from this type of localised corrosion can lead to localised failure of the treatments. The resulting small cracks that develop in the protective coating can become water traps and produce ideal conditions for destructive corrosion of the reinforcement.

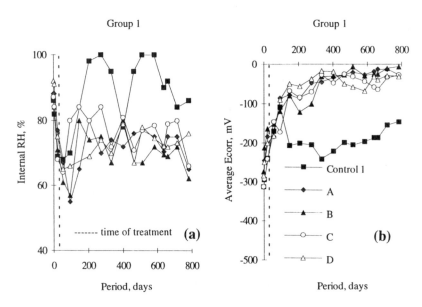

Figure 7 *Average Ecorr (SCE) of the steel and internal RH of concrete in chloride- contaminated balcony railings*

The results obtained so far are relatively short-term. It is hoped that monitoring of these specimens over a longer period of time can provide better qualified conclusions about the long term behaviour of these types of treatments in controlling reinforcement corrosion.

4 CONCLUSIONS

This is a continuing research programme. The specimens are to be monitored for a period of about five years, by which time it should be possible to draw firm conclusions about the ability of certain surface coatings to control reinforcement corrosion and prolong the service-lives of structures in a cost-effective way.

The conclusions drawn so far are based on a limited number of results available after less than two years and are subject to confirmation. They can be summarised as follows:

The three elastomeric, crack-bridging surface treatments and the siloxane treatment used in this investigation appear to be able to reduce the corrosion activity of the reinforcing steel in carbonated concrete structures exposed to natural weathering, as a consequence of the decrease in moisture content indicated by the reduced internal RH of the concrete.

The same treatments were not always successful in maintaining a crack-free, coherent coating on chloride-contaminated concrete elements which had suffered from significant reinforcement corrosion prior to the application of treatment. Cracking on some of the treatments allowed penetration of moisture into the concrete and an increase in the corrosion activity of the reinforcement.

Acknowledgements

This work is supported by contributions from EPSRC, DoE, Aston Material Services Ltd, Aston University, British Cement Association, Building Research Establishment, Transport Research Laboratory, Fosroc International Ltd, SBD Ltd, Stirling Lloyd Polychem Ltd, Pullar Strecker Consultancy Ltd, Taywood Engineering Ltd, L G Mouchel & Partners and Scott Wilson Kirkpatrick & Partners through the LINK CMR Programme.

References

1. G. Sergi, S. E. Lattey and C. L. Page, 'Influence of surface treatments on corrosion rates of steel in carbonated concrete', 'Corrosion of Reinforcement in Concrete', (Eds. C. L. Page, K. W. J. Treadaway & P. B. Bamforth), Elsevier Applied Science, London, 1990, 409.
2. C. Alonso, A. M. Gracia and C. Andrade, 'Some studies on the use of concrete coatings to protect rebars from corrosion', Proc. 5th Int. Conf. on 'Durability of Building Materials and Components', (Eds. J. M. Baker, P. J. Nixon, A. J. Majumdar & H. Davies), Chapman & Hall, London, 1990, 691.
3. P. R. Vassie, 'Concrete coatings: do they reduce on-going corrosion of reinforcing steel', 'Corrosion of Reinforcement in Concrete', (Eds. C. L. Page, K. W. J. Treadaway & P. B. Bamforth), Elsevier Applied Science, London, 1990, 456
4. S. Tanikawa and R. N. Swamy, 'Protection of steel in chloride contaminated concrete using an acrylic rubber surface coating', Proc. Int. Conf. on 'Corrosion and Corrosion Protection of Steel in Concrete', (Ed. R. N. Swamy), Sheffield Academic Press, Sheffield, 1994, 1055.
5. L. J. Parrott, 'Assessing carbonation in concrete structures', Proc. 5th Int. Conf. on 'Durability of Building Materials and Components', (Eds. J. M. Baker, P. J. Nixon, A. J. Majumdar & H. Davies), Chapman & Hall, London, 1990, 575.
6. BS 3781, 'Specification for laboratory humidity ovens (non-injection type)', British Standards Institution, London, 1964.
7. M. Stern and A. L. Geary, 'A theoretical analysis of the shape of polarization curves', *J. Electrochem. Soc.*, 1957, **104(1)**, 56.
8. M. Stern, 'A method for determining corrosion rates from linear polarization data', *Corrosion*, 1958, **14**, 60.
9. M. G. Grantham and J. Broomfield, 'The use of linear polarization corrosion rate measurements in aiding rehabilitation options for the deck slabs of a reinforced concrete underground car park', Proc. 6th Int. Conf. on 'Structural Faults and Repair', (Ed. M. C. Forde), Engineering Technics Press, London, 1995, 199.

ADMIXTURES AND HIGH PERFORMANCE CONCRETE

ELECTROCHEMICAL PROPERTIES OF CONCRETE ADMIXTURES

J. Vogelsang and G. Meyer

Sika Chemie GmbH
Kornwestheimer Str. 107
70439 Stuttgart, Germany

1 INTRODUCTION

In concrete production plants concrete admixtures are widely used for many different purposes. Of them, accelerators and corrosion inhibitors are the most interesting. The influence of the admixtures on the concrete properties is important and has to be considered during the development of a new product. Compression strength, hardening velocity and hydration kinetics are the most relevant properties, but pore content and heat insulation also have to be considered in some cases.

On the other hand admixtures can sometimes cause corrosion of reinforcing steel although iron is passive in alkaline media. To study the corrosive potential of admixtures, it is necessary to perform electrochemical corrosion tests in alkaline solution as well as in mortar samples. Such tests use a wide range of different measuring techniques, e.g. potentiostatic, galvanostatic, linear sweep of current or potential and impedance measurements.

For the reason that quite different measuring techniques (with respective sensitivities) are required, the admixtures have to be split into two classes, the inhibitors and all other types of additives.

Most admixtures were added to almost pure mortar or concrete and here we only have to check the corrosive potential which could be caused by the chemical content of the admixture. The corrosive potential of the well known $CaCl_2$ for example is an unintentional side effect of this accelerator. It will be shown below, that a new class of accelerators was developed to reduce this corrosive potential while retaining the positive influence on the concrete properties[1]. So the necessity for special testing methods is evident. The most interesting question is whether corrosion could be affected by the admixture or not. This relatively simple distinction has to be verified experimentally.

Normally, the high pH value of the cement paste protects reinforcements against corrosion. But in the presence of chloride the passivity of the steel could be disrupted. Deicing salts, chloride-containing sands or aggregates and chloride contamination of the mixing water possibly combined with the carbonation of structures can cause corrosion of the reinforcing steel[2]. To reduce chloride induced corrosion of rebars, corrosion inhibitors may be added to the mixing water[3]. A second possibility is given by surface impregnation of corroding structures with a migrating inhibitor, which diffuses to the reinforcements and reduces the corrosion velocity[4]. For both inhibitor applications we have to investigate a corrosion system with much higher corrosion currents and fluctuations resulting from pitting corrosion in comparison to "ordinary" admixtures in uncorroding media.

In this context a great variety of different electrochemical tests are available. Potentiostatic, galvanostatic and linear polarization techniques are well established. Macrocell tests using three iron rods in mortar cubes are described in ASTM G 109-92.

For all purposes outlined above it is much easier to perform measurements in simulated pore solutions than in mortar samples, because of fluctuations in the ohmic resistance of the mortar. This leads to unreproducible results due to relatively high local current densities in some places. Nevertheless, for better comparability with practical requirements, measurements using mortar samples cannot be neglected.

In this paper we report our experiments to elaborate the corrosive behaviour of accelerators and, secondly, our screening procedure for inhibitor testing. This procedure was choosen to test a large number of substances with respect to their corrosion inhibition potential. Of course, we are not able to give the results in detail, we only want to demonstrate our action and show some general results.

2 EXPERIMENTAL

All measurements were carried out in aerated solutions without oxygen exclusion. Oxygen exclusion is not always appropriate in corrosion rate measurements, because passivators such as chromate or nitrite need oxygen to form a proper passivation layer. For all samples ordinary German construction steel with the strength of St 37 type from the same batch was used.

The measurements were performed in solution with a classical three electrode arrangement using Haber Luggin capillary and Pt sheet counter electrode. The samples were conditioned for one hour at their open circuit potential in the measuring solution.

The working electrode was prepared as follows: the diameter of a steel rebar (St 37) was reduced to 12 mm and then cut into slices with a thickness of 2 mm. These steel disks were ground and polished to 1μm using an automatic polishing machine and diamond paste. Finally, the electrode was degreased with acetone and ethanol and was mounted in a PTFE holder with an electrical contact at the back. Due to the ohmic drop, the distance between the iron surface and the capillary entrance is a sensitive parameter and responsible for variations in the maximum value of currents and the position on the potential scale, when potentiodynamic experiments with higher current densities are performed. The distance between the working electrode and the Luggin capillary must only be adjusted if the differences become too pronounced.

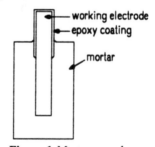

Figure 1 *Mortar samples*

The mortar samples were prepared with a working electrode embedded in the mortar and sealed with an epoxy primer as shown in Figure 1. The mortar was prepared as described in the European Standard DIN EN 196 part 1[5]. All admixtures were applied to the cement content. The samples for the potentiostatic test (see below[6]) were conditioned for one day in saturated $Ca(OH)_2$ solution before measurement and the other mortar samples were kept for four weeks in a humid chamber. The potential was controlled and recorded with a CAMEC II station (7 independent measurements at the same time). The purity of the chemicals was of analytical grade (p. a. grade Merck, Darmstadt), if they were commercially available in this purity.

3 RESULTS

3.1 Accelerators in Solution

We performed two different types of electrochemical tests in solution, the galvanodynamical and the potentiodynamical polarizations with linear cyclic scanning of current or potential. These methods are characterized in Figure 2 (galvanodynamic) and Figure 3 (potentiodynamic).

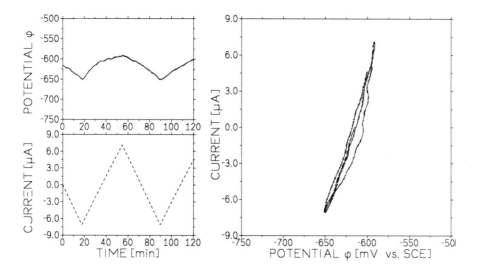

Figure 2 *Galvanodynamic polarisation, GAIN1, applied current (dashed) and measured curve (solid)*

At a pH value of 10 with 0.1M KNO_3, added as a conducting salt, iron remains passive, if no additional potential is overlayed, which can be seen in the Pourbaix diagram. A conducting salt is required to obtain nearly equal conductive properties, if nonionic and ionic chemicals are to be compared. Under these conditions nitrate is a relatively noncorrosive salt, so it should be possible to judge the commencement of corrosive behaviour, caused by the admixtures.

We found that galvanodynamic measuring methods in solutions are not appropriate for distinguishing between accelerators of different corrosive potentials. The conducting salt was corrosive and caused a higher corrosion rate, as can be detected with this method. The poorly informative galvanodynamic polarization curves of $CaCl_2$ and two typical accelerators are shown in Figure 4.

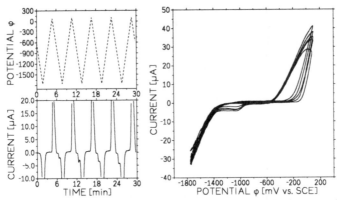

Figure 3 *Fast potentiodynamic polarisation, CV; applied potential (dashed) and measured current curve (solid)*

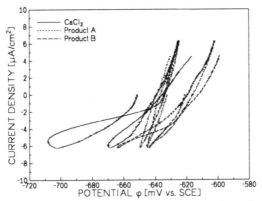

Figure 4 *Galvanodynamic polarisation with steel in alkaline solutions of accelerators*

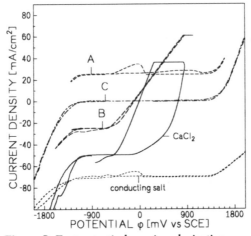

Figure 5 *Fast potentiodynamic polarisation*

However, slow cyclic voltammetry (which could be called fast potentiodynamic polarization, Figure 3) with a scan speed of 10 mV/s showed more mechanistic details and allowed us to distinguish between corrosive and noncorrosive admixtures. Figure 5 shows CV measurements of solutions with conducting salt, $CaCl_2$ or accelerators. It can be seen that in the anodic scan the current densities of corrosive admixtures rise at much lower potentials. Sometimes the anodic part of the polarization curve increases at the beginning of the active region to such a high current density that no passive behaviour remains. Due to these differences, classification into three types of behaviour seems to be useful: non corrosive (the conducting salt and product C), increased active current with repassivation (product A) and severe corrosion without passive region ($CaCl_2$ and product B).

3.2 Accelerators in Mortar

To obtain proper measurements in mortar samples we have to select methods other than linear polarization due to the ohmic drop error referred to above. The ohmic resistance of mortar samples varies strongly with the surface of the mortar, more than with the surface of the working electrode. We observed resistances between 300 and 10^4 Ohm. The ohmic drop was 3 mV to 100 mV, with a current of 10 μA, but 10^4 Ohm is only found when the samples are dried at a relative humidity below 50%. The conductivity of the mortar increases during conditioning in the measuring device. Therefore, the real resistance will be below 1000 Ohm and the absolute potential error will be below 10 mV. These reflections allowed us to perform galvanodynamical measurements with a maximum current of 6 μA. Figure 6 shows the results of $CaCl_2$ and product B as corrosive and product A as noncorrosive accelerators. A sample without admixtures is shown for comparison.

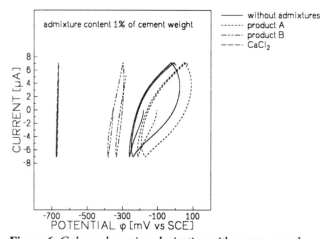

Figure 6 *Galvanodynamic polarization with mortar sample*

Except for $CaCl_2$, all the curves show a hysteresis effect which can be explained by nonstationary states, caused by a relatively high scan speed. Under corrosive conditions ($CaCl_2$) the equilibration at the steel surface must be faster because no hysteresis is observed. Perhaps the corrosive properties of chloride on the passive layer could also be responsible for this effect.

Galvanostatic and potentiostatic tests were also carried out with mortar samples. It turned out that the galvanostatic step method allows good distinction between the different kinds of corrosive properties of the admixtures. Two regions are marked in Figure 7, the shaded area at 600 mV represents curves from samples without admixtures and the lower area (-200 mV) is obtained from samples containing 2% sodium chloride. Samples containing 2% $CaCl_2$ show lower potential curves, and samples with product B reach the upper area considerably later than the other samples. For both concentrations of product A no corrosion was detected, although the higher concentration of product A led to a curve below the lower concentrated sample.

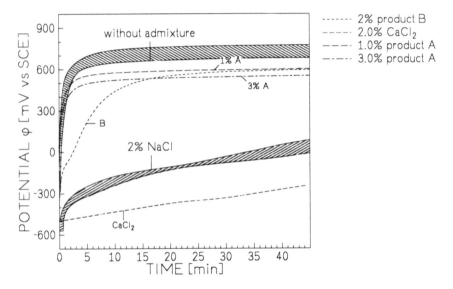

Figure 7 *Galvanodynamic step (10 $\mu A/cm^2$)*

Potentiostatic step tests according to the IfBt recommendations[6] give practical guidelines for the maximum allowable current density of freshly prepared mortar samples (1 day curing, 1 day in saturated $Ca(OH)_2$ solution). The preparation and the parameters for this test have been described elsewhere[6]. A potential of 500 mV vs. NHE has to be applied for 24 hours and the current density is recorded. The current density is not allowed to exceed 10 $\mu A/cm^2$ between one hour and 24 hours after the beginning of the experiment. A further limitation is that the nature of the current density versus time plot for samples with admixtures has to be the same as for samples without additives. In Figure 8 a semilogarithmic plot of current vs. time shows some differences between the products. Both products are able to meet the demands of the IfBt test if only the current density limit is regarded. Product A shows almost the same behaviour as the sample without admixture. The current fluctuations of product B indicate pitting corrosion and the curve deviates from the curve without admixture. Therefore product B could not be certified according to the requirements of IfBt. The other tests have already indicated this failure. The logarithmic current scale leads to a pronounced noise level at small currents (0.1 - 0.02 μA). This is caused by the restricted range of the potentiostat (CAMEC II).

All the methods discussed above only show the short time behaviour. As long as no practical experience over a longer period is available, these short time tests are the only way to investigate corrosive properties.

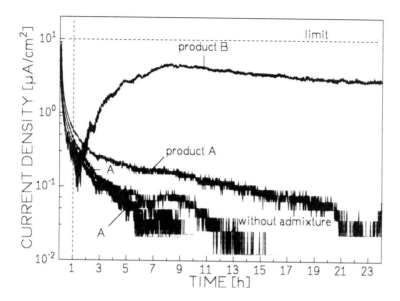

Figure 8 *Potentiostatic step (500 mV vs. SHE)*

3.3 Screening of Inhibitors to be Applied in Concrete

Inhibitors against chloride induced corrosion of reinforcements have to be checked in an environment comparable to practical conditions. Therefore we developed a two day immersion screening test, using a ground and degreased steel plate of 4 cm x 4 cm x 2 mm in a petri dish containing 0.09 molar NaCl solution with a pH value[7] of 10. After two days of immersion it was possible to compare the steel plates and to find appropriate inhibitors in such media. Stimulators can also be found (more corrosion products than the sample without added substance), and it is also possible to differentiate between crevice corrosion, the corrosion at the edges and on the top of the steel sheet (mostly pitting corrosion). This phenomenological approach for evaluation of corrosion inhibitors can be checked with two types of electrochemical test in solution, galvanodynamic and fast potentiodynamic polarization. Slow linear polarization at 200 mV/h gave almost the same results as the 180 times faster method. So the error due to faster cycling can not be too significant. Nevertheless, there are some problems left. For example, pitting corrosion is very difficult to detect. Figure 9 compares the curves obtained from steel in a 0.09 m/l NaCl solution with pH of 10 which are measured with the two scan speeds. It is evident that the faster measurement has sufficient accuracy for the screening of several substances.

Measurements at three pH values and at least five sodium chloride concentrations were performed to gain complete information about the influence of pH and [Cl⁻]. A higher [Cl⁻]/[OH⁻] ratio effects an increase of current density and corrosion rate.

In all cases of very small corrosion traces on the immersed samples we also found an anodic shift of the current density potential plot, if compared to a sample without inhibitor. This was confirmed by open circuit potential measurements using the galvanodynamical method. For lower distortions of the steel surface we reduced the maximum current to approximately 0.7 µA/cm². This type of measurement makes it possible to determine the open circuit potential and to judge the curvature at the origin (polarization resistance). Here we found a strong pH dependence as would be expected from the Pourbaix diagram. Figure 10 shows measured curves at three pH values. With increasing pH the curves are shifted to less negative potentials and the slope is reduced. This indicates a lower corrosion rate.

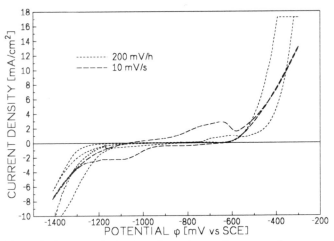

Figure 9 *Linear polarization, dependence of scan speed*

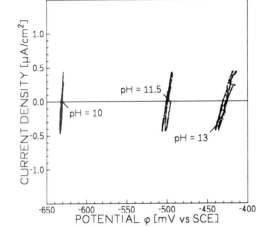

Figure 10 *Galvanodynamic polarization, influence of pH, [NaCl] = 0.09 m/l*

If substances which are expected to work as inhibitors, are added to the alkaline NaCl solution, and produce a positive shift in the steel potential (more than 50 - 100 mV), then the inhibitor can be called effective. The effectiveness can not be quantified, but a relative ranking of different substances according to their shift is clearly appropriate to select the most promising inhibitor. Figure 11 shows the curves of the inhibitive admixture and the impregnation inhibitor measured in a 0.09 m/l NaCl solution with a pH value of 10.

Another ranking can be achieved in a relatively short time by measuring the breakpoint potential during the anodic scan of fast polarization measurements. The higher the breakpoint potential the lower the pitting corrosion tendency of the system. Figure 12 shows three curves for the same substances as Figure 11. It can be seen that both rankings are equivalent, the highest open circuit potential combined with the lowest corrosion rate and, on the other hand, the highest break point potential is correlated with the same type of inhibitor. All results presented here are in accordance with the theoretical expectations.

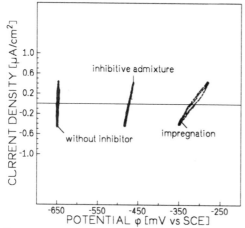

Figure 11 *Galvanodynamic polarization, influence of inhibitor, pH 10, 0.09 m/l NaCl*

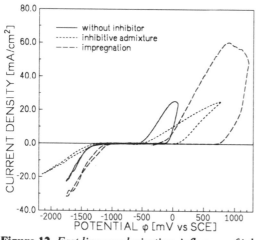

Figure 12 *Fast linear polarisation, influence of inhibitor, pH 10, 0.09 m/l NaCl*

The advantage of these screening methods is the saving of time, because with the CAMEC II station it is possible to perform seven measurements independently at the same time by computer controlled data capture. All the methods described above can be performed with this machine.

4 DISCUSSION, CONCLUSION

The determination of the corrosive potential of concrete admixtures in short time tests is well established, but the requirement for meaningful short time measurements which are highly correlated with long time behaviour is evident. Some practical experiences are given when the IfBt test is applied. However, for inhibitor testing no really powerful test for long time prediction is available, especially for application in concrete structures. Powerful means that a large number of test substances can be checked experimentally in a relatively short time. The macro cell corrosion test ASTM G 109-92 is cheap and easy to do, but time consuming with samples that are heavy and too big.

Of course, there are highly sophisticated experimental methods to measure the polarization resistance with proper consideration of the mortar resistance and electrolyte conductivity, but these methods are not really appropriate if time is a limiting factor.

References

1. Patent pending, see also Sika manual "Technologie und Konzepte zur Herstellung von frühhochfestem Beton" 1.95.
2. C.L. Page and K.W. Treadaway, *Nature,* 1982, **297,** 109.
3. D.F. Griffin, *ACI SP-49,* 1975, 95.
4. U. Mäder, *2nd Regional Concrete Conference - Concrete Durability in the Arabian Gulf,* March 1995, Manama, Bahrain.
5. Deutsches Institut für Normung, DIN, Normenausschuß Bauwesen, *DIN 196 part 1 from March 1990,* Beuth Verlag, Berlin.
6. Y. Efes, Institut für Bautechnik, *Mitteilungen IfBt 1/1990* : "Erläuterungen zu den Prüfrichtlinien und Überwachungsrichtlinien für Betonzusatzmittel" 6.89.
7. E. Ludmann, Thesis, 1992, Fachhochschule für Farbe u. Druck, Stuttgart.

INVESTIGATION AND DEVELOPMENT OF A PROPOSED EUROPEAN STANDARD TEST METHOD FOR ASSESSING THE INFLUENCE OF CEMENT ADMIXTURES ON REINFORCEMENT CORROSION

D. G. Jarratt-Knock, A. Bromwich and J. M. Dransfield

Fosroc International Technology Centre
Birmingham B6 7RB
UK

1 INTRODUCTION

The performance of admixtures for cementitious systems is specified in the new European standard EN 934. Part 2 of EN 934[1] relates specifically to admixtures for concrete and in it, performance requirements are specified according to standard test methods, which are specified either in EN 480,[2] or in ISO standards where possible. Besides specific performance requirements for particular admixture types, admixtures of all types must fulfil a set of general requirements. These include such properties as chloride content and infra-red spectrum (compared with a reference spectrum), and also a property defined as 'corrosion behaviour'. This latter requirement was included at the drafting stage, because CEN Technical Committee TC104/SC3, responsible for EN 934-2, believed an assurance that an admixture would not enhance corrosion was desirable. However, current draft versions of the standard do not refer to a test method, because agreement could not be reached on a suitable means of assessing corrosion risk.

The intention was that an accelerated test should be used, to allow rapid evaluation and thus demonstrate the harmlessness of the admixture. The first proposal was to use potentiostatic anodic polarisation; however insufficient experimental evidence could be produced by the proponents, and an alternative method based on galvanostatic anodic polarisation was proposed. Currently neither method has been accepted for use, because of insufficient data to confirm or deny suitability.

The current work was carried out to compare the two proposed methods and, if possible, suggest performance criteria against which admixtures could be tested. Much of the work was carried out as part of a project for a Diploma in Advanced Concrete Technology, and was supported by funding from the Department of the Environment.

2 APPARATUS

2.1 Galvanostatic Anodic Polarisation

The experimental set-up described in prEN 480 Part XXX[3] 'Galvanostatic Electro-

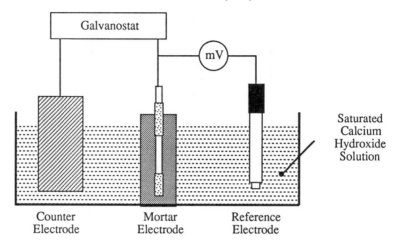

Saturated
Calcium
Hydroxide
Solution

| Counter | Mortar | Reference |
| Electrode | Electrode | Electrode |

Figure 1   *Schematic diagram for Galvanostatic Anodic Polarisation*

chemical Test for the Measurement of Corrosion of Steel in Concrete' was used, and the procedures described therein were taken as a starting point. Some of the procedures were modified as part of the test programme and this is described later. The test cell was as shown in Figure 1. The galvanostat used was an 8-channel Multistat II from Sycopel Scientific used in galvanostatic mode, driving a constant current density through the test cell and monitoring the potential attained by the mortar electrode. It was connected to an IBM compatible PC which was used to run the control and data-acquisition software. The mortar electrodes consisted of smooth, polished, low carbon mild steel bars embedded in mortar. The bars were cleaned by washing in 1:1 hydrochloric acid containing 3 g/litre of hexamine as an inhibitor.

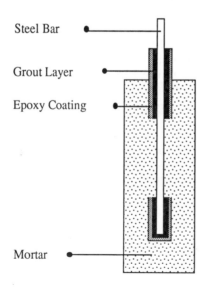

Steel Bar

Grout Layer

Epoxy Coating

Mortar

They were then washed, dried and checked for visible signs of corrosion, repeating the process if necessary until all surface rust had been removed; they were then stored under white spirit until use. The bars were coated in an OPC/water grout of 0.5 w/c ratio, and an epoxy resin to leave a known area of the bar exposed to the test. The prepared bars were then cast into cylindrical mortar specimens, using a 3:1 sand:cement mortar, w/c ratio 0.5, in accordance with EN 196 Part 1. The cement is required to have a C_3A content below 5%, and SRPC ex

Figure 2   *Cross-section of mortar electrode (not to scale)*

Figure 3 *General view of experimental set-up showing multi-channel potentiostat*

Castle Cement (Ketton) with a C_3A content in the range 1.8 - 2.6% was used. Figure 2 shows a schematic diagram of the 'mortar electrode'.

The counter electrode in all cases was a rectangular plate of dimensions 70 x 80 mm, made from type 316 austenitic stainless steel.

A general view of the experimental set-up is shown in Figure 3.

The test regime specified in the standard is as follows:
- Curing period of mortar from mixing until commencement of electrochemical test: 7 days
- Applied current density: $0.1 \pm 0.005 \, \mu A/mm^2$
- Duration of polarisation: 24 hours

2.2 Potentiostatic Anodic Polarisation

The circuit diagram for the potentiostatic test is shown in Figure 4. The test method is described in prEN 480 Part YYY,[4] 'Potentiostatic Electrochemical Test for the Measurement of Corrosion of Steel in Concrete'.

The equipment for the potentiostatic test was essentially the same as for the galvanostatic test, except that the potentiostat was run in potentiostatic mode, applying a controlled potential and monitoring corrosion current.

The test regime specified in the standard is as follows:
- Curing period of mortar from mixing until commencement of electrochemical test: 7 days
- Applied potential: $+260 \pm 5 \, mV_{SCE}$
- Duration of polarisation: 24 hours

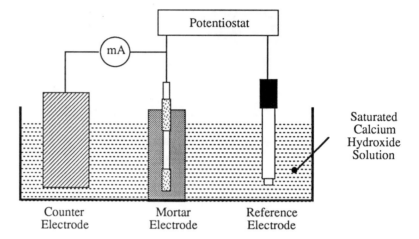

Figure 4   *Schematic diagram for Potentiostatic Anodic Polarisation*

3 BASIS OF TESTING

EN 934 Part 2 would probably require the admixture being assessed to be tested at maximum dosage. However, in order to evaluate the sensitivity of the method and to set compliance limits for new admixtures, the ability of the method to induce corrosion at chloride levels down to 0.1% and up to 1.0% was measured. This range was chosen to span the maximum levels of chloride considered to present negligible corrosion risk in reinforced concrete (0.4% Cl⁻ by weight of cement) and prestressed concrete (0.2% Cl⁻ by weight of cement) according to prEN 206 'Concrete - Performance, production, placing and compliance criteria'.

The chloride, in the form of calcium chloride ($CaCl_2.2H_2O$ - Technical grade), was dissolved in the mixing water and added to the mortar over the above range of dosages.

A curing period of seven days is specified in both electrochemical draft standards. Tests were also carried out using curing periods of one, two and three days. The samples were cured in their moulds, at 20±2 °C, and RH ≥ 90%. Tests were carried out in groups of eight, the eight specimens being cast from the same mix and tested simultaneously using the eight channels of the potentiostat.

4 RESULTS

4.1 Form of Results

4.1.1 Galvanostatic Test. The galvanostatic test maintains a constant current density through the mortar electrode and monitors variations in electrode potential with time. Thus the results can be expressed in the form of a graph of electrode potential as a function of time. In the absence of pitting corrosion, the graph is of the form shown by the solid line of Figure 5, where the potential steadily increases until a plateau value is reached. If pits develop, the graph is of the form shown by the dotted line in Figure 5, where the potential rises initially

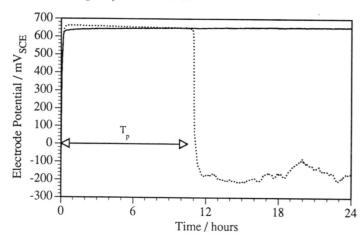

Figure 5 *Typical galvanostatic test results*

but then falls off in a characteristic fashion.[5] The graph is therefore clearly indicative of whether or not pitting corrosion is occurring. The graph can also be analysed further to obtain values for time to onset of pitting T_p .

4.1.2 Potentiostatic Test. The potentiostatic test polarises the bar to a constant potential and monitors the current flowing in the cell. The results are therefore expressed in the form of a graph of current density against time. In the absence of pitting corrosion, the graph will be of the form shown by the solid line in Figure 6, where the current density is small or zero. If pitting corrosion occurs, the current density will tend to rise above that of the control,[6] as shown by the dotted line of Figure 6.

The method suggests that values can be obtained from the graph for current density, J_1, at 1 hour, J_{24} at 24 hours, and the maximum current density obtained during the test, J_{max}. Only J_{max} has been considered in the present study.

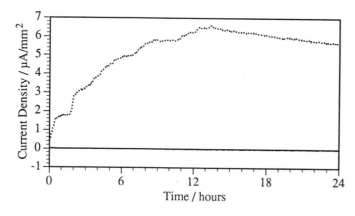

Figure 6 *Typical potentiostatic test results*

5 DISCUSSION OF RESULTS

5.1 Variability

Previous authors[7,8] have commented on the variability to be expected in electrochemical corrosion studies of reinforced cementitious systems and this was certainly experienced in this work. Figure 7 shows a typical set of galvanostatic results on eight replicate specimens, where three initiated pitting almost immediately, one initiated after 12 hours, and four remained passive for the full 24 hours. Similarly, Figure 8 shows a comparable series for the potentiostatic method, where three specimens showed no corrosion behaviour and the other five showed maximum current densities between 1 and 7 $\mu A/mm^2$.

Both methods are designed to detect pitting corrosion. Initiation of pitting depends on

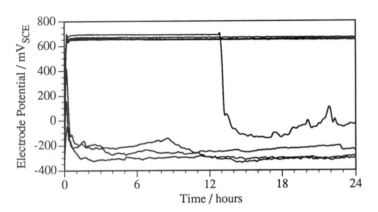

Figure 7 *Illustration of variability of galvanostatic method - eight replicate specimens from same mix, 1% CI^-, one day cure*

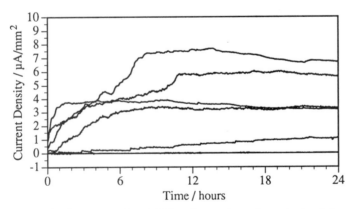

Figure 8 *Illustration of variability of potentiostatic method - eight replicate specimens from same mix, 0.3% CI^-, one day cure*

having the appropriate chemical environment in the vicinity of the steel[7,9] and, because of the inhomogeneous structure of the developing cement matrix, the likelihood of this occurring in any mortar specimen of a given overall composition will vary. Thus it is not unexpected that there should be considerable variability in the results. This does, however, make it difficult to use a test of this nature to specify absolute performance levels for an admixture. However, previous work[10,11] suggests that, notwithstanding the overall variability, the likelihood and severity of corrosion should increase with the level of aggressive species.

5.2 Galvanostatic Results

Figure 9 shows a plot of average T_p against chloride dosage, for various curing periods. To obtain this graph, the results for each test condition were examined and the average period T_p before initiation of pitting was calculated. If no pitting occurs, T_p will equal 24 hours under

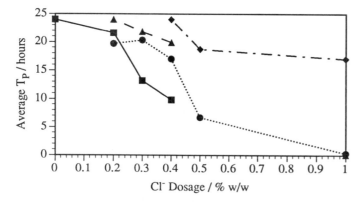

Figure 9 *Galvanostatic Test - T_p as a function of chloride dosage*
■ *1D Cure,* ● *2D Cure,* ▲ *3D Cure,* ◆ *7D Cure*

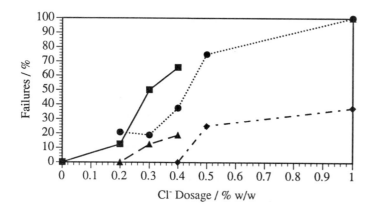

Figure 10 *Galvanostatic Test - percentage of failures as a function of chloride dosage*
■ *1D Cure,* ● *2D Cure,* ▲ *3D Cure,* ◆ *7D Cure*

the regime of this test. There is still some anomalous behaviour, but clear trends towards decreasing T_p with increasing chloride dosage, and increasing T_p as curing time increases, are seen.

The galvanostatic results were also analysed in terms of the proportion of failures observed for a given test condition. Figure 10 shows this information as a plot of the percentage of failures (a failure being defined as a specimen where corrosion behaviour is observed) against chloride dosage, for each of the curing periods. Again, there is a reasonable indication of increasing incidence of failure with increasing chloride dosage, and decreasing incidence as curing time increases.

5.3 Potentiostatic Results

A similar treatment was carried out on the potentiostatic results. Figure 11 shows a plot of average J_{max} (maximum current density averaged over eight replicate specimens) against chloride dosage, for different curing periods. In this case, little corrosion activity is seen for curing periods above two days, but for one and two days' curing, J_{max} increases with increasing chloride content and decreases with increasing curing time.

Figure 12 shows percentage of failures ('failure' has here been arbitrarily defined as occurring when $J_{max} > 0.5\ \mu A/mm^2$) against chloride content and curing time, in an analogous manner to Figure 10. Here there is considerable scatter in the results, with only the one day curing period giving any indication of a trend towards increasing failure incidence with increasing chloride dosage. There is some suggestion of a trend towards decreasing failures as curing time increases.

As stated earlier, it is to be expected that the incidence and severity of corrosion should increase as chloride content increases, due to the increased likelihood of chloride/steel contact in a suitable environment for corrosion to occur. It is also expected that incidence and severity might decrease as curing time increases, since continuing hydration means that the cement matrix will become denser with time, reducing the availability of oxygen at the steel surface,[9] and also inhibiting any polarity-induced flow of chloride ions to the steel with time.

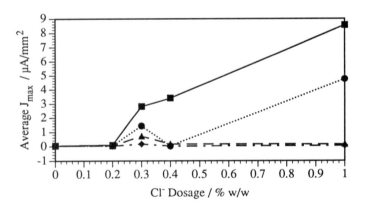

Figure 11 *Potentiostatic Test - J_{max} as a function of chloride dosage*
■ *1D Cure,* ● *2D Cure,* ▲ *3D Cure,* ◆ *7D Cure*

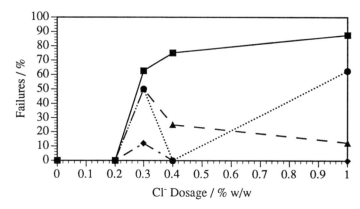

Figure 12 *Potentiostatic Test - percentage of failures as a function of chloride content*
■ 1D Cure, ● 2D Cure, ▲ 3D Cure, ◆ 7D Cure

5.4 General Observations

From the present study the galvanostatic test appears more sensitive, in the sense that corrosion behaviour was observed at lower levels of chloride than with the potentiostatic test. The trends observed with increasing chloride content and curing time were also more consistent with the galvanostatic test. It is not currently understood why this should be the case, although Page and Treadaway[9] suggest that galvanostatic polarisation may be more reliable than potentiostatic or potentiodynamic methods when small current densities are applied; however the results reported here would tend to favour the galvanostatic method on the grounds of sensitivity to low chloride levels and it is acknowledged that the galvanostatic test is more severe than the potentiostatic, as a higher potential is applied to the steel.

A problem which has been reported with accelerated laboratory tests is that appreciably higher chloride levels may be required to induce corrosion than are known to cause problems in practice.[9,12] Certainly the seven day curing period suggested in the drafts makes the methods appear relatively insensitive and could usefully be shortened. Ideally one would hope to see an increasing incidence/severity of corrosion through the critical range of 0.2-0.4% chloride; this was only seen with short curing periods, whilst after seven days' curing, significant corrosion only occurred above 0.4% chloride (and this only in galvanostatic mode). The authors favour the use of a two day curing period under the galvanostatic regime. If the potentiostatic method were to be used, the results would suggest that sufficient sensitivity would only be obtained after a one day cure.

Existing literature suggests that a range of factors can affect corrosion behaviour, including pore structure, permeability, cement type and composition, mix quality, metal surface variations and the intimacy of the contact between the steel and the aqueous phase of the cementitious matrix.[5,9,13-16] Given this, and also given the observed inherent variability of the results, the authors feel it is not possible to define an absolute performance limit in terms of percentage failures or average T_p, which would indicate 'harmlessness' of an admixture. It might, however, be possible to relate the performance of an admixture under test to that of calcium chloride, and to require that the admixture gives an improved performance over a given level of chloride. To compare relative performance, it is of course still necessary to take into account the variability of the test method.

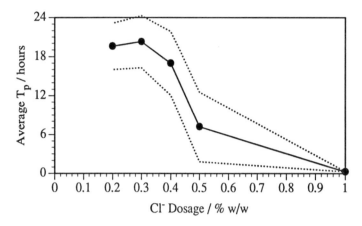

Figure 13 *Galvanostatic Test - Tp as a function of chloride dosage,*
showing 95% confidence limits (two day-cured samples)
———— *Mean,* ········ *95% Confidence*

6   PROPOSED PERFORMANCE EVALUATION

6.1 Variability of Two Day Cured Specimens

Figure 13 shows a plot of T_p against chloride dosage for samples cured for two days; the mean values are those shown previously in Figure 9. For each dosage level, 95% confidence limits have been calculated. In practice, the mean T_p value for an admixture under test would be compared with the 95% confidence interval for 0.4% chloride and to pass the test, the admixture's performance would need to be no worse than that of the chloride.

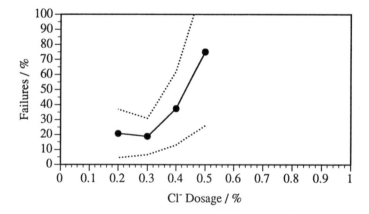

Figure 14 *Galvanostatic Test - percentage of failures as a function of chloride*
dosage, showing 95% confidence limits (two day-cured samples)
———— *Mean,* ········ *95% Confidence*

6.2 Test Procedure

From the present results, it can be seen that the confidence limits are wide, and this inherent variability needs to be allowed for when defining performance criteria. A hypothetical test regime might be as follows:

The admixture under test would be dosed at the manufacturer's recommended maximum dosage. Equivalent mixes would be carried out, using the same batches of cement, sand and steel as the test mix, with no admixture, and with 0.4% chloride. All tests would be replicated a specific number of times, at least eight based on the present study. For each set of results, the mean time before pitting occurs (T_p) or the percentage of failures would be calculated.

The values obtained for the test admixture would be compared with those for the chloride, to ascertain whether or not it showed impaired performance. The variability of both sets of results would need to be assessed in order to ensure that any differences were statistically significant. The control series with 0% chloride would be required to show a mean T_p of 24 hours, or 0% failures. Any reduction in Tp or increased incidence of failures would indicate possible problems with the equipment or procedure.

7 FURTHER WORK

Further work is required to define performance criteria in detail. For a standard method, it would be necessary to choose whether to evaluate on the basis of T_p or percentage of failures, and more extensive and rigorous work is required on reproducibility and repeatability, although the conclusions drawn in this paper are based on more than 400 individual tests. Certain admixture types might require modifications to the procedures; for example, a retarding admixture at the maximum dosage would significantly delay setting, resulting in an effective difference in curing time.

8 CONCLUSIONS

- Neither the galvanostatic nor the potentiostatic method is suitable as it stands to allow specification of a safe level of performance for an admixture under test, but it is believed that modifications could be made to the procedures and results analysis to render them usable.
- The galvanostatic method proved more sensitive and it is recommended that further work be based on this method.
- The curing time for specimens should be reduced to two days.
- The inherent variability of the corrosion process means that it will be necessary to test multiple specimens and express performance in terms of the proportion of those specimens showing corrosion behaviour, or in terms of an average duration of test before initiation of corrosion. The performance would then be compared with that of calcium chloride at a specified level. To achieve a pass, the admixture would have to show improved performance compared with the appropriate level of chloride, by a given margin of safety.
- Further work is required before performance criteria can be specified, including a rigorously planned statistical study of the method's repeatability and reproducibility.

References

1. prEN 934 -2 - Admixtures for concrete, mortar and grout. Part 2: Concrete admixtures. Definitions, specifications and conformity criteria.
2. prEn 480 - Admixtures for concrete, mortar and grout. Test methods (various parts).
3. prEN 480 Part XXX - part number not yet assigned.
4. prEN 480 Part YYY - part number not yet assigned.
5. K. W. J. Treadaway and A. D. Russell, *Highways and Public Works*, 1968, **36(1705)**, 40.
6. W. Manns and W. R. Eichler, *Betonwerk + Fertigteil-Technik*, 1982, **3**, 154.
7. C. L. Page, 'The corrosion of reinforcing steel in concrete: its causes and control', paper presented at the Institution of Corrosion Science and Technology / Institution of Civil Engineers Symposium, University of Strathclyde, on 'The Corrosion and Degradation of Structural Materials', 1979.
8. C. L. Page and J. Havdahl, *Matériaux et Constructions*, 1985, **18**, 41.
9. C. L. Page and K. W. J. Treadaway, *Nature*, 1982, **297**, 109.
10. P. Schießl and M. Schwarzkopf, *Betonwerk + Fertigteil-Technik*, 1986, **10**, 626.
11. Building Research Establishment Digest No. 263, 1982.
12. P. Lambert, C. L. Page and P. R. W. Vassie, *Matériaux et Constructions*, 1991, **24**, 351.
13. P. S. Mangat and B. T. Molloy, *Cement and Concrete Research*, 1991, **21**, 819.
14. ACI Committee 222, *ACI Journal*, 1985, **January-February**, 3.
15. Rasheeduzzafar, *Cement and Concrete Research*, 1991, **21**, 777.
16. W. López and J. A. Gonzalez, *Cement and Concrete Research*, 1993, **23**, 368.

SUPERCOVER CONCRETE: A COST EFFECTIVE METHOD OF PREVENTING REINFORCEMENT CORROSION

C. Arya[a], F.K. Ofori-Darko[a], G. Pirathapan[a] and P.R.W. Vassie[b]

[a]South Bank University
Wandsworth Road
London SW8 2JZ, U.K.

[b]Transport Research Laboratory
Crowthorne
Berkshire RG11 6AU, U.K.

1 ABSTRACT

The paper describes a new design method called Supercover Concrete which should significantly reduce or eliminate the incidence of reinforcement corrosion in concrete structures. It involves using glass fibre reinforced plastic (GFRP) rebars at nominal cover depths to control surface crack widths while conventional steel reinforcement, situated at deep cover depths, will provide the tensile strength and will not be subject to corrosion.

Investigations on single span, simply supported reinforced concrete beams made using this design show that GFRP rebars are effective in reducing surface crack widths to values recommended in BS8110 and Eurocode 2. Moreover, it would appear that introducing GFRP rebars in the concrete cover does not significantly affect either the structural strength or the deflections which occur. The structural strength, design crack width and deflections of the concrete beams can be predicted using the procedures given in BS8110. Life cycle costing studies on beams show that it is a cost effective option for preventing reinforcement corrosion.

2 INTRODUCTION

Corrosion of reinforcement in concrete remains a significant problem. Structures especially vulnerable to this form of deterioration are those exposed to chloride ions from sea water and de-icing salts.

Many methods have been proposed to combat this problem including treating the surface of the concrete with silanes or siloxanes, replacing a portion of the cement with ground granulated blast furnace slag or pulverised fuel ash and using stainless steel or coated reinforcement. Another approach is simply to increase the depth of concrete cover to the reinforcement. Since the penetration of chlorides into hardened concrete roughly follows a square root time function,[1] if the cover to the reinforcement was increased from 40 mm to, say, 100 mm i.e. an increase of 60 mm above the nominal cover specified in Eurocode 2,[2] the time taken for carbonation or chlorides to reach the level of the reinforcement will increase by a factor of $(100/40)^2$ i.e. approximately six fold.

Field experience shows that many concrete highway structures in the U.K. show signs of deterioration despite being only twenty years old[3] and therefore providing cover depths of 100 mm or more should eliminate the possibility of reinforcement corrosion occurring during the design life of most structures. Increasing the cover to the reinforcement increases surface crack widths which may be undesirable from the viewpoint of durability, and is often unacceptable for reasons of appearance. However, with a limited amount of glass fibre reinforced plastic (GFRP) rebars introduced in the cover zone at a nominal depth of say 30 mm it may be possible to control surface cracking within acceptable limits. This additional reinforcement can be easily attached to the main steel and offset with spacers.

The use of GFRP for the reinforcement of concrete is relatively new. Therefore before describing the laboratory and desk studies which are being carried out on beams designed using the new design method, the following section briefly summarises the properties of GFRP rebars.

3 GFRP REBARS

GFRP rebars are normally made of fibres of E-glass embedded in a thermosetting resin produced from vinyl esters or epoxies.

GFRP rebars are commonly manufactured by the pultrusion process. This involves drawing several spools of fibres in the form of strands through a resin bath. They are then pulled through a heated die which consolidates the fibre/resin composite in the form of smooth rods and also cures the resin. The rods may be left smooth or roughened by, for example, wrapping the rod with an additional strand of resin soaked glass fibre in a 45 degree helical pattern or coated with a layer of sand particles.

There are currently no standards for the size of GFRP rebars. They are produced in various lengths e.g. 6 m but generally with the same diameters as conventional steel reinforcement. Unlike steel reinforcement, GFRP rebars cannot be bent into different shapes on site; any hooks or bends etc must be manufactured in the plant before the resin has cured. Electrical type nylon ties can be used for fixing GFRP rebars.

It is difficult to be precise about the cost of GFRP rebar as it varies between producers because of the range of materials and manufacturing methods used. But by way of an example, the prices obtained from one supplier suggests that GFRP will cost between 2 and 4 times of conventional steel.

Research on the use of GFRP in concrete has mainly focused on using it as a substitute for conventional steel reinforcement. In North America and Europe, the principal reason for this is to find a way of preventing reinforcement corrosion.[4] Tension tests performed on GFRP bars have shown that it does not exhibit any signs of plasticity. Moreover, its tensile strength is almost three times higher than that of steel reinforcement. Despite this, flexural tests on concrete beams reinforced with equal amounts of steel and GFRP have shown that the beams reinforced with GFRP have significantly lower strengths, greater deflections and larger surface crack widths compared with steel reinforced beams.[5,6] This is due to the fact that the modulus of elasticity of GFRP is approximately 25% that of steel and that the bond strength of GFRP bars with concrete is also significantly lower, being approximately 65% that of the steel bars to concrete strength.[7]

Other studies which have been carried out have suggested that the strength and stiffness of GFRP bars may be adversely affected due to excessive moisture absorption, exposure to alkaline environments such as that in concrete, and exposure to high temperatures.[8]

Nevertheless, many of these difficulties can be avoided and the others marginalised if GFRP bars are used in combination with conventional steel reinforcement in the aforementioned manner. The following section describes the experimental work undertaken to:

1) assess the effect on structural behaviour of providing deep concrete cover to the steel reinforcement
2) assess the effect of GFRP in crack control
3) compare actual and predicted crack widths and deflections using the procedures in BS8110.[9]

4 EXPERIMENTAL PROCEDURE

Four reinforced concrete beams, 300 mm wide, 400 mm deep and 3100 mm long, were cast in timber moulds. Figure 1 shows the reinforcement details. Mix details are given in Table 1. Two batches, each consisting of 500 kg of concrete, were used to make each beam. An electric vibrator was used to compact the concrete. Three 100 mm cubes were cast from each batch of concrete to measure the compressive strength.

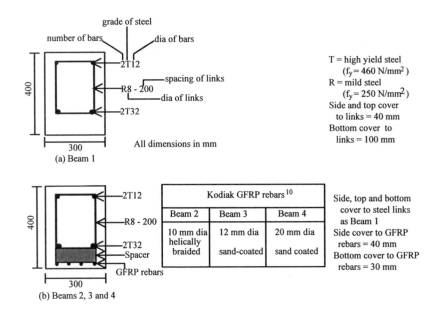

Figure 1 *Reinforcement details*

Table 1 *Mix proportions (by weight)*

OPC	1.00
Sand (Zone 3)	1.78
10 mm aggregate (flint gravels)	1.31
20 mm aggregate (flint gravels)	1.96
Water	0.43

The beams were allowed to cure in their moulds for two days. Thereafter they were removed from the mould and wrapped in damp hessian and polythene, and allowed to cure for a further 26 days.

Prior to testing, the beams were painted white in order to ease monitoring of crack development during loading. The beams were tested in bending using a four point loading arrangement as shown in Figure 2. The effective span of the beam was 2.9 m. The load was applied at a rate of 0.03 kN/sec, generally in increments of 12.5 kN or 25 kN up to a maximum of 210.5 kN in order to produce a design moment of 100 kNm in the 1 m central pure bending region of the beam.

At each load increment, the crack widths were measured using a portable microscope. The position and extent of individual cracks was also recorded. In addition, the mid-span defections of the beams were measured by means of a linear displacement transducer.

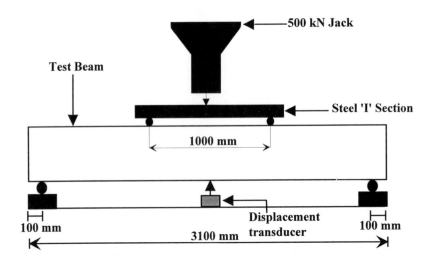

Figure 2 *Bending test*

5 RESULTS AND DISCUSSION

5.1 Concrete Compressive Strengths

Table 2 summarises the average 28 day compressive strengths for the concrete used in Beams 1-4.

Table 2 *Average 28 day compressive strengths*

	Beam 1	Beam 2	Beam 3	Beam 4
Compressive strength (N/mm²)	47	53	53	52

5.2 Crack Widths

Figure 3 shows the pattern of cracking and maximum surface crack widths obtained on Beams 1-4 at a load of 210.5 kN. Beam 1, containing only steel reinforcement, had a total of six cracks, four of which exceeded a surface width of 0.3 mm. The maximum crack width on Beam 1 was 0.5 mm. Beam 2, containing four 10 mm diameter braided GFRP rebars, had a total of seven cracks, one of which exceeded a surface crack width of 0.3 mm. The maximum crack width on Beam 2 was 0.4 mm. Beam 3, with four 12 mm diameter sand coated GFRP rebars, had a total of eleven cracks and the maximum crack width was 0.3 mm. Beam 4, containing four 20 mm diameter sand coated GFRP rebars, had a total of six cracks and the maximum crack width was 0.25 mm.

Beams 3 and 4 complied with BS8110 as none of the crack widths exceeded 0.3 mm. Beam 2 also complied with BS8110 as more than 80%, actually 85%, of the cracks had a surface width of 0.3 mm or less. Beam 1, on the other hand, fell outside the recommendations for crack control in BS8110.

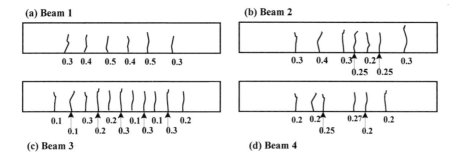

Figure 3 *Crack patterns and maximum surface crack widths (mm) at a load of 210.5 kN*

Figure 4 shows the predicted and experimental crack widths as a function of design moment for Beams 1-4. The predicted surface crack widths, w, were calculated using equation 1, taken from BS8110:[9]

$$w = \frac{3 a_{cr} \, \epsilon_m}{1 + 2\left(\dfrac{a_{cr} - c_{min}}{h - x}\right)} \tag{1}$$

where a_{cr} is the distance from the point considered to the surface of the nearest longitudinal bar; ϵ_m the average strain at the level where cracking is being considered; c_{min} is the minimum cover to the tension steel/GFRP rebar; h is the overall depth and x is the depth of the neutral axis.

As can be seen from Figure 4a, the predicted crack widths for Beam 1 always exceed the experimental values. Generally, there is good agreement between the experimental and predicted crack widths. With Beams 2-4, it was found that if the presence of the GFRP rebars was ignored in calculating the depth of neutral axis and stress in the steel, good agreement existed between the predicted and experimental crack widths (Figures 4b-4d). This suggests that the GFRP rebars do not significantly affect the moment capacity of the section, which might otherwise result in an over-reinforced failure. The GFRP rebars are largely ineffective structurally probably because they are not tied to the shear reinforcement.

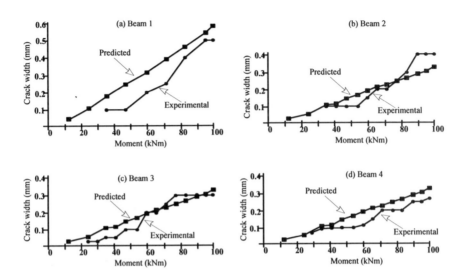

Figure 4 *Predicted and experimental crack widths*

5.3 Deflection

The maximum mid-span deflections for Beams 1-4 were, respectively, 4.7 mm, 5.3 mm, 5 mm and 4.0 mm. The predicted deflection for Beam 1 was 5.9 mm and was calculated using equations 2 and 3, taken from BS8110:[9]

$$a = K L^2 \frac{1}{r_b} \qquad (2)$$

where L is the effective span of the member; K is a constant that depends on the shape of the bending moment diagram; and $1/r_b$ is the curvature at mid-span and is given by:

$$\frac{1}{r_b} = \frac{f_c}{xE_c} = \frac{f_s}{(d - x)E_s} \qquad (3)$$

in which f_c is the design service stress in the concrete; E_c is the short term modulus of the concrete; f_s is the estimated design service stress in tension reinforcement; d is the effective depth of the section; x is the depth of the neutral axis and E_s is the modulus of elasticity of the reinforcement.

The maximum permissible deflection is 11.6 mm and therefore all the beams comply with the deflection requirements in BS8110.[9] Figure 5 compares the deflections occurring in Beams 1-4. Generally, it can be seen that the sections containing GFRP rebars showed similar deflection characteristics to Beam 1 with no GFRP and therefore the existing approach for estimating deflection in BS 8110 can be used, if the presence of the GFRP is ignored.

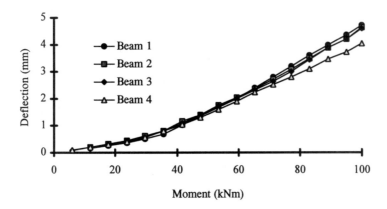

Figure 5 *Deflections in Beams 1-4*

5.4 Life Cycle Costing

The use of GFRP will increase the construction cost of any structure due to the high cost of the reinforcement. However, it should eliminate the need for maintenance arising from reinforcement corrosion, the primary cause of deterioration of reinforced concrete. A method for quantifying the effects of increased initial costs and reduced maintenance costs is life time costing.[11]

Life time costing has been carried out for the three beams shown in Figure 6. Typical prices have been used for the steel and GFRP reinforcing bars, the concrete and for the repairs required by beam A which contains only steel reinforcement. It has been assumed that each beam is to be exposed to a saline environment similar to a crosshead beam situated under a leaking expansion joint on a bridge.

The maintenance free life (MFL) for beam A (steel reinforcement only) is taken as 20 years which is typical for a crosshead under a leaking expansion joint on a bridge. The MFL of beam B (GFRP rebars only) is taken to be 120 years, the same as the design life of the bridge, because there will be no corrosion. The MFL of beam C (steel and GFRP reinforcement) will be determined by the time to corrosion of the steel bars. Typically the bars in a crosshead beam are at a cover depth of about 40 mm. The transport mechanism for chloride ions in concrete in this situation is predominantly by diffusion obeying Fick's Second Law under semi-infinite and linear boundary conditions.[12] The solution to Fick's law shows that there is a square root time relationship between time and depth such that doubling the cover depth should quadruple the time for chloride concentrations around the steel reinforcement to reach a value sufficient to initiate corrosion. Thus increasing the depth of cover to the steel bars from 40 mm in beam A to 100 mm in beam C should increase the MFL by a factor of $(100/40)^2$ i.e. by more than six times, to achieve a value of 120 years. This is similar for the time to corrosion caused by carbonation of the concrete in beams with steel bars at 40 mm cover.

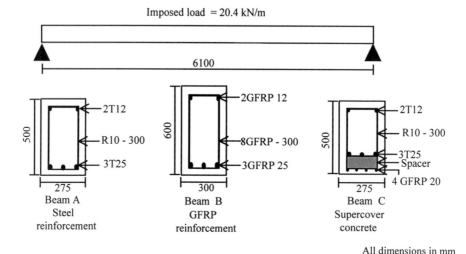

Figure 6 *Loading and reinforcement arrangement for beams A, B and C*

The results of the life cycle costing study are shown in Table 3. The 'whole life cost' (WLC) row in the table represents the values calculated using the quoted values of MFL. The 'risk analysis/mean WLC' row represents the mean value of the frequency distribution of whole life costs derived from a Monte Carlo simulation risk analysis to take account of uncertainty in the values of the inputs, MFL and discount rate. The results clearly show that the lifetime costs of the beams containing GFRP are significantly lower than the lifetime cost for the beam containing only steel reinforcement. It is also evident that the beam which contains GFRP for crack control purposes only, has the lowest lifetime cost. This beam does not need maintenance and its cost is only about 20% more than the construction cost of the all steel reinforced beam, whereas the construction cost of the all GFRP beam is about 100% more.

The current costing exercise does not include any contribution from user delay costs incurred during maintenance work since this is not relevant for individual beams. However if beams of similar type were scaled up to a bridge deck it would be justified to include user delay costs which can sometimes amount to a factor of several times the engineering cost. Under these circumstances the advantages of GFRP rebars would be even more marked.

5.5 Overall Remarks

From the foregoing it can be seen that many of the problems which are currently hampering the use of GFRP rebars in concrete structures can be avoided if GFRP is used in combination with conventional steel. Moreover, it would seem that the existing procedures in BS8110 can be used to predict the behaviour of concrete beams reinforced with steel and GFRP bars. Life cycle costing has shown that using steel and GFRP reinforcement is cost effective.

Work now in hand seeks to assess the applicability of the system, which the authors have termed Supercover Concrete,[13] to other elements of structure and also addresses some of the outstanding issues such as long term durability, fire and creep.

Table 3 *Life time costing for beams A, B and C*

	Beam A	Beam B	Beam C
Load carrying reinforcement	Steel 40 mm cover	GFRP 40 mm cover	Steel 100 mm cover
Crack control reinforcement	-	-	GFRP 30 mm cover
Maintenance Free Life (Years)	20	120	120
Construction cost (£)	449	924	550
Maintenance cost (£)	1022	nil	nil
Whole Life Cost (£)	1471	924	550
Risk analysis/ mean value of WLC cumulative frequency distribution	1776	947	573

Note: All costs discounted at 8% per annum; period of analysis 120 years; repair method:- concrete cut out and backfill

6 CONCLUSIONS

This paper has described a novel method called Supercover Concrete for preventing reinforcement corrosion in concrete structures using GFRP rebars. The method involves using conventional steel reinforcement together with concrete covers in excess of 100mm, with a limited number of GFRP rebars at a nominal depth of, say, 30 mm. Preliminary results on single span, simply supported reinforced concrete beams made using this method show:

1. Introducing GFRP rebars in the concrete cover does not significantly affect the structural performance of the member.
2. The GFRP rebars are effective in controlling surface cracking and can be used to reduce surface crack widths to values currently recommended in BS8110 and Eurocode 2.
3. Deflections under working loads are not significantly affected as a result of introducing the GFRP rebars in the cover zone.
4. Existing procedures given in BS8110 can be used to assess structural strength, design crack width and deflections.

7 ACKNOWLEDGEMENTS

The authors thank International Grating Inc., Huston, Texas, for donating the GFRP rebars used in this work.

8 REFERENCES

1.   Comite Euro-International du Beton (1992), 'Durable Concrete Structures', Thomas Telford, London.
2.   British Standards Institution (1992), 'DD ENV 1992-1-1: Eurocode 2: Part 1: 1992: Design of Concrete Structures: General rules and rules for buildings', BSI, London.
3.   G. Maunsell and Partners (1989), 'The performance of concrete bridges', Her Majesty's Stationery Office, London.
4.   A. Nanni, 'Fiber-reinforced-plastic (FRP) reinforcement for concrete structures: properties and applications: Developments in Civil Engineering 42', Elsevier, Amsterdam, 1993.
5.   J. Wines and G. C. Hoff, 'Laboratory investigation of plastic-glass fiber reinforcement for reinforced and prestressed concrete', Reports 1 and 2, U.S. Army Corps of Engineers, Paper 6-779, 1966.
6.   V. L. Brown and C. L. Bartholomew, *ACI Materials Journal*, 1993, **90**, 1, 34.
7.   O. Chaallal and B. Benmokrane, Materials and Structures, 1993, **26**, 157, 167.
8.   M. R. Ehsani, 'Alternative materials for the reinforcement and prestressing of concrete' (ed. J. L. Clarke), E & FN Spon, London, 1993, p34.
9.   British Standards Institution, 'BS8110: Structural use of concrete. Part 1: Code of Practice for Design and Construction; Part 2: Code of Practice for Special Circumstances'. BSI, London, 1985.
10. International Gratings Inc, 'Kodiak: Fiberglass-reinforced plastic rebar data sheet', Huston, Texas, 1992.
11. Life Cycle Costing - A Radical Approach, CIRIA, Report No. 122, 1991.
12. J. Crank, 'Mathematics of Diffusion', Oxford University Press, Oxford, 1956.
13. International Publication No. WO95/29307, Publication date: 2.11.1995.

EFFECT OF MICROCRACKS ON DURABILITY OF ULTRA HIGH STRENGTH CONCRETE

Bendt Aarup

Aalborg Portland A/S
Rørdalsvej 44, P.O.Box 165, DK-9100 Aalborg

Oskar Klinghoffer

FORCE Institutes
Park Allé 345, DK-2605 Brøndby

1 INTRODUCTION

High strength concretes or high performance concretes are used increasingly in construction today. Often these materials are preferred not because of their high strength but because they exhibit a much higher durability than conventional concretes. Examples of this class of materials are the so-called DSP-materials, developed in 1978, with contents of microsilica based on weight of cement of 20-25%, with water/powder ratios down to 0.15 and with compressive strengths in the range of 150-400 MPa.[1]

These materials are very dense, and also extremely durable, even without air entrainment, and early tests have demonstrated excellent performance against freeze-thaw and also good protection of reinforcement against corrosion.[2]

2 COMPACT REINFORCED COMPOSITE

The good protection of reinforcement became even more important as Compact Reinforced Composite (CRC) was developed in 1986.[3] DSP is - as are most high strength concretes - relatively brittle, and in order to be able to utilise the high strength in structures, ductility was provided through the incorporation of steel fibres - typically in contents of 6% by volume. This high ductility makes it possible to utilise even large contents of reinforcement without excessive cracking, providing a very high strength also in bending. This is demonstrated in Figure 1, where a beam with prestressing strands is loaded in 4-point bending. The strands have not been prestressed, but are simply used because of their high strength.

Even with deformations of 70 mm for a 2 metre beam only small cracks are observed, but as bending stresses of only approximately 30 MPa are obtained for unreinforced beams, it is possible that microcracks are developed at the point indicated on the figure, at a load of 40 kN where a change in stiffness of the beam is observed. It is important to determine the effect of any such microcracks with regard to corrosion, especially as the thickness of cover of the reinforcement is as small as 10 mm in this case and will typically be in the range 8-20 mm for CRC.

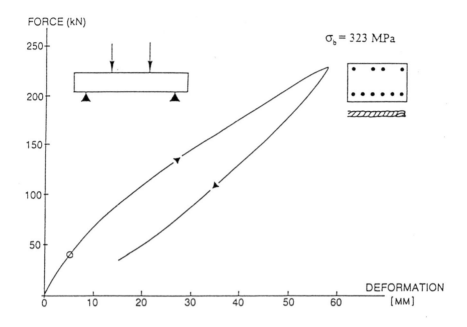

Figure 1   *Load-deformation behaviour of CRC beam with prestressing wires as tensile reinforcement. The circle on the graph indicates a change in stiffness*

This small thickness of cover is used partly to better utilise the main reinforcement, but also to ensure an optimal distribution of microcracks, which prevents larger cracks from developing.

3 INVESTIGATION OF CORROSION

The corrosion of reinforcement is a result of two processes, namely initiation and propagation, which take place if the initially passive conditions are destroyed.[4] Therefore, in order to investigate the reinforcement corrosion, the factors responsible for both initiation and propagation must be defined and then created in the performed test.

The initiation is the period when the aggressive ions, such as chloride, permeate through the concrete cover. The corrosion starts if a critical chloride concentration is built up on the reinforcement.

The propagation is the period after initiation, when the corrosion process continues if some necessary conditions are present. It requires the availability of oxygen and water. At the same time the concrete resistivity must be maintained at a level at which the electro-chemical reactions can take place.

The investigation program for CRC concerning its resistance against corrosion, has been prepared in such a way that both parameters with influence on initiation and propagation are investigated.

Earlier investigations on durability have been performed on DSP materials without steel fibres.[2] It was unclear what effect these fibres - and the high content of main reinforcement - would have on corrosion. On the one hand the steel may use up all the available oxygen very rapidly, thus stifling active corrosion. On the other hand, if the electrical conductivity of the material is sufficient and oxygen and water are available, very high corrosion rates can be expected.

Also, most tests on the corrosion of reinforcement have been carried out on un-loaded specimens, and in order to take advantage of CRC in structures it is likely that service loads will be in a range where microcracks could develop. It was therefore very important to test specimens under high loads and determine whether this would accelerate the intrusion of chloride ions.

This was accomplished with a test rig as shown in Figure 2. Specimens with a 400 mm span were mounted in this flexure rig and loaded to constant centre deflections of 0.2, 1 or 2 mm. Then specimens were exposed to alternating drying and wetting in a sodium chloride solution - a test which was initiated in March 1991 and has been carried out for more than four years. A cross section of one of the small beams is shown in Figure 3, along with the load-deflection curve for such a beam. Deflections of 0.2, 1 and 2 mm are indicated on the curve, corresponding to bending stresses of appoximately 10, 40 and 75 MPa.

A number of specimens have also been exposed without being mounted in the deflection rig. These are reference specimens with exactly the same composition as the beams exposed under flexural load, beams where steel fibres have not been included in the matrix, and finally a few specimens that have been loaded so heavily that visible cracks have developed prior to exposure. These last specimens are simply termed "cracked".

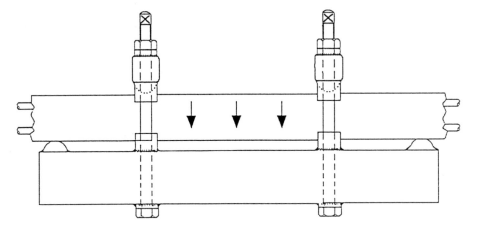

Figure 2 *Mounting rig, where specimens are loaded to a constant center deflection before being exposed to chlorides*

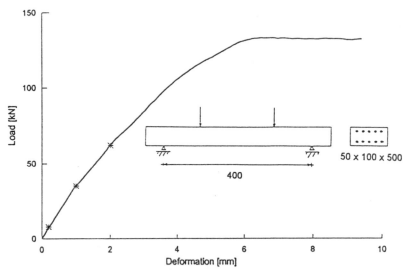

Figure 3   *Load-deformation curve for specimens exposed to chlorides. Deflection of 0.2, 1 and 2 mm are shown on the curve. Measurements are given in mm on the small figure*

3.1 Chloride Ingress through CRC Matrix

Chloride profiles were obtained by measuring the chloride content at different depths in samples exposed to cyclic wet/dry conditions in a concentrated sodium chloride solution.

The measured chloride concentration is plotted against the penetration depth. The effective diffusion coefficients are then calculated by means of Fick's second law and non-linear curve fitting.[5,6] A typical chloride profile is shown in Figure 4.

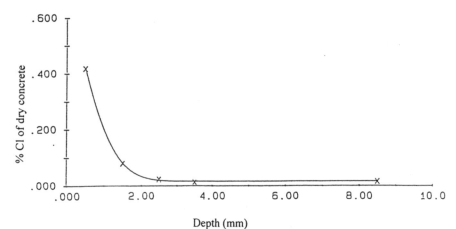

Figure 4 *Chloride profile measured on sample with 1 mm deflection after 385 days of exposure*

Table 1 *Results of chloride diffusion test*

Deflection (mm)	Exposure Duration (days)	D_{eff} 10^{-14} (m²/s)		C_s (w/w%)		Depth at which chloride concentration is below 0.1 w/w% (mm)	
		Surface	Form side	Surface	Form side	Surface	Form side
0.2	185	2.10	2.35	0.42	0.50	1.35	1.35
0.2	953	2.35	2.18	0.60	0.69	2.40	3.00
1	385	1.19	0.78	0.72	0.43	1.50	1.50
2	567	0.81	1.08	0.80	0.65	2.15	2.25
none	1315	0.92	0.74	0.27	0.47	3.25	1.50

The calculated diffusion coefficients for five different samples are presented in Table 1.

The ingress of chloride is limited to the outer 2 mm, and the estimated diffusion coefficients are extremely low (2.35 to 0.075 x 10^{-14}m²/s). For comparison the specified upper limit for concrete, produced for the Great Belt Link with estimated 100 years service life, is 60 x 10^{-14} m²/s.

When all the chloride profiles are taken into account, there is some indication that the measured diffusion coefficients are decreasing with exposure time. This is in accordance with the information published previously,[7] and it also confirms observations from the field.[8] However, the number of measurements in this case is too low to draw any conclusions with regard to the time dependence of the diffusion coefficients.

It seems that there is no increase in the diffusion coefficients with flexural load, as in fact the lower values are obtained with higher deflections. It thus appears that either there is no difference in the number of microcracks present under different loads, or the microcracks have no influence on the penetration of chloride ions.

This very low and well contained chloride diffusion can be explained by the following properties of CRC:

• very low porosity with disconnected pores
• self-desiccation with time
• self-closing of the microcracks, which may appear due to the applied deflection, because of the large number of unhydrated cement particles in the matrix.

These properties are characteristic for DSP materials such as CRC, which contain a large amount of microsilica, and at the same time a small amount of water. The result is that chloride penetrating from the environment is concentrated on the surface of the material, while its diffusion through the bulk is very slow.

In order to destroy the passivity and to initiate corrosion, the critical chloride concentration must be reached at the reinforcement.[9] This concentration is normally between 0.05% and 0.1% w/w of concrete. This range of chloride content is not observed in any investigated samples at a depth of 5 mm below the cover. Therefore, the possibility that active corrosion will be initiated on the rebar surface is, in practice, very low (in test samples the depth of cover to the reinforcement is 10 mm).

If the corrosion is initiated as local pitting, the continuation of this process (propagation) will be difficult due to high electrical resistivity and content of metallic fibres in the matrix. These metallic fibres will consume the oxygen permeating inside the matrix, thus limiting the possibility of continuation of the cathodic reaction in the corrosion process. Measurements indicate that there is very little oxygen available in the matrix. Also, because of the dense matrix, water is not easily available. These factors combine to indicate that even if the critical chloride concentrations usually ascribed to concrete are reached after a long time, corrosion will not be initiated due to other limiting factors. This aspect has been investigated by the Eduardo Torroja Institute, Spain where chlorides have been introduced in the mixing water, and these results will be published at a later date.

In order to determine the range of corrosion rate, after almost four years of exposure in the concentrated sodium chloride solution, linear polarization resistance measurements have been performed.

3.2 Determination of Corrosion Rate by Means of Linear Polarization Technique

The linear polarization resistance is a commonly used electrochemical technique, especially suitable for determination of corrosion rate in the laboratory.

This technique, well described in the literature,[10] is characterized by applying a small polarization to the reinforcement close to its corrosion potential. The polarization should remain in the linear range, which is typically approximately ± 20 mV from the free corrosion potential.

Figure 5 illustrates the principle of the polarization resistance measurements, also known as the "Stern-Geary" method.

The corrosion rate is calculated from the Stern-Geary formula:

$$I_{corr} = B/R_p \tag{1}$$

where I_{corr} is corrosion current expressed in mA/cm^2, which value yields the corrosion rate in $\mu m/year$. R_p is polarization resistance in ohm, and B is an empirical constant determined to be 25 mV for actively corroding steel and 50 mV for passive steel.

The results of polarization resistance measurements, performed on samples representing different deflection loads, are collected in Table 2.

As the polarized area of reinforcement is approximately 100 cm^2, only the average corrosion rate is obtained, which may not reflect the very local corrosivity.

In spite of many years exposure in concentrated NaCl under cyclic wet/dry conditions, the majority of the calculated average corrosion rates are low, under 10 $\mu m/year$.

Corrosion rates under 10 $\mu m/year$ should be classified as negligible. It is remarkable, that the cracked sample, and the one with applied 2 mm deflection, show even lower corrosion rates than the remaining specimens. The explanation of this behaviour reflects the material properties mentioned above. Even with high deflection load, only small cracks are developed in the matrix, and these defects have a self-closing tendency with time.

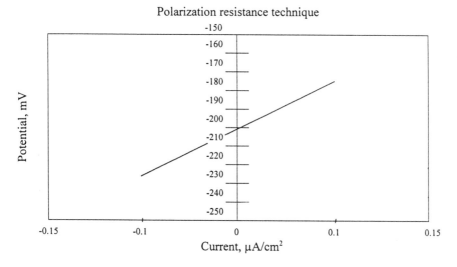

Figure 5 *Principle of polarization resistance measurements. The slope of the potential/ current curve through zero applied potential is the polarization resistance, Rp*

Table 2 *Results of polarization resistance measurements*

Deflection (mm)	Exposure Duration (days)	E_{coor} (mV) vs SCE	R_p (Kohm x cm²)	I_{corr} (µA/cm²)	Corrosion rate (µm/year)
none	975	-551	76	0.66	7.66
	1172	-460	100	0.50	5.80
	1390	-425	242	0.21	2.44
0.2	975	-556	53	0.94	10.90
	1172	-370	150	0.33	3.83
	1390	-567	42	1.18	13.70
1.0	975	-570	62	0.81	9.40
	1172	-381	36	1.38	16.00
	1390	-463	43	1.16	13.45
2.0	975	-551	500	0.10	1.16
	1172	-372	136	0.37	4.29
	1390	-385	539	0.09	1.04
cracked	975	-532	480	0.10	1.16
	1172	-336	165	0.30	3.48
	1390	-494	178	0.28	3.25

The first attempts at thin section analysis have not shown differences in microcracking due to differences in flexural load, but this analysis will now be repeated with a slightly different preparation technique. However, it is apparent that the CRC matrix has a larger number of microcracks than is normally observed in conventional concrete. These defects have limited influence on the chloride ingress,[8] and, therefore, also on the initiation of the corrosion process.

3.3 Supplementary Investigation of the Corrosion State

For samples exposed in concentrated NaCl under cyclic wet/dry conditions, the corrosion state was tested by permanent recording of electrochemical potentials and electrical resistance between rebars. Both measurements have confirmed the results obtained by means of the polarization resistance technique.

Parallel with exposure in concentrated NaCl, the effect of accelerated carbonation in a chamber with relative humidity between 55 and 60%, and CO_2 content approximately 50 times higher than in the atmosphere, has been investigated. So far (after approximately 2 years of exposure) it is not possible to detect any sign of carbonation in the CRC specimens, while a good quality reference concrete exhibited a clear carbonation front. This test will be continued for another few years.

The described results indicate that the CRC matrix has an excellent resistance to chloride induced corrosion and carbonation. Therefore, the small thickness of cover (typically 10 mm) to the reinforcement can be permitted.

These excellent properties have now resulted in application of CRC for 40,000 drain-covers under the tunnel of the Great Belt Link and other similar applications are being considered.

4 CONCLUSIONS

♦ CRC exhibits an excellent resistance to chloride induced corrosion and carbonation, even with high deflection load applied on samples exposed in a very aggressive environment.

♦ This behaviour is due to the composition of CRC, which results in the beneficial properties regarding corrosion prevention. In spite of high deflection loads, only very fine cracks are developed in the matrix.

♦ Diffusion of chloride through the CRC matrix is extremely low and limited to a few millimetres below the surface. The corrosion rates, determined after four years of exposure in a very aggressive environment, are still below the range of significance.

♦ These low corrosion rates make it possible to use a thickness of cover as small as approximately 10 mm to the rebar, which also gives better utilization of the main reinforcement and ensures an optimal distribution of microcracks, thus preventing larger cracks from occurring.

♦ The density of the matrix makes it likely that corrosion will not occur, even with high concentrations of chlorides at the rebars.

♦ These properties of CRC and its excellent performance in the highly corrosive environment open the possibility of using this material as a better alternative than steel in various applications.

5 ACKNOWLEDGEMENTS

These investigations have been carried out partly under the EUREKA programme (Contract EU264) with support from the Danish Department of Trade and Industry, and partly under the Brite/EuRam programme (Contract BRE4-0351) with support from the European Commission.

References

1. H. H. Bache, "Densified Cement/Ultrafine Particle-Based Materials", CBL Report 40, Aalborg Portland, 1981, 33.
2. C. M. Preece, H. Arup and T. Frølund, "Electrochemical Behaviour of Steel in Dense Silica-Cement Mortar", Proceedings of ACI Symposium, Fly Ash, Silica Fume, Slag & Other Mineral By-Products in Concrete, Detroit, 1983, 2.
3. H. H. Bache, "Compact Reinforced Composite, Basic Principles", CBL Report 41, Aalborg Portland, 1987, 87.
4. K. Tuutti, "Corrosion of Steel in Concrete", CBI Research, Stockholm, 1982, 4.
5. H. Sørensen and J. M. Frederiksen, "Testing and Modelling of Chloride Penetration into Concrete", Proceedings of Nordic Concrete Research Meeting, Trondheim, Nordic Concrete Research, Research Project 1990.
6. T. F. Pedersen and O. Klinghoffer, "Factors Influencing the Uncertainity in the Determination of Diffusion Coefficients by Non-linear Curve Fitting", Nordic Miniseminar, Chalmers University of Technology, Gøteborg, 1993.
7. E. Poulsen, "On a Model of Chloride Ingress into Concrete having Timedependent Diffusion Coefficient", Nordic Miniseminar, Chalmers University of Technology, Gøteborg, 1993.
8. P. S. Mangat and K. Gurusamy, "Chloride Diffusion in Steel Fibre Reinforced Marine Concrete", *Cement and Concrete Research*, 1987, **17**, 385.
9. C. M. Hansson and B. Sørensen, "The Threshold Concentration of Chloride in Concrete for the Initiation of Reinforcement Corrosion", Proceedings of ASTM Symposium: Corrosion Rates of Steel in Concrete, Baltimore, 1988.
10. M. Stern and A. L. Geary, "Electrochemical Polarization I". A Theoretical Analysis of Shape of Polarization Curves, *Journal of Electrochemical Society*, 1957, **104**, 56.

SUPPLEMENTARY CORROSION PROTECTION

CORROSION BEHAVIOUR OF WELDED STAINLESS REINFORCED STEEL IN CONCRETE

U. Nürnberger

Research and Testing Institute
Baden-Württemberg
Pfaffenwaldring 4
70569 Stuttgart
Germany

1 INTRODUCTION

In general steel in concrete is sufficiently protected against corrosion. Problems only arise if
- the concrete cover and the concrete quality is - by design or otherwise - reduced relative to the necessary values for the surrounding environmental conditions (e.g. by extreme filigree elements);
- special structures have to be erected, e.g. connections between precast and cast in place elements or heat insulated joints between the structure and external structural elements (e.g. balconies);
- non-dense or dense lightweight concrete is designed to reach a required thermal insulation as well as low ownweight;
- higher corrosion conditions occur e.g. in parking decks due to the use of deicing salts.

Under these conditions reinforcement has to be protected against corrosion. Different types of protection systems have been used in practice, but performance is often unsatisfactory.[1, 2] Therefore studies on stainless steel reinforcement have been carried out to prevent steel corrosion under extreme environmental conditions and for low density concrete.

2 MATERIALS AND TESTING PROCEDURE

Table 1 shows the stainless steels used (materials 1 to 7) together with a comparative unalloyed steel (material 8). The alloyed materials are subdivided into the high quality austenitic and ferritic-austenitic steels and a series of ferritic steels with different chromium contents between 7 and 17% by mass. The sensitivity against pitting corrosion is characterized by the 'effective sum' (% Cr + 3.3 % Mo).

The different materials were tested welded and unwelded. In general, the welding line was not treated; this means that the tests were performed with fixed weld seam and a thin oxidised layer on the surface of the steel bar within the area of the weld. In special cases the weld seam was removed by a corrosive paste.

Electrochemical and exposure tests were carried out. The electrochemical potentio-

Table 1 *Listing of the applied materials*

no.	material		structure	% Cr + 3.3 % Mo	
	chemical comp.	material no.1)		RS	WM
1	X2CrNiMoN 17-13-5	1.4439	A	33.5	
2	X2CrNiMoN 22-5-3	1.4462	A-F	31.9	28
3	X6CrNiMoTi 17-12-2	1.4571	A	23.6	
4	X6Cr 17	1.4016	F	17	
5	X20Cr 13	1.4021	F	13	12
6	X2Cr 11	1.4003	F	11	
7	X10CrAl 7	1.4713	F	7	
8	unalloyed	1.0466	F	0	0

1)European standard   RS reinforcing steel   F ferritic
                           WM welding materials   A austenitic

static investigations were performed in the first instance on cold deformed and welded specimens with a plain surface. On four steels tests were carried out on the final product of unwelded and welded reinforcing bars. During production these steel bars were hot rolled to plain round bars, annealed and then cold deformed with a reduction in cross section of 20 to 30 %. Table 2 contains the different diameters and mechanical properties.

The performance of the above mentioned stainless steel types (Table 1) in the welded as well as the unwelded state has been investigated under the following conditions:

(I)   carbonated concrete,
(II)   alkaline, chloride containing concrete,
(III) carbonated, chloride containing concrete.

Different concentrations of 1, 3 and 5 % chloride (related to the cement content) were added. One half of the alkaline mortar electrodes (without and with chloride) were artificially carbonated in 3 Vol-% content air. The "carbonated" state is possible for normal-weight concrete of inadequate quality and also for structural lightweight concrete during the period of use. Non-dense lightweight concrete is either not high-alkaline or

Table 2 *Mechanical properties of the deformed bars*

no.	structure	D	$R_{p0,2}$	R_m	A_{10}
		mm	N/mm²	N/mm²	%
2	ferritic-austenitic	7,0	870	934	13,1
3	austenitic	10,0	456	599	39,3
6	ferritic	8,0	518	608	15,6
8	ferritic (unalloyed)	8,0	533	596	11,6

Table 3 *Composition and main properties of the used concrete mixtures*

type of concrete	nominal strength	cement type	content kg/m³	w/c	aggregates degradation/ type	content kg/m³		density t/m³	compr. strength N/mm²	porosity Vol-%
normal weight concrete	B 25	FAZ 35 F	270	0.78	0-2 0-8 0-16	655 578 694	1927	2.30	37.0	15
open structural concrete	LB 8	-	230	0.52	Liapor 4 Liapor sand 0-4 sand 0-2	410 52 250	712	1.10	9.0	52
gas concrete	GB 3.3	PC 35 F or 45 F	130 cem. +65 line	0.50 to 0.70	fine silicon sand 70% < 90 µm	400		0.65	3.4	74

tends to carbonate quickly. The state 'alkaline plus chloride content' can be expected most realistically   for chloride attacked normal-weight and structural lightweight concrete.

Additionally, reinforced concrete elements of normal-weight concrete and light-weight concrete (dense and non-dense) have been stored in the open air. Table 3 presents the composition and main properties of the mixtures used. The concrete beams were reinforced with unwelded and welded reinforced bars of the quality in Table 2. Chlorides such as NaCl were mixed in concrete or deicing salts were sprayed on. The spraying procedure represents the situation of concrete structures exposed to the splash-zone of highway-constructions. Some other samples were artificially carbonated. More details are given in 'reference 3'.

3 RESULTS

3.1 Electrochemical investigations

The results of the potentiostatic tests [1, 3] on the welded plain bars are presented in Figures 1 and 2. The following essential conclusions can be drawn:

– The pitting corrosion potential decreases with decreasing effective sum of the steel types.
  In particular three groups appear:
  . the austenitic and ferritic-austenitic types 1.4439 (mat.1), 1.4462 (mat. 2) and 1.4571 (mat.3);
  . the ferritic types 1.4016, 1.4021 and 1.4003 (mat. 4 to 6) with chromium content > 10 % by mass;
  . the ferritic types 1.4713 and 1.4066 (mat. 7 and 8) with chromium content ≤ 7 % by mass.

The resistance against pitting corrosion decreases gradually for these three steel types. Stainless steel bars with chromium contents ≥ 11 % by mass are less susceptible than

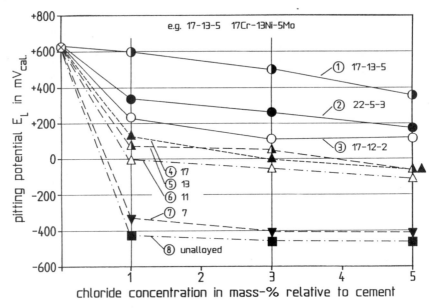

Figure 1 *Pitting potential of plain welded steel specimens in PC-mortar electrodes depending on the steel-type and chloride content; potentiostatic test (t=24h), oxygen evolution E=+600 mV_cal*

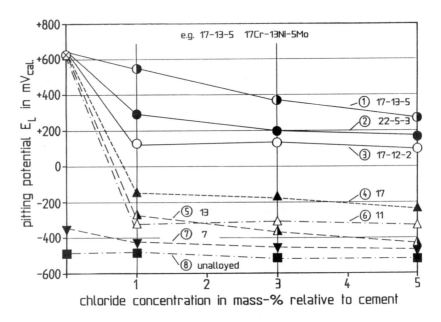

Figure 2 *Pitting potential of plain welded steel specimens in carbonated PC-motar electrodes depending on the steel-type and chloride content; potentiostatic test (t=24h), oxygen evolution E=+600 mV_cal*

unalloyed reinforced steel.
- In carbonated, chloride free concrete the welded steel bars with chromium content ≥ 11 % by mass (for all types of stainless steel) are found to be passive.
- The pitting potential becomes more negative as the chloride content in concrete increases. This reduction of potential is more pronounced between 0 and 1 % by mass chloride than between 1 and 5 % by mass.
- In carbonated, chloride containing concrete the pitting potential is more negative than in alkaline, chloride containing concrete. This is especially significant in the case of ferritic steel bars with chromium content ≥ 11 % by mass, which shows a drop of the pitting potential of approx. 200 to 300 mV.

According to these results on welded specimens with a plain surface the types 1.4462 (ferritic-austenitic), 1.4571 (austenitic), 1.4003 (ferritic) and 1.0466 (ferritic, unalloyed) were chosen for further investigations on welded and unwelded deformed reinforcing bars. Figure 3 summarises the results from the ribbed steels 1.4571, 1.4003 and unalloyed. In addition to those from specimens with a plain surface, the following conclusions can be drawn:
- In the welded state (deformed) reinforcing bars show a less favourable behaviour than plain bars. This is more pronounced for the ferritic steel 1.4003 than for the austenitic steel 1.4571 and more distinct in alkaline than in carbonated concrete.
- Unwelded deformed stainless steel reinforcing bars in concrete with chlorides show a more positive pitting corrosion potential than welded bars. For unalloyed material no difference between welded and unwelded bars was observed.

In total the ribs of the bars and the application of welding have a negative influence on the pitting corrosion behaviour of stainless steel. The ribs and the scratches produced during cold deforming have a disadvantageous influence only together with the welding, if the steel surface contains additional oxide films due to welding.

Corrosion of reinforced structures in open air can only be expected, if the corrosion potential is more negative than -100 mV_{cal} (see the horizontal axis in the top part of Figure 3); in this case the necessary condition $E_L < E_{corr}$ is fulfilled. According to this definition and the results presented in Figure 3, for the following materials and mediums corrosion can be excluded:
- Unalloyed steel (unwelded and welded): alkaline, chloride free concrete.
- Ferritic steel 1.4003 (welded): alkaline and carbonated, chloride free concrete.
- Ferritic steel 1.4003 (unwelded): alkaline concrete, chloride contaminated; carbonated concrete, chloride free.
- Austenitic steel (unwelded and welded): all possible corrosive conditions.

Also in the deformed unwelded and welded state the ferritic-austenitic quality 1.4462 shows still better performance than the steel 1.4571.

Brushing of the welded area and removal of the welding cinder only slightly increases the pitting corrosion potential. The treatment of welded specimens with a pickle paste resulted in an unusual improvement. In the pickled state also a welded ferritic steel 1.4003 showed no tendency for pitting corrosion in a chloride - containing carbonated concrete.

3.2 Field tests

Table 4 shows a summary representation of the extensive results.[3, 4] Corrosion intensities are assigned to the amount of metal wastage and depth of pitting corrosion.

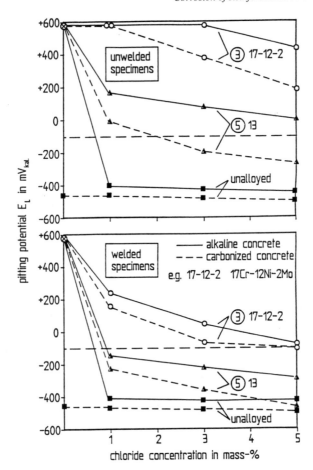

Figure 3 *Pitting potential of deformed steel specimens in mortar (potentiostatic test)*

Welded and non welded areas are distinguished.

According to the expectation the unalloyed steel corroded in carbonated and/or chloride containing concrete. The strongest attack was in carbonated plus chloride contaminated concrete.

The unwelded ferritic chromium steel 1.4003 behaved more favourably than unalloyed steel: in carbonated or alkaline concrete with chlorides no corrosion occured. In the sphere of weld, chlorides favour marked pitting corrosion. The pit-depth increases with the rising chloride content and was in chloride containing carbonated concrete more intensive than in alkaline concrete. But in the ferritic chromium steel 1.4003 the corrosion was more intensive at welds than in the unalloyed steel. In carbonated concrete with chlorides an unwelded ferritic steel 1.4003 can also corrode slightly.

Accordingly ferritic chromium steels 1.4003 are safe in carbonated concrete. In the unwelded state corrosion resistance exists in chloride concrete, when the concrete is not carbonated at the same time. Nevertheless, the steels are suspect at welds in chloride-containing concrete. The ferritic steels have a lower tendency to concrete spalling than

Table 4 *Corrosion behaviour of steel in concrete (survey)*

steel	concrete	normal-weight-concrete				non-dense-concrete	
		alkaline			carbon.	carbonated	
	Cl⁻ M.-%¹⁾	0	0.12	0.3	0	0	0.3–0.7
unalloyed	unwelded		moderate	severe	moderate	moderate	very severe
	welded		moderate	severe	moderate	moderate	very severe
ferritic 11Cr	unwelded						moderate
	welded	moderate	severe				very severe
austenitic 17Cr-12Ni-2Mo	unwelded						
	welded						

¹⁾ chloride content in concrete

☐ none ▨ moderate ▧ severe ■ very severe corrosion

the unalloyed steels, because of the lack of general corrosion.

The austenitic Cr-Ni-Mo-steel 1.4571 does not corrode in the welded or in the unwelded condition. This is valid for all conditions of concrete.

Summarising the results, austenitic steel 1.4571 represents a high security against corrosion, in both the welded and unwelded condition.

References

1. U. Nürnberger, 'Corrosion and Corrosion Protection in Civil Engineering', Bauverlag, Wiesbaden, 1995.
2. T. Mansour, 'Corrosion Protection of Reinforcement in Lightweight Concrete', Dissertation, University of Stuttgart, 1995.
3. U. Nürnberger, W. Beul and G. Onuseit, Otto-Graf-Journal, 1993, *4*, 225.
4. U. Nürnberger, W. Beul and G. Onuseit, Bauingenieur 1995, *70*, 73.

THE INFLUENCE OF STEEL GALVANIZATION ON REBARS BEHAVIOUR IN CONCRETE

R. Fratesi[(I)], G. Moriconi[(I)] and L. Coppola[(II)]

[(I)] Department of Materials and Earth Sciences
   University of Ancona, 60131 Ancona, Italy
[(II)] Engineering Concrete
   31027 Spresiano, Italy

1 INTRODUCTION

Concrete is commonly considered to be an easily produceable and readily available material. However, this presumption has meant that the application of this material is often not supported by an adequate understanding of its properties and of the effective limits to its use in relation to durability problems not only in particularly aggressive environments.

When correctly applied, the present level of knowledge relating to concrete technology, makes it possible to predict the average life of a reinforced concrete structure, the major limitation to durability being the corrosion of the steel reinforcement.

Due to the rising costs of restoration, there is increasing interest in the prevention of corrosion in reinforced concrete structures exposed to aggressive environments. For concrete structures exposed to chlorides or carbonation induced corrosion risk, the most effective approach to producing durable structures is to manufacture a low porosity, high quality concrete (corresponding to a water/cement of less than 0.50 according to ENV 206)[1] with an adequate cover thickness (at least 40-50 mm as recommended by Eurocode 2).[2]

However, even when the concrete is correctly designed, defects may still occur due to lack of control, poor workmanship or accidental causes. Therefore, in aggressive environments, where the risk of corrosion is greatest, the durability of the structure can only be guaranteed by providing additional protection to the steel reinforcement, and at the same time manufacturing a concrete of high quality, which is sufficiently impermeable to the penetration of aggressive agents.

Among the possible methods for improving the corrosion resistance of reinforcement in concrete, new consideration has recently been given to the use of galvanized rebars, in view of their relatively low cost when compared to other protection systems. However, the benefits of using galvanized steel in reinforced concrete structures, is still uncertain due to contradictory data on their effectiveness. Indeed, although successful practical results have been reported in the literature,[3-6] laboratory test results remain fairly controversial.[7-11] Swamy[12] recently stated that the results of laboratory tests must be viewed with caution due to the fact that the simulated environment does not fully match the actual conditions. In general, however, it is recognized that, in comparison to the severe attacks observed on uncoated steel, zinc coatings are effective in delaying corrosion initiation of reinforcing steel.

For structures without macrodefects, Yeomans[13] has clearly illustrated the generally accepted conceptual model to describe the mechanism of corrosion of galvanized steel reinforcement in concrete. This model shows that steel galvanization prolongs the average service life in comparison to bare steel reinforcement, thus increasing the durability of reinforced concrete structures.

An additional hazard arises when cracks appear in concrete structures as a consequence of plastic settlement, shrinkage, creep, thermal stresses, dynamic loads, etc. In such cases carbon dioxide and/or chloride ions can penetrate through the cracks and corrosion can occur even with a low porosity cementitious matrix and thick concrete covers.[14-18]

The effect of cracks or damage to the concrete on corrosion of the reinforcement has not been seriously considered until recently. Beeby[19] has addressed this issue and pointed out that the results obtained by simulating corrosion in atmospheric conditions cannot be extended to corrosion processes occurring underwater. In fact, in immersed structures a large area of embedded steel may act as a cathode, with the very small area of exposed steel at the bottom of the crack acting as the anode. Beeby also claims that there is no basis for assuming a direct relationship exists between corrosion and crack width, even if larger cracks might initiate corrosion in a shorter time.

The role played by cracks is one of the most controversial subjects concerning the corrosion of reinforcement in concrete structures exposed to aggressive environments. The effectiveness of zinc coating in protecting steel rebars in the presence of cracks is also subject to some controversy, and contradictory results[12,20-23] have been presented by different researchers. Galvanization has however been consistently observed to delay the onset of steel corrosion, with this delay being considerable in the case of uncracked concrete.

Another controversial issue is that of the bond strength between galvanized rebars and the concrete. Indeed, conflicting results and opinions are reported in the literature[24-26] regarding the loss of adherence between galvanized rebars and concrete. This loss of adhesion is considered to be due to hydrogen evolution on the rebar surface resulting from the attack of the zinc coating by the hydroxyl ions released into the concrete pore solution during cement hydration.

Passivation of the zinc surface by chromate ions is a valid preventive method of inhibiting hydrogen evolution. It should be pointed out that chromates, which are the most effective passivating agents for zinc, are hazardous to human health and may cause a contact allergy (chromium eczema). As a consequence, a serious effort is being made to find alternative and equally effective passivating agents which could replace chromates. Unfortunately, these efforts have been unsuccessful so far. For health and environmental reasons the addition of chromates to the concrete mix must also be considered unacceptable. In fact, the European Community is currently working on a directive that will impose a reduction in the content of hexavalent chromium ions in concrete to very low levels (< 2 ppm).

The aim of the present work is to investigate the corrosion mechanism and the electrochemical performance of galvanized steel reinforcement embedded in cracked concrete which is immersed in seawater. A secondary aim is to investigate the adhesion of galvanized steel to concrete and the interaction of the two materials. For this purpose, the experimental program presented here included pull-out tests on bare, galvanized and chromated galvanized smooth steel bars.

2 EXPERIMENTAL

Ninety prismatic reinforced specimens (100 x 100 x 400 mm, Figure 1) were produced using CE IV/A 42.5 cement with a water/cement equal to 0.50.

Some of the specimens (type A in Figure 1) were reinforced with a single steel plate (340 x 40 x 1 mm), either bare, hot dip galvanized (80 μm thick zinc coating) or hot dip galvanized and chromated. The plate was embedded 30 mm from the specimen side containing a preformed notch (10 mm deep). The choice of steel plates as reinforcement was justified by the need to enhance the macrocell effect, and to control the crack width induced in the specimens by a flexural stress.

The remaining specimens (type B in Figure 1) were reinforced with two steel plates, which were not in contact with each other; one (80 x 40 x 1 mm) was placed as described above, centered with respect to the notch; the second (340 x 40 x 1 mm) was embedded 40 mm deeper, in order to obtain the same concrete cover on both plates. Various combinations of bare steel, galvanized, and chromated galvanized plates in the upper and lower positions were employed.

A few galvanized plates were scratched to bare steel in order to investigate the corrosion behaviour of galvanized reinforcement damaged during placement.

Before casting, insulated cables were attached to each steel plate for electrical connection to the measuring apparatus.

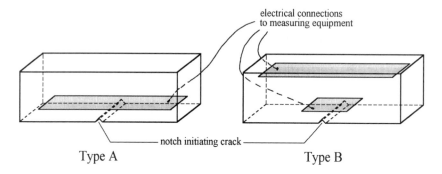

Figure 1 *Prismatic concrete specimens reinforced by one (Type A) or two (Type B) steel plates*

After about one month of air curing, some sound specimens of both types A and B were directly exposed to an aggressive environment to provide reference results for uncracked concrete. All the other specimens were stressed by bending, by loading the surface opposite to the notch, to initiate the development of a crack. Crack widths equal to 1.0 mm were obtained with sufficient accuracy by slowly varying the applied load.

Exposure in an aggressive environment was simulated by continuous full immersion of the specimens in natural seawater at a marine site in the Ancona harbour. Monitoring of the potential with respect to a calomel electrode was carried out on all the specimens; in addition, short circuit currents were measured for the specimens reinforced with two steel plates using a null resistance ammeter. After completion of these measurements, some specimens of each type were split and the steel plates were disbonded at fixed times to allow visual observation and to evaluate the extent of corrosive attack.

Pull-out tests were carried out according to CEB Recommendations RC 6[27] to investigate the extent of adherence between rebars and concrete. About one hundred concrete specimens were made employing CE IV/A 42.5 cement and bare or galvanized or chromated galvanized smooth steel reinforcing bars. After two months air curing, half of the specimens were partially immersed (about 6 cm deep) in tap water, while the other half were placed in a 3% aqueous solution of sodium chloride. Pull-out tests were performed at predetermined times, by recording the tensile force as a function of the relative displacement and transforming the tension forces into bond stresses.

3 RESULTS AND DISCUSSION

3.1 Reinforcing Steel Protection by Galvanization

The free corrosion potential of bare and galvanized steel reinforcement was monitored from the time of concrete casting. Bare steel reinforcement in fresh concrete showed potential values of -350 mV (SCE), which rose to about -130 mV (SCE) after 30 days air curing. At the same time, galvanized and chromated galvanized steel reinforcements showed average potential changes from -1200 mV (SCE) to -500 mV (SCE) for galvanized steel and from -850 mV (SCE) to -580 mV (SCE) for chromated galvanized steel. It should be pointed out that for the galvanized steel, a potential variation of 600 mV occurred within the first five days of curing, causing the potential to change from an active state of corrosion to a passive one. This phenomenon can be explained by the precipitation of calcium hydroxyzincate on the zinc surface which then becomes passivated. The results of tests carried out on zinc electrodes immersed in an aerated saturated $Ca(OH)_2$ solution support this hypothesis. In these tests it was possible to observe the formation of a continuous coating of calcium hydroxyzincate crystals on the zinc surface along with a sharp variation in potential, similar in magnitude and duration to the one described for the reinforced concrete specimens.

The potential values of bare, galvanized, and chromated galvanized steel plates are shown as a function of the immersion time in seawater in Figures 2a and 2b. Figure 2a shows the results for the plates embedded in sound concrete, while the data in Figure 2b refers to the plates embedded in cracked concrete. In the presence of cracks, which allow the immediate exposure of the reinforcement to seawater, a rapid decrease in the potential of the steel plates is produced which is independent of the surface conditions, and thus causes each reinforcement type to reach an active state of corrosion. When the concrete is sound, this change of state is delayed and no trace of corrosion is detectable after one year of immersion in seawater, regardless of the type of reinforcement used. In spite of the apparently more rapid activation of the galvanized steel plate with respect to the bare one, a tendency of zinc to increase its potential value with time is observed (Figure 2a).

In cracked concrete, the potential value of the galvanized steel plates was not influenced by chromation or by scratching of the galvanization (Figure 2b). In fact, during the first immersion, all the potential values settled in the -1200 to -1000 mV (SCE) range and no variation from this range was observed during the entire testing period. In spite of this, the zinc corrosion rate at the crack apex remained low and there was no evidence of penetration of the corrosive attack, in contrast to the results obtained under the same experimental conditions for the bare steel plates. In particular, observation by optical microscopy of the cross sections of the galvanized plates corresponding to the crack apex, indicated a progressive reduction (sometimes baring the underlying steel after one year or more of immersion) of the zinc coating in this area. No iron corrosion products were observed for either the chromated or the non-chromated galvanized steel plates. Furthermore, no traces of steel corrosion were detected on the scratched galvanized areas, thus supporting the idea that the zinc can act cathodically to protect these bare steel zones. In all cases a thick, white, compact deposit was observed wherever zinc dissolution had occurred. This deposit, which appeared well adherent to the steel plate and sealed the corroded area, was identified by X-ray diffraction as calcium hydroxyzincate.

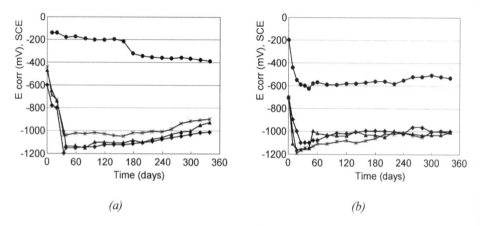

(a)   *(b)*

Figure 2 *Average values of free corrosion potential for continuous full immersion in seawater of uncracked (a) and cracked (b) type A specimens:*
● *bare steel,* ▲ *galvanized steel,* ✻ *scratched galvanization,* ◆ *chromated galvanization*

Generalized corrosion was also observed on galvanized steel plate surfaces, and was considered to be due to the chemical reaction between zinc and the hydrolysis lime produced by cement hydration. However, after one year's immersion in seawater, no steel corrosion products were detected on any of the specimens. Chromated galvanized steel plates showed a similar behaviour, but exhibited lower generalized corrosion on the surfaces.

Measurements of the short circuit potential and current were carried out by electrically coupling the two steel plates embedded in type B specimens, with the smaller plate acting as the anode, and the larger plate acting as the cathode. The experimental program considered the following combinations of plates respectively: (anode (a) and cathode (c)): bare steel (a) - bare steel (c), galvanized steel (a) - galvanized steel (c), galvanized steel (a) - bare steel (c) and chromated galvanized steel (a) - bare steel (c). The short circuit potential of the plates was about -600 mV (SCE) in the double bare steel coupling, and about -1000 mV (SCE) in all the other cases. These results indicate that the smaller steel plate reached by the crack is able to polarize the other plate which is embedded in sound concrete when the two are electrically connected. In fact, the circuit opening resulted in potential changes in all the steel plates which were acting as cathodes and embedded in sound concrete, except for the case of the double galvanized steel coupling; in this case the potential remained constant at -1000 mV (SCE) even after the plates were electrically disconnected.

The short circuit currents measured in the cracked concrete specimens reinforced by two steel plates are shown in Figure 3 as a function of time. The measured values and the flow of current indicate the formation of a macrocell for the cases of bare steel (a) - bare steel (c), galvanized steel (a) - bare steel (c) and chromated galvanized steel (a) - bare steel (c), where the steel plates reached by cracks act as anodes. Zero or insignificant currents with changes in direction were observed when the coupled steel plates were both galvanized. However, in all the cases considered, the higher currents were observed within the first few weeks of immersion and tended to decrease with time, reaching very low values after one year of immersion.

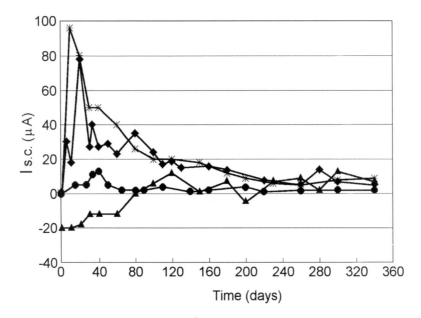

Figure 3 *Average values of short circuit current measured between the steel plates embedded in continuously and fully immersed type B cracked specimens:*
● *Fe(a) - Fe(c),* ▲ *Zn(a) - Zn(c),* ✳ *Zn(a) - Fe(c),* ◆ *Zn-Cr(a) - Fe(c)*

The highest short circuit current values were measured in the cracked specimens when bare steel plates were used as cathodes, and galvanized or chromated galvanized steel plates as anodes. These currents initially reached values of 80 to 100 μA, but decreased to less than 10 μA after one year of immersion. A similar trend was observed for the double bare steel coupling, even though the short circuit current values were lower than those obtained for the galvanized steel (a) - bare steel (c) coupling, due to the lower electromotive force in the macrocell. However, the corrosive attack on the bare steel anodes was shown to be penetrating and localized in the very narrow crack area. Conversely, as observed in the type A specimens, the galvanized anodes showed a generalized corrosion over a much larger area in the crack zone, leading to the higher values of current measured. Finally, if the higher values measured at the beginning of the monitoring phase are neglected, the magnitudes of the short circuit current over the surface area of the steel plates are in good agreement with the data relating to oxygen flow through saturated concrete reported in the literature.[28-30]

Due to the presence of calcium hydroxyzincate and the generalized attack on the zinc surface, caused by the interaction between zinc and the cement paste, it was not possible to determine the weight loss of zinc at the apex of the crack area. The weight loss could only be determined on bare steel plates, where after splitting of type A and B specimens the zones reached by the cracks showed localized, more or less intensely corroded areas covered with black corrosion products which were easily removed by acid pickling. Immediately after steel plate disbonding and before the removal of corrosion products, the pH was measured at the center of the corroded area. These measurements indicated acidic values (usually about 5) sometimes falling down to 3-4. The weight losses were then determined and the related currents were calculated using the Faraday relationship to produce average values of around 15 µA. The short circuit current measured by the null resistance ammeter was found to be much lower (about 3 µA, Figure 3) than the one calculated from the weight loss. Since the macrocell current is due only to oxygen reduction and its value was lower than that calculated from the effective weight loss, another cathodic reaction must occur, which contributes to the overall corrosion process. Such a reaction might involve the reduction of hydrogen ions originating at the crack apex as a result of iron chloride acid hydrolysis. This mechanism, already suggested in the literature,[30-32] seems to be supported by the acidic pH values measured just inside the corroded areas after splitting of the specimens.

By referring to the Pourbaix diagram, for the potential values between -500 and -600 mV (SCE) exhibited by the steel reinforcement in the presence of concrete cracks, it can be seen that the system representation falls below the hydrogen reduction line, and therefore there is no thermodynamic barrier preventing the cathodic reduction of hydrogen.

3.2 Adhesional Behaviour of Galvanized Rebars

Figure 4 shows the tensile force as a function of the relative displacement during pull-out tests on bare, galvanized and chromated galvanized steel reinforcing bars before exposure to the testing environments. The values of the ultimate bond stress are reported in Figures 5a and 5b as a function of the time of exposure to the different experimental conditions.

The highest initial resistance to pull-out (measured after two months air curing before partial immersion) was observed for the galvanized rebars, and the lowest resistance was noted for the chromated galvanized rebars, whereas intermediate values were recorded for the bare steel rebars. During the first year of partial immersion in a 3% sodium chloride aqueous solution (Figure 5a), an increase in the bond stress was observed for all the reinforcement types. Both the total increase in bond stress and its development over time, vary according to the type of reinforcement. The results obtained for the galvanized reinforcement were encouraging, since its resistance to pull-out, and therefore its adherence to the concrete, were always high, particularly after exposure to an aggressive environment. As a general observation, the galvanized rebars usually showed concrete fragments attached to the surface after disbonding, while the surfaces of the chromated galvanized rebars generally appeared bright and perfectly clean, provided that the chromate film had not dissolved.

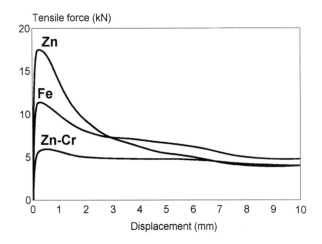

Figure 4 *Tensile load as a function of the relative displacement for bare, galvanized and chromated galvanized rebars*

The results of the bond tests on specimens exposed to tap water are much more ambiguous than those for the sodium chloride solution, due to the much higher scatter observed in the values of the ultimate bond stress measured after longer exposure times (Figure 5b).

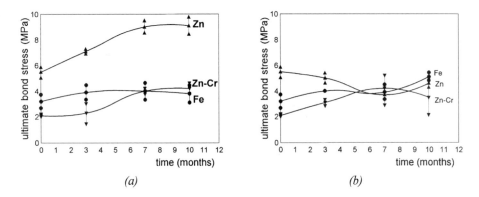

(a) *(b)*

Figure 5 *Ultimate bond stress versus exposure time in sodium chloride solution (a) and in tap water (b)*

The significant increase in ultimate bond stress between galvanized rebars and concrete immersed in sodium chloride aqueous solution (Figure 4) could be explained by densification of the transition zone between the zinc coating and the cement paste, as a result of the penetration of the non-expansive zinc corrosion products into the concrete pores near the interface. This process would contribute to sealing the pores and microvoids in the interfacial zone, causing the formation of bridges between the metal and the concrete, as already suggested by some models reported in the literature.[31,33] This phenomenon could be enhanced by the presence of chloride ions, since they are able to facilitate the zinc dissolution. This interpretation is supported by the increase in the ultimate bond stress observed for the chromated galvanized rebars. Initially the value of the ultimate bond stress is at its lowest (Figure 4) due to the presence of chromates which inhibit the corrosion of zinc. The bond stress starts to increase only after dissolution of the chromate film is complete, when, due to the appearance of zinc on the rebar surface, the mechanism described above is initiated.

3.3 Microscopic Observation of the Rebar/Concrete Transition Zone

Cross sections of some specimens were examined after their disbonding in order to observe the rebar/concrete interface with optical and scanning electron microscopes.

SEM observations of concrete fragments attached to the surface of the galvanized rebars showed a continuously dense transition zone (Figure 6a, where A and B indicate the zinc coating, and C and D the interfacial zone), while the EDXA analysis carried out during the microscopic observations confirmed that the cement paste (E in Figure 6a) surrounding the galvanized rebars had been widely penetrated by zinc corrosion products.

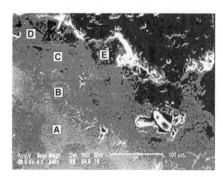   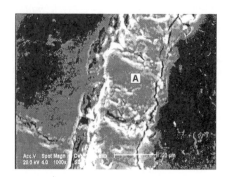

Figure 6a *SEM analysis of galvanized rebar/concrete interface*   **Figure 6b** *SEM analysis of bare steel rebar/concrete interface*

Magnification of the galvanized rebar/concrete transition zone where the zinc coating was slightly corroded revealed the formation of calcium hydroxyzincate crystals perpendicular to the rebar surface (Figure 7). These crystals act as bridges between the metal and the concrete, providing further confirmation of the mechanism hypothesized to explain the increased adherence of the galvanized rebars.

Where formed, iron corrosion products may also diffuse into the cement paste, but in this case, because of their expansive nature, they contribute to the formation of a disjoined layer of corrosion products between the reinforcement and the concrete (A in Figure 6b).

After about one year of exposure to an aqueous solution of sodium chloride, the surface of the zinc coating on the disbonded bars appeared dark grey on the non-chromated rebars; but on the chromated rebars, the surface was still bright even when it had been reached by the chloride ions. Dark zones were observed only on portions of the chromated surface, indicating the disappearance of the η phase layer. This indicates that when the chromate film breaks down in presence of chloride ions, the formation of calcium hydroxyzincate is inhibited.

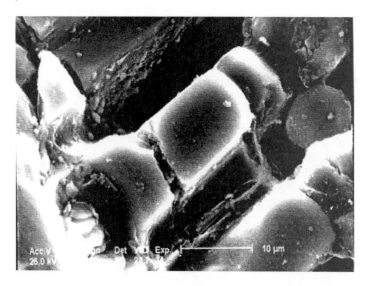

Figure 7 *Calcium hydroxyzincate crystals formed at the galvanized rebar/concrete interface*

4 CONCLUSIONS

In cracked reinforced concrete immersed in seawater, localized corrosion around the cracked area occurred when a bare steel reinforcement was used. There was experimental evidence for a process of hydrogen reduction contributing to the corrosion mechanism occurring under these conditions.

Using galvanized rebars to prevent corrosion, appeared to be an effective solution for cracked concrete, after at least one year of exposure to an aggressive environment. The zinc coating appeared to be able to cathodically protect the rebars when the steel reinforcement was locally uncovered. The galvanized steel embedded in cracked concrete gave the worst results when it was electrically coupled with bare steel embedded in sound concrete.

Regarding the adhesional behaviour of the galvanized rebars, the excellent results obtained do not always agree with the data reported in the literature by other researchers, and thus require further confirmation. For instance, it is suggested that the pull-out tests should be repeated using different cements and employing a wider range of aggressive environments. Above all, much more effort should be put into trying to replace chromates as passivating agents, due to their toxic nature. Finally, some full scale applications should be encouraged in buildings to verify the experimental results.

References

1. European Prestandard ENV 206, 'Concrete, Performance, Production, Placing and Compliance Criteria', 1992.
2. Eurocode N°2, 'Design of Concrete Structure', Part 1, 1991.

3. D. Stark and W. F. Perenchio, 'The Performance of Galvanized Reinforcement in Concrete Bridge Decks', Final Report Project No.2E-206, Constr. Technol. Lab., 80, 1975.

4. J. E. Slater, *Mater. Performance,* 1979, **18**(6), 34.

5. D. Stark, 'Corrosion of Reinforcing Steel in Concrete', (Ed: D. E. Tonini and J. M. Gaidis), ASTM STP 713, American Society for Testing and Materials, Philadelphia, 1980, 132.

6. K. W. J. Treadaway, B. L. Brown and R. N. Cox, 'Corrosion of Reinforcing Steel in Concrete', (Ed: D. E. Tonini and J. M. Gaidis), ASTM STP 713, American Society for Testing and Materials, Philadelphia, 1980, 102.

7. I. Cornet and B. Bresler, 'Galvanized Reinforcement for Concrete - II', International Lead Zinc Research Organization, New York, 1981, 1.

8. C. Andrade, A. Macias, A. Molina and J. A. Gonzales, 'Advances in the study of the corrosion of galvanized reinforcements embedded in concrete', 'Technical Symposia - Corrosion 85, Boston, 25-29 March 1985', NACE, Houston, 1985, Paper N°270.

9. C. Andrade and A. Macias, 'Surface Coatings-2', (Ed: A. D. Wilson, J. W. Nicholson and H. J. Prosser), Elsevier Applied Science, London, 1988, 137.

10. G. Sergi, N. R. Short and C. L. Page, *Corrosion*, 1985, **41**, 618.

11. E. Maahn and B. Sorensen, *Corrosion*, 1986, **42**, 187.

12. R. N. Swamy, 'Corrosion of Reinforcement in Concrete Construction', (Ed: C. L. Page, K. W. J. Treadaway and P. B. Bamforth), Elsevier Applied Science, London, 1990, 586.

13. S. R. Yeomans, 'Corrosion and Corrosion Protection of Steel in Concrete', (Ed: R. N. Swamy), Proceedings of International Conference held at the University of Sheffield, 24-28 July 1994, Sheffield Academic Press, Sheffield, 1994, Vol. II, 1299.

14. J. I. De Lange, K. Verhulst and R. Breeuwer, 'Behaviour of Offshore Structures', Elsevier Science Publishers B.V., Amsterdam, 1985, 513.

15. K. Suzuki, Y. Ohno, S. Praparntanatorn and H. Tamura, 'Corrosion of Reinforcement in Concrete Construction', (Ed: C. L. Page, K. W. J. Treadaway and P. B. Bamforth), Elsevier Applied Science, London, 1980, 19.

16. F. Rendell and W. Miller, 'Corrosion of Reinforcement in Concrete Construction', (Ed: C. L. Page, K. W. J. Treadaway and P. B. Bamforth), Elsevier Applied Science, London, 1980, 167.

17. C. Collepardi, R. Fratesi, G. Moriconi, L. Coppola and G. Corradetti, 'Admixtures for Concrete - Improvement of Properties', (Ed: E. Vázquez), Chapman and Hall, London, 1990, 279.

18. C. Branca, R. Fratesi, G. Moriconi and S. Simoncini, 'Proceedings of Eurocorr 91', (Ed: I. Karl and M. Bod), Budapest, 21-25 October 1991, GTE, Budapest, 1991, Vol. II, 739.

19. A. W. Beeby, 'Cracking and Corrosion', Concrete in the Oceans Technical Report No.1, 1978.

20. U. Nürnberger and W. Beul, *Werkst. Korros.*, 1991, **42**, 537.

21. G. Rehm, U. Nürnberger and B. Neubert, 'Chloridkorrosion von Stahl in gerissenem Beton', Deutscher Ausschuss für Stahlbeton, Berlin, 1988, Heft 390, 43.

22. H. Shimada and S. Nishi, 'Corrosion of Reinforcement in Concrete Construction', (Ed: A. P. Crane), Ellis Horwood, Chichester, 1985, 407.

23. J. Satake, M. Kamamura, K. Shirakawa, N. Mikami and N. Swamy, 'Corrosion of Reinforcement in Concrete Construction', (Ed: A. P. Crane), Ellis Horwood, Chichester, 1985, 357.

24. C. E. Bird, *Nature*, 1962, **194**(4830), 768.

25. C. E. Bird, *Corros. Prev. Control*, 1964, **11**(7), 17.

26. G. Rehm and A. Lammke, *Betonstein Feitung*, 1970, **360**, 36.

27. RILEM/CEB/FIP Recommendations on Reinforcement Steel for Reinforced Concrete, RC 6, 'Bond Test for Reinforcement Steel: 2. Pull-out Test', CEB News N°73, 1983.

28. O. E. Gjorv, O. Vennesland and A. H. S. El-Busaidy, *Mater. Performance*, 1986, **25**, 39.

29. C. L. Page and P. Lambert, *J. Mater. Sci.*, 1987, **22**, 942.

30. E. Otero, J. A. Gonzales, S. Feliu, W. Lopez and C. Andrade, 'Innovation and Technology Transfer for Corrosion Control', Proceedings of the 11th International Corrosion Congress, Florence, 2-6 April 1990, Associazione Italiana di Metallurgia, Milan, 1990, Vol. 2, 459.

31. G. Arliguie, J. Grandet and R. Duval, Proceedings of the 7th International Congress on the Chemistry of Cement, Paris, June 30 - July 5, Editions Septima, Paris, 1980, Vol. III, VII-22.

32. R. Fratesi, G. Moriconi and S. Simoncini, *Oebalia*, 1993, Vol. XIX (Suppl.), 595.

33. O. E. Gjørv, P. J. M. Monteiro and P. K. Mehta, *ACI Mater. J.*, 1988, **87**(6), 573.

ACCELERATED TESTING OF EPOXY COATED REINFORCING STEEL-PART II: AMBIENT TEMPERATURE AQUEOUS EXPOSURE AND ELECTROCHEMICAL IMPEDANCE SPECTROSCOPY

S. K. Lee and W. H. Hartt

Center for Marine Materials
Department of Ocean Engineering
Florida Atlantic University
Boca Raton, FL 33431, U.S.A.

J. F. McIntyre

Advanced Polymer Sciences, Inc.
951 Jaycox Road
Avon, OH 44011, U.S.A.

1 INTRODUCTION

Deterioration of concrete structures due to the corrosion of embedded reinforcing steel has become a serious and costly problem for infrastructure in the United States and other countries. Based on early research by National Institute for Standards and Technology[1] and Federal Highway Administration,[2] epoxy coated reinforcing steel (ECR) has been used in North America for approximately twenty years as a corrosion control method to guard against chloride-induced corrosion of concrete structures exposed to deicing salts or marine environments. However, premature corrosion failure of ECRs in substructure members of several bridges in the Florida Keys was observed after about eight years,[3] and other cases of unsatisfactory performance of ECR have since also been reported.[4] Since ECR is a relatively new technology, unexpectedly poor performance of ECR in concrete for some service situations has led to research to identify the exact mechanism(s) involved in such failure.

The utility of hot water testing (HWT) of ECR at 80° C in conjunction with Electrochemical Impedance Spectroscopy (EIS) monitoring was proposed as an accelerated technique for coating qualification and in-plant quality control.[5] The present paper describes companion experiments that were performed in four different aqueous test solutions at ambient temperature, with periodic EIS scans being made during the immersion testing. Monitoring of defect development and of open circuit potential (OCP) for these specimens was also performed as a function of time.

2 EXPERIMENTAL

2.1 Materials

Epoxy coated reinforcing steel (16 mm in diameter) was obtained from ten sources, including two which were not North American, and which included four types of fusion bonded epoxy powder. All the bars were claimed by the suppliers to conform to appropriate industry standards. ECR specimens were given an identification letter according to the source as **A, B, C, D, E, F, J, N, T** and **U**. Three substrate deformation patterns were represented among the ten sources. Test specimens were prepared in 15.2 cm lengths by placing the bar in question in a clear plastic tube to avoid coating damage from gripping and cutting with a power hacksaw. A lead wire was attached by a screw after drilling and tapping a hole in the end of each specimen. Care was taken to avoid coating damage during the specimen preparation process. The approximate surface area of these specimens was 66 cm². Subsequent to cleaning with an ethanol soaked laboratory tissue, a thick, two part epoxy resin was applied to both cut ends. The end without the lead wire was further protected by an epoxy filled PVC plug.

2.2 Coating Characterization

All ECR specimens were characterized according to coating thickness, defects (holidays) and hardness prior to exposure testing. Coating thickness was measured according to ASTM G 12[6] at various locations along the specimen surface using a digital magnetic coating thickness tester and recorded as the average of three readings per location. Macroscopic coating damage was identified by visual examination, and holiday detection was performed according to ASTM G 62.[7] Coating defects were classified by type and location. For bars from two sources excessive mechanical damage was present, and holiday detection was abandoned because indicator signals from bare areas were so extensive. Coating hardness measurements were made according to ASTM D 3363[8] using a hardness pencil test kit. Since this procedure is destructive, measurement locations were limited to free space between deformations at the end portion of specimens or upon ECR samples not scheduled for testing. A minimum of ten readings was obtained for each source.

2.3 Electrochemical Impedance Spectroscopy

The experimental setup consisted of a frequency analyzer (Schlumberger 1260 Gain Phase Analyzer), a potentiostat (EG&G Princeton Applied Research 273A) and a PC computer. Instrument control and data acquisition were performed using commercial software (ZPLOT®, Scribner Associates). Impedance measurements were made using a three electrode cell arrangement that employed a Ti-mesh counter electrode and a Ag/AgCl reference electrode, as shown in Figure 1. For EIS measurements in low conductivity environments, the Ag/AgCl electrode was fitted with a Pt wire at its tip, and the two were coupled through a 0.1 µF capacitor. Occasionally, different scan parameters were employed depending on the condition of ECRs. For example, in the case of badly deteriorated ECR specimens an amplitude of 25 to 50 mV was employed so as not to accelerate the deterioration process. On the other hand, for defect-free specimens with high impedances the voltage perturbation was up to 150 mV. Scans were generally conducted between 65 kHz and 10 mHz.

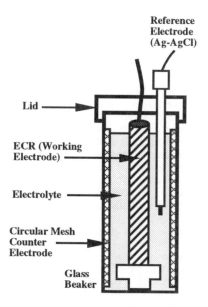

Figure 1 *Schematic of test cell*

A commercially available software package (ZFIT®, Scribner Associates) that employs a CNLS (Complex Non-Linear Least Squares) routine was used to model the selected EIS data, and four equivalent circuit models were used. One of the main electrical components was a resistor, R; and different resistors represented particular resistive components in the equivalent circuits such as solution resistance, $R\Omega$, pore resistance, R_{po}, and charge transfer resistance, R_{ct}. In addition, constant phase elements (CPEs) were employed to fit non-ideal pseudocapacitance and diffusion-related responses.

2.4 Chemical Immersion Test (CIT) Concurrent With EIS

ECR specimens were exposed under ambient conditions to four aqueous test solutions which included distilled water (DW) 3.5 weight percent NaCl solution (NaCl) simulated pore water with KCl (3.44 weight percent) and without addition of KCl (SP w/KCl and SP w/o KCl, respectively). The latter two solutions consisted of 2.63 weight percent NaOH, 1.07 weight percent KOH and 0.22 weight percent $Ca(OH)_2$. Since the acquisition of ECRs from the various sources was made at different times during the test program, there were three groups of test specimens which were exposed to specific environments at three different times. Two ECR specimens per source were immersed in each environment and were subjected to periodic open circuit potential (OCP) measurements and holiday detections. In addition, EIS scanning took place at approximately two to three month intervals. These EIS data were compared with the corresponding baseline data, the majority of which were performed in tap water (TW) and the remaining in DW. This latter environment was abandoned in favor of TW because the low conductivity of DW sometimes caused instability problems. The first round of EIS scans was performed in the respective immersion solutions after two months for the first group and some of the second group of specimens. During initial EIS scans, however, it was found that when coating defects were present, the impedance response was influenced by the type of electrolyte, especially in the basic solutions; and so a comparison of data from the different solutions was not possible. Subsequent EIS scans were performed in TW only. Exposure to TW lasted until the impedance was obtained, and then the specimen was returned immediately to the appropriate test environment. The time required for this ranged from one to two hours.

3 RESULTS AND DISCUSSION

3.1 Coating Characterization

Average coating thickness values for all sources at different locations on ECR specimens indicated that the coating was thickest on the top portion of lug deformations and that the top portion of rib deformations was thicker than valley areas between lug deformations. Conversely, areas close to a rib base were the thinnest. According to the average coating thickness at valleys per source, there was about a 100 µm difference between the thinnest coating (source **J**) and the thickest (source **N**); however, bars from all sources conformed to the ASTM specification (minimum thickness 125 µm and maximum thickness 305 µm).[9,10] In most coating operations the epoxy powder is carried by an air stream through a spray gun and directed by electrostatic forces onto the heated substrate surface. The complexity of the substrate geometry can influence the contact angle of the powder stream and subsequent movement of powder particles. Consequently, more particles are deposited at locations of reduced obstruction to air flow which results in a thicker coating than at complicated locations such as the base of deformations.

Examination of coatings using a stereomicroscope revealed four types of defects based on physical appearance. The first type was a bare area which was attributed to mechanical damage of the coating. The three remaining defect types had non-mechanical origins and included pin-holes, cracks and burrs. Bare areas or mechanical damage are caused by rough handling after the coating is applied, while non-mechanical defects are associated more with the quality of coating application. In general, ECRs with cross deformations (**A**, **B** and **N** sources) had more non-mechanical defects compared to bars with the other (non-crossing) deformation patterns.

ECRs from two sources (**E** and **F**) possessed a uniform hardness value of 2B, while the ECRs from the other eight sources exhibited values of B or HB which indicated that they were slightly harder than the former ECRs.

3.2 Defect Development During CIT

Figure 2(a) shows the total number of defects developed as a function of time for bars from each of eight source during CIT, regardless of solution type. Data for **C** and **T** sources were not included on this plot due to the presence of a large number of initial coating defects. As the plot indicates, **J** source specimens developed the largest number of defects over 500 days in the four environments, but this was due to one particular bar exposed to NaCl solution. Next were specimens from **N**, **B** and **A** sources in that order. Among the eight **D** source specimens only one (**D1** in SP w/KCl) developed a defect (evident at 175 days). Even though immersion durations were shorter for **U**, **E** and **F** source specimens, it can be inferred that their performance was more satisfactory than for specimens from other sources that were immersed for longer times. The legend to Figure 2(a) includes the type of deformation for each source. It can be seen from this that ECR sources with cross deformations had higher frequencies of defect development. It is reasoned that cross deformations have more potential sites for defect development due to the inherent complexity of the geometry, which makes it more likely that imperfect or thin spots in the coating are present after its application. Figure 2(b) demonstrates the effect of solution type on defect development and shows that ECR specimens exposed to chloride ion containing solutions (NaCl and SP w/KCl) were more susceptible. This indicates that aggressive ions like Cl⁻ play a role in defect development, regardless of whether the pH is near neutral or comparable with that of pore water. The highest number in the NaCl category resulted from a single specimen from the **J** source, as mentioned above. If this observation is considered as an exception and disregarded, then SP w/KCl was the most defect promoting medium. Since flaw-free organic coatings have low permeability to Cl⁻, pathways had to develop for the anions to migrate to the coating/substrate interface in order to account for the observed influence of this species.

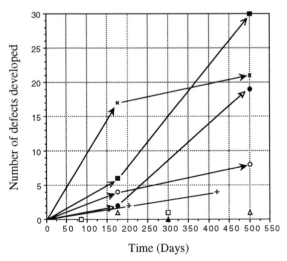

Figure 2(a) *Total number of defects developed in various solutions as a function of immersion time (*O *source A (cross);* ● *source B (cross);* Δ *source D (inclined);* ▲ *source E (inclined);* □ *source F (parallel);* ■ *source J (parallel);* × *source N (cross);* + *source U (inclined))*

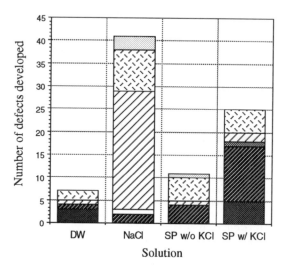

Figure 2(b) *Type of solutions versus number of defects developed (☐ source U; ◫ source N; ▨ source J; ☐ source F; ☐ source E; ▦ source D; ▨ source B; ▮ source A)*

3.3 Open Circuit Potentials

Figures 3(a) - (d) show changes in OCP of ECR specimens with time in the four different aqueous solutions at room temperature. The legend for each plot includes the total number of defects identified during CIT. This indicates that DW was the least aggressive of the solutions. As shown in Figure 3(a), each specimen in DW exhibited relatively positive OCPs (greater than - 0.1 V), except one (**T8,** potential about - 0.3 V) which developed local rust spots during the test.

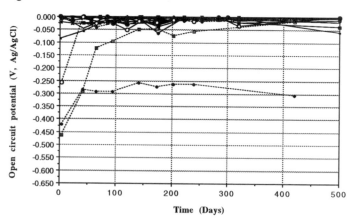

Figure 3(a) *Change of open circuit potentials for ECR specimens exposed to distilled water (*———□——— *A4 (1);* ———◆——— *A8 (2);* ———○——— *B4 (0);* ———◇——— *B8 (1);* ———■——— *C4 (0);* ———□——— *C8 (0);* ———▲——— *D4 (0);* ———△——— *D8 (0);* ———■——— *E4 (0);* ———┼——— *E6 (0);* ———●——— *F11 (0);* ———✕——— *F12 (0);* ———✕——— *J4 (0);* ———⊟——— *J8 (1);* ———■——— *N4 (1);* ————— *N8 (1);* ---□--- T4 (excessive); ---◆--- T8 (excessive); ---○--- U4 (0); ---◇--- U8 (0))*

Conversely, the OCPs in the NaCl solution, as shown in Figure 3(b), tended to fall into one of two groups with a distinctive separation of about 0.5 V between them. The OCP of two specimens (**B3** and **N3**) did, however, exhibit intermediate behavior in the later stages of immersion. One defect was found on each of these two specimens after 284 days of exposure and two on **B3** at 500 days. When the coating remained intact without defect development, ECRs in NaCl solution maintained a positive OCP similar to specimens in the DW environment. This observation is supported by the behavior of defect-free specimens such as **A7, B7, D3, D7, E7** and **E14** in Figure 3(b). The specimens from **A, B** and **D** sources were exposed for 500 days and **E** source specimens for 300 days. However, specimen **F10** did not develop any detectable defects from the initial defect-free condition but still behaved as a defect containing specimen as shown by its relatively negative OCP. For specimens that started with initial defects, such as **T3, T7** and **U3**, the OCP was negative at the beginning of exposure and maintained approximately the same potential throughout the test. For specimens without initial defects which developed defects during immersion, both types of behavior (relatively positive versus negative potential) were observed, as exemplified by specimens **C3** and **C7** in Figure 3(b). Initially, no defects were detected using a holiday detector, and initial OCPs were relatively positive. By day 42 a number of coating cracks at the base of lug deformations were visually identifiable and confirmed by the holiday detector. The corresponding OCPs shifted from an initial value of - 0.05 V to below - 0.5 V (Figure 3(b)). The detrimental effect of chloride ions was evident by the appearance of rust spots and an OCP that changed in the active direction, as observed for defect containing ECR specimens.

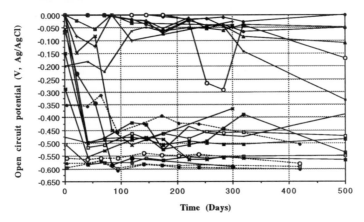

Figure 3(b) *Change of open circuit potentials for ECR specimens exposed to NaCl solution (———◻— A3 (1); ——◆— A7 (0); ——○— B3 (2); ——◇— B7 (0); ——■— C3 (excessive); ——◻— C7 (excessive); ——▲— D3 (0); ——▲— D7 (0); ——■— E7 (0); ——+— E14 (0); ——●— F7 (1); ——✕— F10 (0); ——✱— J3 (30); ——◧— J7 (2); ——▪— N3 (1); ————— N7 (8); ---◧--- T3 (excessive); ---◆--- T7 (excessive); ---○--- U3 (1); ---◆--- U7 (3))*

Figure 3(c) shows OCP behavior with time in a high pH solution with chlorides (SP w/KCl). In general, OCP fluctuated with time. The OCPs of ECR specimens containing excessive coating defects exhibited behavior that was similar to those with many defects in NaCl solution in that specimens **C2, C6, T2** and **T6** had OCPs near - 0.6 V. The change of OCP with time in the chloride-free alkaline solution (SP w/o KCl) is shown in Figure 3(d). Here most ECR specimens behaved similarly regardless of the density of coating defects and exhibited OCPs throughout the test duration that were more positive than - 0.2 V. A possible reason for such noble potentials is formation of a stable passive film on bare areas in this alkaline, Cl⁻ free environment.

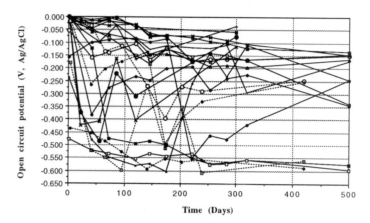

Figure 3(c) *Change of open circuit potentials for ECR specimens exposed to simulated pore water solution with KCl (———□——— A1 (3); ———◆——— A5 (3); ———○——— B1 (2); ———◇——— B5 (10); ———■——— C1 (excessive); ———□——— C5 (excessive); ———▲——— D1 (1); ———▲——— D5 (0); ———■——— E8 (0); ———+——— E10 (0); ———●——— F6 (0); ———✕——— F9 (0); ———✕——— J1 (1); ———□——— J5 (2); ——— N1 (5); ——— N5 (0); - - -□- - - T1 (excessive); - - -◆- - - T5 (excessive); - - -○- - - U1 (0); - - -◇- - - U5 (0))*

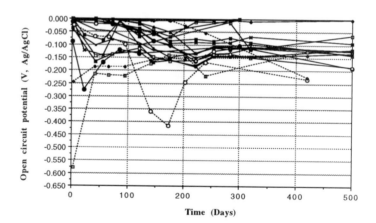

Figure 3(d) *Change of open circuit potentials for ECR specimens exposed to simulated pore water solution without KCl (———□——— A2 (0); ———◆——— A6 (0); ———○——— B2 (1); ———◇——— B6 (3); ———■——— C2 (excessive); ———□——— C6 (excessive); ———▲——— D2 (0); ———▲——— D6 (0); ———■——— E11 (0); ———+——— E12 (0); ———●——— F4 (1); ———✕——— F8 (0); ———✕——— J2 (0); ———□——— J6 (1); ——— N2 (2); ——— N6 (4); - - -□- - - T2 (excessive); - - -◆- - - T6 (excessive); - - -○- - - U2 (1); - - -◇- - - U6 (0))*

Good performance of specimens from **D** and **E** sources is anticipated because of the noble OCPs of these and the fact that they developed the least number of defects throughout the CIT. In this case performance appeared to be independent of the environment.

Overall OCP behavior with time in the four different test solutions can be summarized as follows: in chloride-free environments such as DW and SP w/o KCl, the OCPs of ECR specimens were not influenced significantly by the presence of coating defects. In addition, ECR specimens in the DW environment exhibited more positive OCPs than those in SP w/o KCl solution; in chloride containing environments the OCPs were sensitive to the presence of coating defects and were more negative for specimens containing many coating defects, especially in the NaCl solution; and the fluctuation of OCPs with time was more evident in high pH solutions, especially the chloride containing one. Presence and change with time of active-passive cells may have been a factor here. It can be concluded that coating defects (size and density) are the key factor influencing OCP and the subsequent degradation process due to corrosion in the different chemical environments investigated.

3.4 EIS Response

Before the major CIT portion of the program was launched, a **J** source specimen was immersed in DW at room temperature. Later, this specimen was added to the main CIT program and designated **J8**. Periodic EIS scan results for this specimen are shown in Figure 4 as Nyquist and Bode plots. According to the conventional holiday detection test, **J8** did not contain any initial defects which is consistent with the baseline impedance response having been purely capacitive. However, on day 49 the impedance response showed the presence of a second time constant in the low frequency region indicating coating breakdown. At 100 days a visible local defect was noted, and the corresponding EIS scan revealed a fully developed second time constant and subsequent EIS scans at 275 and 507 days showed more distinctive second time constants (Figure 4).

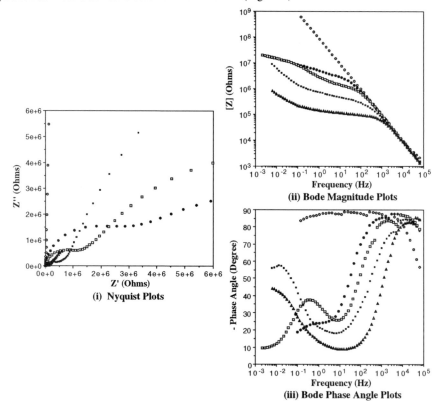

Figure 4 *Nyquist and Bode plots for J source ECR exposed to distilled water for 500 days at room temperature (○ baseline; ● 49th day; □ 100th day; ■ 275th day; ▲ 507th day)*

EIS proved to be a useful technique for monitoring coating breakdown processes where defects were not yet visible. These results indicated that coating breakdown could occur in the absence of chloride ions.

In general, EIS responses from initially defect-free ECRs exposed to CIT conformed to one of three categories: in type 1 the impedance remained unchanged with time (similar to the baseline response in Figure 4); in type 2 there was a reduction of pore resistance (R_{po}) from initial capacitive behavior to a finite magnitude (the 49th day response in Figure 4) and in type 3 there was appearance and progressive growth of a second time constant (as noted for the 100th, 275th and 507th day responses in Figure 4). The third category corresponded to the development of defects and formation of rust at the defect sites. Sometimes a diffusion tail was present in the impedance spectrum. Such responses were basically the same as those observed in HWT environments.[4,5] The only difference is that the HWT condition could produce these responses within one day compared to several months or longer at room temperature during CIT.

By adding specific circuit elements which represented specific changes in the coating/substrate system, a general coating breakdown sequence was simulated. The quality of the fitting was judged by degree of similarity between the experimental and calculated curves and the minimum value of PDAV (the average of the absolute values of the relative standard deviations) or PDRMS (root mean square average of PDAV). For well fitting cases a PDRMS of 0.01 to 0.05 was achieved. Examples of curve fitting results are shown in Figures 5(a) through (d). These examples were randomly selected based on the impedance response and type of circuit model. Most of the fitting results agreed well with the corresponding experimental data, and the examples above do not necessarily represent the best fit. The circuit models can be interpreted as follows: circuit model 1 represents an intact coating/substrate system that should correspond to defect-free ECRs. Upon exposure to a test environment the coating absorbs water and mass transport takes place. Circuit model 2 is intended to represent this, and EIS responses from good performance ECR specimens routinely conformed to this (Figure 5(b)). The impedance response for diffusion can be simulated by a nonuniformly distributed RC transmission line analog (CPE (3)). Kendig et al.[11] introduced an element, C_d, in their circuit model as the capacitance of the diffuse double layer at the coating/substrate interface of a defect-free coating system. They suggested that C_d became a conventional double layer capacitance, C_{dl}, with progressive disbonding in a corrosive environment. It seems that C_d has virtually the same meaning as CPE (3) in circuit model 2 of this study. Charged species or ions, if present, also migrate into the coating; but their penetration rate should be much lower than that of water or oxygen; and they probably reach the interface at holidays and subsequently migrate laterally beneath the coating.

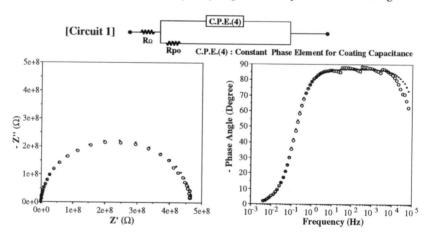

Figure 5(a) *Example of curve fitting using circuit model 1 (specimen J4, ○ experimental; ■ calculated)*

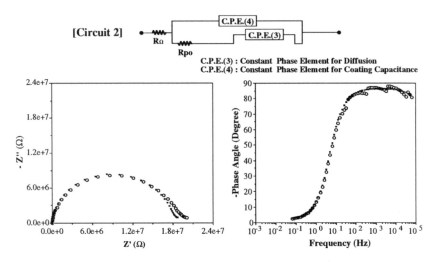

Figure 5(b) *Example of curve fitting using circuit model 2 (specimen D3, ○ experimental;*
▪ calculated)

Once water and oxygen reach the substrate, electrochemical reactions can begin. Circuit model 3 (Figure 5(c)) is representative of this degradation process. Whenever the rate of reaction is controlled by diffusion, the impedance response corresponding to circuit model 4 is applicable. For ECRs containing initial defects the impedance response starts with the elements comprising circuit model 2 or 3, depending on the size and density of defects. For initially defect-free ECR specimens which performed poorly, all of circuit models 1 through 4 could be successively applied during the period of immersion. These circuit models were also applicable to EIS responses obtained from HWT.[4,5]

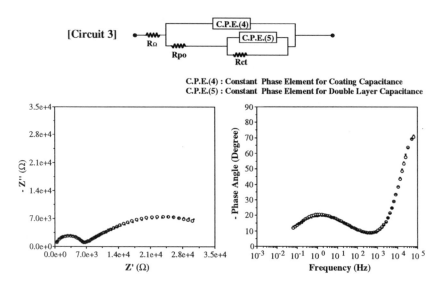

Figure 5(c) *Example of curve fitting using circuit model 3 (specimen T1, ○ experimental;*
▪ calculated)

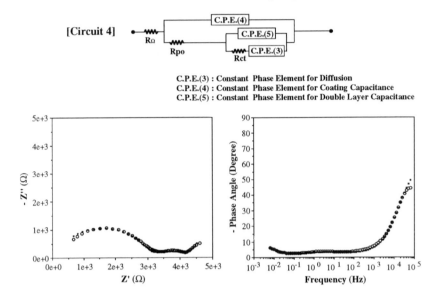

Figure 5(d) *Example of curve fitting using circuit model 4 (specimen T5, ○ experimental; ■ calculated)*

4 CONCLUSIONS

Coating thickness of ECRs acquired from ten sources, including two non-North American sources, indicated that regardless of source the coating at the top of lug deformations was thicker than in the area between the deformations and the coating was thinnest in the base of rib deformations. However, the average coating thickness in the valleys for all ECR specimens conformed to the current ASTM specification. In addition to bare areas induced by mechanical damage, three types of non-mechanical coating defects were identified on virgin ECRs and categorized according to physical appearance. These were pin-holes, cracks and embedded burrs. Virgin ECRs from eight sources exhibited B or HB pencil coating hardness, while ECRs from the two other sources were softer with a hardness of 2B.

The frequency of defect development was influenced by the nature of the test solution. That is, ECR specimens developed more defects in chloride containing solutions than in chloride-free ones regardless of solution pH. Furthermore, ECRs having cross deformations or a relatively thin coating were particularly susceptible to defect development.

Regardless of pH, the OCP of ECR specimens in chloride containing solutions was sensitive to the presence of coating defects and became more negative with time as new defects developed or corrosion at existing defects progressed. On the other hand, the OCP of the specimens exposed to chloride-free solutions was not significantly influenced by defects and remained relatively positive. OCP measurements can be employed to monitor coating defect development in Cl^- containing solutions.

For poorly performing defect-free ECR specimens in CIT, EIS data indicated rapid water and oxygen absorption with these species reaching the coating/substrate interface and causing coating degradation. This process was sometimes controlled by diffusion. This progressive degradation was simulated by four circuit models. For high-performing ECR specimens, the impedance response, in general, showed no signs of active electrochemical reactions but, instead, a diffusion-related response which is indicative of a low rate of corrosion.

5 REFERENCES

1.   J. R. Clifton, H. F. Beeghly and R. G. Mathey, 'Corrosion of Metals in Concrete', SP 49, American Concrete Institute, Detroit, 1975, 115.
2.   K. C. Clear and Y. P. Virmani, 'Corrosion of Non-Specification Epoxy-Coated Rebars in Salty Concrete', Paper No. 114, Corrosion/83, Anaheim, 1983.
3.   D. Gustafson, *Civil Engineering*, 1988, **58**, 10.
4.   K. C. Clear, W. H. Hartt, J. F. McIntyre and S. K. Lee, 'Performance of Epoxy-Coated Reinforcing Steel in Highway Bridges', Final Report 370 submitted to National Cooperative Highway Research Program (Project No. 10-37), Transportation Research Board, National Academy Press, Washington D.C., 1995.
5.   S. K. Lee, J. F. McIntyre and W. H. Hartt, 'Accelerated Testing of Epoxy Coated Reinforcing Steel - Part I: Hot Water Exposure/Electrochemical Impedance Spectroscopy', Paper No. 11, Corrosion/95, Orlando, 1995.
6.   American Society of Testing Materials, 'Nondestructive Measurements of Film Thickness of Pipeline Coatings on Steel', G12, ASTM, Philadelphia, 1987.
7.   American Society of Testing Materials, 'Holiday Detection in Pipeline Coatings', G 62, ASTM, Philadelphia, 1987.
8.   American Society of Testing Materials, 'Film Hardness by Pencil Test', D 3363, ASTM, Philadelphia, 1987.
9.   American Society of Testing Materials, 'Epoxy-Coated Reinforcing Steel', D 3963, ASTM, Philadelphia, 1987.
10. American Society of Testing Materials, 'Epoxy-Coated Reinforcing Steel Bars', A 775, ASTM, Philadelphia, 1987.
11. M. W. Kendig, S. Jeanjaquet and J. Lumsden, 'Electrochemical Impedance: Analysis and Interpretation', (Eds, J. R. Scully, D. C. Silverman and M. W. Kendig), STP-1188, American Society of Testing Materials, Philadelphia, 1993, 407.

A DISCUSSION OF METHODS FOR PREVENTING REINFORCEMENT CORROSION IN BRIDGES

P. R. Vassie

Transport Research Laboratory
Old Wokingham Road
Crowthorne
Berkshire, RG45 6AU
UK

1   INTRODUCTION

Sodium chloride in the form of rock salt is spread on most roads in the UK during the winter months to prevent the formation of ice and maintain safe transport communications. Chloride ions from the deicer can be dissolved in condensation or rain and then enter the concrete on contact. The layer of concrete between the surface of the bridge and the reinforcement, which is typically about 40 mm thick, provides a physical barrier to the chloride ions. Concrete contains pores, therefore the barrier is not continuous and chlorides can move through these pores, reaching the reinforcing steel when the bridge is about 10 to 30 years old. When the chloride ions have reached the reinforcement in sufficient numbers corrosion is initiated and cracking, spalling or delamination of the concrete and loss of steel section occur within a few years. This correlates well with the established model for reinforcement corrosion where the time to corrosion initiation is much greater than the time needed for corrosion to damage the structure.[1] Consequently the most effective way of preventing or delaying reinforcement corrosion is by increasing the time to initiation. This paper discusses the feasibility, cost and effectiveness of a range of methods for increasing the time to corrosion.

2   REINFORCEMENT CORROSION IN BRIDGES

Corrosion normally only occurs in those locations which regularly come into contact with salt solution. Locations which are most vulnerable to reinforcement corrosion are:-
   a)   crosshead beams and piers under leaking expansion joints
   b)   the top surface of bridge decks if the waterproofing membrane is defective
   c)   the lower parts of the vertical faces of piers, columns and abutments that are in close proximity to the road.
   In general other parts of bridges receive little exposure to salt and consequently their rate of deterioration due to salt ingress and corrosion is negligible. Carbonation is another cause of corrosion, but the cover depths and concrete quality used in bridges generally result in a time to corrosion due to carbonation of more than 100 years. Therefore corrosion caused by carbonation will not be considered further in this paper.
   There are two basic mechanisms whereby chloride ions are transported into

concrete, diffusion[2] and capillary action.[3] Diffusion involves the movement of chloride ions through an aqueous solution from regions of high chloride concentration near the concrete surface to regions of low concentration in the bulk of the concrete. Capillary action involves the bulk movement of an aqueous solution containing chloride ions into dry pores in concrete. On a bridge diffusion is the predominant mechanism in locations (a) and (b) whereas capillary action is dominant in location (c). The mechanism of chloride ingress can be important since some preventative methods may only inhibit one mechanism.

Observations have shown that the time to corrosion depends on the location and the mechanism of chloride ingress. In locations (a) and (b) the time to corrosion is typically 15 to 20 years and in location (c) it is typically 30 to 40 years. These differences are probably due in part to the different durations of exposure that occur in the different locations. In locations (a) and (b) the exposure duration is close to 100% of the time after breakdown of waterproofing function on joints and decks. The exposure duration in location (c) is determined by periods of rainfall and condensation and is only about 10%.

Bridges are designed for a serviceable life of 120 years although to achieve this life periodic maintenance work will be needed. To minimise the whole life cost of a bridge calculated using discounted cash flow at a rate of 8% per annum[4] it would be necessary to increase the age of the bridge at which major maintenance is first required to about 40 years. Thus if corrosion due to chlorides is the main cause of deterioration then the time to corrosion should be increased to about 40 years. The aim is therefore to find methods of improving durability that will give a maintenance free life of more than 40 years without significantly increasing the cost of construction. Any increase beyond 40 years, although desirable, would only be justified economically if no additional construction cost was incurred.

3   METHODS FOR PREVENTING/DELAYING REINFORCEMENT CORROSION

There are many different approaches to improving durability:-

i   )   use alternative deicers that do not include chloride ions
ii)   improve the performance of joints and waterproofing membranes
iii)   minimise the use of joints
iv)   use barriers at the concrete surface to prevent contact between chlorides and the concrete
v)   increase the depth of cover of concrete to the reinforcement
vi)   reduce the rate of transport of chlorides through the concrete
vii)   use a protective coating on the reinforcing steel
viii)   use non-corrodible reinforcement

Each approach will be considered in terms of its feasibility, cost and effectiveness.

3.1 Alternative Deicers

Non corrosive deicers such as calcium magnesium acetate[5] and urea[6] are available. They can be spread using existing plant although some modifications may be required to pre-wet the material to prevent it being blown off the road. These alternative deicers will not cause corrosion of the reinforcement and do not appear to have any deleterious effect on concrete, however urea is a less effective deicer than rock salt. The use of urea may also lead to environmental problems in water courses.

Rock salt deicer costs about £2M per annum (raw material cost) to deice trunk roads. In recent years the cost of concrete repairs to corroding trunk road bridges has been about £40M per annum, although this should fall when the backlog of corroding structures has been dealt with. The comparative costs of CMA and urea would be about £40M and £20M per annum respectively.[7] Hence the cost of preventing corrosion using non-corrosive deicers is very high. It is sometimes claimed that the cost penalty of using alternative deicers can be partly overcome by applying rock salt to the roads and CMA or urea to the bridges. However, this is unlikely to be effective in preventing bridge corrosion since there is considerable carry over of rock salt from road to bridge on the tyres of vehicles.

3.2 Improving the Performance of Joints and Waterproofing Membranes

If joints and waterproofing membranes did not leak then reinforcement corrosion would be avoided in locations (a) and (b) where currently the time to corrosion is lowest. The performance of waterproofing has improved considerably over the last few years following the advent of sprayed membranes. It is expected that these membranes will achieve an effective maintenance free life of 50 years providing they are not damaged during resurfacing operations. The cost of sprayed membranes is only marginally more than for the older type of membrane at about £30 m^{-2} and they are the most appropriate and cost effective treatment for protecting the top surfaces of bridge decks.

Many attempts have been made to design expansion joints that will not leak, but success has been limited. Although there are examples of joints which have performed well for long periods without leaking a substantial proportion of joints do develop leaks[8] and presently no joint has been produced that can be relied upon not to leak. Problems due to leaking joints can in part be overcome by providing drainage beneath the joint, but this will only be effective if the drains are cleaned regularly which is costly and difficult to ensure. Although drainage under joints should prevent the salt solution from contaminating crosshead beams and bearing shelves the deck ends will still be exposed to salt solution.

3.3 Minimise the Use of Expansion Joints

Since expansion joints are the source of many of the corrosion problems on concrete bridges minimising the number of joints used will have obvious benefits. Traditionally the most common form of construction for concrete bridges has involved simply supported spans where an expansion joint is used at the ends of each deck span. Another form of construction that is appropriate in most circumstances is to use continuous decks[9] where expansion joints are only required at the two ends of the deck. Continuous decks avoid a number of joints where the overall length of the bridge is sufficient to necessitate a number of spans. There is no overall increase in cost associated with continuous construction since increased reinforcement costs are balanced by savings on joints and bearings. The joints at the abutments of the bridge can be eliminated in many situations by using integral construction where the abutments and deck are continuous.[9] The use of integral construction should not increase construction costs although structural maintenance and improvements such as widening could be more difficult.

Alternative methods for dealing with leaking joints at abutments are:

A)   to modify the design of the abutment to form an abutment gallery that is well drained and uses sprayed waterproofing to prevent contact between salt solution and concrete.

or

B)   to use sprayed waterproofing on vulnerable areas under the joints.

Both these alternatives should be effective although their long term durability is not known. Alternative (B) should be cheaper since it only introduces an additional waterproofing cost.

3.4 Barriers at the Concrete Surface

The purpose of barriers at the concrete surface is to prevent contact between salt solution and concrete. Barriers are likely to be most effective in location (c). Waterproofing membranes can be effective for locations (a) and (b) and they have been considered in Sections 3.2 and 3.3.

Barriers for location (c) include cladding,[10] renders, waterproofing,[11] paints,[12] and impregnants.[13] Cladding can be decorative and effective in preventing chloride ingress, but it is expensive and may need to be removed in order to carry out bridge inspections. Renders can also be used, at a lower cost, to improve the aesthetics and provide some protection against chloride ingress. However, they can debond and it is not possible to inspect the underlying concrete. Sprayed waterproofing membranes bond well to concrete and are likely to provide better protection than renders or paints. They can also be produced with a range of colours. Paint systems also combine corrosion protection and improved aesthetics. Impregnants such as silane provide corrosion protection by generating a hydrophobic layer on the surface of the concrete pores and have no effect on the appearance of the concrete.

The most appropriate barriers are either impregnants, paints or waterproofing membranes. In terms of cost impregnants are cheapest and membranes the most expensive. Membranes are likely to be the most effective and paint systems the least effective. In practical terms impregnants will have no effect on surface reflectivity and can be re-applied with minimal surface preparation. Their durability is uncertain but they have no known breakdown mechanism. Paints can affect surface reflectivity and they typically have a life of 10 to 15 years. Considerable surface preparation is needed to replace paint treatments. Sprayed waterproofing membranes are expected to have a life of at least 25 years, but considerable surface preparation would be needed in order to replace a membrane.

3.5 Increased Cover

If steps are not taken to prevent contact between chlorides and the concrete it will be necessary to increase the time taken for chloride ions to pass through the concrete to the reinforcing steel or to use protected or non-corrodible reinforcement. An obvious way of increasing the transport time is to increase the depth of cover. In engineering terms this is easy to achieve either by a small increase in dimensions or by using some additional reinforcement. In either case the additional cost of construction is small. The benefits of increasing the cover depth are substantial. In locations (a) and (b) where diffusion is the dominant ingress mechanism, the time to corrosion is related to the square of the cover depth by the semi infinite linear solution to Fick's Law of Diffusion.[14] Thus a 20% increase in cover depth should result in a 44% increase in the time to corrosion.

The extent of chloride ingress due to capillary action alone in location (c) of bridges is typically less than 20 mm depth. Thus if the cover depth is less than about 25 mm the time to corrosion may be virtually independent of cover depth. If the cover depth is more than about 20 mm the time to corrosion will depend on both ingress mechanisms and is probably roughly proportional to $(x - 20)^2$ where x is the cover depth in mm.

The benefits of increased cover could also be obtained by improving workmanship and supervision during construction since there is evidence that the cover depths achieved are frequently less than those specified.

3.6 Reducing the Rate of Transport of Chloride Ions

There are three main ways to reduce the rate of transport of chloride ions through concrete:
- reduce the water/cement ratio of the concrete
- use cement replacement materials
- use permeable formwork

Reducing the water/cement ratio will result in a reduction in concrete porosity and the number of interconnected pores and consequently should retard chloride ingress by both the diffusion and capillary action mechanisms. Reducing the water cement ratio has little effect on the cost of construction but a ratio of about 0.4 is the limit for conventional mixes. Further reductions can be achieved by employing superplasticiser at a small additional cost.[15]

The cement replacement materials PFA (pulverised fuel ash)[16] and GGBS (ground granulated blastfurnance slag)[17] are known to reduce the rate of diffusion of chloride ions in concrete, but they do not appear to reduce chloride ingress due to capillary action. They do not add to the cost of construction, but if high replacement levels are employed additional curing may be needed. Replacement materials are also known to increase chloride binding and thus decrease the chloride ion concentration.

Silica fume[18] is another cement replacement material and although there is little information to date on its effect on chloride ingress in bridges there are expectations that it could retard both mechanisms.

Permeable formwork allows bleed water to drain more effectively from the concrete surface after placing. This results in the surface layer of concrete having a higher density and being less permeable to water and chlorides.[19] The long term benefits of permeable formwork are not yet known. The use of permeable formwork results in a small increase in the cost of construction. It can be used in most situations but cannot be used on top horizontal surfaces or where architectural features are required.

3.7 Protecting the Reinforcing Steel

This can be achieved by metallic coatings such as zinc which corrode sacrificially to the steel or by non-metallic coatings such as electrostatically applied epoxy coatings. Galvanised reinforcement can be bent effectively after galvanising, is robust and not easily damaged during construction, but its life is limited in saline environments.[20] Epoxy coatings can also be bent after coating, but bending can result in reduced adhesion and protection. They are less robust than galvanised steel and it is generally necessary to repair damage to the coating and cut ends of the reinforcement before casting the concrete. In general epoxy coatings provide good protection against corrosion[21] although there have been a few instances where corrosion has initiated at a holiday in the coating and then spread along the bar undercutting the coating.[22] Both types of coating add significantly to the cost of construction. Epoxy coatings are generally the more expensive.

3.8 Non Corrodible Reinforcement

There are three common types of non-corrodible reinforcement
- stainless steel grade 316
- carbon fibre
- glass fibre

Each of these types of reinforcement should be effective in preventing corrosion. They all add substantially to the cost of construction; carbon fibres are the most expensive and glass fibres the least expensive. It is possible for stainless steel to corrode although this is very unlikely in chloride contaminated concrete.[20] Carbon[23] and glass fibres[24] cannot corrode but it is possible that the adhesive used to bond the fibres into the form of a bar may deteriorate in the alkaline concrete environment. The glass fibres may also be susceptible to alkali.

The mechanical properties of stainless steel are similar to high yield steel hence its use should not present any practical difficulties. Carbon and glass fibres have a lower modulus of elasticity than steel and their ultimate failure could be brittle. Therefore the use of carbon or glass fibre reinforcement would necessitate design modifications.

To overcome the possible difficulties associated with the cost and mechanical properties of non-metallic reinforcement a design has been suggested[25] which uses high yield steel at a cover depth of about 100 mm to carry the load and a small quantity of glass fibre reinforcement at a depth of about 25 mm to control the crack width. The large cover depth to the steel should prevent corrosion.

4   DECIDING WHICH OPTIONS ARE ECONOMICALLY VIABLE

In order to decide whether the increased construction costs arising from the use of the durability options described in Section 3 will ultimately be offset by reduced maintenance costs it is necessary to carry out a whole life costing exercise where the economic viability of options will relate to factors such as ease of access and the effect of maintenance work on the flow of traffic as well as the effectiveness of the option in delaying the initiation of corrosion. Nevertheless, the information in Section 3 can be used to short list those durability options that offer the greatest benefits for the lowest cost.

The only options for which there is a reasonable prospect of achieving 40 years of maintenance free life under joints are: the avoidance of joints by using continuous decks and integral abutments, and the use of sprayed waterproofing membranes on elements under expansion joints. The better option for most bridges in terms of cost and effectiveness is the avoidance of joints. This is applicable to most bridges; the exceptions are bridges of long total length (>60 m) and bridges located in regions of potential subsidence. The use of integral abutments may limit the scope for future widening of the carriageway.

For a concrete distant from joints there are a number of possible viable options
- increase cover
- impregnants
- cement replacements
- reduced water/cement ratio
- permeable formwork
- sprayed waterproofing membrane

There could be benefits from combining options such as increased cover and silane impregnant whereby access to chloride is restricted by the hydrophobic layer and any chloride that does enter the concrete will take longer to penetrate to the steel because of the longer path.

5   REFERENCES

1. K. Tuutti, 'Corrosion of steel in concrete', Research Report 4/82 Swedish Cement and Concrete Research Institute, Stockholm, 1982.
2. S. Sergi, S. W. Yu and C. L. Page, 'Diffisuion of chloride and hydroxyl ions in cementitious materials exposed to a saline environment', *Magazine of Concrete Research*, 1992, **44**, 63.
3. C. Hall, 'Water sorptivity of mortars and concretes', *Magazine of Concrete Research*, 1989, **41** (147), 51.
4. M. Spackman, 'Discount rates and rate of return in the public sector': Economic issues, Govt. Econ. Service Working Paper No. 113, HM Treasury, London, 1991.
5. C. E. Locke and M. D. Bora, 'The effect of CMA on corrosion of reinforcing steel in Portland Cement Concrete', Corrosion 87 Symposium on corrosion of metals in concrete, San Fransisco paper No. 129, NACE, 1987.
6. Department of Transport, 'Repair and maintenance of Midland Links Viaducts': Working party report, West Midland Regional Office, 1988.
7. 'The ice man goeth', *New Civil Engineer*, 12 October 1989, 52.
8. I. D. Johnson and S. P. McAndrew, 'Research into the condition and performance of bridge deck expansion joints', TRL Report PR9, Transport Research Laboratory, Crowthorne, 1993.
9. Department of Transport, 'Design for Durability', Departmental Standard BD 57/94 and BA 57/94, DOT, London, 1995.
10. S. Hay, 'Overcladding and rainscreens', The aesthetics and refurbishment of bridges, Institution of Civil Engineers, London, 1995.
11. A. R. Price, 'Waterproofing of concrete bridge decks: site practice and failures', Transport Research Laboratory Report RR 317, TRL, Crowthorne, 1991.
12. R. Barton, 'State of the art report: coatings and renders', The aesthetics and refurbishment of bridges, Institution of Civil Engineers, London, 1995.
13. M. B. Leeming, 'CIRIA review and future of surface treatments', Permeability of concrete and its control, Concrete Society, 1985.
14. J. Crank. 'Mathematics of diffusion', Oxford University Press, Oxford, 1956.
15. J. C. Payne and J. M. Dransfield, 'The influence of admixtures and curing on permeability', Permeability of concrete and its control, Concrete Society, 1985.

16. M. N. Hague, The use of flyash to produce durable concrete in the Arabian Gulf, Proceedings of the 2nd Int. Conf. on deterioration and repair of reinforced concrete, Bahrain 1987, 357.

17. D. Higgins, Reducing the ingress of deicing salts into concrete, *Construction repairs and maintenance*, 1986, 12.

18. K. Byfors, 'Influence of silica fume and flyash on chloride diffusion and pH values in cement paste', *Cement and concrete research*, 1987, **17(1)**, 115.

19. M. Marosszeky, M. Chew, M. Arioka and P. Peck, 'Textile form method to improve concrete durability, Concrete International', 1993, **15(11)**, 37.

20. K. W. J. Treadaway, R. N. Cox and B. L. Brown, 'Durability of corrosion resisting steels in concrete', *Proc. Institution of Civil Engineers, Part 1*, 1989, **86**, 305.

21. C. D. Lawrence, 'Permeability and protection of reinforced concrete, Part 1 Degredation processes', *Industrial Corrosion*, 1986, 6.

22. K. Clear, W. H. Harrt, J. McIntyre and S. K. Lee, 'Performance of epoxy coatings on reinforcing steel in highway bridges', NCHRP Report 370, Transportation Research Board, 1995.

23. J. L. Clarke, 'Alternative materials for the reinforcement and prestressing of concrete', Chapman and Hall, London, 1993.

24. A. Nanni, 'Fibre reinforced plastic (FRP)', Reinforcement for concrete structures, Elsevier, 1993.

25. C. Arya, F. K. Ofori-Darko, S. Pirathapan, 'FRP rebars and the elimination of reinforcement corrosion in concrete structures', Proceedings of 2nd International RILEM Symposium on Non-Metallic Reinforcement for Concrete Structures, (Ed: L. Taerwe), E & F N Spon, Ghent, 1995, 227.

THE LONG TERM PERFORMANCE OF AUSTENITIC STAINLESS STEEL IN CHLORIDE CONTAMINATED CONCRETE

R.N. Cox[1] and J.W. Oldfield[2]

[1]Building Research Establishment
Garston, Watford WD2 7JR

[2]Cortest Laboratories Ltd, Consultants to NiDI
23 Shepherd Street, Sheffield S3 7BA

1 INTRODUCTION

In the early 1970s it became apparent that a major factor contributing to the deterioration of concrete was corrosion of steel reinforcement in both general buildings[1] and bridge decks.[2-4] The deterioration was associated with the presence of chloride in quantities large enough to initiate and propagate corrosion and/or to the loss of alkalinity of the concrete by carbonation down to the depth of the reinforcing steel. Corrosion inducing conditions had been attributed to the injudicious use of chloride-bearing set-accelerators, chloride contamination of some mix materials, the ingress of chloride from the external environment and/or neutralisation of the passivating alkalinity at the steel surface. Although these causes can be, and often are, attributed to lack of control and poor workmanship there are occasions where exposure to a chloride-bearing environment or inherently chloride-contaminated concrete are inevitable and place the reinforcing steel at higher than normal corrosion risk.

To investigate this increased corrosion risk an extensive natural weathering study of the performance of corrosion resistant and corrosion protected reinforcing steel embedded in concrete was instigated in the early 1970s. The materials studied were plain carbon steel, galvanised steel, weathering steel, ferritic stainless steels and austenitic stainless steels.

The performance of galvanised steel after 5 years was reported in 1980[5] and after 10 years in 1989,[6] and the performance of austenitic stainless steel ferritic stainless steel and weathering steel after 10 years.[6] Whilst the programme was originally designed to cover a 10 year exposure period it was decided to retain some of the austenitic stainless steel reinforced specimens on exposure for a longer period as there was no evidence of deterioration of the rebar in the concrete. Prisms reinforced with type 316 stainless steel and beams reinforced with types 302, 315 and 316 stainless steel were retained. After 22 years these specimens have been removed and destructively examined under sponsorship from the Nickel Development Institute.

2 EXPERIMENTAL

Twelve prisms reinforced with type 316 stainless steel and twenty-three beams reinforced with either type 316 or types 304 and 315 stainless steel bars formed part of an extensive test programme and were exposed to natural weathering for 22 years. Some of the beams and the prisms were contaminated with chloride at manufacture. The prisms and most of the beams were exposed on an industrial site at Beckton, East London and three beams were exposed in the splash zone on a marine site at Hurst Castle on the south coast of England.

3 PRACTICE

3.1 Prisms

Descaled 10 mm diameter type 316 stainless steel bar of composition give in Table 1 was cut to length, drilled, degreased, weighed and assembled on to the supporting mild steel frame (Figure 1). The stainless steel was electrically isolated from the frame.

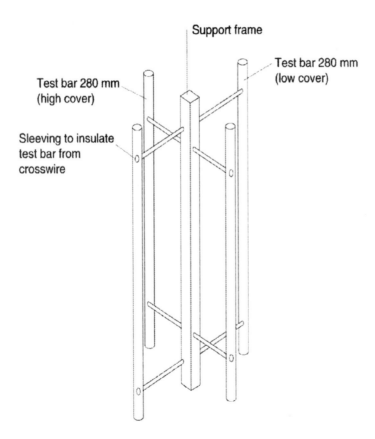

Figure 1 *Arrangement of steel on support frame for prisms*

Table 1   *Chemical analysis of austenitic stainless steels*

Steel	C%	Si%	N%	Mn%	P%	S%	Cr%	Ni%	Mo%	Cu%
302	0.096	0.47	0.02	0.78	0.021	0.023	17.76	8.80	0.18	0.098
315	0.056	0.29	0.030	1.64	0.30	0.009	16.96	10.10	1.42	0.01
316	0.041	0.46	0.22	1.87	0.030	0.007	17.28	12.35	2.14	0.154

Table 2   *Concrete mixes designed for specimens*

	Prism Concrete		Beam concrete
	High permeability (HP)	Low permeability (LP)	
Cement/aggregate ratio	0.125	0.183	0.2
Approximate cement content kg/m^3	220	290	350
Water/cement ratio	0.75	0.60	0.55
Fine aggregates (Ham river 5 mm down) %	40	40	40
Coarse aggregate (Ham river 10 mm) %	60	60	60

The prisms were cast vertically and the correct covers of 10 and 20 mm were achieved by locating the steel assembly within the mould. The concrete mixes used are given in Table 2. Calcium chloride was added to the mixes to give chloride contamination levels, with respect to the cement, of 0, 0.32, 0.96, 1.92 and 3.20%. The prisms were cured under wet hessian for 24 hours at 20°C, demoulded and stored under wet hessian for 27 days at 20°C. They were then stored indoors until placed on exposure. In all, 12 prisms were exposed, each containing 4 stainless steel bars, 2 bars with high cover and 2 bars with low cover. Half the prisms were made with high permeability (HP) concrete, the other half with low permeability (LP) concrete.

Throughout the exposure period the prisms were inspected regularly for rust staining and spalling and, at the end of the exposure period, they were removed for destructive examination. The depth of carbonation was determined using phenolphthalein indicator. The bars were cleaned manually to remove adherent concrete and chemically cleaned in 30% nitric acid at room temperature. The cleaned and washed bars were dried and reweighed and subsequently visually examined for type and degree of deterioration.

3.2 Beams

Twenty-three beams were exposed, each containing 2 stainless steel bars with the same depth of cover which was either 15 mm or 30 mm. Three different stainless steel reinforcements were used, types 302, 315 and 316; their chemical analysis is given in Table 1. Of the 23 beams, 13 were reinforced with 2 bars of 316 stainless steel; in the remainder, each contained 1 bar of 302 and 1 bar of 315 stainless steel. The concrete mix used is given in Table 2; calcium chloride was added to the mix to give chloride contamination levels, with respect to the cement, of 0, 0.32 and 0.96. The beams were cured under wet hessian for 7 days and then they were cured under water for 20 days. After casting and curing, the beams were transported to the exposure site and cracked by fastening pairs of beams of similar cover back to back across roller pivots.

Two of the beams exposed at Hurst Castle contained 316 stainless reinforcement, the other one contained 302 and 315 stainless steel. No chloride had been added to these 3 beams.

Prior to destructive examination, the free corrosion potential and the corrosion current density of each bar were determined using the linear polarisation resistance technique. In using the LPR technique the second bar in the beam was used as the auxiliary electrode. In doing this one can convert the total current into a current density using the embedded area of the bar.

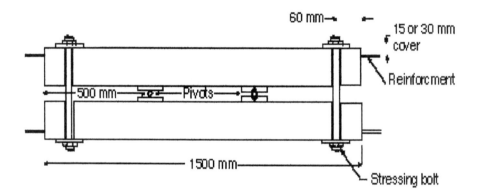

Figure 2 *Assembly of Stressed beams*

4 RESULTS

4.1 Examination of Prisms

None of the prisms exhibited cracking that could be attributed to the stainless steel reinforcement. The supporting mild steel frame did however corrode and cause cracking of some of the 1.92% and 3.20% chloride ion containing cover concrete.

Measurements of depth of carbonation are given in Figure 3. These show that carbonation did not penetrate to the surface of the stainless steel. Visual examination of the stainless steel showed no evidence of corrosion. The weight loss was measured for each bar and found to be less than 0.05% in all cases. This is assumed to be insignificant when considered in conjunction with the visual observation.

4.2 Examination of Beams from Hurst Castle

No cracking of the concrete cover was found on any of the 3 beams that could be associated with the deterioration of the reinforcing bars. The transverse cracks induced at the start of the exposure remained visible.

LPR measurements indicated current densities in the range 0.02 to 0.2 $\mu A/cm^2$. These are considered to be typical of passive currents and an indication that no corrosion was occurring. Potential measurements ranged between -27 mV and +124 mV vs SCE, again indicating that the stainless steel was passive.[8]

On destructive examination the embedded regions of the bars were free from any signs of corrosion. There was, however, evidence of minor pitting at the concrete-bar-atmosphere interface. A maximum depth of attack of 20 μm was measured. Carbonation had only penetrated 1 or 2 mm into the concrete.

To examine the ingress of chloride from the atmosphere the concrete was sampled at various depths and the chloride content determined and the results are summarised in Table 3.

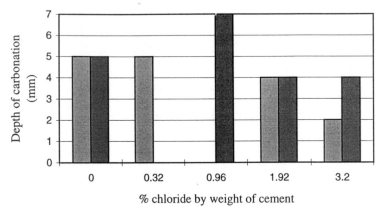

Figure 3 *Depth of carbonation*

□ Low permeability ■ High permeability

Table 3 *Chloride penetration in the Hurst Castle beams (weight % with respect to the cement content)*

Stainless steel	Cover	Sample depth	% chloride
316	15	1-5 6-10 11-15 16-25 26-35	0.79 0.89 0.78 0.49 0.49
316	30	1-10 11-20 21-30 31-40 41-50	1.07 0.94 0.79 0.60 0.29
302/315	30	1-10 11-20 21-30 31-40 41-50	1.12 0.94 0.79 0.75 0.60

These data show that the chloride at the depth of the bar was such that plain carbon steel would be at moderate risk of corrosion.

4.3 Examination of Beams from Beckton

There was no cracking of the concrete cover which could be associated with deterioration of the reinforcing steel. The transverse crack induced at the start of the exposure period remained visible.

LPR measurements indicated current densities in the range 0.002 to 0.6 $\mu A/cm^2$. These are considered to be typical of passive currents and an indication that no corrosion was occurring. Potential measurements ranged between -40 and 216 mV indicating that the stainless steel was passive.

Destructive examination showed that the embedded region of the reinforcing steel bars had suffered no corrosion. Carbonation had only penetrated 1 to 2 mm into the concrete.

The type 316 stainless steel bars embedded in four beams with 30 mm cover had light green deposits on their surfaces. The concrete was contaminated with 0.96% chloride with respect to the cement content. Qualitative analysis using EDAX under a scanning electron microscope indicated these deposits consisted of iron and sulphur suggesting they were deposits of ferrous sulphate.

5 DISCUSSION

Of the twelve prisms containing 48 stainless steel bars, and the twenty three beams containing 46 bars, only three bars showed any sign of corrosion after 22 years exposure. Minor pitting attack occurred at the point at which the reinforcing bar passed through the concrete/air interface on one bar each of type 304, 315 and 316 stainless steel in beams subject to a marine splash zone exposure at Hurst Castle. The affected areas were very small and could only be seen clearly under a microscope. The position of the minor pits (less than 20μm deep) and the environment in which they occurred suggested that the most likely cause was crevice corrosion, since a crevice exists between the steel bar and concrete at the point at where it protrudes from the concrete.

There was no pitting attack of the steel in the crevice created by local loss of bond along the bar adjacent to the intersection of the induced cracks and the bar. It is probable that autogenous healing of cracks restricted the ingress of chloride down the crack and there was insufficient concentration of chloride, even with the chloride diffusing through the concrete to create suitable conditions for crevice corrosion at the base of cracks. There was no pitting where the steel was fully embedded in concrete. There was no evidence of similar pitting attack of steel protruding from beams exposed at Beckton even when chloride had been added to the concrete. This difference in performance between beams exposed at Beckton and beams exposed at Hurst Castle suggest that the pitting is associated with the environment to which the beams were exposed rather than the contamination of the concrete.

In laboratory studies Sorensen[8] indicated that the free corrosion potential for austenitic stainless steel should be between 0 and -200 mV if no corrosion is to occur. Nurnberger[7] suggested that for all practical conditions corrosion of stainless steel can be expected if the potential falls below -50 mV. All the potentials in the present study were more positive than -40 mV and away from the levels where corrosion of the covered bar might occur.

6 CONCLUSIONS

1. After 22 years of exposure at an industrial site at Beckton, East London, types 302, 315 and 316 stainless steel reinforcing bar have not corroded in concrete contaminated with up to 3.2% chloride ion with respect to the cement concrete. This applies to both prism and cracked beam samples. In all cases carbonation had not reached the concrete bar interface.

2. After 22 years of exposure at a marine site at Hurst Castle, types 302, 315 and 316 stainless steel reinforcing bar in cracked concrete with no added chloride contamination but subjected to the splash zone showed no signs of corrosion. In all cases carbonation had not reached the concrete bar interface.

3. The results reported here indicate that austenitic stainless steels are a suitable reinforcement for both chloride contaminated cement/concrete exposed to chloride containing atmospheres.

Reference

1. H. G. Midgley, J. W. Figg and M. J. Mclean, *Concrete*, 1973,7, **1**, 24.
2. P. G. Caralier and P. R. Vassie, *Proc. Instn. Civil Engrs Part 1*, 1981, **70**, 461.
3. P. R. Vassie, *Proc. Instn. Civil Engrs. Part 1*, 1984, **76**, 712.
4. J. E. Slater, 'Corrosion of Metals in association with concrete bridges' ASTM, STP.,

Philadelphia Pa. No 818, 1983, p.53.

5. K. W. J. Treadaway, B. L. Brown and R. N. Cox, ASTM, STP., Philadelphia Pa. 712, 1980, 102.
6. K. W. J. Treadaway, R. N. Cox and B. L. Brown, *Proc Inst Civil Engrs Part 1*, 1989, **86**, 1, 305
7. U. Nurnberger, W. Beul and G. Onuseit, *Otto Graf Journal*, 1993, **4**, 527
8. B. Sorensen, P. B. Jensen and E. Maahn, Corrosion of Reinforcement in Concrete C. L. Page, K. W. J. Treadaway and P. B. Bamforth, Elsevier Applied Science, London, 601

Index of Contributors

Subject Index